T0297760

CAMBRIDGE LIBRARY COLLECTION

Books of enduring scholarly value

Botany and Horticulture

Until the nineteenth century, the investigation of natural phenomena, plants and animals was considered either the preserve of elite scholars or a pastime for the leisured upper classes. As increasing academic rigour and systematisation was brought to the study of 'natural history', its subdisciplines were adopted into university curricula, and learned societies (such as the Royal Horticultural Society, founded in 1804) were established to support research in these areas. A related development was strong enthusiasm for exotic garden plants, which resulted in plant collecting expeditions to every corner of the globe, sometimes with tragic consequences. This series includes accounts of some of those expeditions, detailed reference works on the flora of different regions, and practical advice for amateur and professional gardeners.

The Miscellaneous Botanical Works of Robert Brown

The botanist Robert Brown (1773–1858) is regarded as one of the most significant figures in the advancement of plant science in the nineteenth century. After studying at Aberdeen and Edinburgh, he made the acquaintance of Sir Joseph Banks via William Withering, and in 1801 was appointed naturalist in Matthew Flinders' expedition to Australia. Brown made extensive collections of animals and minerals, but his 3,400 plant specimens from Australia, Tasmania and Timor were the foundation of his work for the rest of his life, as an active member of the Linnean Society, as Banks's librarian, and as an under-librarian in the British Museum. This two-volume collection of his 'miscellaneous botanical works', edited by John J. Bennett, Brown's assistant at the British Museum, was published in 1866–7. It has not been possible to reissue the accompanying quarto volume of plates. Volume 1 contains 'Geographico-Botanical Memoirs' and 'Structural and Physiological Memoirs'.

Cambridge University Press has long been a pioneer in the reissuing of out-of-print titles from its own backlist, producing digital reprints of books that are still sought after by scholars and students but could not be reprinted economically using traditional technology. The Cambridge Library Collection extends this activity to a wider range of books which are still of importance to researchers and professionals, either for the source material they contain, or as landmarks in the history of their academic discipline.

Drawing from the world-renowned collections in the Cambridge University Library and other partner libraries, and guided by the advice of experts in each subject area, Cambridge University Press is using state-of-the-art scanning machines in its own Printing House to capture the content of each book selected for inclusion. The files are processed to give a consistently clear, crisp image, and the books finished to the high quality standard for which the Press is recognised around the world. The latest print-on-demand technology ensures that the books will remain available indefinitely, and that orders for single or multiple copies can quickly be supplied.

The Cambridge Library Collection brings back to life books of enduring scholarly value (including out-of-copyright works originally issued by other publishers) across a wide range of disciplines in the humanities and social sciences and in science and technology.

The Miscellaneous Botanical Works of Robert Brown

Volume 1

Edited by John Joseph Bennett

CAMBRIDGE
UNIVERSITY PRESS

University Printing House, Cambridge, CB2 8BS, United Kingdom

Cambridge University Press is part of the University of Cambridge.
It furthers the University's mission by disseminating knowledge in the pursuit of education, learning and research at the highest international levels of excellence.

www.cambridge.org
Information on this title: www.cambridge.org/9781108076814

This edition first published 1866
This digitally printed version 2015

ISBN 978-1-108-07681-4 Paperback

Selected botanical reference works available in the CAMBRIDGE LIBRARY COLLECTION

al-Shirazi, Noureddeen Mohammed Abdullah (compiler), translated by Francis Gladwin: *Ulfáz Udwiyeh, or the Materia Medica* (1793) [ISBN 9781108056090]

Arber, Agnes: *Herbals: Their Origin and Evolution* (1938) [ISBN 9781108016711]

Arber, Agnes: *Monocotyledons* (1925) [ISBN 9781108013208]

Arber, Agnes: *The Gramineae* (1934) [ISBN 9781108017312]

Arber, Agnes: *Water Plants* (1920) [ISBN 9781108017329]

Bower, F.O.: *The Ferns (Filicales)* (3 vols., 1923–8) [ISBN 9781108013192]

Candolle, Augustin Pyramus de, and Sprengel, Kurt: *Elements of the Philosophy of Plants* (1821) [ISBN 9781108037464]

Cheeseman, Thomas Frederick: *Manual of the New Zealand Flora* (2 vols., 1906) [ISBN 9781108037525]

Cockayne, Leonard: *The Vegetation of New Zealand* (1928) [ISBN 9781108032384]

Cunningham, Robert O.: *Notes on the Natural History of the Strait of Magellan and West Coast of Patagonia* (1871) [ISBN 9781108041850]

Gwynne-Vaughan, Helen: *Fungi* (1922) [ISBN 9781108013215]

Henslow, John Stevens: *A Catalogue of British Plants Arranged According to the Natural System* (1829) [ISBN 9781108061728]

Henslow, John Stevens: *A Dictionary of Botanical Terms* (1856) [ISBN 9781108001311]

Henslow, John Stevens: *Flora of Suffolk* (1860) [ISBN 9781108055673]

Henslow, John Stevens: *The Principles of Descriptive and Physiological Botany* (1835) [ISBN 9781108001861]

Hogg, Robert: *The British Pomology* (1851) [ISBN 9781108039444]

Hooker, Joseph Dalton, and Thomson, Thomas: *Flora Indica* (1855) [ISBN 9781108037495]

Hooker, Joseph Dalton: *Handbook of the New Zealand Flora* (2 vols., 1864–7) [ISBN 9781108030410]

Hooker, William Jackson: *Icones Plantarum* (10 vols., 1837–54) [ISBN 9781108039314]

Hooker, William Jackson: *Kew Gardens* (1858) [ISBN 9781108065450]

Jussieu, Adrien de, edited by J.H. Wilson: *The Elements of Botany* (1849) [ISBN 9781108037310]

Lindley, John: *Flora Medica* (1838) [ISBN 9781108038454]

Müller, Ferdinand von, edited by William Woolls: *Plants of New South Wales* (1885) [ISBN 9781108021050]

Oliver, Daniel: *First Book of Indian Botany* (1869) [ISBN 9781108055628]

Pearson, H.H.W., edited by A.C. Seward: *Gnetales* (1929) [ISBN 9781108013987]

Perring, Franklyn Hugh et al.: *A Flora of Cambridgeshire* (1964) [ISBN 9781108002400]

Sachs, Julius, edited and translated by Alfred Bennett, assisted by W.T. Thiselton Dyer: *A Text-Book of Botany* (1875) [ISBN 9781108038324]

Seward, A.C.: *Fossil Plants* (4 vols., 1898–1919) [ISBN 9781108015998]

Tansley, A.G.: *Types of British Vegetation* (1911) [ISBN 9781108045063]

Traill, Catherine Parr Strickland, illustrated by Agnes FitzGibbon Chamberlin: *Studies of Plant Life in Canada* (1885) [ISBN 9781108033756]

Tristram, Henry Baker: *The Fauna and Flora of Palestine* (1884) [ISBN 9781108042048]

Vogel, Theodore, edited by William Jackson Hooker: *Niger Flora* (1849) [ISBN 9781108030380]

West, G.S.: *Algae* (1916) [ISBN 9781108013222]

Woods, Joseph: *The Tourist's Flora* (1850) [ISBN 9781108062466]

THE

RAY SOCIETY.

INSTITUTED MDCCCXLIV.

This volume is issued to the Subscribers to the RAY SOCIETY *for the Year* 1866.

LONDON.

MDCCCLXVI.

THE MISCELLANEOUS BOTANICAL WORKS

OF

ROBERT BROWN, ESQ., D.C.L., F.R.S.,

FOREIGN ASSOCIATE OF THE ACADEMY OF SCIENCES OF THE INSTITUTE OF FRANCE, ETC., ETC., ETC.

VOL. I.

CONTAINING

I. GEOGRAPHICO-BOTANICAL,

AND

II. STRUCTURAL AND PHYSIOLOGICAL MEMOIRS.

LONDON:
PUBLISHED FOR THE RAY SOCIETY BY
ROBERT HARDWICKE, 192, PICCADILLY.

MDCCCLXVI.

J. E. ADLARD, PRINTER, BARTHOLOMEW CLOSE.

PREFACE.

(BY THE EDITOR.)

THE present volume contains the first portion of the works of the distinguished author, now for the first time collected in England, and reprinted from the originals, without change, in accordance with his express desire. It had been his intention to reprint them himself with annotations, but, unfortunately for science, this intention was never carried out, and it remained for the Editor simply to superintend a verbatim reprint.

The Memoirs are arranged in three divisions—1st. Geographico-Botanical; 2nd. Structural and Physiological; 3rd. Systematic. Of course this arrangement is in some degree arbitrary, inasmuch as observations relating to both of the other divisions are continually occurring in the Memoirs referred to each of them; but, on the whole, it has appeared to be the most convenient for reference. The present volume contains the first two of these divisions; the second will be devoted to Systematic Memoirs and Miscellaneous Descriptions of Plants; and a separate volume, in large 4to, will contain the illustrative figures to both.

JOHN J. BENNETT.

DECEMBER 30TH, 1865.

CONTENTS.

PART I.

GEOGRAPHICO-BOTANICAL MEMOIRS.

PART II.

STRUCTURAL AND PHYSIOLOGICAL MEMOIRS.

PART I.

GEOGRAPHICO-BOTANICAL MEMOIRS.

GENERAL REMARKS,

GEOGRAPHICAL AND SYSTEMATICAL,

ON THE

BOTANY OF TERRA AUSTRALIS.

BY

ROBERT BROWN, F.R.S. LIBR. L.S.,

ACAD. REG. SCIENT. BEROLIN. CORRESP.

NATURALIST TO THE VOYAGE OF H.M.S. INVESTIGATOR, COMMANDED BY CAPTAIN FLINDERS.

[*Reprinted from a Voyage to Terra Australis, by Matthew Flinders.*]

LONDON.

1814.

GENERAL REMARKS, &c.

THE coasts of the great South Land commonly [533* called New Holland have been discovered partly by Dutch and partly by English navigators. Captain Flinders, considering it therefore unjust towards the English to retain a name for the whole country which implies its discovery to have been made by the Dutch alone, has thought proper to recur to its original name Terra Australis; under which he includes the small islands adjacent to various parts of its coasts, and the more considerable southern island called Van Diemen's Land.

In this extended sense I shall use Terra Australis in the following observations, but when treating of the principal Land separately, shall continue to employ its generally received name New Holland; that I may be more readily understood by botanists, for whom these observations are intended, and preserve consistency with the title of a work, part of which I have already published, on the plants of that country.

In the following pages I have endeavoured to collect such general, and at the same time, strictly botanical, observations on the vegetation of Terra Australis, as our very limited knowledge of this vast country appears already to afford. To these observations are added descriptions of a few remarkable plants, which have been selected for publication, from the extensive and invaluable collection of drawings made by Mr. Ferdinand Bauer in New Holland, chiefly during the voyage of the Investigator.

* These figures throughout the volume correspond with the paging in the original.

The materials for the present essay were acquired principally in the same voyage, from Captain Flinders's account of which a general notion of the opportunities afforded for observation may be gathered. It seems necessary, however, 534] to present in one view the circumstances under which our collections were formed, both in the Investigator's voyage, and subsequently, during a stay of eighteen months in New South Wales and Van Diemen's Island; as also to state other sources from which additional materials have been obtained. By this means the reader will be better enabled to judge how far I am entitled to make those observations of a more general nature which he will find in the following pages.

The first part of New Holland examined in Captain Flinders's voyage was the South Coast, on various and distant points of which, and on several of its adjacent islands we landed, in circumstances more or less favorable for our researches. The survey of this coast took place from West to East, and our first anchorage was in King George Third's Sound, in 35° S. lat. and 118° E. lon. In this port we remained for three weeks, in the most favorable season for our pursuits; and our collection of plants, made chiefly on its shores and a few miles into the interior of the country, amounts to nearly 500 species, exclusive of those belonging to the class Cryptogamia, which, though certainly bearing a small proportion to phænogamous plants, were not, it must be admitted, equally attended to. At our second anchorage, Lucky Bay of Captain Flinders's chart, in 34° S. lat. and about 4° to the eastward of King George's Sound, we remained only three days, but even in that short time added upwards of 100 species to our former collection.

Goose-Island Bay, in the same latitude and hardly one degree to the eastward of the second anchorage, where our stay was also very short, afforded us but few new plants; and the remaining parts of the South Coast, on five distant points of which we landed, as well as on seven of its adjacent Islands, were still more barren, altogether producing only 200 additional species. The smallness of this number is to be accounted for, partly, no doubt, from the less favorable season in which this part of the coast was examined; but it appeared to depend also in a considerable degree on its greater sterility, and especially that of its islands.

Of New South Wales, or the East Coast of New Holland, scarcely any part beyond the tropic was examined in the voyage; our first landing after leaving Port Jackson being at Sandy Cape, in nearly 25° S. lat. Between this and 21° S. lat. we had many, and upon the whole, favorable opportunities for observation, especially at Port Curtis, Keppel Bay, Port Bowen, Strong-tide Passage, Shoal- [535
water Bay, and Broad Sound, the survey of which was completed; we landed also on two of the Northumberland and on one of the Cumberland Isles.

On the North Coast we landed on Good's Island, one of the Prince of Wales' Isles of Captain Cook; for a few hours at Coen River on the east side of the Gulf of Carpentaria; and in more favorable circumstances on many of the islands and some points of the mainland on the west side of this Gulf. Several of the group called the Company's Islands in the chart, the shores of Melville Bay, of Caledon Bay, and a small part of Arnhem Bay were also examined.

We then left the coast, owing to the decayed state of the ship, which, on our return to Port Jackson, was surveyed and pronounced unfit for the prosecution of the voyage.

Captain Flinders having, in consequence of this, determined to repair immediately to England, for the purpose of obtaining another vessel to complete the objects of the expedition, Mr. Bauer and myself agreed to remain in the colony of New South Wales until his return, or, if that should not take place, for a period not exceeding eighteen months. During this time we added very considerably to our collections of plants, within the limits of the Colony of Port Jackson and its dependent settlements; the banks of the principal rivers and some part of the mountains bounding the colony were examined; I visited also the north and south extremities of Van Diemen's Land, remaining several months in the vicinity of the river Derwent; and repeatedly landed on Kent's Islands, in Bass' Strait, on the shores of which the principal part of the Submarine Algæ contained in our collections were found.

The reader of Captain Flinders's narrative is already acquainted with the unfortunate circumstances that prevented his revisiting Port Jackson within the expected period, soon after the expiration of which we embraced an opportunity of returning to England, where we arrived in October, 1805, with the greater part of our collections, and without having absolutely lost any one species; though many of our best specimens of the South Coast, and all the living plants collected in the voyage perished in the wreck of the Porpoise.

The collection of Australian plants thus formed amounts to nearly 3900 species. But before embarking in the voyage
536] of Captain Flinders, I enjoyed no common advantages, through the liberality of Sir Joseph Banks, in whose Herbarium I had not only access to nearly the whole of the species of plants previously brought from Terra Australis, but received specimens of all those of which there were duplicates. Of these plants, exceeding 1000 species, the far greater part were collected by Sir Joseph Banks himself, in the voyage in which New South Wales was discovered. The rest were found at Adventure Bay in Van Diemen's Land, by Mr. David Nelson, in the third voyage of Captain Cook; at King George's Sound on the south-west coast of New Holland, by Mr. Menzies, in Captain Vancouver's voyage; and in the colony of New South Wales by several botanists, especially the late Colonel Paterson and Mr. David Burton. Since my return from New Holland I have had opportunities of examining, in the same Herbarium, many new species, found in New South Wales by Mr. George Caley, an acute and indefatigable botanist, who resided nearly ten years in that colony: and have received from the late Colonel Paterson several species discovered by himself within the limits of the colony of Port Dalrymple; which was established under his command.

I have also examined, in the Sherardian Herbarium at

Oxford, the greater part of the plants brought from Shark's Bay by the celebrated navigator Dampier, and have seen a few additional species from that and other parts of the West Coast of New Holland, collected in the voyage of Captain Baudin.

The additional species obtained from all these collections are upwards of 300; my materials, therefore, for the commencement of a Flora of Terra Australis amount to about 4200 species; a small number certainly for a country nearly equal in size to the whole of Europe, but not inconsiderable for the detached portions of its shores hitherto examined.

In Persoon's Synopsis, the latest general work on phænogamous plants, their number is nearly 21,000. The cryptogamous plants already published, by various authors, exceed 6000; and if to these be added the phænogamous plants that have appeared in different works since the publication of Persoon's Synopsis, and the unpublished species of both classes already existing in the collections of Europe, the number of plants at present known may be estimated at 33,000, even exclusive of those peculiar to Terra Australis.

The observations in the present essay being chiefly on extensive tribes of plants, they are necessarily arranged [537
according to the natural method.

Of this method the primary classes are Dicotyledones, Monocotyledones, and Acotyledones.

These three divisions may be admitted as truly natural, and their names, though liable to some exceptions, appear to me the least objectionable of any hitherto proposed.

Of the Australian plants at present known, upwards of 2900 are Dicotyledonous; 860 Monocotyledonous; and 400 Acotyledonous, Ferns being considered as such.

It is well known that Dicotyledonous plants greatly exceed Monocotyledonous in number; I am not however aware that the relative proportions of these two primary divisions have anywhere been given, or that it has been inquired how far they depend on climate. Into this subject I can enter only very generally in the present essay.

According to the numbers already stated the Dicotyledones of Terra Australis are to be Monocotyledones as rather more than 3 to 1, or somewhat less than 7 to 2.

In Persoon's Synopsis, to which, as the latest general work, I again refer, these two classes are to each other nearly as 11 to 2. But, from the nature of this compilation, it may be assumed that certain difficult and extensive orders of Monocotyledones, especially Gramineæ and Cyperaceæ, are considerably under-rated; an addition of 500 species to Monocotyledones would make the relative numbers of the two classes as 9 to 2, which I am inclined to think an approximation to the true proportion.

With a view to determine how far the relative proportions of these two classes are influenced by climate, I have examined all the local catalogues or Floras which appeared most to be depended on, and have likewise had recourse to unpublished materials of great importance in ascertaining this point. The general results of this examination are, that from the equator to 30° of latitude, in the northern hemisphere at least, the species of Dicotyledonous plants are to Monocotyledones as about 5 to 1; in some cases considerably exceeding, and in a very few falling somewhat short of this proportion; and that in the higher latitudes a
538] gradual diminution of Dicotyledones takes place, until in about 60° N. lat. and 55° S. lat. they scarcely equal half their intratropical proportion.

In conformity with these results the Dicotyledones should be to the Monocotyledones of Terra Australis as nearly 9 to 2; whereas the actual proportion as deduced from our materials is hardly 7 to 2: but it appears, on arranging these materials geographically, that the relative proportions of the different regions of Terra Australis itself, are equally at variance with these results. About half the species of Australian plants at present known have been collected in a parallel included between 33° and 35° S. lat.; for this reason, and for one which will hereafter appear, I shall call this the *principal parallel*. At the eastern extremity of this parallel, within the limits of the colony of Port Jackson, where our materials are the most perfect, the proportion of Dicotyledones to Monocotyledones does not exceed 3 to 1. At the western extremity of the same parallel, in the vicinity of King George's Sound, the proportion is but little different from that of Port Jackson, being nearly as 13 to 4. At the south end of Van Diemen's Island in 43° S. lat., it is fully 4 to 1. And with this proportion that of Carpentaria, and I may add the whole of the equinoctial part of New Holland, hitherto examined, very nearly agrees.

I confess I can perceive nothing either in the nature of the soil or climate of Terra Australis, or in the circumstances under which our collections were formed, to account for these remarkable exceptions to the general proportions of the two classes in the corresponding latitudes of other countries.

With regard to the proportion of Acotyledones in Terra Australis, it is necessary to premise that I consider my collections of some of the Cryptogamous order, especially of Fungi, as very imperfect. If, however, 300 species were added to the 400 actually collected, I believe it would give an approximation to the true proportions, which on this supposition, would be of Phænogamous to Cryptogamous plants as nearly 11 to 2. But the general proportion of these two great divisions, as deduced from the published materials, is very different from this, being nearly 7 to 2.

If we inquire in what degree these proportions are dependent on climate, we find that in the more northern parts of Europe, as in Lapland and even in Great Britain, Cryptogamous plants somewhat exceed the Phænogamous in number. In the south of Europe, even making allow- [539
ance for its being at present less perfectly examined, these proportions seem to be inverted. And within the tropic, unless at very great heights, Cryptogamous plants appear to form hardly one fifth of the whole number of species. But their proportion in Terra Australis is still smaller than the assumed intratropical proportion: for this, however, in the northern parts of New Holland at least, the comparative want of shade and moisture, conditions essential to the vegetation of several of these tribes, will in some measure

account; for at the southern extremity of Van Diemen's Island, where the necessary conditions exist, the relative proportion of Cryptogamous plants is not materially different from that of the south of Europe.

In that which I have called the principal parallel of New Holland, however, Cryptogamous plants appear to be much less numerous than in the corresponding latitudes of the northern hemisphere; and within the tropic they probably do not form more than one twelfth of the whole number of species.

In several of the islands of the Gulf of Carpentaria, having a Flora of Phænogamous plants exceeding 200 species, I did not observe a single species of Moss.

From the three primary classes of plants already treated of I proceed at once to those groups called NATURAL ORDERS or Families; for the intermediate divisions are too much at variance with the natural series to be made the subject of such general remarks as have been already offered on the primary classes, and which are equally admissible with respect to the natural families.

A methodical, and at the same time a natural, arrangement of these families is, in the existing state of our knowledge, perhaps impracticable. It would probably facilitate its future attainment, if at present, entirely neglecting it, attention were turned to the combination of these orders into Classes equally natural, and which, on a thorough investigation, might equally admit of being defined. The existence of certain natural classes is already acknowledged, and I have, in treating of the Australian natural families, ventured to propose a few that are perhaps less obvious, still more, however, might have been suggested had this been the place for pursuing the subject.

540] The natural orders in the Genera Plantarum of Jussieu are exactly 100; subsequent observations of Jussieu himself and of other botanists have considerably increased their numbers, so that in the lately published *Théorie Elémentaire de la Botanique* of Decandolle they amount to 145.

The plants of Terra Australis are referable to 120

natural orders, some of which are not included in Decandolle's list.

On such of these as either contribute largely to form the mass or the striking peculiarities of the Australian vegetation, I proceed to offer a few observations, chiefly on their geographical distribution, and more remarkable points of structure: taking them nearly in the same series in which they are given by Decandolle in the work already referred to.

MALVACEÆ. The Malvaceæ may be considered as a class including several orders, namely, *Malvaceæ* of Jussieu,[1] *Sterculiaceæ* of Ventenat,[2] *Chlenaceæ* of Du Petit Thouars,[3] *Tiliaceæ* of Jussieu,[4] and an order very nearly related to the last, and perhaps gradually passing into it, but which I shall in the mean time, distinguish under the name of *Buttneriaceæ*.

Of the *Malvaceæ* strictly so called, upwards of fifty species have been observed in Terra Australis, where the maximum of the order appears to be within the tropic. In the principal parallel Malvaceæ are more abundant at its eastern than its western extremity; and at the south end of Van Diemen's Island two species only have been observed. There is nothing very peculiar in the structure or appearance of the New Holland plants of this family; most of them belong to genera already established, and several of the species are common to other countries.

BUTTNERIACEÆ.[5] The Australian portion of *Buttneriaceæ* consists of Abroma, Commersonia, Lasiopeta- [541
lum, and several unpublished genera, intermediate to the last two.

[1] *Gen. pl.* 271.
[2] *Malmais.* 91.
[3] *Plant. des isles d'Afrique,* 46.
[4] *Gen. pl.* 289.

[5] BUTTNERIACEÆ. *Calyx* 1-ph. 5-fid. æqualis, marcescens, æstivatione valvata. *Petala* 5: vel basi saccata superne variè producta; vel minuta squamuliformia; quandoque nulla. *Stamina* hypogyna, definita: *Filamenta antherifera* cum laciniis calycis alternantia, simplicia, vel 2-3 connata; *sterilibus* quandoque alternantibus. *Ovarium* 3-5 loculare, loculis 2-polyspermis, ovulis erectis: *Styli* 3-5, sæpius connati: *Stigmata* simplicia. *Capsula* 3-5-loc. *Semina:* umbilico strophiolato. *Embryo* erectus, in axi albuminis carnosi cujus dimidio

The greater part of the order exists in the principal parallel, very few species have been observed within the tropic, and one only in Van Diemen's Island.

Lasiopetalum, the most extensive genus of the family, was established by Dr. Smith,[1] who considered it to belong to Ericeæ. Ventenat,[2] taking a different view of its structure, has assigned some plausible reasons for referring it to Rhamneæ. From both these orders it appears to me sufficiently distinct, and it is certainly more nearly related to the genera with which I have placed it.

DILLENIACEÆ. It was first, I believe, proposed by Mr. Salisbury to separate Dillenia, Wormia, Hibbertia, and Candollea from the Magnoliæ of Jussieu, and to form them into a distinct order, which he has called *Dilleneæ.*[3] It is remarkable that Decandolle,[4] who has adopted this order, should also limit it to these genera, Jussieu[5] having previously suggested the separation of Dillenia from Magnoliæ and its combination with Tetracera and Curatella, genera which certainly belong to Dilleniaceæ, as do also Pleurandra of Labillardière[6] and Hemistemma of Du Petit Thouars.[7]

The Dilleniaceæ appear to be more abundant in Terra Australis than in any other part of the world, nearly seventy Australian species having already been observed; most of these belong to Hibbertia and Pleurandra, both of which are very generally diffused, their maximum, however, is in the principal parallel, to the western extremity of which Candollea seems to be limited. Hemistemma, Wormia,
542] and an unpublished genus remarkable for its thickened filaments and flat leafless stems, are found only within the tropic. The remaining genera of the order have not yet been observed in New Holland.

longior. Frutices *rarò* Arbores, *pube sæpe stellari.* Folia *alterna, simplicia, stipulata, sæpius dentata.* Pedunculi *subcymosi, oppositifolii; pedicellis utplurimum bracteatis.*

[1] *Linn. soc. transact.* 4, p. 216.

[2] *Malmais.* 59. *Dec. gen. nov. p.* 7.

[3] *Paradis. Lond.* 73.

[4] *Annales du mus.* 17, *p.* 400.

[5] *Annales du mus.* 14, *pp.* 129-130.

[6] *Plant. Nov. Holl.* 2, *p.* 5.

[7] *Gen. nov. Madagasc. n.* 61.

Magnoliaceæ and Dilleniaceæ appear to me to form two orders of one natural class. These orders are sufficiently distinct from each other in most cases, both in fructification and habit; they are not, however, easily defined. The ovaria, which are indefinite in number, in the greater part of Magnoliaceæ, are also so in certain Dilleniaceæ; there are likewise examples in both orders, in which they are reduced to unity; and the stipulation of Magnoliaceæ exists in Wormia.

PITTOSPOREÆ.[1] Authors have generally been disposed to consider Pittosporum, Bursaria, and Billardiera, as belonging to Rhamneæ or Celastrinæ, from both of which they are certainly widely different; and they appear to me to constitute, along with some unpublished Australian genera, a very distinct natural family. PITTOSPOREÆ form a small tribe chiefly belonging to Terra Australis, where most of them have been observed in the principal parallel; but certain species of all the published genera exist at the south end of Van Diemen's Island, and both Pittosporum and Bursaria are found within the tropic. Pittosporum, the only genus of the order which is not confined to Terra Australis, has the most extensive range in that country, and has been found in many other parts of the world, namely, New Zealand, Norfolk Island, the Society and Sandwich Islands, the Moluccas, in China, Japan, and even Madeira. It has not, however, been observed in any part of America.

POLYGALEÆ.[2] The curious observation of Richard,

[1] PITTOSPOREÆ. *Calyx* 5-ph. (raò 1-ph. 5-fid.) æstivatione imbricata *Petala* 5: unguibus conniventibus, nunc cohærentibus; laminis patulis, æstivatione imbricatis. *Stamina* 5, hypogyna, distincta, cum petalis alternantia. *Ovarium* loculis placentisve 2-5 polyspermis: *Stylus* 1: *Stigmata* numero placentarum. *Pericarpium* capsulare vel baccatum, loculis polyspermis quandoque incompletis. Embryo minutus, prope umbilicum, inclusus albumine carnoso. Frutices *vel* Arbores. Folia *simplicia, alterna, exstipulata.* Flores *terminales, vel axillares, quandoque polygami.*

[2] POLYGALEÆ. *Calyx* 5-ph. raro 5-fid. æstivatione imbricata: sæpius irregularis: foliolis 2 lateralibus interioribus majoribus quandoque petaloideis;

543] that the arillus of the seed, whether general or partial, is never found in the Dicotyledonous orders with monopetalous flowers, seems to have determined Jussieu[1] and other French botanists to remove Polygala, remarkable for its caruncula umbilicalis, from Rhinanthaceæ, with which they had placed it, and to consider it, along with some nearly related genera, as forming a distinct polypetalous order. They appear to me, however, not to have taken so correct a view of the structure of its Corolla as Adanson,[2] who very justly observes that both in this genus and Securidaca, which he rightly associates with it, the apparently monopetalous corolla is made up of three petals, united by means of the cohering filaments, the external sutures remaining visible; but Adanson himself has not observed the minute rudiments of two additional petals in Securidaca, the existence and position of which assist in explaining the nature of the irregularity in Polygala, where no such rudiments are found, but in which the corolla is in every other respect very similar. A much nearer approach to regularity, however, takes place in an unpublished genus, having five petals, which, though irregular, are of nearly equal size and similarly connected by the cohering filaments, likewise five in number. The essential characters of the order Polygaleæ to which Krameria, Monnina, Salomonia, and several unpublished genera also belong, consist in the hypogynous insertion of its corolla, which is always irregular, and frequently reduced to three petals, connected together by the cohering filaments, whose antheræ are simple and bursting only at the top.

reliquorum duobus anterioribus (respectu spicæ) tertio postico. *Petala* 3-5, mediante tubo stamineo connexa, raró distincta. *Stamina* hypogyna, 8 (nunc 3, 4 vel 5): filamentis infernè connatis in tubulum hinc apertum inde petala connectentem: *Antheræ* simplices, basi insertæ, poro apicis dehiscentes. *Ovarium* 2-loc. (quandoque 1-3 loc.) ovulis solistariis pendulis: *Stylus* 1: *Stigma* sæpe bilabiatum. *Pericarpium* sæpius capsulare, biloculare, bivalve, valvis medio septigeris: nunc Drupa vel Samara. 1-2 sperm. *Semina* pendula, umbilico (in capsularibus) strophiolato vel comoso. *Embryo* in axi albuminis carnosi vix longioris, quandoque (præsertim in pericarpiis clausis) deficientis. Herbæ *vel* Frutices, *utplurimum glabri.* Folia *simplicia indivisa alterna exstipulata.* Flores *spicati sæpius terminales.*

[1] *Annales du mus.* 14, *p.* 386, *et seq.* [2] *Fam. des Plantes*, 2, *p.* 348.

About thirty species of this order are found in Terra Australis; these are either Comespermæ or Polygalæ, with a single species of Salomonia of Loureiro, a genus [544
which is certainly not monandrous, as that author affirms, but has four connected filaments with distinct unilocular antheræ, and consequently half the number of stamina usually found in the order. Most of the Comespermæ exist in the principal parallel, and equally at both its extremities; several, however, are found beyond it, and in both directions; the genus extending from Arnhem's Land to Adventure Bay. The greater part of the Polygalæ and the genus Salomonia exist only within the tropic.

TREMANDREÆ.[1] The genus Tetratheca of Dr. Smith and one very nearly related to it, which I shall hereafter publish under the name of *Tremandra*, constitute a small tribe of plants peculiar to Terra Australis. For this tribe I prefer the name *Tremandreæ* to that of Tetrathecaceæ, as it is more distinctly, and at the same time more correctly descriptive of the structure of stamina in both genera; the four distinct cells in the ripe state of the antheræ not existing in Tremandra, nor even in all the species of Tetratheca. In the quadrilocular antheræ of the latter genus there is indeed nothing peculiar, that being the original structure of all those antheræ which are commonly described as bilocular; and the difference in this case depending on the mode of bursting, which, when lateral, necessarily obliterates two of the septa, but when terminal, as in Tetratheca, admits of their persistence. It is remarkable that both Dr. Smith and Labillardière have mistaken the fungous appendix of the apex of the seed for an umbilical

[1] TREMANDREÆ. *Calyx* 4-5 ph. æqualis, æstivatione valvata. *Petala* 4-5, æqualia: æstivatione involuta stamina includentia. *Stamina* 8-10, hypogyna, distincta: *Antheræ* 2-4 loculares, basi insertæ, poro tubulove apicis dehiscentes. *Ovarium* 2-loc. loculis 1-3-spermis, ovulis pendulis: *Stylus* 1: *Stigmata* 1-2. *Capsula* bilocularis, bivalvis, valvis medio septigeris. *Semina* umbilico nudo: extremitate opposita appendiculata; albuminosa. *Embryo* in axi albuminis carnosi cujus dimidio longior: *radicula* umbilicum spectante. Fruticuli *ericoides.* Folia *sparsa vel verticillata, exstipulata.* Pedunculi *axillares, uniflori.*

caruncula, a mistake involving a second, that of considering the seeds erect in the capsule, and which has led Labillardière into a third error, namely, describing the radicule of the embryo as pointing towards this supposed umbilical appendix.

545] The Tremandreæ are in several respects nearly related to Polygaleæ; they appear to me, however, sufficiently distinct, not only in the regularity of the flower, and in the structure of antheræ, but in the æstivation of both calyx and corolla, in the appendix of the seed being situated at its apex, and not at the umbilicus, and, I may also add, in a tendency to produce an indefinite number of ovula in each cell of the ovarium.

The greater number of Tremandreæ are found in the principal parallel of New Holland, they extend also to the south end of Van Diemen's Island, but none have been observed within the tropic.

DIOSMEÆ. To this natural order, in addition to the Australian genera hereafter to be mentioned, and the south African genus from which its name is derived, I refer Fagara, Zanthoxylon, Melicope, Jambolifera, Euodia, Pilocarpus, Empleurum, and Dictamnus: and four genera of equinoctial America, namely, Cusparia of Humboldt and Bonpland, Ticorea and Galipea of Aublet, and Monnieria, if not absolutely of this order, belong at least to the same natural class.

Both Ruta and Peganum may be annexed to Diosmeæ, but neither of them are calculated to give a clear idea of the order, from the usual structure and habit of which they deviate in some important points; I have therefore proposed to derive the name of the family from one of its most extensive and best known genera. The first section of Jussieu's Rutaceæ is sufficiently different to admit of its being considered a distinct order, which may be named Zygophylleæ.

Diosmeæ are numerous in Terra Australis, and form, at least in its principal parallel and more southern regions, a striking feature in the vegetation. Nearly seventy species

have been observed, of which the greater part are referable to Boronia, Correa, Eriostemon, and Zieria, of Dr. Smith, and Phebalium of Ventenat. Of these genera *Boronia* is both the most extensive and the most widely diffused, existing within the tropic, and extending to the South end of Van Diemen's Island; like the others, however, its maximum is in the principal parallel, at both extremities of which it is equally abundant. *Correa,* though extending to the south end of Van Diemen's Island, is not found within the tropic, nor was it observed at the western extremity of the principal parallel; in the intermediate part of which, however, where many of the peculiarities in the [546
vegetation of the parallel are less remarkable, or entirely wanting, it may be said to abound.

Eriostemon, which appears to be most abundant at the eastern extremity of the principal parallel, has not been observed either at its western extremity or intermediate part; it extends, however, to the south end of Van Diemen's Island on the one hand, and within the tropic as far as Endeavour River on the other.

Phebalium, very nearly related to Eriostemon, has like that genus its maximum at the eastern extremity of the principal parallel, it is found also at the western extremity of this parallel, and as far as the south end of Van Diemen's Island, but it has not been observed within the tropic.

Zieria seems to be limited to the eastern extremity of the principal parallel, and the more southern regions.

The most remarkable plant of the order with regard to structure, is that imperfectly figured and described in Dampier's voyage.[1] Of this genus, which may be named DIPLOLÆNA, I have examined Dampier's original specimen in the Sherardian Herbarium at Oxford, and others recently collected, also at Shark's Bay, in the voyage of Captain Baudin, and have ascertained that what appear to be calyx and corolla in this singular plant, are in fact a double involucrum containing many decandrous flowers, whose stamina and pistilla exactly agree with those of the

[1] Vol. 3, p. 110, tab. 3, f. 3.

order, but of which the proper floral envelopes are reduced to a few irregularly placed scales.

Another Australian genus of Diosmeæ differs from the rest of the order in having a calyx with ten divisions, an equal number of petals, and an indefinite number of stamina with evidently perigynous insertion.

MYRTACEÆ.[1] This is one of the most extensive tribes in Terra Australis, in which considerably above 200 species have already been observed, and where the order is also more strikingly modified than in any other part of the world. It is very generally spread over the whole of Australia, but its maximum appears to be in the principal parallel. Many observations might here with propriety be introduced on the more remarkable structures which occur among the Australian Myrtaceæ; I must, however,
547] confine myself to a few remarks on the distribution of the most extensive genera.

Of *Eucalyptus* alone nearly 100 species have been already observed; most of these are trees, many of them of great and some of enormous dimensions. Eucalyptus globulus of Labillardière and another species peculiar to the south end of Van Diemen's Island, not unfrequently attain the height of 150 feet, with a girth near the base of from 25 to 40 feet. In the colony of Port Jackson there are also several species of great size, but none equal to those of Van Diemen's Island; and no very large trees of this genus were seen either on the south coast or in the equinoctial part of New Holland. Mr. Caley has observed within the limits of the colony of Port Jackson nearly 50 species of Eucalyptus, most of which are distinguished, and have proper names applied to them, by the native inhabitants, who, from differences in the colour, texture, and scaling of the bark, and in the ramification and general appearance of these trees, more readily distinguish them than botanists have as yet been able to do. Eucalyptus, although so generally spread over the whole of Terra Australis, and so abundant as to form at least four fifths of its forests, is

[1] Myrti. *Juss. gen.* 322.

hardly found beyond this country. I am acquainted with one exception only, in an additional species which is said to be a native of Amboyna.

Next to Eucalyptus in number, is the beautiful genus *Melaleuca*, of which upwards of 30 Australian species have already been observed, exclusive of Tristania, Calothamnus, Beaufortia, and an unpublished genus which I separate from it. The maximum of Melaleuca exists in the principal parallel, but it declines less towards the south than within the tropic, where its species are chiefly of that section which gradually passes into Callistemon, a genus formed of those species of Metrosideros that have inflorescence similar to that of Melaleuca, and distinct elongated filaments. With the exception of two species of this section, namely, Melaleuca leucadendron, and M. Cajeputi, the genus Melaleuca appears to be confined to Terra Australis.

Leptospermum, of which, nearly 30 Australian species have been observed, exists also in New Zealand and in the Moluccas. In Terra Australis its maximum is decidedly in the principal parallel, and like Melaleuca, it is much more abundant in the southern regions than within the tropic.

Bæckia, to which I refer Imbricaria of Dr. Smith, as [548
well as the opposite-leaved Leptospermums, is also an extensive Australian genus, having its maximum in the principal parallel, extending like the two former genera to the highest southern latitude, and hardly existing within the tropic: one species, however, has been found in New Caledonia, and that from which the genus was formed is a native of China.

COMBRETACEÆ.[1] I have formerly[2] made some

[1] COMBRETACEÆ. *Calyx* superus: limbo 4-5-fido, æquali. *Petala* 4-5 vel nulla. *Stamina* 8-10; quandoque laciniis calycis æqualia et cum iisdem alternantia. *Ovarium* uniloculare, ovulis 2-4, ab apice loculi pendulis absque receptaculo communi vel columna centrali: *Stylus* 1: *Stigma* 1. *Pericarpium* monospermum, clausum, figura et textura varium, Drupa v. Samara. *Semen* exalbuminosum. *Embryo* cotyledonibus sæpius involutis: plumula inconspicua.

Arbores *vel* Frutices. Folia *simplicia, integra, exstipulata, alterna nunc opposita, raro punctato-pellucida.* Flores *spicati, axillares.*

[2] *Prodr. fl. Nov. Holl.* 351.

remarks on the structure and limits of Combretaceæ, one of whose principal characters consists in the unilocular ovarium with two or more ovula simply pendulous from the upper part of its cavity, and not inserted, as in Santalaceæ, into a central receptacle or column. Guiera of Jussieu, having the same structure, and also leaves dotted with pellucid glands, appears to connect this order with Myrtaceæ.

The Australian Combretaceæ, which belong to Terminalia, Chuncoa, and Laguncularia, are not numerous, and all of them are found within the tropic.

CUNONIACEÆ.[1] This order, several of whose genera have been referred to Saxifrageæ, is more readily distinguished from that family by its widely different habit, than by any very important characters in its fructification; like
549] Saxifrageæ also it comprehends genera with ovarium superum and inferum.

The genera strictly belonging to *Cunoniaceæ* are Weinmannia, Cunonia, Ceratopetalum, Calycomis, and Codia. To this order Bauera may also be referred, but it must form a separate section from the genera already mentioned. Of these *Weinmannia*, *Ceratopetalum* and *Calycomis* are found in Terra Australis, and hitherto only at the eastern extremity of its principal parallel, where also *Bauera* is most abundant; but this genus is found beyond the parallel in one direction, extending to the southern extremity of Van Diemen's Island.

RHIZOPHOREÆ.[2] The genera Rhizophora, Bruguiera,

[1] CUNONIACEÆ. *Calyx* 1-ph. 4-5-fidus, semisuperus vel inferus. *Petala* 4-5; raró nulla. *Stamina* perigyna, definita, 8-10. *Ovarium* biloculare, loculis 2-polyspermis: *Stylus* 1-2. *Pericarpium* biloculare, capsulare vel clausum. *Embryo* in axi albuminis carnosi.

Arbores *vel* Frutices. Folia *opposita, composita vel simplicia, sæpius stipulata stipulis interpetiolaribus.*

[2] RHIZOPHOREÆ. *Calyx* superus 4-5-fidus, æstivatione valvata. *Petala* 4-5. *Stamina* perigyna, 8-15. *Ovarium* 2-loc. loculis 2-polyspermis ovulis pendulis: *Stylus* 1. *Pericarpium* clausum, monospermum. *Semen* exalbuminosum. *Embryo* sæpe germinans et pericarpium semisuperum perforans.

Arbores. Folia *opposita, simplicia, stipulis interpetiolaribus.*

and Carallia, all of which are found in the equinoctial part of New Holland, form a distinct natural order which may be called *Rhizophoreæ.* This order agrees with Cunoniaceæ in its opposite leaves and intermediate stipulæ, and with great part of them in the æstivation of its calyx, and in the structure and cohesion of the ovarium. From these it differs chiefly in the want of albumen and greater evolution of its embryo. Jussieu[1] has combined Rhizophora and Bruguiera with Loranthus and Viscum, neglecting some very obvious, and, as they appear to me, important differences in the flower, and probably never having had an opportunity of comparing the very distinct structures of their ovaria; the affinity too of Rhizophoreæ to Cunoniaceæ is unquestionable, and it will hardly be proposed to unite both these tribes with Loranthus, which I consider as even more nearly related to Proteaceæ.

HALORAGEÆ. The greater part of the genera of which this order is composed, have been referred to Onagrariæ, to certain parts of which they no doubt very nearly approach; but it must appear rather paradoxical to unite Fuchsia in the same family with Myriophyllum and even Hippuris, and it would be in vain to attempt a definition [550
of an order composed of such heterogenous materials. By the separation of the order here proposed it becomes at least practicable to define Onagrariæ. It is still, however, difficult to characterise Halorageæ, which will probably be best understood by considering as the type of the order the genus Haloragis, froan which all the others differ by the suppression of parts or separation of sexes. Thus *Meionectes*, an unpublished genus of New Holland, is reduced to half the number of parts both of flower and fruit. *Proserpinaca* is deprived of petals and of one fourth of all the other parts. *Myriophyllum,* which is monœcious, has the complete number of parts in the male flower, but in the female wants both calyx and corolla; what several authors have described as petals being certainly bracteæ.

[1] *Annales du mus.* 12, *p.* 288.

Serpicula differs from Myriophyllum in having only half the number of stamina in the male flower, and in its unilocular four-seeded ovarium.

Hippuris, though retaining the habit of Myriophyllum, yet having a monandrous hermaphrodite flower without petals, and a single-seeded ovarium, is less certainly reducible to this order: and it may appear still more paradoxical to unite with it *Callitriche*, in which, however, I am inclined to consider what authors have denominated petals as rather analogous to the bracteæ in the female flower of Myriophyllum and Serpicula, and to both these genera Callitriche in the structure of its pistillum, and even in habit, very nearly approaches.

The Australian genera of this order are Haloragis, Meionectes, Myriophyllum, and Callitriche.

Of *Haloragis*, many new species have been observed in Terra Australis, in every part of which this genus is found, most abundantly, however, at both extremities of the principal parallel.

That *Gonocarpus* really belongs to the same genus, I am satisfied from an examination of original specimens sent by Thunberg himself, to Sir Joseph Banks, for in these I find not only petals, but eight stamina and a quadrilocular ovarium.

LEGUMINOSÆ.[1] This extensive tribe may be considered as a class divisible into at least three orders, to
551] which proper names should be given. Of the whole class about 2000 species are at present published, and in Terra Australis, where this is the most numerous family, considerably more than 400 species have already been observed.

One of the three orders of Leguminosæ which is here for the first time proposed may be named MIMOSEÆ. It consists of the Linnean Mimosa, recently subdivided by Willdenow into five genera, along with Adenanthera and Prosopis.

[1] *Juss. gen.* 345.

This order is sufficiently distinguished from both the others by the hypogynous insertion and valvular æstivation of its corolla, which being perfectly regular differs in this respect also from the greater part of Lomentaceæ and from all the Papilionaceæ.

Nearly the whole of the Australian species of the Linnean genus Mimosa belong to *Acacia* of Willdenow, as it is at present constituted; and about nine tenths of the Acaciæ to his first division of that genus, described by him as having simple leaves, but which is in reality aphyllous; the dilated foliaceous footstalk performing the functions of the true compound leaf, which is produced only in the seedling plant, or occasionally in the more advanced state in particular circumstances, or where plants have been injured.

The great number of species of Acacia having this remarkable economy in Terra Australis forms one of the most striking peculiarities of its vegetation. Nearly 100 species have already been observed; more than half of these belong to the principal parallel, at both extremities of which they appear to be equally abundant; they are, however, very generally diffused over the whole country, existing both on the north coast of New Holland, and at the south end of Van Diemen's Island. But though the leafless Acaciæ are thus numerous and general in Terra Australis, they appear to be very rare in other parts of the world; none of the Australian species are found in other countries, and at present I am acquainted with only seven additional species, of which five are natives of the intratropical Islands of the southern hemisphere; the sixth was observed in Owhyhee, and is said to be the largest tree in the Sandwich Islands; the seventh is *Mimosa stellata* of Loureiro, upon whose authority it entirely rests.

The second order, LOMENTACEÆ or CÆSALPINEÆ, comprehends all the genera having perigynous stamina, a corolla whose æstivation is not valvular, and which though [552
generally irregular is never papilionaceous. To these characters may be added the straight embryo, in which they

agree with Mimoseæ, but differ from all the Papilionaceæ except Arachis and Cercis.

The Lomentaceæ of New Holland are not numerous, and consist chiefly of the genus Cassia, the greater part of whose species grow within the tropic. On the east coast they probably do not extend beyond 35° lat.; and on the south coast only one species has been observed, it was found in 32° S. lat. and is remarkable in being aphyllous, with dilated footstalks exactly like the Acaciæ already noticed.

The third order, PAPILIONACEÆ, which comprehends about three fourths of the whole class at present known, includes also nearly the same proportion of the Australian Leguminosæ.

Papilionaceæ admit of subdivision into several natural sections, but in Terra Australis they may be divided almost equally, and without violence to natural affinities, into those with connected and those with distinct stamina.

The decandrous part of the whole order bears a very small proportion to the diadelphous, which in Persoon's synopsis is to the former as nearly 30 to 1, while in Terra Australis, as I have already stated, the two tribes are nearly equal.

This remarkably increased proportion of Decandrous Papilionaceous plants, forms another peculiarity in the vegetation of New Holland, where their maximum exists in the principal parallel. They are not so generally spread over the whole of Terra Australis, as the leafless Acaciæ, for although they extend to the southern extremity of Van Diemen's Island, they are even there less abundant, and very few species have been observed within the tropic. Papilionaceous plants with distinct stamina do not in fact form a very natural subdivision of the whole order, though those of New Holland, with perhaps one or two exceptions, may be considered as such: this Australian portion, however, forms nearly three fourths of the whole section, at present known; the remaining part, consisting of genera, most of which are very different, both from each other and from those of Terra Australis, are found at the Cape of

Good Hope, in equinoctial and North Africa, in the different regions of America, in New Zealand, in India, very sparingly in North Asia, and lastly in the South of Europe, where, [553 however, only two species have been observed, namely, Anagyris fœtida, and Cercis siliquastrum; but the latter having a straight embryo and a habit approaching to that of Bauhinia, rather belongs to Lomentaceæ.

Among the Diadelphous genera of Terra Australis the most remarkable in habit and structure, namely Platylobium, Bossiæa, Hovea, Scottia, and Kennedia, are found chiefly in the principal parallel and higher latitudes; within the tropic the greater part of these cease to exist, and most of the genera which there occur are common to other countries, especially India.

ATHEROSPERMEÆ.[1] Jussieu, in his excellent memoir on Monimieæ[2] has referred *Pavonia* of Ruiz and Pavon and *Atherosperma* of Labilardière to that order, from the other genera of which, namely, Ambora, Monimia, and Ruizia, they appear to me very different, not only in the insertion of the seed, the texture of the albumen, and relative size of the embryo, but in having antheræ similar to those of Laurinæ. I separate them therefore into a distinct family with the name of ATHEROSPERMEÆ. The propriety of this separation is confirmed by the discovery of two New Holland plants, evidently belonging to this family, but which have hermaphrodite flowers; a structure not likely to occur in Monimieæ, in which what has been termed calyx is more properly an involucrum.

[1] ATHEROSPERMEÆ. *Flores* diclines vel hermaphroditi. *Calyx* monophyllus, limbo diviso: laciniis sæpe duplici serie, interioribus omnibusve semipetaloideis: *Squamulæ faucis* in femineis et hermaphroditis. *Corolla* nulla. *Stamina* in masculis floribus numerosa, fundo calycis inserta, squamulis aucta; in hermaphroditis pauciora, fauce imposita: *Antheræ* adnatæ, biloculares, loculis valvula longitudinali a basi ad apicem dehiscenti. *Ovaria* uno plura, sæpius indefinita, monosperma, ovulo erecto: *Styli* simplices, nunc laterales v. basilares: *Stigmata* indivisa. *Pericarpia* clausa seminiformia, stylis persistentibus plumosis aristata, tubo aucto calycis inclusa. *Embryo* erectus brevis, in basi albuminis carnosi mollis.

Arbores. Folia *opposita simplicia exstipulata.* Pedunculi *axillares, uniflori.*

[2] *Annales du museum,* 14, *p.* 116.

The place of Atherospermeæ in the natural series is not very easily determined. It is singular that differing so widely as they certainly do in most parts of their structure from Laurinæ they should notwithstanding agree with them in the economy of their antheræ, and very remarkably with some of them in their sensible qualities. Of the
554] three Australian plants of this order two are found in the colony of Port Jackson, the third through the whole of Van Diemen's Island. Pavonia of the Flora Peruviana (Laurelia of Jussieu), a native of a similar climate, and possessing the same sensible qualities, is more nearly related to Atherosperma than is generally supposed, differing from it merely in the oblong form and regular bursting of its female calyx.

RHAMNEÆ. Into this order I admit such genera only as have ovarium cohering more or less with the tube of the calyx, of which the laciniæ have a valvular æstivation; stamina equal in number to these laciniæ, and alternating with them; an ovarium with two or three cells and a single erect ovulum in each; an erect embryo generally placed in the axis of a fleshy albumen, or entirely without albumen; the petals, which are opposite to the stamina, and inclose the antheræ in their concave laminæ, are in some cases wanting.

With these characters Rhamnus, Ziziphus, Paliurus, Ceanothus (from which Pomaderris is hardly distinct), Colletia, Cryptandra, Phylica, Gouania, Ventilago, and probably Hovenia correspond. In comparing this description of Rhamneæ with that of Buttneriaceæ formerly given, they will be found to coincide in so many important points, that the near relationship of these two orders cannot be doubted, and thus an unexpected affinity seems to be proved between Rhamneæ and Malvaceæ.

In Terra Australis upward of thirty species of Rhamneæ belonging to Ziziphus, Ceanothus, Pomaderris, Colletia and Cryptandra, have been observed, and chiefly in its principal parallel or southern regions.

CELASTRINÆ.[1] This order comprehends the greater part of the first two sections of the Rhamni of Jussieu; it is obviously different from the more limited order of Rhamneæ, which I have already attempted to define, and in many respects so nearly approaches to the *Hippocra-* [555 *ticeæ* of Jussieu,[2] that it may be doubted whether they ought not to be united.

In New Holland the Celastrinæ are not numerous, nor do they form any part of its characteristic vegetation; their distribution is somewhat different from that of Rhamneæ, for they are found either in the principal parallel, or within the tropic.

STACKHOUSEÆ.[3] Stackhousia of Dr. Smith,[4] and an unpublished genus, exactly agreeing with it in flower, but remarkably different in fruit, form a small tribe of plants, sufficiently distinct from all the natural orders hitherto established. I have placed it between Celastrinæ and Euphorbiaceæ; to both of which, but especially to the former, it seems to be related in a certain degree.

The Stackhouseæ are peculiar to Terra Australis, and though found chiefly in its principal parallel, extend more sparingly both to the southern extremity of Van Diemen's Island, and to the North coast of New Holland.

[1] CELASTRINÆ. *Calyx* 4-5-partitus, æstivatione imbricata. *Petala* 4-5. *Stamina* totidem, cum petalis alternantia, insertione ambiguè perigyna. *Ovarium* liberum, 2-4-loculare loculis 1-polyspermis, ovulis erectis (raro pendulis): *Stylus* 1-4. *Pericarpium* capsulare, vel clausum (Baccatum, Drupaceum vel alatum.) *Semina* in capsularibus arillata. *Embryo* fere longitudine albuminis carnosi, axilis.

Frutices *vel* Arbores. Folia *simplicia* (*raro composita*) *alterna vel opposita, stipulata stipulis sæpius minutis, quandoque nullis.*

[2] *Annales du mus.* 18, *p.* 486.

[3] STACKHOUSEÆ. *Calyx* 1-ph. 5-fidus, æqualis: tubo ventricoso. *Petala* 5, æqualia, summo tubo calycis inserta: unguibus cohærentibus in tubum calyce longiorem; laminis angustis stellato-patulis. *Stamina* 5, distincta, inæqualia (duo alterna breviora), fauci calycis inserta. *Ovarium* liberum, 3-5-lobum, lobis discretis monospermis, ovulis erectis: *Styli* 3-5, nunc basi cohærentes: *Stigmata* indivisa. *Pericarpium* 3-5-coccum, coccis evalvibus, apteris v. alatis; columna centrali persistenti. *Embryo* erectus axilis, longitudine fere albuminis carnosi.

Herbæ. Folia *simplicia, integerrima, sparsa, quandoque minuta:* Stipulæ *laterales minutissimæ.* Spica *terminalis; floribus tribracteatis.*

[4] *Linn. soc. transact.* 4, *p.* 218.

EUPHORBIACEÆ.[1] This is an extensive and very general family, of which about 100 species have already been observed in Terra Australis. Of these the greater part exist within the tropic, but the order extends to the southern extremity of Van Diemen's Island, and the greater number of the genera peculiar to this country are found in the principal parallel or higher latitudes.

556] The species of *Euphorbia* are not numerous in Terra Australis, most of them are intratropical plants, and all of them are referable to one section of the genus. It appears to me that the name of the order ought not to be taken from this genus, which is so little calculated to afford a correct idea of its structure that authors are still at variance in the names and functions they assign to several parts of the flower. The view I take of the structure of *Euphorbia* is, in one important particular at least, different from those given by Lamarck,[2] Ventenat,[3] Richard?[4] and Decandolle,[5] though possibly the same that Jussieu has hinted at;[6] so briefly, however, and I may add obscurely, that if his supposition be really analogous to what I shall presently offer, he has not been so understood by those who profess to follow him in this respect.

With all the authors above quoted, I regard what Linneus has called calyx and corolla in *Euphorbia* as an involucrum, containing several male flowers which surround a single female. By some of these authors the male flowers are described as monandrous, and in this respect, also, I agree with them; but the body, which all of them describe as a jointed filament, I consider to be made up of two very distinct parts, the portion below the joint being the footstalk of the flower, and that above it the proper filament; but as the articulation itself is entirely naked, it follows that there is no perianthium; the filiform or laciniated scales which authors have considered as such, being on this supposition analogous to bracteæ. The female flower, in conformity with this supposition, has also its pedunculus, on

[1] *Jus. gen.* 384.
[2] *Encyclop. botan.* 4, *p.* 413.
[3] *Tableau,* 3, *p.* 487.
[4] *In Michaux. fl. bor. Amer.* 2, *p.* 209.
[5] *Flor. Franc.* 3, *p.* 329.
[6] *Gen. pl.* 386.

the dilated, and in a few cases obscurely lobed, apex of which the sessile ovarium is placed. If this be a correct view of the structure of Euphorbia, it may be expected that the true filament or upper joint of what has commonly been called filament, should, as in other plants, be produced subsequent to the distinct formation of the anthera, which consequently will be found at first sessile on the lower joint or peduncle, after that has attained nearly its full length; and accordingly this proves to be the case in such species as I have examined. Additional probability is given to this view by the difference existing between the surfaces [557
of the two joints in some species. I consider it, however, as absolutely proved by an unpublished genus of this order, having an involucrum nearly similar to that of Euphorbia, and like it, inclosing several fasciculi of monandrous male flowers, surrounding a single female; but which, both at the joint of the supposed filament, and at that by which the ovarium is connected with its pedicellus, has an obvious perianthium, regularly divided into lobes.

UMBELLIFERÆ.[1] This order may be considered as chiefly European, having its maximum in the temperate climates of the northern hemisphere; in the corresponding southern parallels it is certainly much less frequent, and within the tropics very few species have been observed. In Terra Australis the Umbelliferæ, including a few Araliæ, which belong at least to the same natural class, exceed 50 species. The greater part of these are found in the principal parallel, in which also those genera deviating most remarkably from the usual structure of the order occur. The most singular of these is *Actinotus* of Labillardière,[2] which differs from the whole order in having a single ovulum in the unimpregnated ovarium. A second genus, which I shall hereafter publish with the name of *Leucolæna*, is worthy of notice on account of the great apparent differences of inflorescence existing amongst its species; which agree in habit,

[1] *Jus. gen.* 218.
[2] *Nov. Holl. pl. spec.* 1, *p.* 67, *t.* 92. Eriocalia, *Smith exot. bot.* 2, *p.* 37.

in the more essential parts of fructification, and even in their remarkable involucella. Of this genus, one species has a compound umbel of four many-flowered radii; a second has an umbel of three rays with two or three flowers in each; several others, still retaining the compound umbel, which is proved by the presence of their involucella, have from four to two single-flowered rays: and lastly one species has been observed, which is reduced to a single flower; this flower, however, is in fact the remaining solitary ray of a compound umbel, as is indicated by the two bracteæ on its footstalk, of which the lower represents the corresponding leaf of the general involucrum, while the upper is evidently similar to the involucellum of the two-rayed species of the genus.

558] COMPOSITÆ.[1] Of this family, which is the most extensive among Dicotyledones, upwards of 2500 species have been already described. About 300 are at present known in Terra Australis, in which therefore the proportion of Compositæ to its Dicotyledonous plants is considerably smaller than that of the whole order to Dicotyledones generally, and scarcely half that which exists in the Flora of South Africa. It is also inferior in number of species to Leguminosæ, like which it seems expedient to consider it as a class including several natural orders. Of these orders *Cichoraceæ* and *Cinarocephalæ* are comparatively very rare in Terra Australis, not more than ten species of both having hitherto been observed.

The class therefore chiefly consists of *Corymbiferæ*, which are very generally diffused; they are however evidently less numerous within the tropic, and their maximum appears to exist in Van Diemen's Island. Corymbiferæ may be subdivided into sections and the greater part of the genera peculiar to Terra Australis belong to that section which may be named *Gnaphaloideæ*, and exist either in the principal parallel or higher latitudes.

The whole of *Compositæ* agree in two remarkable points

[1] *Adans. fam.* 2, *p.* 103. *Decand. Theor. elem.* 216.

of structure in their corolla; which, taken together at least, materially assist in determining the limits of the class. The first of these is its valvular æstivation, this, however, it has in common with several other families. The second I believe to be peculiar to the class, and hitherto unnoticed. It consists in the disposition of its fasciculi of vessels, or nerves; these, which at their origin are generally equal in number to the divisions of the corolla, instead of being placed opposite to these divisions and passing through their axes, as in other plants, alternate with them; each of the vessels at the top of the tube dividing into two equal branches running parallel to and near the margins of the corresponding laciniæ, within whose apices they unite. These, as they exist in the whole class, and are in great part of it the only vessels observable, may be called primary. In several genera, however, other vessels occur, alternating with the primary and occupying the axes of the laciniæ: in some cases these secondary vessels, being most distinctly visible in the laciniæ, and becoming gradually fainter as they descend the tube, may be regarded as recurrent; originating from the united apices of the primary branches; but [559
in other cases where they are equally distinct at the base of the tube, this supposition cannot be admitted. A monopetalous corolla not splitting at the base is necessarily connected with this structure, which seems also peculiarly well adapted to the dense inflorescence of Compositæ; the vessels of the corolla and stamina being united, and so disposed as to be least liable to suffer by pressure.

As this disposition of vessels is found in Ambrosia and Xanthium, they ought not to be separated from Compositæ as Richard[1] has proposed; and as it does not exist in Brunonia I prefer annexing that genus to Goodenoviæ, with which it agrees in the peculiar indusium of the stigma.

GOODENOVIÆ.[2] This order I have formerly separated from Campanulaceæ, considering the peculiar membranous cup surrounding the stigma, along with a certain irregu-

[1] *Annales du mus.* 8, *p.* 184. [2] *Prodr. fl. Nov. Holl.* 573.

larity in the corolla, as sufficient distinguishing characters, especially as these are accompanied by other differences which appear to me important. In Goodenoviæ I have not included Lobelia, which, however, has also an irregular corolla, and although it wants the peculiar indusium of the stigma, has in its place a fasciculus or pencil of hair surrounding that organ. This structure has been regarded by Jussieu and Richard, in a very learned memoir, more recently written on the subject,[1] as analogous to the indusium of Goodenoviæ, to which they have therefore added Lobelia and derived the name of the order from this, its most extensive and best known genus. To the opinion of these authors I hesitate to accede, chiefly for the following reasons:

1st. In Goodenoviæ the deeper fissure of the tube of the corolla exists on its inner or upper side; a circumstance readily determined in those species having single spikes. In Lobelia, on the other hand, the corresponding fissure is on the outer or lower side, a fact, however, which can only be ascertained before the opening of the corolla, the flowers in the greater number of species becoming resupinate in the expanded state, a circumstance that does not appear to have been before remarked. The relation therefore not only of the corolla but of the calyx and stamina to the axis of inflorescence, is different in these two tribes.

560] 2ndly. In Goodenoviæ the greater part of the tube of the corolla is formed by the cohesion of five laciniæ, the distinct inflected margins of which are in most cases visible nearly to its base; these laciniæ are in some cases unconnected, as in *Diaspasis*, and more remarkably still in *Cyphia*, which is actually pentapetalous. I have observed no such structure in Lobelia.

3rdly. At the period of bursting of the antheræ the stigma in Lobelia is almost completely evolved, and capable of receiving impregnation from the pollen of the same flower; the function therefore of its surrounding pencil, is similar to that of the hairs which are almost equally obvious

Annales du mus. 18, *p.* 1.

in many Compositæ, especially Cinarocephalæ. On the contrary, in Goodenoviæ the stigma at the same period is hardly visible, and is certainly not then capable of receiving impregnation from the pollen of its proper flower; it is therefore either impregnated by the antheræ of different flowers, or in some cases at a more advanced stage by the pollen of its own antheræ, which is received and detained in the indusium. To these arguments for the exclusion of Lobelia I may add that in the greater part of Goodenoviæ with dehiscent fruit the dissepiment is parallel to the valves of the capsule, in which respect they differ equally from Lobelia and the valvular-fruited Campanulaceæ; and lastly, that many species of Lobelia as well as Campanulaceæ contain a milky juice of which there is no instance in Goodenoviæ. If, therefore, in Lobelia the pencil surrounding the stigma and the irregularity of the corolla, which, however, in some species is hardly perceptible, be considered as characters sufficient to separate this extensive genus from Campanulaceæ, it may form a separate order, admitting, perhaps, of subdivision into several distinct genera.

I have formerly observed[1] that in two genera of Goodenoviæ, namely, *Euthales* and *Velleia*, the base of the corolla coheres with the ovarium while the calyx remains entirely distinct. This structure I had stated as being peculiar to these genera, and as in some degree invalidating one of Jussieu's arguments for considering the floral envelope of Monocotyledones as calyx rather than corolla. The fact, however, seems not to be admitted by Richard, who in the dissertation already quoted[2] describes what has hitherto been called calyx in Velleia as bracteæ; a view of the structure which in those species of that genus having triphyllous calyx may appear plausible, but of which the probability is [561
diminished even in those with pentaphyllous calyx, and still more in Euthales, where the calyx is also tubular. But a stronger argument for the part usually denominated calyx being in these genera really such may be derived from certain species of Goodenia, in which it will be admitted that both calyx and corolla are present, and where, though

[1] *Prod. fl. Nov. Holl.* 580. [2] *Annales du mus.* 18, *p.* 27.

both these envelopes adhere to the ovarium, they may be separately traced to its base; the coloured corolla being plainly visible in the interstices of the foliaceous calyx.

Goodenoviæ, whose maximum exists in the principal parallel of New Holland, are nearly but not absolutely confined to Terra Australis; the only known exceptions to this consist of the genus *Cyphia*, which is peculiar to Africa, and chiefly occurs at the Cape of Good Hope; of some species of *Scævola* which are found within the tropics; and of *Goodenovia littoralis*, which is common to the shores of Terra Australis and New Zealand, and according to Cavanilles is also a native of the opposite coast of South America.

STYLIDEÆ.[1] This order, consisting of Stylidium, Levenhookia, and Forstera, I have formerly separated from Campanulaceæ, on account of its reduced number of stamina, and the remarkable and intimate cohesion of their filaments with the style, through the whole length of both organs. It differs also both from Campanulaceæ and Goodenoviæ in the imbricate æstivation of the corolla, and where its segments are unequal in the nature of the irregularity. In the relation which the parts of its flower have to the axis of inflorescence, and in the parallel septum of its capsule, it agrees with Goodenoviæ and differs from Lobelia, which, however, in some other respects it more nearly resembles.

Very different descriptions of the sexual organs in this tribe, and especially of the female, have been given by several French botanists. According to Richard the lateral appendices of the labellum in *Stylidium* are the real stigmata, the style being consequently considered as cohering with the tube of the corolla, and the column as consisting of stamina only. This view of the structure demands particular notice, not only from the respect to which its author is himself entitled, but because it has also been adopted by Jussieu,[2] whose arguments in support of it, and against the
562] common opinion, may be reduced to three. 1st. Were the common opinion admitted, the difficulty of conceiving so wide a difference in what he terms insertion of stamina

[1] *Prod. fl. Nov. Holl.* 565. [2] *Annales du mus.* 18, *p.* 7.

in two orders so nearly related as Campanulaceæ and Stylideæ obviously are: 2ndly. The alleged non-existence of the stigma, which preceding authors had described as terminating the column: and lastly, the manifest existence of another part, which, both from its appearance and supposed origin is considered as capable of performing the function of that organ.

In opposition to these arguments it may be observed, that the real origin of the stamina is in both orders the same, the apparent difference arising simply from their accretion to the female organ in Stylideæ, a tendency to which may be said to exist in Lobelia. The inability to detect the stigma terminating the column in Stylidium must have arisen from the imperfection of the specimens examined, for in the recent state, in which this organ is even more obvious than in Goodenoviæ at the time of bursting of the antheræ, it could not have escaped so accurate an observer as Richard; and were it even less manifest in Stylidium, its existence would be sufficiently confirmed from the strict analogy of that genus with Levenhookia, whose stigma, also terminating the column, consists of two long capillary laciniæ, which are in no stage concealed by the antheræ.

With respect to the part considered as stigma by Richard I have formerly observed that it is obsolete in some species of Stylidium and entirely wanting in others,[1] and there is certainly no trace of anything analogous to it in Forstera.

The greater part of the Australian *Stylideæ* exist at the western extremity of the principal parallel, several species are found at the eastern extremity of the same parallel, and a few others occur both within the tropic and in Van Diemen's Island. Beyond Terra Australis very few plants of this order have been observed; two species of Stylidium, very similar to certain intratropical species of New Holland, were found in Ceylon and Malacca, by Kœnig; and of the only two known species of Forstera, one is a native of New Zealand, the other of Terra del Fuego, and the opposite coast of Patagonia.

[1] *Bauer illustr. tab.* 5.

563] RUBIACEÆ.[1] As this order is now constituted it appears to me impracticable to distinguish it from Apocineæ, by characters taken from the fructification alone; and even if the Stellatæ or Asperuleæ be excluded, and the remarkable stipulation of its remaining sections be taken into account, it will not then, perhaps, admit of a definition entirely free from exceptions. It must also, I think, be allowed that Rubiaceæ, Apocineæ, Asclepiadeæ, and certain genera at present referred to Gentianeæ, form but one great natural class. In this class the leaves are uniformly simple, perfectly entire, and, with a very few exceptions, occurring in Asclepiadeæ and Apocineæ, also opposite; while in the parts of fructification there are hardly any characters that are not liable to exceptions, unless the monopetalous regular corolla, and stamina alternating with its laciniæ and not exceeding them in number.

The order *Rubiaceæ*, admitting it as it is at present established, is chiefly equinoctial. In Terra Australis its maximum is also within the tropic, where, however, it is not very numerous; and the most remarkable Australian part of the order, consisting of *Opercularia* and *Pomax*, is chiefly found in the principal parallel. Jussieu is very unwilling to admit these two genera into Rubiaceæ, and is rather disposed to consider them as a distinct family; chiefly on account of their single-seeded ovarium. To prove that this character alone, however, is not of such importance as to separate plants into different natural orders, it is sufficient to advert to Proteaceæ, Amaranthaceæ, and Epacrideæ, all of which contain genera with one, two, and even an indefinite number of seeds; and as Operculariæ entirely agree with many genera of Rubiaceæ in other points of structure of fructification, in habit, and especially in their remarkable stipulation, I think there can be no doubt that they ought to be referred to the same order, of which they form a section, characterised not only by its single-seed ovarium, but by the peculiar dehiscence of its compound fruit.

[1] *Juss. gen.* 196.

APOCINEÆ.[1] I have already observed[2] that this order is very nearly related to Rubiaceæ and Gentianeæ; the [564 former appearing to differ chiefly in its remarkable stipulæ, the latter in its minute embryo. If these characters be admitted, certain New Holland genera which I have placed with Gentianeæ will either be transferred to Rubiaceæ, or, as I have formerly proposed,[3] may, with some others, constitute a family intermediate to Rubiaceæ and Apocineæ.

This order or section, which may be named LOGANEÆ, will consist of Logania, Geniostoma (from which Anasser of Jussieu is not distinct), Usteria, Gærtnera of Lamarck,[4] Pagamea of Aublet, and, perhaps, Fagræa. Of these, the only New Holland genus is *Logania*, the greater part of whose species are found in the principal parallel. In this genus, which admits, however, of subdivision, the importance of stipulation seems to be entirely lost, for it contains species agreeing in this respect with Rubiaceæ, others in which the stipulæ are lateral and distinct, and one species, at least, in which they are entirely wanting.

There is an evident affinity between certain species of Logania and *Mitrasacme*, which I had therefore placed in Gentianeæ. Mitrasacme is very general in Terra Australis, but its maximum is within the tropic; it is not absolutely confined to New Holland, for I have observed in the Sherardian Herbarium two species collected at Cheusan, by Mr. Cunningham.

Among the true Apocineæ of New Holland, which are chiefly found within the tropic, the most remarkable genus is Alyxia, in which the albumen and embryo agree with those of the very different family Annonaceæ.

ASCLEPIADEÆ.[5] These plants differ from Apocineæ solely in the peculiar structure of their genitalia, a character, however, which appears to me fully sufficient to justify their separation. They are not very numerous in New

[1] *Prodr. fl. Nov. Holl.* 465. Apocinearum pars, *Juss. gen.* 143.
[2] *Werner. soc. transact.* 1, *p.* 12.
[3] *Prodr. fl. Nov. Holl.* 455. [4] *Illustr. gen. tab.* 167.
[5] *Werner. soc. transact.* 1, *p.* 12; *Prodr. fl. Nov. Holl.* 458.

Holland, where they are found chiefly within the tropic, and I have not observed any plant of the order in that country in a higher latitude than 34° S.

EPACRIDEÆ.[1] The abundance of this family in Terra
565] Australis constitutes one of the peculiarities of its vegetation. About 140 species have already been observed, the greater part of which are found in the principal parallel; the order, however, continues numerous at the south end of Van Diemen's Island, where several genera appear that have not been met with in other parts; within the tropic very few species have been observed, and none with capsular fruit.

Epacrideæ, with the exception of two species found in the Sandwich Islands, are confined to the southern hemisphere; several species have been observed in New Zealand, a few in the Society Islands, and even in the Moluccas; the only species with capsular fruit found within the tropic is *Dracophyllum verticillatum*, observed by Labillardière in New Caledonia; and the only plant of the family known to exist in America is an unpublished genus, also with capsular fruit, found by Sir Joseph Banks in Terra del Fuego.

The sections into which I have divided this order differ from each other in two remarkable points of structure. The *Stypheleæ*, as they may be called, having a valvular or very rarely a plaited æstivation of the corolla, and a definite number of seeds; while the *Epacrideæ*, strictly so called, have along with their indefinite number of seeds and capsular fruit a corolla with imbricate æstivation. I have formerly[2] pointed out what seems to be the natural subdivision of this section, depending more on the differences of insertion in its leaves than on characters derived from the parts of fructification.

LABIATÆ and VERBENACEÆ appear to me to form one natural class, the two orders of which gradually pass into each other. Terra Australis contains several remarkable genera of both orders, and chiefly in its principal pa-

1 *Prodr. fl. Nov. Holl.* 535. Ericearum genera. *Juss. gen.* 160.
2 *Prodr. fl. Nov. Holl.* 536.

rallel. *Chloanthes*[1] is the most singular among Verbenaceæ, having, with the fruit of that order, entirely the habit of Labiatæ.

Westringia and Prostanthera, with the genera nearly related to each of these, are the most worthy of notice among Labiatæ; all of them are limited to Terra Australis, and they are found chiefly in its principal parallel, but Westringia and Prostanthera abound also in Van Diemen's Island, and extend, though more sparingly, in the opposite direction as far as the tropic. *Prostanthera* is remarkable [566
in the appendages to its antheræ, in the texture of its fruit, and in the remains of albumen existing in the ripe seeds of several of its species. *Westringia*, and its related genera Microcorys and Hemigenia, differ from the rest of the order in having verticillate leaves, and from the greater part in the structure of antheræ, particularly in the order in which these organs become abortive. *Westringia*, according to Dr. Smith, has resupinate corolla, a term which in this case cannot allude to a mere inversion in the form of its lips, for this does not exist; and if it mean an absolute change in the relation of its parts to those of the calyx or to the included organs, it cannot, I apprehend, be admitted either in this genus or in any other of the order. The fact which I formerly stated[2] against the resupination of corolla in Labiatæ is the uniformity of its æstivation in this order, in which the upper lip always covers the lower. To those who do not consider this as a sufficient proof, the following, drawn from another equally uniform point of structure, may perhaps appear more satisfactory. In Labiatæ, as well as in several other orders with irregular flowers, the deviation from the usual quinary division of calyx and corolla in Dicotyledones, does not consist in an absolute suppression of parts, but merely in their confluence, a fact indicated by the disposition of vessels; thus the upper lip of the corolla, which in this order generally consists of one piece, either entire or more or less deeply bifid, is always furnished with two longitudinal nerves equidistant from its axis, which is without vessels; while each of the

[1] *Bauer illustr. tab.* 4. [2] *Prodr. fl. Nov. Holl.* 499.

three laciniæ usually forming the lower lip has a single nerve passing through its axis; the upper lip is therefore to be considered, even when entire, as made up of two confluent laciniæ; and if this test be allowed to be conclusive, and applied to the corolla of those genera of Labiatæ in which it is supposed to be resupinate, the opinion will be found to be erroneous.

MYOPORINÆ.[1] The principal characters in the fructification of this order, by which it is distinguished from Verbenaceæ, are the presence of albumen in the ripe seed, and the direction of the embryo, whose radicule always points towards the apex of the fruit. The first of these characters, however, is not absolute, and neither of them can
567] be ascertained before the ripening of the seed; for previous to the complete development of the embryo the fluid albumen or liquor amnios equally exists in both orders; and although all the genera of Verbenaceæ have an embryo whose radicule points towards the base of the fruit, yet many of them have pendulous seeds, and consequently a radicule remote from the umbilicus. Hence *Avicennia*,[2] which I formerly annexed to Myoporinæ, should be restored to Verbenaceæ, with which also it much better agrees in habit.

Myoporinæ, with the exception of Bontia, a genus of equinoctial America, and of two species of Myoporum found in the Sandwich Islands, has hitherto been observed only in the southern hemisphere, and yet neither in South Africa nor in South America beyond the tropic. Its maximum is evidently in the principal parallel of Terra Australis, in every part of which it exists; in the more southern parts of New Holland, and even in Van Diemen's Island it is more frequent than within the tropic. The genus Myoporum is also found in New Zealand, Norfolk Island, New Caledonia, and the Society Islands.

PROTEACEÆ.[3] I have formerly[4] offered several obser-

[1] *Prodr. fl. Nov. Holl.* 514.
[2] *Prodr. fl. Nov. Holl.* 518.
[3] *Ibid.*
[4] *Lin. soc. transact.* 10, *p.* 15.

vations both on the geographical distribution and on some of the more remarkable points of structure of this order of plants. I shall now therefore confine myself to a few of the most important facts on each of these subjects.

Proteaceæ are chiefly natives of the southern hemisphere, in which they are most abundant in a parallel included between 32° and 35° lat., but they extend as far as 55° S. lat. The few species found in the northern hemisphere occur within the tropic.

Upwards of 400 species of the order are at present known; more than half of these are natives of Terra Australis, where they form one of the most striking peculiarities of the vegetation. Nearly four fifths of the Australian Proteaceæ belong to the principal parallel, in which, however, they are very unequally distributed; the number of species at its western extremity being to those of the eastern as about two to one, and, what is much more remarkable, the number even at the eastern extremity being to that of the middle of the parallel as at least four to one. From the principal parallel the diminution of the order in number of species is nearly equal in both directions; but while no genus has been met with [568
within the tropic which does not also exist in the principal parallel, unless that section of *Grevillea* having a woody capsule[1] be considered as such, several genera occur at the south end of Van Diemen's Island which appear to be peculiar to it.

No Australian species of this order has been observed in any other part of the world, and even all its genera are confined to it, with the exception of Lomatia, of which several species have been found in South America; and of Stenocarpus, the original species of which is a native of New Caledonia.

The genera of Terra Australis that approach most nearly to the South African portion of the Proteaceæ exist in the principal parallel, and chiefly at its western extremity; those allied to the American part of the order are found either at the eastern extremity of the same parallel or in Van Diemen's Island.

[1] Cycloptera, *Lin. soc. transact.* 10, *p.* 176; *Prodr. fl. Nov. Holl.* 380.

There is no species of Proteaceæ common to the east and west coasts of New Holland, and certain genera abound at one extremity of the principal parallel which at the opposite extremity are either comparatively rare or entirely wanting.

I have formerly remarked that in this order no instance of deviation from the quaternary division of the perianthium has been observed; a fact which is the more remarkable as this is itself a deviation from the prevailing quinary number in the floral envelopes of Dicotyledonous plants.

There is a peculiarity in the structure of the stamina of certain genera of Proteaceæ, namely, Simsia, Conospermum, and Synaphea, in all of which these organs are connected in such a manner that the cohering lobes of two different antheræ form only one cell.

Another anomaly equally remarkable exists in Synaphea, the divisions of whose barren filament so intimately cohere with the stigma as to be absolutely lost in its substance, while the style and undivided part of the filament remain perfectly distinct.

SANTALACEÆ. I have formerly[1] proposed and at-
569] tempted to define this natural order, one of whose most remarkable characters consists in its unilocular ovarium, containing more than one, but always a determinate number of ovula, which are pendulous and attached to the apex of a central receptacle. This receptacle, which varies in its figure in the different genera, in some being filiform, in others nearly filling the cavity of the ovarium, had not been previously noticed in any plant of the order.

The greater part of the Santalaceæ of Terra Australis are found in the principal parallel, to which several genera, namely, *Leptomeria*, *Corethrum*, and *Fusanus*, are nearly limited; *Santalum*, on the other hand, is found chiefly within the tropic.

I have added *Exocarpus* and *Anthobolus* to this order, with certain genera of which they agree in habit and many points of structure, both of the flower and fruit; but they

[1] *Prodr. fl. Nov. Holl.* 350.

are readily distinguishable from the whole order by their fructus superus, and they may possibly differ also in the internal structure of their ovarium, which has not yet been satisfactorily ascertained.

The genus *Exocarpus* is most abundant in the principal parallel and southern parts of Terra Australis, but it is not unfrequent even within the tropic. *Exocarpus cupressiformis* is not only the most common species of the genus, but the most general tree in Terra Australis, being found in nearly the whole of the principal parallel, in every part of Van Diemen's Island that has been visited, and even within the tropic. I am acquainted with only three plants that have in that country an equally extensive range. These are *Anthistiria australis*, the most valuable grass as well as the most general plant in Terra Australis; *Arundo Phragmites*, less frequent than the former, but which extends from the southern extremity of Van Diemen's Island to the north coast of New Holland; and *Mesembryanthemum æquilaterale*, which occurs on almost every part of the sandy sea shores of both these islands.

Exocarpus is not absolutely confined to Terra Australis, for M. Bauer has discovered a very remarkable species bearing its flowers on the margins of dilated foliaceous branches, analogous to those of Xylophylla; and *Xylophylla longifolia*, which was taken up by Linnæus from Rumphius,[1] [570 appears more probably, both from the description and figure of that author, to be also a species of Exocarpus.

There is so great a resemblance between the enlarged fleshy receptacle of Exocarpus and the berry of Taxus, that some botanists have been led to compare these plants together in other respects. A complete coincidence in this part of their structure would not indeed prove the affinity of these two genera, any more than it does that of Exocarpus to Anacardium or Semecarpus, with which also it has been compared; and to determine their agreement even in this respect it is necessary to understand the origin of the berry of Taxus, of which very different accounts

[1] Xylophyllos ceramica, *Herb. amb.* 7, *p.* 19, *t.* 12.

have been given. According to Lamarck[1] it consists of the enlarged ovarium itself, perforated by the seed soon after impregnation; while Mirbel[2] considers it as formed of the scales of the female amentum, immediately surrounding the organ, named by him *cupula;* and considered as containing the pistillum, but which most other authors have regarded as the pistillum itself. My observations differ from both these accounts, for on examining the female fructification of Taxus before impregnation I find the rudiments of the future berry, consisting at that period of a narrow fleshy ring, surrounding the base only of the cupula of Mirbel, and very similar to the annular hypogynous nectarium of many flowers. If this cupula, therefore, were the pistillum itself, the berry of Taxus would have an origin analogous to that of Balanites,[3] as it has been very lately described by Mirbel; and, on the other hand, if that author's view of the female fructification of Taxus and Coniferæ generally be adopted, it might then to a certain degree be compared with the external cupula of Dacrydium, which will be more particularly noticed hereafter; but from this it would still be very distinct both in its texture and in its not enclosing in the early stage the cupula; on neither supposition, however, does its origin agree with that of the berry of Exocarpus, which in some respects more nearly resembles the fleshy receptacle of Podocarpus.

I have annexed *Olax* to Santalaceæ,[4] not, however, considering it as absolutely belonging to the same family, but
571] as agreeing with it in some important circumstances; especially in the internal structure of its ovarium, and that of its pericarpium and seed; but as in Olax there appears to be a double floral envelope, as its antheriferous stamina alternate with the segments of the inner envelope, and its ovarium does not cohere with either, there are sufficient grounds for regarding it, with Mirbel, as a distinct family.

[1] *Encyclop. botan.* 3, *p.* 228. [2] *Nouv. bulletin des scien.* 3, *p.* 73.
[3] *Delile in mem. sur l'Egypte,* 3, *p.* 326. Ximenia ægyptiaca, *Linn.*
[4] *Prodr. fl. Nov. Holl.* 357.

CASUARINEÆ. The genus *Casuarina* is certainly not referable to any natural order of plants at present established; and its structure being now tolerably understood, it may be considered a separate order, as Mirbel has already suggested.[1]

The maximum of Casuarina appears to exist in Terra Australis, where it forms one of the characteristic features of the vegetation. Thirteen Australian species have already been observed; the greater number of these are found in the principal parallel, in every part of which they are almost equally abundant; in Van Diemen's Island the genus is less frequent, and within the tropic it is comparatively rare; no species except *Casuarina equisetifolia* having been observed on the north coast of New Holland. Beyond Terra Australis only two species have been found, namely, *C. equisetifolia*, which occurs on most of the intratropical islands of the Southern Pacific, as well as in the Moluccas, and exists also on the continent of India; and *C. nodiflora*, which is a native of New Caledonia.

In the male flowers of all the species of Casuarina, I find an envelope of four valves, as Labillardière has already observed in one species, which he has therefore named *C. quadrivalvis*.[2] But as the two lateral valves of this envelope cover the others in the unexpanded state, and appear to belong to a distinct series, I am inclined to consider them as bracteæ. On this supposition, which, however, I do not advance with much confidence, the perianthium would consist merely of the anterior and posterior valves, and these, firmly cohering at their apices, are carried up by the anthera, as soon as the filament begins to be produced, while the lateral valves or bracteæ are persistent; it follows from it also that there is no visible perianthium in the female flower, and the remarkable economy of its lateral bracteæ may, perhaps, be considered as not only affording an additional argument in support of the view now taken [572 of the nature of the parts, but also as in some degree again approximating Casuarina to *Coniferæ*, with which it was formerly associated.

[1] *Annales du mus.* 16, *p.* 451.

[2] *Plant. Nov. Holl.* 2, *p.* 67, *t.* 218.

The outer coat of the seed or caryopsis of Casuarina consists of a very fine membrane, of which the terminal wing is entirely composed; between this membrane and the crustaceous integument of the seed there exists a stratum of spiral vessels, which Labillardière, not having distinctly seen, has described as an "integumentum arachnoideum;" and within the crustaceous integument there is a thin proper membrane closely applied to the embryo, which the same author has entirely overlooked. The existence of spiral vessels, particularly in such quantity, and, as far as can be determined in the dried specimens, unaccompanied by other vessels, is a structure at least very unusual in the integuments of a seed or caryopsis, in which they are very seldom at all visible, and have never, I believe, been observed in such abundance as in this genus, in all whose species they are equally obvious.

CONIFERÆ.[1] The structure of the female parts of fructification in Coniferæ having, till very lately, been so little understood, and certain facts concerning it being still unpublished, I shall prefix a few observations on this subject to the remarks I have to offer on the Australian part of the order.

In the late essays of Mirbel and Schoubert on *Coniferæ*[2] that part of the female fructification which had previously been considered as the pistillum, having a perforated style, is described as a peculiar organ enclosing the ovarium, and in most cases also the stigma. This organ, which they have named cupula, they regard as more analogous to an involucrum than to a perianthium, which, according to them, also exists, cohering, however, with the body of the ovarium. Without absolutely adopting this latter part of their statement, it appears to me that impregnation really takes place in the manner these authors describe. Their principal argument is derived from the genus Ephedra, in which both the stigma and a considerable part of the style project beyond this cupula, without cohering with its aperture. In further confirmation of their opinion it may be

[1] *Juss. gen.* 411. [2] *Nouv. bulletin des scien.* 3, *p.* 73, 85, *et* 121.

observed that I have found a projection of the stigma, [573
though certainly in a much less obvious degree, both in Agathis[1] and in a species of Podocarpus.

Towards this discovery, as extending to the Coniferæ more strictly so called, an important step was made in *Pinus*, by the accurate Schkuhr,[2] who first correctly described and figured the cupula of that genus, but who considered it as the ovarium itself and the two processes of its aperture as stigmata. Mr. Salisbury, who seems to have been unacquainted with Schkuhr's observations, published a few years afterwards,[3] the same opinion, which continued to be generally received till the appearance of the essays, already quoted, of Mirbel and Schoubert.

But these authors do not seem to be aware that certain plants of the order are even furnished with a double cupula. This is most remarkable in *Podocarpus*, in which the drupa is formed of this external cupula, whose aperture exists not at the apex, but very near its base or point of insertion. The inner cupula in this genus is in every stage entirely enclosed in the outer, and is in like manner inverted.

That this is the real structure of Podocarpus seems to be proved by that of the nearly related genus *Dacrydium*, hitherto so imperfectly understood. This genus has also a double cupula, the outer in the young state enclosing the inner, and both of them at this period being inverted, as in Podocarpus; but the inner in a more advanced stage acquires nearly an erect position, by rupturing one side of the external cupula, which, not continuing to increase proportionally in size, forms a cup surrounding the base only of the ripe fruit.

Three species of *Podocarpus* are found in Terra Australis; two of these exist in the colony of Port Jackson, the third was observed on the summit of the Table Mountain in Van Diemen's Island. *Podocarpus asplenifolia* of Labillardière[4] is certainly not a Podocarpus, but either forms a distinct

[1] *Salisbury in Linn. soc. transact.* 8, *p.* 311. Pinus Dammara, *Lamb, pin. p.* 61, *t.* 38. [2] *Botan. handb.* 3, *p.* 276, *t.* 308.
[3] *Linn. soc. transact.* 8, *p.* 308.
[4] *Plant. Nov. Holl.* 2, *p.* 71, *t.* 221.

genus, as Richard has already supposed,[1] or it may possibly be a species of Dacrydium; a conjecture which I have no means of verifying, having never seen the female fructification of this remarkable plant.

574] *Callitris* of Ventenat[2] is peculiar to Terra Australis, where it exists very generally, but most abundantly in the principal parallel; it consists of several species, differing from each other chiefly in the form of their fruit.

Araucaria excelsa, which was first observed in Norfolk Island and New Caledonia, is found also on the east coast of New Holland, immediately within the tropic; it is there, however, a tree of very moderate dimensions, and never of that enormous size which it not unfrequently attains in Norfolk Island.

ORCHIDEÆ.[3] The Australian species of this order already known amount to 120; many of these, however, are of very rare occurrence, and none of them appear to be produced in abundance.

The maximum of the order exists in the principal parallel, a considerable part extends to Van Diemen's Island, and very few have been observed within the tropic.

The greater part form genera nearly or entirely peculiar to Terra Australis, and most of these genera belong to that division of the order having farinaceous pollen, with an anthera which is inserted, but not deciduous, and either parallel to the stigma or terminating the column. The two sections of this division with parallel and terminal anthera are found in New Holland to pass very gradually into each other, and several genera belonging to the former are, in that country, remarkable for the great expansion of the lateral lobes of the column. These lateral lobes I have considered as barren stamina, which, like those of Philydrum, are occasionally, though indeed very rarely, furnished with rudiments of antheræ. This structure, as well as that of Cypripedium, in which the lateral lobes are antheriferous, while the middle is barren, approximates the flower of Orchideæ to what

[1] *Annales du mus.* 16, *p.* 299. [2] *Dec. gen. nov.* 10.
[3] *Prodr. fl. Nov. Holl.* 309.

may be called the type of Monocotyledones, that is, a regular flower with ternary division of its envelope, stamina, and cells or placentæ of the fruit.

I have attempted a similar approximation of true *Scitamineæ*,[1] whose processes crowning the ovarium, and usually two in number, form the complement of the stamina.

Maranteæ or *Canneæ*,[2] an order at present referred to [575
Scitamineæ, may also be reduced to this type; they differ, however, from Scitamineæ in the mutual relation of their barren and fertile stamina, somewhat as Cypripedium does from the other genera of Orchideæ; except that in Maranteæ the imperfection is greater, a single lobe only of one of the lateral stamina having the appearance of an anthera and producing pollen.

It is remarkable that so very few Orchideæ of Terra Australis belong to that section of the order with angular elastic pollen and adnate anthera; this section being not only the most numerous in Europe, but existing in an equal proportion, though singularly modified, at the Cape of Good Hope.

Of another section of the order, formerly comprehended under the Linnean genus Epidendrum, most of which, though not properly parasitical, grow upon trees, several species, chiefly belonging to Dendrobium, are found in New Holland. In the northern hemisphere very few plants of this section that grow on trees have been observed beyond the tropic. The only exceptions to this, that I am acquainted with, consist of two species of a genus related to Dendrobium, discovered by Dr. Buchanan, in Upper Nepaul;[3] of *Dendrobium moniliforme*, observed by Kæmpfer and Thunberg, in Japan, near Nagasaki; and of *Epidendrum conopseum*,[4] which, according to Mr. William Bartram, grows in East Florida, in lat. 28° N.

In some parts of the southern hemisphere this section appears to have a more extensive range. On the east coast of New Holland several species of Dendrobium and

[1] *Prodr. fl. Nov. Holl.* 305. [2] *Loc. citat.* 307.

[3] Epidendrum præcox and Epidendrum humile. *Smith exot. bot. tabb.* 97 and 98. [4] *Hort. Kew, ed.* 2, *vol.* 5, *p.* 219.

Cymbidium are found in 34° S. lat.; but this is probably about their southern limit in that country, no species having been met with on any part of its south coast. They have, however, been observed in a considerably higher latitude in New Zealand, in the northern island of which several species were collected by Sir Joseph Banks, in about 38° S. lat., and *Epidendrum autumnale* of Forster grows in the neighbourhood of Dusky Bay, in upwards of 45° S. lat.

I am not acquainted with the limit of this section in South America; but in South Africa, at the Cape of Good Hope, none of those, at least, that are parasitical on trees, have been observed.

576] ASPHODELEÆ.[1] In this order I include the greater part, both of Asphodeleæ and Asparageæ of Jussieu, distinguishable from each other only by texture and dehiscence of fruit; differences which, as they separate Stypandra from Dianella, and Eustrephus from Luzuriaga, cannot be admitted to be of more than generic importance.

I confess myself unable to point out satisfactory distinguishing characters for this order, in my description of which, however, I have noted two circumstances, neither of them indeed peculiar to the order, but both of them appearing to extend through the whole of it; namely, the reduction of stamina from six to three, which occasionally occurs, constantly taking place by the suppression of those opposite to the outer series of the perianthium; and the existence of the black crustaceous testa or outer integument of the seed. It is probable I have given too much weight to this latter circumstance, in combining, partly on account of it, genera so very dissimilar as Anthericum, Xanthorrhœa, and Astelia.

Xanthorrhœa, which I have included in Asphodeleæ, is in habit one of the most remarkable genera of Terra Australis, and gives a peculiar character to the vegetation of that part of the country where it abounds. This genus is most frequent in the principal parallel, but it extends to the

[1] *Prodr. fl. Nov. Holl.* 274.

south end of Van Diemen's Island, and is also found within the tropic.

A plant of a very similar habit to Xanthorrhœa, agreeing with it in its caudex and leaves, having, however, a very different inflorescence, was observed abundantly at King George's Sound, but with fructification so decayed and imperfect that I have not been able to determine the structure either of its flower or fruit. This plant is introduced by Mr. Westall in the view of King George's Sound published in Captain Flinders's account of his voyage.

I had annexed *Hypoxis* and *Curculigo* to the Asphodeleæ, chiefly on account of a similarity in the testa of the seed; but they differ so much from this order in other parts of their structure, and from Amaryllideæ both in this respect and in the singular umbilicus of the seed, as well as in habit, that it is better to consider them as forming a separate family.

Of this family, which may be called HYPOXIDEÆ,[1] only five species have been observed in Terra Australis, four of [577
these belong to Hypoxis, which is chiefly an extratropical genus, the fifth is a Curculigo very like those of India.

PALMÆ. Only six species of this order have been observed in New Holland, and of two of these the fructification is at present unknown.

The New Holland Palms exist chiefly within the tropic, but one species is found in 34° S. lat.; it seems, however, that this is nearly the southern limit of the order in that country, no species having been seen on any part of the South coast.

In New Zealand a species of *Areca* was observed by Sir Joseph Banks, in about 38° S. lat., which is probably nearly the limit of Palms, in the southern hemisphere. In the northern hemisphere their extent is not materially different from this: in North America, indeed, they do not appear

[1] HYPOXIDEÆ. *Perianthium* superum: limbo sexpartito, regulari, æstivatione imbricata. *Stamina* sex, imis lacinis inserta. *Ovarium* 3-loc. loculis polyspermis. *Capsula* evalvis, nunc baccata, polysperma. *Semina* umbilico laterali rostelliformi; testa atra crustacea. *Embryo* in axi albuminis carnosi: *radicula* vaga.

to grow beyond 36° lat.; but in Europe *Chamærops humilis* extends as far as the neighbourhood of Nice.

It is remarkable that no species of Palm has been found in South Africa, nor was any observed by Mr. Leschenault,[1] on the West coast of New Holland, even within the tropic.

JUNCEÆ. We are now in possession of so many links connecting together the Monocotyledonous orders with regular flowers, that in attempting to define several of them, we are obliged to have recourse to differences, many of which may appear, and some of which unquestionably are, of but secondary importance. Of this kind may be considered the characters by which I have endeavoured to distinguish *Junceæ* from Asphodeleæ, namely the difference in the texture of the perianthium, and in that of the testa of the seed, in the consistence of the albumen, and in the order of suppression of the stamina; these when reduced to three in number being always placed opposite to the three outer leaves of the perianthium: in this respect and
578] in the more important character of the position of the embryo Junceæ differ also from Restiaceæ, to which they more nearly approach in habit.

Three very remarkable genera, which I have referred to Junceæ, are peculiar to Terra Australis. Of two of these, *Calectasia* and *Dasypogon*, each consisting of only one species, figures and descriptions are annexed to this essay.

Of the third, *Xerotes*, 24 species have already been observed. This genus is somewhat more abundant in the principal parallel than in other parts; but it is very generally extended, and is more frequent within the tropic than in Van Diemen's Island. *Xerotes*, in the structure and appearance of its flowers and in the texture of albumen, has a considerable resemblance to Palms, but it wants the peculiar characters of the seed and also the habit of that remarkable order.

Flagellaria, which I have added to Junceæ, differs from Xerotes chiefly in its pericarpium, and in the form and relation of its embryo to the albumen, which is also of a

[1] *Annales du mus.* 17, *p.* 87.

different texture; in all these respects it approaches to Cyperaceæ, with some of whose genera it has even a certain resemblance in habit. This genus has usually been found only within the tropics, but in New Holland it extends as far as 33° S. lat.

Philydrum, which I have annexed to Junceæ, has always appeared to me an insulated genus, yet though not referable to any established natural order, it may be compared with several in certain respects. In the structure of its stamina it may in one point of view be said to be intermediate between Scitamineæ and Orchideæ; in that of its pericarpium and even of its seeds it has some affinity to the latter order; yet it differs from both of them in almost every other respect. In general appearance, it bears a considerable resemblance to Cartonema, which belongs to Commelineæ. In some parts of its structure it may be compared with Xyris, and perhaps with Burmannia; a genus which I have likewise annexed to Junceæ, but whose real affinities are equally obscure.

Philydrum pygmæum differs in so many respects from *P lanuginosum* that it may probably hereafter be considered as a distinct genus; and a very few additions to this tribe of plants would sanction their formation into a separate natural order.

RESTIACEÆ. The principal character distinguish- [579
ing this family from Junceæ and Cyperaceæ consists in its lenticular embryo being placed at the extremity of the seed opposite to the umbilicus; from Junceæ it also differs in the order of suppression of its stamina, which when reduced to three are opposite to the inner laciniæ of the perianthium; and most of its genera are distinguishable from both these orders as well as from Commelineæ by their simple or unilocular antheræ.

With the exception of Eriocaulon, Tonina, and Xyris, the order appears to be confined to the Southern hemisphere. In Terra Australis its maximum is in the principal parallel, but it extends to the southern extremity of Van Diemen's Island, where it is even in considerable abundance, and exists, though much more sparingly, within the tropic.

Restiaceæ are almost equally numerous at the Cape of Good Hope as in the principal parallel of New Holland. One species only of the order has been observed in New Zealand, and hitherto none in South America.

CYPERACEÆ. In Terra Australis this is a very extensive order, consisting already of more than 200 species. It contains, however, fewer peculiarities in structure than several other orders that are much less numerous. Its maximum appears to be in the principal parallel; but the species observed solely within the tropic exceed one third of the whole number. Cyperaceæ, in many respects, are nearly related to Restiaceæ, and when furnished with a true perianthium are distinguishable from the monospermous genera of that order, solely by the different position of the embryo in the seed. But in the greater part of the order the perianthium is either entirely wanting or merely setaceous. Fuirena, Lepidosperma and Orcobolus, all of them natives of New Holland, are almost the only genera in which it is found of nearly the usual appearance.

What I have formerly termed perianthium in Carex, Diplacrum, and Schœnus nemorum, ought, perhaps, rather to be considered as internal bracteæ, analogous to those of Lepyrodia, of Irideæ, and, perhaps, to the upper valve of the inner envelope of grasses.

I have formerly remarked that the Perianthium of Hypœ-
580] lyptum consists merely of the squamæ of a spicula, similar to that of Kyllinga, but reduced to two valves.

GRAMINEÆ. This order comprehends, at least, one fourth of the whole of Monocotyledones, and in Terra Australis, where upwards of 200 species have already been observed, it bears the same proportion to that primary division.

I have formerly, in arranging the Australian genera of *Gramineæ,* endeavoured to explain what I conceived to be the natural subdivision of nearly the whole order into two great tribes. The reasons which I then assigned for this arrangement appear, however, either not to have been comprehended, or to have been considered too hypothetical. With a view of removing the supposed obscurity and strengthening my former arguments, I shall preface what I have now to say on the subject, by a few observations common to both tribes.

The natural or most common structure of *Gramineæ* is to have their sexual organs surrounded by two floral envelopes, each of which usually consists of two distinct valves: but both of these envelopes are in many genera of the order subject to various degrees of imperfection or even suppression of their parts.

The outer envelope or *gluma* of Jussieu, in most cases, containing several flowers with distinct and often distant insertions on a common receptacle, can only be considered as analogous to the bracteæ or involucrum of other plants.

The tendency to suppression in this envelope appears to be greater in the exterior or lower valve, so that a gluma consisting of one valve may, in all cases, be considered as deprived of its outer or inferior valve. In certain genera with a simple spike, as Lolium and Lepturus, this is clearly proved by the structure of the terminal flower or spicula, which retains the natural number of parts; and in other genera not admitting of this direct proof, the fact is established by a series of species showing its gradual obliteration, as in those species of Panicum which connect that genus with Paspalum.

On the other hand, in the inner envelope or *calyx* of
Jussieu, obliteration first takes place in the inner or [581
upper valve; but this valve having, instead of one central nerve, two nerves equidistant from its axis, I consider it as composed of two confluent valves, analogous to what takes place in the calyx and corolla of many irregular flowers of other classes; and this confluence may be regarded as the first step towards its obliteration, which is complete in many species of Panicum, in Andropogon, Pappophorum, Alopecurus, Trichodium, and several other genera.

With respect to the nature of this inner or proper envelope of grasses, it may be observed that the view of its structure now given, in reducing its parts to the usual ternary division of Monocotyledones, affords an additional argument for considering it as the real perianthium. This argument, however, is not conclusive, for a similar confluence takes place between the two inner lateral bracteæ of the greater part of Irideæ; and with these, in the relative insertion of its valves, the proper envelope of grasses may be supposed much better to accord, than with a genuine perianthium. If, therefore, this inner envelope of grasses be regarded as consisting merely of bracteæ, the real perianthium of the order must be looked for in those minute scales, which in the greater part of its genera are found immediately surrounding the sexual organs.

These scales are in most cases only two in number, and placed collaterally within the inferior valve of the proper envelope. In their real insertion, however, they alternate with the valves of this envelope, as is obviously the case in Ehrharta and certain other genera; and their collateral approximation may be considered as a tendency to that confluence which uniformly exists in the parts composing the upper valve of the proper envelope, and which takes place also between these two squamæ themselves, in some genera, as Glyceria and Melica. In certain other genera, as Bambusa and Stipa, a third squamula exists, which is placed opposite the axis of the upper valve of the proper envelope, or, to speak in conformity with the view already taken of the structure of this valve, opposite to the junction of its two component parts. With these squamæ the stamina in triandrous grasses alternate, and they are consequently opposite to the parts of the proper envelope; that is, one stamen is opposed to the axis of its lower or outer valve, and the two others are placed opposite to the two nerves of the upper valve. Hence, if the inner envelope be
562] considered as consisting of bracteæ and the hypogynous squamæ as forming the perianthium, it seems to follow, from the relation these parts have to the axis of inflorescence that the outer series of this perianthium is wanting, while its corresponding stamina exist, and that the whole or part

of the inner series is produced while its corresponding stamina are generally wanting. This may, no doubt, actually be the case, but as it would be, at least, contrary to every analogy in Monocotyledonous plants, it becomes in a certain degree probable that the inner or proper envelope of grasses, the calyx of Jussieu, notwithstanding the obliquity in the insertion of its valves, forms in reality the outer series of the true perianthium, whose inner series consists of the minute scales, never more than three in number, and in which an irregularity in some degree analogous to that of the outer series generally exists.

It is necessary to be aware of the tendency to suppression existing, as it were, in opposite directions in the two floral envelopes of grasses, to comprehend the real structure of many irregular genera of the order, and also to understand the limits of the two great tribes into which I have proposed to subdivide it.

One of these tribes, which may be called PANICEÆ, comprehends Ischæmum, Holcus, Andropogon, Anthistiria, Saccharum, Cenchrus, Isachne, Panicum, Paspalum, Reimaria, Anthenantia, Monachne, Lappago, and several other nearly related genera; and its essential character consists in its having always a locusta of two flowers, of which the lower or outer is uniformly imperfect, being either male or neuter, and then not unfrequently reduced to a single valve.

Ischæmum and Isachne are examples of this tribe in its most perfect form, from which form Anthenantia, Paspalum, and Reimaria, most remarkably deviate in consequence of the suppression of certain parts: thus Anthenantia (which is not correctly described by Palisot de Beauvois) differs from those species of Panicum that have the lower flower neuter and bivalvular, in being deprived of the outer valve of its gluma; Paspalum differs from Anthenantia in the want of the inner valve of its neuter flower; and from those species of Panicum, whose outer flower is univalvular, in the want of the outer valve of its gluma; and Reimaria differs from Paspalum in being entirely deprived of its gluma. That this is the real structure of these genera may be proved by a series of [583 species connecting them with each other, and Panicum with Paspalum.

Paniceæ have their maximum within the tropics, and they cease to exist in the most northern parts of Europe and the higher southern latitudes. Of this tribe, 99 species have been observed in Terra Australis, 79 of which were found within the tropic, and of these, 66 only within it. There is no Australian genus of this tribe, Neurachne and Hemarthria excepted, which is not chiefly intratropical.

The second tribe, which may be called POACEÆ, is more numerous than Paniceæ, and comprehends the greater part of the European genera, as well as certain less extensive genera peculiar to the equinoctial countries; it extends also to the highest latitudes in which Phænogamous plants have been found, but its maximum appears to be in the temperate climates considerably beyond the tropics. The locusta in this tribe may consist of one, of two, or of many flowers, and the two-flowered genera are distinguished from Paniceæ by the outer or lower flower being always perfect; the tendency to imperfection in the locusta existing in opposite directions in the two tribes. In conformity with this tendency in Poaceæ, the outer valve of the perianthium in the single-flowered genera is placed within that of the gluma, and in the many-flowered locusta the upper flowers are frequently imperfect. There are, however, some exceptions to this order of suppression, especially in Arundo Phragmites, Campulosus, and some other genera, in which the outer flower is also imperfect, but as all of these have more than two flowers in their locusta, they are still readily distinguished from Paniceæ.

In Terra Australis the *Poaceæ* amount to 115 species, of which 69 were observed beyond the tropic and of these 63 only beyond it; but of the 52 species that occur within the tropics 49 belong to genera which are either entirely or chiefly intratropical, and of the remaining three species, two, namely Arundo Phragmites, and Agrostis virginica, are very general and also aquatic plants. The distribution of this tribe, therefore, in Terra Australis agrees with that which obtains in other parts of the world.

FILICES.[1] Of this order nearly 1000 species are described in the fifth volume of Willdenow's edition of the [584 Species Plantarum. In their geographical distribution, Ferns differ from all the other orders of cryptogamous plants, their maximum being in the lower latitudes, probably near, or very little beyond the tropics. Thus, Norfolk Island, situated in 29° S. lat. and only a few leagues in circumference, produces as many species of the order as are described in Dr. Smith's Flora Britannica.

But as shade and moisture are essential conditions to the vegetation of the greater part of Ferns, few species only have been observed in those parts of equinoctial New Holland hitherto examined. The number of species already found, however, in the different regions of Terra Australis exceeds 100, of which nearly one fourth are also natives of other countries.

Among the Australian Ferns there is no genus absolutely confined to that country, except Platyzoma, but this, perhaps, ought not to be separated from Gleichenia.

Only two arborescent Ferns have been observed in Terra Australis, one in the colony of Port Jackson, the second, *Dicksonia antarctica*, is frequent in Van Diemen's Island, at the southern extremity of which its trunk is not unfrequently from 12 to 16 feet in height. An arborescent species of the same genus was found by Forster, in New Zealand, at Dusky Bay, in nearly 46° S. the highest latitude in which tree ferns have yet been observed. It is remarkable that, although they have so considerable a range in the southern hemisphere, no tree fern has been found beyond the northern tropic: a distribution in the two hemispheres somewhat similar to this has been already noticed respecting the Orchideæ that are parasitical on trees.

I have formerly, in treating of the New Holland *Asplenia*, observed that *Cænopteris* does not differ from them in the relation its involucra have to the axis of the frond or pinna but merely in having the ultimate pinna more deeply divided, with one, or, at most, two involucra on each segment,

[1] *Prodr. fl. Nov. Holl.* 145.

towards the margins of which they must necessarily open: hence, the characters of both genera not unfrequently occur in the same frond, and are even exhibited by the same involucrum when it happens to extend below the origin of the segment.

I have observed also, in the same place, that in *Asplenium*
585] when the involucrum originates from the inner branch of a primary vein, which is usually the case, it opens inwards or towards the mid-rib of the frond from which the vein is derived; and that when it arises from the lower or outer branch of a vein it opens outwards, or in an opposite direction, instances of which occur in several species of the genus, in some of those especially where the frond is simple. On the same law also depends the peculiar character of Scolopendrium, in which the involucra are produced in pairs, one of each pair originating from the lower branch of a vein, the other from the upper branch of the vein immediately below it; they consequently open in opposite directions and towards each other. This law, however, in Asplenium is only observed where the vein has but few branches, for when these are more numerous, and especially when, in consequence of their greater number, the vein has a manifest trunk or axis, the involucra of all its branches open towards this axis; the most remarkable instances of this occur in those species of the genus which authors have separated from it, under the name of Diplazium, where, however, another peculiarity exists, depending on the same law. This peculiarity consists in the inner branch of the vein, or that adjoining the mid-rib, appearing to have a relation not only to the axis of the vein but to that of the pinna or frond from which the vein originates; a relation indicated by its having two involucra, one of which bursts towards the axis of the vein, the other towards the adjoining mid-rib. This double involucrum constitutes the character of Diplazium, but as it is confined to the inner branch, all the others being simple, and opening towards the axis of the vein, there do not appear to be sufficient grounds for its separation from Asplenium. I consider the curved involucrum of *Asplenium Filix-Fœmina,* which

exists only on this inner branch of the vein, as somewhat analogous to the double involucrum of Diplazium; but in another point of view it may be regarded as an approach to the structure of Nephrodium, to which this plant has been improperly referred.

THERE are some other Australian natural families of [586 plants to which, either as containing distinct and peculiar genera, or a considerable number of species, similar remarks might be extended; but I have already exceeded the limits prescribed for the present essay, which I shall therefore conclude with a few general observations, chiefly deduced from the facts previously stated, and with a very slight comparison of the vegetation of Terra Australis with that of other countries.

I have formerly remarked that nearly half the Australian species of plants, at present known, have been collected in a parallel included between 33° and 35° S. latitude; and it appears, from the preceding observations on the several natural orders, that a much greater proportion of the peculiarities of the Australian Flora exist in this, which I have therefore called the *principal parallel;* and that many of them are even nearly confined to it. But these peculiarities exist chiefly at its western and eastern extremities, and are remarkably diminished in that intermediate part which is comprehended between 133° and 138° E. long.

From the principal parallel most of the characteristic tribes diminish in number of species as well as of individuals, not, however, equally in both directions, but in a much greater degree towards the equator. In Van Diemen's Island the same general aspect of vegetation is retained; but of the natural orders forming the peculiar character of the principal parallel several are very much reduced, while none are augmented in numbers; and the only tribes which enter in nearly the same proportion into the composition of its Flora are *Eucalyptus*, the *Leafless Acaciæ* and, perhaps, *Epacrideæ*. Within the tropic, at least on the East coast, the departure from the Australian character is much more remarkable, and an assimilation nearer to that of India than of any other country takes place. Several of the peculiar orders and extensive genera of the principal parallel are here exceedingly diminished, and none remain in nearly equal proportion except *Eucalyptus* and the *Leafless Acaciæ*.

These two genera are not only the most widely diffused, but, by far the most extensive in Terra Australis, about 100 species of each having been already observed; and if taken together and considered with respect to the mass of
587] vegetable matter they contain, calculated from the size as well as the number of individuals, are, perhaps, nearly equal to all the other plants of that country. They agree very generally also, though belonging to very different families, in a part of their economy which contributes somewhat to the peculiar character of the Australian forests, namely, in their leaves or the parts performing the functions of leaves being vertical, or presenting their margin, and not either surface, towards the stem; both surfaces having consequently the same relation to light. This economy, which uniformly takes place in the Acaciæ, is in them the result of the vertical dilatation of the foliaceous footstalk; while in Eucalyptus, where, though very general, it is by no means universal, it proceeds from the twisting of the footstalk of the leaf.

The plants of Terra Australis at present known, amounting to 4200, are referable, as has been already stated, to 120 natural orders; but fully half the number of species belong to eleven orders.

Of these Leguminosæ, Euphorbiaceæ, Compositæ, Orchideæ, Cyperaceæ, Gramineæ, and Filices are most extensive and very general tribes, which are not more numerous in Terra Australis than in many other countries.

Thus Leguminosæ and Compositæ, which taken together comprehend one fourth of the whole of Dicotyledones, and Gramineæ, which alone form an equal part of Monocotyledones, bear nearly the same proportion to these primary divisions in the Australian Flora.

The four remaining orders are Myrtaceæ, Proteaceæ, Restiaceæ, and Epacrideæ. Of these *Myrtaceæ,* though it is likewise very general, has evidently its maximum in Terra Australis, more species having been already observed in that country than in all other parts of the world; *Proteaceæ* and *Restiaceæ,* which are nearly confined to the southern hemisphere, and appear to be most abundant in the principal parallel of New Holland, are also very numerous at the Cape of Good Hope: and *Epacrideæ,* at least, equally limited to the southern hemisphere, are, with very few exceptions, confined to Terra Australis.

Several other less extensive natural families have also their maximum in this country, especially Goodenoviæ, Stylideæ, Myoporinæ, Pittosporeæ, Dilleniaceæ, Diosmeæ, and Halorageæ; but the only orders that appear to be absolutely confined to Terra Australis are Tremandreæ and Stackhouseæ, both of them very small tribes, which [588
many botanists may be disposed to consider rather as genera than separate families.

A great part of the genera of Terra Australis are peculiar to it, and also a considerable number of the species of such of its genera as are found in other countries.

Of the species at present composing its Flora scarcely more than 400, or one tenth of the whole number, have been observed in other parts of the world. More than half of these are Phænogamous plants, of which the greater part are natives of India, and the islands of the southern Pacific; several, however, are European plants, and a few belong even to equinoctial America. Of the Cryptogamous plants the far greater part are natives of Europe.

In comparing very generally the Flora of the principal parallel of Terra Australis with that of South Africa, we find several natural families characteristic of the Australian vegetation, as Proteaceæ, Diosmeæ, Restiaceæ, Polygaleæ, and also Buttneriaceæ, if Hermannia and Mahernia be considered as part of this order, existing, and in nearly equal abundance, at the Cape of Good Hope; others are replaced by analogous families, as Epacrideæ by Ericeæ; and some tribes which form a considerable part of the Australian peculiarities, as Dilleniaceæ, the leafless Acaciæ and Eucalyptus, are entirely wanting in South Africa.

On the other hand, several of the characteristic South African orders and extensive genera are nearly or entirely wanting in New Holland: thus Irideæ, Mesembryanthemum, Pelargonium, and Oxalis, so abundant at the Cape of Good Hope, occur very sparingly in New Holland, where the South African genera Aloe, Stapelia, Cliffortia, Penæa, and Brunia, do not at all exist. Very few species are common to both countries, and of these the only one which is at the same time peculiar to the Southern hemisphere is *Osmunda barbara*.

We have not sufficient materials for a satisfactory comparison of the Flora of the higher latitudes of South America with that of the Southern parts of Terra Australis. If, however, we may judge from those at present in our possession, it would seem that the general character of the South American vegetation differs much more from the Australian than this does from that of South Africa. Yet several instances occur of the same or of very nearly related genera,
589] peculiar to the southern hemisphere, which are common to Terra Australis and South America, and which do not exist at the Cape of Good Hope. Thus the Pavonia or *Laurelia* of Chili has its nearly related genus *Atherosperma* in Van Diemen's Island; where also a genus that I shall name *Tasmania* occupies the place of the *Wintera* of South America, from which it differs chiefly in having a single ovarium; a species of the *Araucaria* of Chili exists in New Holland as well as in Norfolk Island and New Caledonia; several *Lomatiæ* are found in South America; a species of *Astelia* grows in Terra del Fuego; and *Goodenia littoralis* of the southern shores of Terra Australis is found not only in New Zealand but on the opposite coast of America.

Certain tribes of plants common to South Africa and Terra Australis, and almost equally abundant in both these countries, are either very sparingly produced or cease to exist in South America. Others which abound in South

Africa and are comparatively rare in Terra Australis are in South America entirely wanting; and I am acquainted with no tribe of plants common to South Africa and South America and at the same time wanting in Terra Australis, unless the Compositæ with bilabiate corolla.

The character of the New Zealand Flora, known to us chiefly from the materials collected by Sir Joseph Banks, is to a considerable degree peculiar; it has still, however, a certain affinity to those of the two great countries between which it is situated, and approaching rather to that of Terra Australis, than of South America.

In comparing together the Floras of Terra Australis and Europe, I shall chiefly confine myself to an enumeration of the species common to both countries; the subject at present hardly admitting of many remarks of a more general nature. It may, however, be observed, that none of the great natural orders of Europe are absolutely wanting in Terra Australis; that some of them, as Compositæ, Leguminosæ, Gramineæ and Cyperaceæ, are found even in nearly the same proportion; while others, as Cruciferæ, Ranunculaceæ, Caryophylleæ, Rosaceæ, and Ericeæ are reduced to very few species; and that several of the less extensive European orders, namely, Saxifrageæ, Cistineæ, Berberides, Resedaceæ, Fumariaceæ, Grossularinæ, Valerianeæ, Dipsaceæ, Polemonideæ, Globulariæ, Elæagneæ, and Equisetaceæ in Terra Australis do not at all exist.

The greater number of Australian genera, except the [590
Acotyledonous, differ from those of Europe; there are, however, a few European genera, as Utricularia, Drosera, and Samolus, that appear to have even their maximum in Terra Australis.

From the following list of species, common to Terra Australis and Europe, I have carefully excluded all such as, though now existing in the different settlements, have evidently, or probably, been introduced, and I am satisfied that no naturalised plant will be found in it except, perhaps, *Cynodon Dactylon*.

I have also excluded certain plants, as *Elatine Hydropiper*, *Geum urbanum*, *Oxalis corniculata*, *Lycopus euro-*

pæus, and *Typha angustifolia*, which, though appearing to differ in some respects from those of Europe, are probably not specifically distinct. And if among the Phænogamous plants inserted there be any room for doubt respecting the identity of the Australian and European species, it may possibly be as to *Arenaria marina*, *Zapania nodiflora*, *Atriplex Halimus*, *Potamogeton gramineum*, *Cyperus rotundus* and *Holcus Gryllus*.

The first observation that occurs with regard to this list is, that the relative proportions of the three primary divisions of plants compared with those of the Australian Flora are inverted: for of 2900 Dicotyledones of the Flora only 15 are natives of Europe; while of 860 Monocotyledones 30, and of 400 Acotyledones upwards of 120 appear in the list.

The Phænogamous plants of the list are, with very few exceptions, also natives of North America, and several of them are found even in other parts of the world.

There is nothing peculiar in the apparent structure or economy of the *Dicotyledonous* plants common to countries so remote to account for their more general diffusion; though several of them grow in wet or marshy ground, yet very few are properly aquatic plants; and in the structure of their seeds the only circumstance in which they all agree is in the plumula of their embryo not being evolved.

Of the *Monocotyledones*, on the other hand, a considerable number are aquatic plants; and the greater part of those that are not aquatic belong to the irregular tribes, supposed to have a simpler structure.

Among the *Acotyledonous* or Cryptogamous orders it is
591] remarkable that there should be but a single species of Fern in the list, though those of the Flora exceed 100, of which 28 species are found likewise in other countries. It is also worthy of notice that of the Submersed Algæ not more than one sixth of the whole number found occur in the list; while of the Musci and Hepaticæ one third, and of the Lichenosæ two thirds of those observed are also natives of Europe.

The proportion of European plants in Terra Australis,

though only one twenty-fifth[1] of the whole number observed, appears to be greater than that in the Flora of South Africa. And the vegetation of the Cape of Good Hope, not only in the number of species peculiar to it, but in its general character, as depending on the extensive genera or families of which it is composed, differs almost as widely from that of the northern parts of the same continent, and the south of Europe, as that of the corresponding latitude of Terra Australis does from the Flora of India and of Northern Asia.

Of the proportion of European species in the Flora of South America, which is probably still smaller than that of South Africa, we have very insufficient means of judging; we know, however, from the collections made by Sir Joseph Banks that, at the southern extremity of America, certain European plants, as *Phleum alpinum*, *Alopecurus alpinus*, and *Botrychium Lunaria* exist; and that there is even a considerable resemblance in the general character of the Flora of Terra del Fuego to that of the opposite extremity of America and of the North of Europe.

[1] In the original text the proportion is stated as "one-tenth;" but this obvious mistake was corrected as above, by Mr. Brown—himself in the Banksian copy of 'Flinders's Voyage.' *Edit.*

592] A LIST OF PLANTS, NATIVES BOTH OF TERRA AUSTRALIS AND OF EUROPE.

DICOTYLEDONES.

Polypetalæ.

Potentilla anserina, *Linn.*
Aphanes arvensis, *Linn.*
Lythrum Salicaria, *Linn.*
Portulaca oleracea, *Linn.*
Arenaria marina, *Smith brit.* 480.
Nasturtium amphibium, *Hort. Kew. ed.* 2, *vol.* 4, *p.* 110.
Hydrocotyle vulgaris, *Linn.*

Monopetalæ.

Sonchus oleraceus, *Linn.*
Picris hieracioides, *Linn.*
Zapania nodiflora, *Prodr.* 514.
Verbena officinalis, *Linn.*
Prunella vulgaris, *Linn.*
Calystegia sepium, *Prodr.* 483.
Samolus Valerandi, *Linn.*

Apetalæ.

Atriplex Halimus, *Linn.*

MONOCOTYLEDONES.

Hydrocharideæ.

Vallisneria spiralis, *Linn.*
Lemna minor, *Linn.*
— trisulca, *Linn.*

Alismaceæ.

Potamogeton natans, *Linn.*
— perfoliatum, *Linn.*
— crispum, *Linn.*
— gramineum, *Linn.*
Alisma Plantago, *Linn.*

Aroideæ.

Caulinia oceanica, *Prodr.* 339.
Zostera marina, *Linn.*

Junceæ.

Luzula campestris, *Decand. franc.* 3, *p.* 161.
Juncus maritimus, *Smith brit.* 375.
— effusus, *Linn.*

Cyperaceæ.

Carex Pseudo-cyperus, *Linn.*
— cæspitosa, *Linn.*
Cladium Mariscus, *Prodr.* 236.
Scirpus maritimus, *Linn.*
— triqueter, *Linn.*
— mucronatus, *Linn.*
— lacustris, *Linn.*
Isolepis setacea, *Prodr.* 222.
— fluitans, *Prodr.* 221.
Cyperus rotundus, *Linn.*

Gramineæ.

Glyceria fluitans, *Prodr.* 179.
Arundo Phragmites, *Linn.*
Cynodon Dactylon, *Prodr.* 187.
Panicum crus-galli, *Linn.*
Pennisetum glaucum, *Prodr.* 195.
Imperata arundinacea, *Prodr.* 204.
Holcus Gryllus, *Prodr.* 199.

ACOTYLEDONES. [593

Marsileaceæ.

Marsilea quadrifolia, *Linn.*

Filices.

Hymenophyllum tunbridgense, *Smith brit.* 1141.

Musci.

Hypnum recognitum, *Hedw. sp. musc.* 261.
Leskea complanata, *Hedw. sp. musc.* 231.

Hookeria lucens, *Smith in linn. soc. transact.* 9, *p.* 275.
Neckera pennata, *Hedw. sp. musc.* 200
— heteromalla, *Hedw. sp. musc.* 202.
Bryum capillare, *Hedw. sp. musc.* 182.
— argenteum, *Hedw. sp. musc.* 181.
Bartramia pomiformis, *Hedw. sp. musc.* 164.
Funaria hygrometrica, *Hedw. sp. musc.* 172.
Barbula unguiculata, *Hedw. sp. musc.* 118.
Trichostomum canescens, *Hedw. sp. musc.* 111.
— polyphyllum, *Hedw. suppl.* 153.
Cynontodium capillaceum, *Hedw. sp. musc.* 57.
Fissidens exilis, *Hedw. sp. musc.* 152.
Dicranum purpureum, *Hedw. sp. musc.* 136.
— flexuosum, *Hedw. sp. musc.* 145?
— scoparium, *Hedw. sp. musc.* 126.
Encalypta vulgaris, *Hedw. sp. musc.* 60.
Weisia controversa, *Hedw. sp musc.* 67.
Grimmia pulvinata. Dicranum pulvinatum, *Hedw. suppl.* 1, *p.* 189.
— apocarpa, *Hedw. sp. musc.* 76.
Gymnostomum pyriforme, *Hedw. sp. musc.* 38.
Anictangium ciliatum, *Hedw. sp. musc.* 40.
Phascum muticum, *Hedw. sp. musc.* 25.
Sphagnum capillifolium, *Hedw. sp. musc.* 25.

HEPATICÆ.

Jungermannia tomentella, *Hooker junger.* 36.
— tamarisci, *Linn.*
— complanata, *Linn.*
— bidentata, *Linn.*
— pinguis, *Linn.*
— byssacea, *Hooker junger.* 12.
— furcata, *Linn.*
Targionia hypophylla, *Linn.*
Marchantia polymorpha, *Linn.*
— hemisphærica, *Linn.*
Anthoceros punctatus, *Linn.*
Riccia glauca, *Linn.*
— natans, *Linn.*
— fluitans, *Linn.*

LICHENOSÆ.

Lecidea geographica, *Achar. lichenogr.* 163.
— confluens, *Achar. loc cit.* 174.
— parasema, *loc. cit.* 175.
— luteola, *loc. cit.* 195.
— lurida, *loc. cit.* 219.
Gyrophora polyphylla. G. heteroidea, β, *loc. cit.* 219.
Calicium claviculare, *loc. cit.* 235.
— proboscidea, *loc. cit.* 220.
Verrucaria nitida, *loc. cit.* 279.
Endocarpon hepaticum, *loc. cit.* 298.
Thelotrema lepadinum, *loc. cit.* 312.
Lecanora atra, *loc. cit.* 344.
— fusco-atra, *loc. cit.* 359.
— β dendritica, *loc. cit.*
— parella, *loc. cit.* 370.
— subfusca, *loc. cit.* 393.
— ventosa, *loc. cit.* 399.
— sulphurea, *loc. cit.* 399.
— decipiens, *loc. cit.* 409.
— lepidosa, *loc. cit.* 417.
— microphylla, *loc. cit.* 420.
— gelida, *loc. cit.* 428.
— lentigera, *loc. cit.* 423.
— brunnea β nebulosa, *loc. cit.* 419.
Roccella fuciformis, *loc. cit.* 440.
Evernia prunastri, *loc. cit*, 442.
Sticta crocata, *loc. cit.* 447.
— pulmonacea, *loc. cit.* 449?
— scrobiculata, *loc. cit.* 453.
Parmelia caperata, *loc. cit.* 457.
— olivacea, *loc. cit.* 462.
— parietina, *loc. cit.* 463.
— plumbea, *loc. cit.* 466.
— stellaris, *loc. cit.* 476.
— conspersa, *loc. cit.* 486.
— physodes, *loc. cit.* 492.
Peltidea canina, *loc. cit.* 517.
Cenomyce pyxidata, *loc. cit.* 534.
— coccifera, *loc, cit.* 537.
— deformis, *loc. cit.* 538. [594
— cornuta, *loc. cit.* 545.
— rangiferina, *loc. cit.* 564.
— vermicularis, *loc. cit.* 566.

Stereocaulon paschalis, *loc. cit.* 581.
Sphærophoron coralloides, *loc. cit.* 585.
— compressum, *loc. cit.* 586.
Ramalina fraxinea, *loc. cit.* 602.
— fastigiata, *loc. cit.* 602.
Cornicularia spadicea, *loc. cit.* 611.
— lanata, *loc. cit.* 615.
— pubescens, *loc. cit.* 616.
Usnea florida, *loc. cit.* 620.
Collema nigrum, *loc. cit.* 628.
— fasciculare, *loc. cit.* 639.
— tremelloides, *loc. cit.* 655.
Lepraria flava, *loc. cit.* 663.
— incana, *loc. cit.* 665.
— botryoides, *Achar. meth.* 6.

FUNGI.

Rhizomorpha setiformis, *Pers. syn. fung.* 705.
Tubercularia vulgaris, *Pers. syn. fung.* 112.
Sphæria ophioglossoides, *Pers. syn. fung.* 4.
Clavaria pistillaris, *Linn.*
— coralloides, *Linn.*
Peziza scutellata, *Linn.*
Boletus igniarius, *Linn.*
Agaricus alneus, *Linn.*
— muscarius, *Linn.*
— campestris, *Linn.*

ALGÆ.

Conferva ebenea, *Dillwyn brit. conf.* 101.
— ericetorum, *Dillwyn brit. conf.* 1.
Ulva plumosa, *Huds. ang.* 571.
— lactuca, *Linn.*
Fucus articulatus, *Turner fuci* 2, *p.* 93, *t.* 106.
— obtusus, *Turner fuci* 1, *p.* 44, *t.* 21.
— pinnatifidus, *Turner fuci* 1, *p.* 40, *t.* 20.
— corneus, *Eng. bot.* 1970.
— plicatus, *Turner fuci* 3, *p.* 107, *t.* 180.
— palmatus, *Turner fuci*, *p.* 117, *t.* 115.
— rubens, *Turner fuci* 1, *p.* 89, *t.* 42.
— sinuosus, *Turner fuci* 1, *p.* 74, *t.* 35.

DESCRIPTIONS OF PLANTS FIGURED [595
IN THE ATLAS.

FLINDERSIA.

Ord. Nat. *Cedreleæ.*
Syst. Linn. *Pentandria Monogynia,* inter Cedrelam et Calodendrum.

CHAR. GEN. *Stamina* decem, dorso urceoli hypogyni inserta: alterna sterilia. *Capsula* 5-partibilis; segmentis singulis divisis *dissepimento* longitudinali, demum libero, utrinque dispermo. *Semina* erecta, apice alata.

FLINDERSIA AUSTRALIS. *Tab.* 1.

A tree of moderate size, observed September, 1802, both in flower and with ripe capsules, in the woods and thickets near the head of Broad Sound, on the east coast of New Holland, in about 23° S. lat. The examination of Broad Sound was completed at the same time by Captain Flinders, to commemorate whose merits I have selected this genus from the considerable number discovered in the expedition, of which he was the able and active commander.

DESC. *Arbor,* trunco pro ratione altitudinis mediocris crasso, coma irregulari, ramis patulis, ramulis teretibus umbellatis cortice fusco cinereo rugoso, gemmis foliorum apicibusque ramulorum gummiferis. *Folia* alterna, ad apicem ramuli conferta, exstipulata, petiolata, composita, ternata vel cum impari opposito-pinnata 2-3-juga; *foliola* oblongo-elliptica (in ramulis sterilibus quandoque lanceolata), integerrima glaberrima plana pellucido-punctata, dum 2-3 uncias longa 12-15 lineas lata. *Petiolus* communis angulatus mediocris: partialium laterales brevissimi, terminalis foliorum inferiorum 3-4 lineas æquans. *Paniculæ* terminales confertæ, ramis ramulisque alternis patentibus, pube brevi instructis; *bracteis* parvis subulatis. *Flores* parvi albi, odore debili haud ingrato. *Calyx* brevis 5-fidus, extus pubescens, laciniis æqualibus semiovatis acutis, persistens. *Petala* 5 sessilia oblongo-ovata obtusa plana, extus tenuissime pubescentia, basi disci staminiferi inserta, æstivatione imbricata. *Stamina* decem, infra apicem extus disci hypogyni inserta, petalis breviora. *Filamenta* [596
5 *antherifera* cum petalis alternantia, prope basin disci inserta; 5 *sterilia* petalis opposita, breviora, in disco paulo altius imposita: omnia glabra compresso-filiformia conniventia; *Antheræ* conniventes ovato-cordatæ acuminatæ glabræ flavicantes, juxta basin affixæ, loculis appositis longitudinaliter dehiscentibus: *Pollen* flavum globosum læve. *Discus hypogynus* ovarium laxè circumdans,

brevis glaber cyathiformis decemplicatus subcrenatus. *Ovarium* liberum sessile depresso-globosum viride, tuberculis confertissimis obtusis undique tectum, villisque rarioribus tenuibus pubescens, 5-loculare; *Stylus* simplex erectus glaber obtusè 5-gonus; *Stigma* peltatum altè 5-lobum. *Capsula* lignea oblonga obtusa fere 3-uncialis, basi calyce minimo persistenti subtensa, undique echinata processubus suberoso-ligneis confertis subconicis, 5-partibilis, segmentis cymbiformibus, tandem ab apice semibifidis et siccatione sæpe transversim fissis, basibus ante dehisceentiam adnexis axi centrali brevi demum libero et persistenti. *Placenta* centralis longitudinaliter alte 5-loba, efformans *Dissepimenta* quinque longitudine capsulæ, cum segmentis alternantia ideoque eorum cavitates bipartientia, ante dehiscentiam margine interiore connexa, demum soluta, dimidiato-oblonga plana spongioso-lignea, versus dorsum obtusum sensim crassiora, margine interiore in aciem attenuata, utrinque disperma, et insignita lineis duabus alternis a margine interiore arcuatim descendentibus et paulo intra dorsum desinentibus. *Semina* erecta, funiculo brevi compresso paulo supra basin marginis exterioris inserta, plano-convexa, apice in alam membranaceam planam uninervem ipso nucleo subovato sesquilongiorem desinentia. *Integumentum* simplex coriaceum basi lateribusque spongioso-incrassatis. *Albumen* nullum. *Embryo* dicotyledoneus albus; *Cotyledones* transversæ crasso-foliaceæ aveniæ; *Radicula* prope medium marginis interioris seminis transversa, brevissima, sinu baseos cotyledonum inclusa, ab umbilico remota.

Obs. There can be very little doubt that *Arbor Radulifera* of the Herbarium amboinense (3, p. 201, t. 129,) belongs to Flindersia, not only from the external appearance of the capsule as exhibited in the figure, but from the description given by Rumpf of its dehiscence, as well as of the peculiar dissepiments and the structure of the seeds.

The affinities of this genus are not perhaps very evident. I have referred it to *Cedreleæ*, an order certain genera of which are annexed by Jussieu to *Meliaceæ*, but which I have separated from that family chiefly on account of the structure of the fruit, and of the winged seeds. Flindersia, however, does not agree with the other genera of Cedreleæ either in the insertion of its seeds or dehiscence of its capsule; and it appears to differ from them remarkably in its moveable dissepiments; but these may be considered as the segments of a common placenta, having a peculiar
597] form, indeed, but not being in other respects essentially different from that of Cedreleæ. Flindersia is distinct also from the whole of the order, in having its leaves dotted with pellucid glands, in which respect it seems to connect Cedreleæ with Hesperideæ; and, notwithstanding the absence of albumen, even with Diosmeæ.

EUPOMATIA.

Ord. Nat. *Annonaceæ!*
Syst. Linn. *Icosandria Polygynia*, v. *Monadelphia Polyandria.*

CHAR. GEN. *Operculum* superum integerrimum deciduum (integumentis floralibus præterea nullis). *Stamina* numerosa: *exteriora* antherifera: *interiora* sterilia petaloidea imbricata. *Ovarium* multiloculare, loculis indefinitis (numero et ordine), polyspermis. *Stigmata*; areolæ tot quot loculi, in apice planiusculo ovarii. *Bacca* polysperma.

EUPOMATIA LAURINA. *Tab.* 2.

In woods and thickets in the colony of Port Jackson, especially in the mountainous districts, and on the banks of the principal rivers; flowering in December and January.

DESC. *Frutex* erectus ramosus glaberrimus 5-10 pedes altus, trunco gracili, ramulis teretibus subporrectis. *Folia* alterna, in ramulis bifaria, petiolata exstipulata, impunctata coriacea utrinque nitida nigro-viridia, integerrima plana oblonga acumine brevi, basi acuta æquali, dum 5 uncias longa sesquiunciam lata. *Pedunculi* axillares, uniflori folio breviores ramuliformes, foliis 3-4 alternis nanis instructi. *Perianthium* superum, limbo juxta basin transversim dehiscente: *Operculo* caduco semielliptico, paulo ante dehiscentiam albo-virescenti, e calyce et corolla concretis forsan conflato. *Stamina* margine persistenti limbi perianthii inserta, multiplici serie, basibus connatis; *exteriora* antherifera numerosa patula vel arctè reflexa; *Filamentis* e basi dilatata subulatis; *Antheris* ochroleucis linearibus, adnatis apice filamenti in mucronulum ultro producti, bilocularibus longitudinaliter dehiscentibus; *Polline* globoso lævi; *interiora* sterilia petaloidea, arcte imbricata multiplici serie, basi invicem et antheriferis connexa simulque decidua, interioribus sensim minoribus arctiusque imbricatis. *Ovarium* turbinatum multiloculare, loculis sparsis nec verticillatis, polyspermis, ovulis ellipticis angulo interiori loculi insertis. *Stigma* sessile planiusculum integrum, areolis subrotundis numero loculorum notatum. *Bacca* turbinato-obovata glabra, basi angusta limbi persistentis perianthii coronata, apice truncato areolato. *Semina* solitariè vel quandoque geminatim inclusa cellulis clausis, mutua pressione varie angulata, circumscriptione subovata [598
glabra impresso-punctata, altera extremitate, sæpius obtusiore, affixa; chorda ventrali ab umbilico parvo ducta ad extremitatem oppositam ibique in chalazam integumento interiori adnatam desinente; *Integumentum* duplex; *exterius* membranaceum intus undique emittens processus breves inter rugas albuminis demissos, et secundum chordam ventralem processum continuum altiorem nucleum semibipartientem; *interius* tenuissimum albumine arctissime adnatum: *Albumen* semini conforme carnosum lobatum. *Embryo* in regione umbilici, albumine 5-6ies brevior, dicotyledoneus albicans: *Cotyledones* lineares foliaceæ: *Radicula* teres recta longitudine cotyledonum.

Obs. This genus forms a very unexpected addition to *Annonaceæ*, of which it will constitute a distinct section, remarkable in the manifestly perigynous insertion of its stamina and the cohesion of the tube of its calyx with the ovarium. It has therefore nearly the same relation to the other genera of the order that Nymphæa has to Hydropeltis: and the affinity in both cases is chiefly determined by the structure of the seed.

The operculum of Eupomatia, in which there is no mark of longitudinal division, may be considered as formed either of the calyx alone, or of the confluent calyx and corolla, as appears to be the case at least in several species of Eucalyptus.

A singular part of the structure of Eupomatia consists in its internal, barren, petal-like stamina, which, from their number and disposition, completely cut off all communication between the antheræ and stigmata. This communication appears to be restored by certain minute insects eating the petal-like filaments, while the antheriferous stamina, which are either expanded or reflected, and appear to be even slightly irritable, remain untouched. I have at least not unfrequently seen the barren stamina removed in this way, and, as all the stamina are firmly connected at the base and fall off together, it is difficult to conceive any other mode of exposing the stigmata to the influence of the antheræ.

599] EUDESMIA.

Ord. Nat. ***Myrtaceæ*, inter Eucalyptum et Angophoram.**
Syst. Linn. ***Polyadelphia Polyandria.***

CHAR. GEN. *Calyx* superus, 4-dentatus. *Petala* arcte connata in *Operculum* 4-striatum deciduum. *Stamina* in phalanges quatuor polyandras, cum dentibus calycis alternantes, basi connata. *Capsula* 4-loc. polysperma, apice dehiscens.

EUDESMIA TETRAGONA. *Tab.* 3.

In exposed barren places near the shores, in the neighbourhood of Lucky Bay, on the south coast of New Holland in 34° S. lat. and 123° E. lon.; gathered both in flower and fruit in January, 1802.

DESC. *Frutex* 3-5 pedes altus, ramis patentibus, ramulis 4-gonis angulis marginatis. *Folia* opposita quandoque subopposita, petiolata, sæpius aversa, lanceolata vel oblonga, coriacea compacta, integerrima marginata glauca resinoso-punctata, venis vix emersis anastomozantibus, 3-4 uncias longa, 14-16 lineas lata. *Umbellæ* laterales pauciflorae, pedunculo pedicellisque ancipitibus. *Calyx* turbinatus obtuse 4-gonus cum ovario cohærens, angulis apice productis in dentes breves subinæquales, duobus oppositis paulo majoribus. *Operculum* depresso-hemisphæricum muticum glandulosum albicans, striis quatuor cruciatis parum depressis dentibus calycis oppositis notatum, quasi e petalis quatuor conflatum, caducum. *Stamina* plurima; *Filamenta* in phalanges quatuor petalis oppositas approximata, capillaria glabra alba, interiora sensim breviora; *Antheræ* ovato-subrotundæ incumbentes ochroleucæ, loculis longitudinaliter dehiscentibus. *Ovarium* inclusum tubo adherenti calycis, 4-loculare: *Stylus* 1, cylindraceus; *Stigma* obtusum. *Capsula* inclusa et connata tubo aucto turbinato oblongo ligneo calycis, apice 4-fariam dehiscens.

OBS. There can be no doubt respecting the affinity of this genus, which belongs to Myrtaceæ and differs from Eucalyptus solely in having a striated operculum placed within a distinctly toothed calyx, and in its filaments being collected into bundles. The operculum in Eudesmia, from the nature of its striæ and their relation to the teeth of the calyx, appears to be formed of the confluent petals only; whereas, that of Eucalyptus, which is neither striated nor placed within a distinct calyx, is more probably composed, in several cases at least, of both floral envelopes united. But in many species of Eucalyptus a double operculum has been observed; in these the outer operculum, which generally separates at a much earlier stage, may, perhaps, be considered as formed of the calyx, and [600 the inner consequently of corolla alone, as in Eudesmia: this view of the structure appears at least very probable in contemplating *Eucalyptus globulus,* in which the cicatrix caused by the separation of the outer operculum is particularly obvious, and in which also the inner operculum is of an evidently different form.

Jussieu, in some observations which he has lately made on this subject, (*in Annales du mus.* 19. *p.* 432,) seems inclined to consider the operculum of Eucalyptus as formed of two confluent bracteæ, as is certainly the case with respect to the calyptra of Pileanthus, and of a nearly related genus of the same natural family. This account of its origin in Eucalyptus, however, is hardly consistent with the usual umbellate inflorescence of that genus; the pedicelli of an umbel being always destitute of bracteæ; and in *Eucalyptus globulus,* where the flowers are solitary, two distinct bracteæ are present as well as a double operculum. But a calyptra analogous to that of Pileanthus exists also in most of the species of Eucalyptus, where it is formed of the confluent bracteæ common to the whole umbel, and falls off at a very early period.

CEPHALOTUS.

Ord. Nat. *Rosaceæ.*
Syst. Linn. *Dodecandria Hexagynia.*

CHAR. GEN. *Calyx* coloratus, 6-fidus, æstivatione valvata. *Petala* 0. *Stam.* 12, perigyna: *Antherarum* dorso glanduloso. *Ovaria* 6, distincta, monosperma, ovulo erecto. *Styli* terminales.

CEPHALOTUS FOLLICULARIS. *Tab.* 4.

Cephalotus follicularis. *Labillard. nov. holl.* 2, *p.* 7, *t.* 145.

In marshy ground, in the neighbourhood of King George's Sound, especially near the shores of Princess Royal Harbour, in 35° S. lat. and 118° E. long.; beginning to flower about the end of December.

DESC. *Herba* perennis. *Caulis* abbreviatus vix uncialis, demersus sæpe sub terram divisus. *Folia* in apice vix emerso caulis conferta quasi radicalia, numerosa petiolata, exstipulata, elliptica, integerrima, enervia crassiuscula plana glabra pilisve raris instructa, viridia, 8-16 lineas longa. *Petiolus* folio vix brevior, semiteres basi parum dilatata.

Ascidia foliis intermista, petiolisque similibus porrectis parumve deflexis insidentia, in orbem circa folia digesta, respectu petioli dependentia, quoad pro-
601] priam cavitatem erecta, subovata, operculata, uncialia; *Extus* ornata costis tribus ab ore cristato ortis, valde elevatis et sæpius apice longitudinaliter depresso latiusculo marginibus acutis pilosis; *lateralibus* obliquis dorsum versus tendentibus sensim declinantibus et prope medium lateris desinentibus in lineas cursum oblique anticum servantes pauloque supra basin ventris coalescentes; *ventrali* elevatiore recta, longitudine ascidii, apice semper latiusque depresso: *Intus* nitentia et e majore parte nigro-purpurea, paulo infra annulum costatum

oris aucta margine prominulo integerrimo. *Ore* plusquam semiorbiculari, paulo infra marginem extus cincto annulo tenui, ortum præbente processubus numerosis (19-23), parallelis costæformibus, adnatis, extremitate interiore soluta brevi incurva acuta. *Operculo* e petiolo derivato et postico lateri oris ascidii lata basi inserto, foliaceo orbiculato-obovato emarginato planiusculo viridi, venis nigro-purpureis latis ramosis apice anastomozantibus picto, extus pilosiusculo intus glabro.

Scapus simplicissimus erectus pedalis sesquipedalis villosus pilis simplicibus acutis, instructus bracteis nonnullis alternis remotis deciduis; dimidio inferiore quandoque angulato, superiore semper tereti. *Spica* terminalis erecta biuncialis, composita spiculis numerosis, superioribus confertis, inferioribus sensim remotioribus, omnibus pedunculatis 4-5-floris, floribus subcorymbosis ebracteatis. *Bracteæ* pedunculos spicularum subtendentes subulato-lineares deciduæ. *Flores* parvi. *Calyx* albus, altè 6-fidus simplici serie, regularis æqualis, extus pube adpressa simplici, laciniis ovato-lanceolatis patulis apice denticulo interiore auctis; basi intus incrassata pilisque capitatis brevissimis hispidula. *Stamina* margini tubi calycis inserta, ejusdem laciniis breviora; sex laciniis alternantia longiora et præcociora; *Filamenta* subulata erecto-conniventia glabra purpurascentia: *Antheræ* conniventes subrotundæ didymæ, lobis appositis purpurascentibus longitudinaliter dehiscentibus, connectivo subgloboso fungoso celluloso albo adnatis: *Pollen* flavum. *Pistilla* 6 approximata, staminibus minoribus breviora; *Ovaria* cum laciniis perianthii alternantia subovata parum compressa glabra, margine ventrali truncato dorsali rotundato, monosperma; *Ovulo* erecto, magnitudine fere loculi et intra testam membranaceam continente sacculum magnitudine cavitatis testæ, pendulum: *Styli* terminales teretiusculi: *Stigmata* simplicia. *Fasciculus* laxus albus villorum centrum receptaculi intra ovaria occupat.

Obs. Cephalotus has been referred by its discoverer Labillardière to Rosaceæ, to which, notwithstanding its coloured calyx and the absence of petals, it seems to have the nearest affinity; a knowledge of the fruit, however, is wanting to determine absolutely its place in the natural method. From the structure of its ovulum, even in the unimpregnated state, I entertain no doubt that the radicule of the embryo points to the umbilicus of the seed and therefore downwards; a circumstance in which it would differ from the greater part, but not from all the [602
Rosaceæ; and in other respects it does not appear to belong to any subdivision of that order hitherto proposed.

In the structure of its ascidia it agrees with Nepenthes, with which, however, it has no other point of resemblance.

The ascidia or pitchers of Cephalotus were observed to be in general nearly half filled with a watery fluid, in which great numbers of a small species of Ant were frequently found drowned. This fluid, which had a slightly sweet taste, may possibly be in part a secretion of the pitcher itself, but more probably consists merely of rain-water received and preserved in it. The lid of the pitcher in the full grown state was found either accurately closing its mouth or having an erect position and therefore leaving it entirely open; and it is not unlikely that the position of the lid is determined by the state of the atmosphere, or even by other external causes.

ANTIARIS.

Ord. Nat. *Urticeæ,* inter Brosimum et Olmediam.
Syst. Linn. *Monœcia Tetrandria.*

CHAR. GEN. MASC. *Involucrum* multiflorum, basi orbiculata florifera, apice multifido. *Calyx* 4-ph. *Stam.* 4.

FEM. *Involucrum* uniflorum, urceolatum, apice multifidum. *Calyx* 0. *Ovarium* cum involucro cohærens, monospermum, ovulo pendulo. *Stylus* bipartitus. *Drupa* ex involucro aucto formata. *Semen* exalbuminosum. *Embryonis* radicula supera.

ANTIARIS MACROPHYLLA. *Tab.* 5.

A shrub or very small tree observed in barren stony places, on the shores of the Company's Islands, adjacent to Arnhem's Land, on the north coast of New Holland, in about 12° S. lat.; bearing both flowers and ripe fruit in February, 1803.

DESC. *Frutex* orgyalis ramosissimus glaber lactescens. *Ramuli* teretes. *Folia* alterna, petiolata, stipulata, oblonga cum brevi acumine, basi inæquali subcordata, glaberrima integerrima coriacea, suprà nigro-viridia nitida subtus viridiora, venis fere rectangulis juxta marginem anastomozantibus, venulis
603] divaricatis, dum sex uncias longa ultra tres uncias lata. *Petioli* teretiusculi cinerascentes semunciales. *Stipula* intrafoliacea conduplicata lanceolata acuminata foliacea. *Pedunculi* axillares solitarii, brevissimi, androgyni, pedicellis 6-8 alternis, infimo femineo præcociore, reliquis masculis. MASC. *Involucrum* subcyathiforme apice multifido, laciniis imbricatis acutis ciliatis conniventibus, demum expansum orbiculare marginibus reflexis, diametro quinquelineari. *Flosculi* numerosi densê conferti sessiles. *Calyx* 4-ph. foliolis subspathulatis apice conniventibus. *Corolla* 0. *Antheræ* biloculares: *Pollen* album. Nullum rudimentum pistilli. FEM. *Involucrum* uniflorum ovatum parvum glabrum viride apice multifidum, laciniis numerosis lanceolatis ciliatis conniventibus, nonnullisque dorsalibus sparsis similibus. *Calyx* 0. *Stamina* 0. *Ovarium* accretum et inclusum ventre involucri, monospermum, ovulo pendulo: *Stylus*

profundè bifidus, laciniis filiformibus elongatis albicantibus glabris; *Stigmata* acuta. *Drupa* ex involucro aucto efformata, ovalis glabra, magnitudine pruni domestici minoris, nigro-sanguinea, substantia carnosa crassa lactescente intus flavicante, lacte albo; *putamine* ovato crustaceo tenaci lævi fusco. *Integumentum seminis* præter putamen nullum. *Albumen* nullum. *Embryo* dicotyledoneus albus: *Cotyledones* maximæ amygdalino-carnosæ ovatæ plano-convexæ: *Radicula* supera brevissima.

Obs. When I collected and described this plant on the north coast of New Holland, I had no reason to suppose it had any affinity to the *Upas Antiar* or Poison tree of Java, of which the first satisfactory account has been since published by Mr. Leschenault. There can however be no doubt that the plant of New Holland belongs to the same genus with *Antiaris toxicaria* of that author,[1] notwithstanding some difference between our accounts of the structure of the male flowers; with respect to which I have only to observe that my description was taken from living plants, and I consider its correctness to be very much confirmed by the figure, which was afterwards made from dried specimens, by Mr. Bauer, who was unacquainted with my observations. Antiaris evidently belongs to Urticeæ, and in a natural series will stand between *Brosimum* of Swartz and *Olmedia* of Flora Peruviana, agreeing with the latter in the structure of its male flowers, and more nearly resembling the former in its female flowers and fruit.

FRANKLANDIA. [604

Ord. Nat. *Proteaceæ.*
Syst. Linn. *Tetrandria Monogynia.*

Char. Gen. *Perianthium* hypocrateriforme: *limbo* quadripartito deciduo, æstivatione induplicata: *tubo* persistenti. *Antheræ* inclusæ, perianthio accretæ. *Squamæ* 4, perigynæ, in vaginam 4-fidam connatæ. *Ovarium* monospermum. *Caryopsis* fusiformis pedicellata, apice dilatato papposo.

[1] *Annales du mus.* 16, *p.* 478, *t.* 22.

FRANKLANDIA FUCIFOLIA. *Tab.* 6.

Franklandia fucifolia, *Linn. soc. transact.* 10, *p.* 157. *Prodr. fl. nov. holl.* 370.

In moist heaths near the shores of King George's Sound, on the south west coast of New Holland, found in flower and with ripe seed in December, 1801.

DESC. *Frutex* erectus ramosus 2-3 pedes altus glaber, glandulis pustuliformibus aurantiacis undique conspersus. *Rami* teretes striati, epidermide tenui cinerea. *Folia* alterna triuncialia filiformia, basi per spatium semunciale indivisa, dein dichotoma, laciniarum apicibus fastigiatis, ultimæ dichotomiæ ramulo altero bifido altero simplici. *Spicæ* axillares solitariæ simplicissimæ raræ pedunculatæ erectæ, foliis parum longiores: *pedicellis* alternis basi unibracteatis; *bractea* ovata obtusa concava sesquilineam longa, post lapsum perianthii fructiferi persistenti. *Perianthium* luteum fere biunciale, extus conspersum pustulis rufescentibus: *Ungues* in tubum cylindraceum cohærentes, extra medium pappo caryopsidis expandenti solubiles et decidui; inferne arctius connati indurati persistentes: *Laminæ* tubo breviores, æquales lanceolatæ, disco plano, marginibus adscendentibus parum undulatis vividiusque coloratis, sub æstivatione induplicatis. *Stamina* 4, inclusa, antherarum apicibus faucem semiclaudentibus: *Filamenta* medio tubo perianthii quasi inserta, laciniis opposita et iisdem longitudinaliter arcte cohærentia: *Antheræ* lineares dimidio superiore filamenti in mucronulum ultro producti adnatæ, loculis appositis longitudinaliter dehiscentibus: *Pollen* flavum subglobosum obsoletissimè trigonum læve. *Vaginula* dimidio inferiore tubi perianthii arctissime adnata, ita ut quasi ad eandem altitudinem ac filamenta inserta, supernè soluta quadridentata, demum increscenti caryopsidi quadrifida, laciniis subulato-linearibus cum staminibus alternantibus. *Ovarium* teretiusculum, monospermum: dimidio inferiore barbato pilis strictis copiosis, tenuiore et in pedicellum sensim angustato: superiore fusiformi glabro striato: apice coronatum *Pappo* sessili capillari e pilis strictis acutis formato, ipsum ovarium aliquoties superante. *Stylus* filiformis erectus altitudine staminum lanatus apice glabro. *Stigma*
605] turbinatum indivisum glabrum, apice depresso. *Caryopsis* crustacea, dimidio inferiore persistenti indurato tubi perianthii inclusa, striata apice dilatata in vaginulam brevem subcyathiformem extus pappigeram intus glabram. *Semen* fusiforme, membrana propria tenuissima apice chalaza fusca insignita. *Albumen* nullum. *Embryo* erectus subcylindraceus albus; *Cotyledones* brevissimæ semiorbiculatæ; *Radicula* maxima elongato-turbinata teres acuta; *Plumula* inconspicua.

Obs. Franklandia, though evidently belonging to Proteaceæ, differs from the whole of that family in at least three points of structure, any one of which may equally be assumed as the essential character of the genus; namely, in the antheræ being fixed through their whole length to the laciniæ of the perianthium; in the squamæ which alternate with the stamina so intimately cohering at their base with the lower half of the calyx that they appear to originate from its upper part; and in the induplicate æstivation of the laminæ of the hypocrateriform perianthium. In this last respect the genus presents an exception to what I had formerly considered as one of the most constant distinguishing characters of the order; it does not, however, so materially invalidate this character as a change to any other kind of æstivation would have done; the induplicate and valvular modes passing into each other, merely by an abstraction or addition of the elevated margins of the laciniæ. Instances of the abstraction of these elevated margins, in orders where they are generally present, are met with in Goodenoviæ and Convolvulaceæ, and an instance of their addition as in Franklandia occurs, though less obviously, in Chuquiraga, a genus belonging to Compositæ, in which family the valvular æstivation is as general as in Proteaceæ.

The æstivation of Franklandia may be adduced in support of that opinion which considers the floral envelope of Proteaceæ as corolla rather than calyx; there being, I believe, no instance of a similar æstivation in a genuine calyx, unless that of Nyctagines be regarded as such: but a stronger argument for this envelope being really calyx is afforded also by Franklandia, in which the transition from the footstalk to the perianthium is so gradual as to be externally imperceptible, and is not marked either by any change or interruption of the surface.

The apparently similar origin in Franklandia of the stamina and squamæ affords an argument, in addition to what I had formerly stated,[1] for considering the latter as [606
barren filaments; we may, therefore, expect to find octandrous genera belonging to this family. While the persistence and induration of the lower half of the perianthium in this genus, and the perigynous origin of the squamæ, which in other genera of the order are hypogynous, render it not improbable that plants may hereafter be discovered having a calyx absolutely cohering with the ovarium, which nevertheless it may be necessary to refer to Proteaceæ.

[1] *Linn. soc. transact.* 10, *p.* 159.

Elæagneæ, in which the tendency to cohesion of the calyx and ovarium is still more obvious than in Franklandia, approach very near to Proteaceæ in most respects, and the single difference in fructification between these two orders, consisting in the stamina being opposite to the laciniæ of the calyx in the latter and alternating with them in the former, is not an insuperable objection to their union; for *Drapetes*, which evidently belongs to Thymeleæ, has, in opposition to the rest of that order, its stamina alternating with the divisions of the perianthium.

SYNAPHEA.

Ord. Nat. *Proteaceæ*.
Syst. Linn. *Triandria Monogynia*.

CHAR. GEN. *Perianthium* tubulosum, 4-fidum, ringens. *Antheræ* tres, inclusæ: *inferior* didyma cum lateralibus dimidiatis primo cohærens in vaginam bilocularem, lobis proximis vicinarum loculum unicum constituentibus. *Stigma* filamento superiore sterili connatum. *Nux*.

SYNAPHEA DILATATA. *Tab.* 7.

Synaphea foliis apice dilatatis trilobis: lobis inciso-dentatis, petiolis spicisque villosis, stigmate bicorni. *Lin. soc. transact.* 10, *p.* 156. *Prodr. fl. nov. holl.* 370.
Conospermum reticulatum. *Smith in Rees Cyclopæd.*

In exposed barren situations, near the shores of King George's Sound; gathered in flower and fruit, in December, 1801.

DESC. *Fruticulus* procumbens teres crassitie pennæ corvinæ, subramosus, villis patulis mollibus tomentoque appresso cinereus. *Folia* alterna, elongato-petiolata, adscendentia, cuneata, basi valde attenuata, apice dilatato trifido, lobis
607] incisis, segmentis brevibus acutis mucronulo sphacelato; trinervia, nervis lateralibus margini approximatis simplicissimis et infra apicem folii desinentibus, nec in lacinulas extimas productis; medio trifido, ramis lateralibus subalternis; utrinque pulchrè reticulata, areolis minutis subtus magis conspicuis; adulta glabrata, novella villosa. *Petioli* teretes, basi dilatata oblongo-lanceolata scariosa. *Spicæ* axillares solitariæ simplicissimæ erectæ 3-4-unciales folia vix æquantes, basifloræ, rachi tomentosa, floribus alternis sessilibus unibracteatis; bracteis cucullatis ovatis acutis persistentibus. *Perianthium* coloratum flavum deciduum: *ungues* inferne connati in tubum demum 4-partibilem: *laminæ* in limbum bilabiatum dispositæ, lanceolatæ; dorsali labium superius

constituente latiore; anticarum media lateralibus angustiore. *Stamina* tubo inclusa, supra medium unguium inserta; *Filamenta* quatuor, brevia; *superiore* sterili apice cum stigmato connato; *reliqua* antherifera: *Antheræ laterales* dimidiatæ; *inferior* didyma, lobis cum iis lateralium longitudinaliter connexis, ita ut lobus singulus inferioris cum respondenti laterali loculum unicum tandem bivalvem constituit, nullo vestigio dissepimenti: *Pollen* triangulare. *Squamæ hypogynæ* nullæ. *Ovarium* turbinatum pubescens apiceque ornatum pilis brevibus crassis pellucidis strictis, monospermum: *Stylus* curvatus glaber sursum incrassatus deciduus: *Stigma* dilatatum obliquum, hinc apice filamenti arctissime connatum, inde desinens in cornua duo parallela distantia subulata. *Nux* crustacea obovata striata pilosa, apiceque coronata pappo brevi e pilis strictis crassioribus formato.

Obs. In my general observations on Proteaceæ I have noticed two very remarkable characters of Synaphea, namely, the cohesion of the barren filament with the stigma, which is peculiar to this genus, and the structure and connection of the antheræ, in which it agrees with Conospermum: it is also remarkable that these two nearly related genera should differ in the position of their barren and fertile stamina with relation to the perianthium; plants of the same natural family very generally agreeing in the order of abortion or suppression of these organs; to this, however, some other exceptions are known, and one has been already noticed as occurring in Drapetes.

The genus Synaphea seems to be confined to the south-west coast of New Holland, for it is more likely that *Polypodium spinulosum* of Burmannus (*flor. ind.* 233. *t.* 67. *f.* 1.) which I have formerly referred to this genus, as well as *Adiantum truncatum* of the same author, long since determined to be a species of Acacia, by Mr. Dryander, were brought from that coast to Batavia by one of the Dutch navigators, perhaps by Vlaming, than that they are really natives of Java, from which Burmannus received them.

DASYPOGON. [608

Ord. Nat. *Junceæ* inter Xerotem et Calectasiam.
Syst. Linn. *Hexandria Monogynia*, post Xerophytam.

CHAR. GEN. *Perianthium* duplex: *exterius* tubulosum, trifidum: *interius* triphyllum, foliolis semipetaloideis cochleariformibus. *Stamina* 6, imo perianthio inserta. *Ovarium* superum, uniloculare, trispermum, ovulis erectis. *Stylus* subulatus. *Stigma* simplex. *Utriculus* monospermus, tubo indurato aucto perianthii exterioris inclusus.

DASYPOGON BROMELIIFOLIUS. *Tab.* 8.

Dasypogon bromeliifolius. *Prodr. fl. nov. holl.* 263.

On the shores of King George's Sound; observed in flower and fruit in December, 1801.

DESC. *Planta* suffruticosa sesquipedalis bipedalis, habitu peculiari, ad *Xerotem* aliquatenus accedenti. *Caulis* simplicissimus teres foliatus, pilis strictis brevibus copiosis denticulatis reversis tectus. *Folia* graminea; radicalia conferta; caulina sparsa superioribus remotis, breviora, semiamplexicaulia; omnia mucronata glabra marginibus denticulato-asperis. *Capitulum* terminale solitarium sphæricum, magnitudine nucis juglandis vel pruni minoris, bracteis nonnullis patulis foliiformibus involucratum. *Flores* sessiles conferti, paleis e dilatata basi lanceolatis margine denticulatis distincti, aliisque angustioribus intermistis. *Perianthium exterius* 3-partitum, extus pilis longis strictis denticulatis barbatum; *unguibus* in tubum subovatum leviter cohærentibus; *laminis* distinctis ovatis concaviusculis infernè pallidis, supernè nigricantibus ibique intus pube tenuissima: *Interius* longitudine exterioris, glabrum; *unguibus* angustis distinctis approximatis concaviusculis hyalinis glabris; *laminis* ungue paulo latioribus, subellipticis ciliatis hyalinis, carina nigricanti apice pubescenti. *Stamina* ipsi basi perianthii inserta eoque fere duplo longiora: *Filamenta* æqualia filiformia alba glabra, apice incrassato subclavato cum apiculo brevi setaceo antherifero: *Antheræ* oblongæ pallidè flavæ incumbentes, infra medium affixæ, biloculares, loculis appositis approximatis longitudinaliter dehiscentibus. *Ovarium* subovatum trigonum glabrum albicans, ovulis oblongis: *Stylus* strictus glaber albus, inferne obsoletè trigonus, supernè teres. *Utriculus* membranaceus, inclusus tubo perianthii exterioris incrassato nucamentaceo nitido fusco glabrato. *Semen* subglobosum, integumento simplicissimo connato; *Albumine* carnoso semini conformi. *Embryo* . . .

609]
CALECTASIA.

Ord. Nat. *Junceæ.*
Syst Linn. *Hexandria Monogynia.*

CHAR. GEN. *Perianthium* inferum, tubulosum, hypocrateriforme, persistens: *limbo* petaloideo 6-partito, æstivatione imbricata. *Stamina* 6, fauci inserta: *Antheris* conniventibus, poro duplici apicis dehiscentibus. *Ovarium* uniloculare, trispermum, ovulis erectis. *Stylus* filiformis. *Stigma* simplex. *Utriculus* monospermus, tubo indurato perianthii inclusus.

CALECTASIA CYANEA. *Tab.* 9.

Calectasia cyanea. *Prodr. fl. nov. holl.* 264.

On barren hills, near the shores of King George's Sound; flowering in December.

DESC. *Fruticulus* ramosissimus erectus cæspitem efformans, pedalis sesquipedalis, glaber; caule inferne tereti, basibus persistentibus foliorum squamoso. *Folia* e basibus dilatatis semivaginantibus imbricatis patula, acerosa ancipitia rigida semuncialia, mucrone brevi pungenti terminata, glabra; ramea patula, ramulorum modice patentia confertiora. *Flores* ramulos breves ultimos terminantes solitarii, sessiles, foliis floralibus minoribus confertissimis, intimis albicantibus, infernè cincti. *Perianthium: Tubo* angusto-infundibuliformi subcarnoso viridi striato, extus villosiusculo, intus glabro: *Limbo* stellatim patulo, laciniis lanceolatis brevissime mucronulatis immerse nervosis parum concavis vivide cæruleis, disco extus villosiusculo. *Stamina* 6: *Filamenta* fauci perianthii inserta, limbi laciniis opposita, conniventia curvata cærulea glabra: *Antheræ* approximatæ, liberæ, oblongo-lineares obtusæ, basi emarginatæ affixæ, infernè quadriloculares, supernè biloculares poro duplici apicis dehiscentes. *Ovarium* subcylindraceum utrinque attenuatum dilute viride glabrum, longitudine tubi perianthii, uniloculare, trispermum, ovulis erectis: *Stylus* filiformis glaber cæruleus, basi pallidiore pauloque crassiore, pariter ac filamenta curvatus, staminibus paulo longior: *Stigma* acutum. *Utriculus* tubo indurato perianthii inclusus, tenuis, juxta basin transversim abscedens margine lacero, calyptra apicem seminis maturescentis tegente. *Semen* unicum, maturescens elongato-pyriforme teres tenuiter striatum, basi caudata funiculo capillari affixum. *Integumentum* simplicissimum nucleo arcte cohærens, apice area fusca notatum. *Albumen* semini conforme, dense carnosum, album, apice insculptum cavitate superficiali area fusca incrassata integumenti repleta. *Embryo*. . . . .

CORYSANTHES. [610

Ord. Nat. *Orchideæ*.
Syst. Linn. *Gynandria Monandria*.

Perianthium ringens: *Galea* magna: *Labium inferius* 4-partitum, nanum, occultatum *Labello* maximo cucullato vel tubuloso. *Anthera* terminalis, unilocularis, semibivalvis, persistens: *Massæ Pollinis* 4, pulvereæ.

CORYSANTHES FIMBRIATA. *Tab.* 10.

Corysanthes fimbriata. *Prodr. fl. nov. holl.* 328.

In shady places, especially under rocks and large stones, near Sydney, and in other parts of the colony of Port Jackson.

DESC. *Bulbus* solitarius pisiformis radicem longam teretem fibris nonnullis alternis simplicissimis instructam terminans.

Folium unicum, quasi radicale, sed caulem brevissimum demersum, basi squama unica semivaginanti subovata acuta instructum terminans, subrotundum mucrone brevissimo, basi altè cordata, lobis posticis rotundatis altero alterum equitante, explanatum horizontale, viride subtus dilutius, diametro subunciali, venosum venis dichotomis crebre anastomozantibus in nervum margini approximatum et parallelum desinentibus. *Flos* solitarius, pro ratione plantæ magnus, purpureus; *ovario* intra folium subsessili postice bractea semilanceolata erecta subtenso. *Perianthium* petaloideum sexpartitum ringens: *Foliola tria exteriora*, quorum *Galea* hyalina cum maculis crebris purpureis inæqualibus, e basi erectiuscula arcuata angustiore, superne dilatata obovata magis concava porrecta, apice incurvo, marginibus longitudinaliter nudis; *duo antica* cum lateralibus interiorum labium inferius descendenti-porrectum efformantia, subulata plana alba immaculata, ipsis basibus invicem connatis: *tria interiora*, quorum *duo lateralia* anticis exteriorum similia, e basi brevi porrecta adscendentia. *Labellum* maximum unguiculatum indivisum; *ungue* brevissimo erecto albo: *laminæ dimidio inferiore* adscendenti galeæ basi appresso, marginibus nudis inflexis tubum completum efformante, intus nigro sanguineo sursum dilutiore, paulo infra apicem albo virescenti rugoso subglanduloso; *superiore* dilatato ovato concavo deflexo, dilute purpureo maculis numerosis confluentibus rufo-sanguineis, disco intus paulo infra apicem glandulis sessilibus sparsis ornato marginibus inflexis fimbriatis lacinulis subulatis æqualibus.

Columna fructificationis inclusa, brevissima, adscendens, alba carnosa, basi parum coarctata, apice posticè trifido dentibus lateralibus erectis subulatis in-
611] termedio antherifero. *Anthera* mobilis ovata membranacea purpurascens apice semibifido, unilocularis, apicem columnæ incumbens. *Massæ Pollinis* 4, per paria cohærentes, farinaceæ, apicibus affixæ glandulæ communi emarginaturam stigmatis operienti. *Ovarium* oblongum: *Stylus* cum basi columnæ conferruminatus: *Stigma* solutum, horizontale subrotundum, antice concavum, apice plica duplici coarctatum, antheræ subparallelum.

Obs. The three species of which this genus at present consists agree in their anthera being unilocular after bursting, in the singular relative proportions of the parts of the perianthium, and in habit; but in some points, generally of importance in this order of plants, they differ very remarkably, especially in the form of the labellum, which in one species is even furnished with a double calcar. Corysanthes may therefore be considered as affording a proof, and many others might be adduced, of the superior importance of certain modifications of the anthera to those of the labellum in Orchideæ.

AZOLLA.

Ord. Nat. *Marsileaceæ.*
Syst. Linn. *Cryptogamia Filices.*

CHAR. GEN. *Flores* monoici.

MASC. Gemini, involucro clauso monophyllo membranaceo inclusi (nunc solitarii femineum stipantes), ovati, biloculares, membrana exteriore transversim dehiscenti: *loculo superiore* corpusculis 9 vel 6 angulatis, circa axin perforatum apice demum apertum insertis: *loculo inferiore* sphærico clauso, sub duplici membrana materia fluida (demum pulverea ?) repleto.

FEM. In diversis alis ejusdem frondis solitarii (nunc masculo inferiore stipati): *Involucrum* duplex, utrumque clausum membranaceum: *exterius* marium simile: *interius* ovatum, evalve; includens *Capsulas* numerosas evalves, 6—9-spermas, affixas pedicellis capillaribus e receptaculo communi baseos involucri interioris ortis. *Semina* angulata, radiculis exsertis.

AZOLLA PINNATA. *Tab.* 10.

Azolla fronde circumscriptione triangulari pinnata et semibipinnata; [612 foliolis superioribus papulosis, radicibus longitudinaliter plumosis. *Prodr. fl. nov. holl.* 167.

In lakes and ponds, frequent within the limits of the colony of Port Jackson.

DESC. *Plantula* natans, facie Jungermanniæ. *Radices* axillares solitariæ perpendiculares hyalinæ, primo aspectu simplicissimæ, per lentem plumosæ, novellæ calyptra glabra subulata tectæ. *Frons* semuncialis: *Ramis* distichis alternis approximatis parallelis teretiusculis; infimis haud raro pinnatis; superioribus sæpe instructis gemmulis ramulorum nonnullis axillaribus teretibus. *Folia* alterna undique imbricata: in *latere superiore frondis* trapezoideo-ovata, crassiuscula cellulosa, viridia passim rubicunda, margine exteriore submembranaceo, supra convexiuscula papuloso-scabra, subtus lævia: in *latere inferiore* tenuiora lævia, subconformia vix tamen angulata. *Perichætia* in superficie inferiore frondis, prope basin pinnæ solitaria.

Obs. Mr. Bauer's very satisfactory figure and the generic character already given, will in a great measure supersede any farther description of the singular structure of this genus; on which, however, it appears necessary to subjoin a few remarks.

Admitting the parts of fructification to be accurately described, it is not easy to understand in what manner the male influence is communicated to the female organ. In one instance the turbid fluid, which usually fills the cavity of the lower cell of the supposed male organ, was found converted into a powder, and it is not improbable that this change ultimately takes place in all cases where the organ attains perfection. This powder may be supposed either to be discharged by the lateral rupture of the double coat of the containing cell, or a communication may at length be opened between this cell and the tubular axis of the upper cell, which, after the separation of its outer membrane, is open at the top; in this case the ejection of the pollen, or even of a fluid matter, may possibly be aided by the pressure or action of the angular solid bodies which surround this axis, and its dispersion would, no doubt, be assisted by the increased surface of its divided apex.

But whatever supposition may be formed respecting the economy of this part, it appears to me that as it is found in a second species of the genus, and of essentially the same
613] structure, though slightly modified, the angular bodies of the upper cell being only six in number, there can remain little doubt of its being really the male organ.

The genus Azolla was founded by Lamarck on specimens of the South American species entirely destitute of fructification, the remains of which only appear to have been seen more recently by Willdenow, who describes it as "a Capsula unilocularis polysperma."

REFERENCES TO TAB. 10.

AZOLLA PINNATA.

1. Plant of the natural size.
2. — magnified.

3. Leaves, magnified.
4. Male involucrum, containing two flowers, magnified.
5. — empty.
6. Two male flowers.
7. A male flower divided longitudinally.
8. — deprived of its Calyptra, 9.
10. Lower cell of a male flower.
11 and 12. Different views of the contents of the upper cell.
13. Longitudinal section of the upper cell.
14. Inner female involucrum.
15. Capsules, with their footstalks arising from the base of the involucrum.
16. A capsule more highly magnified.
17. — opened transversely to show the position of the seeds.
18. — — empty.
19. Seeds.

LIST

OF

NEW AND RARE PLANTS,

COLLECTED IN

ABYSSINIA

DURING THE YEARS 1805 AND 1810,

ARRANGED ACCORDING TO THE LINNÆAN SYSTEM.

[*Reprinted from 'A Voyage to Abyssinia,' by Henry Salt, Esq., F.R.S., &c. Append., pp.* lxiii—lxv.]

LONDON.

1814.

LIST

[Append. p. lxiii

OF

NEW AND RARE PLANTS, &c.

"THE plants having Br. MSS. annexed form new genera, described in the manuscripts of Mr. Brown. To this gentleman's kindness I am indebted for the list, which he made out from a collection of dried specimens brought by me into the country, and now in the possession of Sir Joseph Banks. The names without reference are considered by Mr. Brown as applying to new species; and for the few that have been published already, contracted references are given to the works in which they occur, namely, Willdenow's 'Species Plantarum'; Forskal's 'Flora 'Ægyptiaco-Arabica'; Vahl's 'Symbolæ Botanicæ'; and the Appendix to the Travels of Mr. Bruce."

DIANDRIA.

Jasminum abyssinicum.
Hypoestis Forskalii (Justicia Forskalii, *Willd. sp. pl.*)
Justicia cynanchifolia.
— bivalvis. *Willd. sp. pl.*
Meisarrhena tomentosa. *Br. MSS.*
Salvia abyssinica.
Stachytarpheta cinerea.

TRIANDRIA.

Geissorhiza abyssinica.
Commelina hirsuta.
— acuminata.
Cyperus involutus.
— laxus.
— scirpoides.
Cyperus melanocephalus.
— densus.
Cenchrus tripsacoides.
Pennisetum villosum.
Aristida ramosa.
Eleusine (?) stolonifer.
Panicum ovale.

TETRANDRIA.

Pavetta congesta.
— reflexa.
Canthium lucidum.
Buddlea acuminata. (Umfar. *Bruce*).
— foliata.
Nuxia congesta.
— dentata.
Dobera glabra. (Tomex glabra, *Forsk.*)
Fusanus alternifolia.

PENTANDRIA.

Heliotropium gracile.
— cinereum.
— ellipticum.
— ? dubium.
Lithospermum ? ambiguum.
Anchusa affinis.
lxiv] Ehretia obovata.
— abyssinica.
Cordia ovalis.
— abyssinica (Wanzey, *Bruce.*)
Plumbago eglandulosa.
Convolvulus cirrhosus.
— congestus.
— pilosus.
Neurocarpæa lanceolata, *Br. MSS.* (Manettia lanceolata, *Vahl.*)
Solanum cinereum.
— uncinatum.
Erythræa compar.
Strœmia longifolia.
— farinosa, *Willd. sp. pl.*
— rotundifolia, *Willd. sp. pl.*
Rhamnus inebrians, (called in Tigré "Sadoo")
Celastrus serrulatus.
— glaucus.
Impatiens tenella.
Paronychia sedifolia.
Saltia abyssinica, *Br. MSS.*
Carissa abyssinica.
— edulis, *Willd. sp. pl.*
Kanahia laniflora. (Asclepias laniflora, *Willd. sp. pl.*)
Pentatropis cynanchoides, *Br. MSS.*
Petalostemma chenopodii, *Br. MSS.*
Breweria evolvuloides.
Taxanthemum attenuatum.
Crassula puberula.

HEXANDRIA.

Loranthus lætus.
— congestus.
— calycinus.

OCTANDRIA.

Combretum ovale.
— molle.
Amyris Gileadensis, *Willd. sp. pl.*
— Kataf, *Willd. sp. pl.*
Polygonum sinuatum.

DECANDRIA.

Cassia pubescens.
Pterolobium lacerans, *Br. MSS.* (Kantuffa, *Bruce.*)
Fagonia armata.
Terminalia cycloptera.
Dianthus abyssinicus.

DODECANDRIA.

Calanchoe pubescens.
Sterculia abyssinica.
Reseda pedunculata.

ICOSANDRIA.

Rosa abyssinica.
Rubus compar.

POLYANDRIA.

Corchorus gracilis.

DIDYNAMIA.

Nepeta azurea.
Satureja ovata.
— punctata.
Ocymum cinereum.
— monadelphum.
Leucas quinquedentata.
— affinis.
Molucella integrifolia.
— scariosa.
— repanda.
Linaria gracilis.
— hastata.
— propinqua.
Buchnera orobanchoides.
Dunalia acaulis, *Br. MSS.*
Bignonia discolor.
Sesamum pterospermum.
Barleria brevispina.
— macracantha.
— eranthemoides.
— grandiflora.
— mollis.
— parviflora. [lxv
Acanthus tetragonus.
Thunbergia angulata.
Lantana polycephala.
Clerodendrum myricoides.

TETRADYNAMIA.

Mathiola elliptica.
Cleome Siliquaria. (Siliquaria glandulosa, *Forsk. Ægypt.* 78.)
— Roridula (Roridula, *Forsk. Ægypt.* 35.
— parviflora.
— paradoxa.

Monadelphia.

Pelargonium abyssinicum.
Geranium compar.
Sida acuminata.
— gracilis.
— pannosa.
Hibiscus parvifolius.
— erianthus.
Urena mollis.
— glabra.

Diadelphia.

Polygala linearis.
— abyssinica.
Erythrina tomentosa.
Crotalaria Saltiana.
— propinqua.
— farcta.
Onobrychis simplicifolia.
Indigofera albicans.
— diffusa.

Syngenesia.

Bracheilema paniculatum, *Br. MSS.*
Teichostemma fruticosum, *Br. MSS.*
Cacalia abyssinica.
Pulicaria involucrata.
— viscida.
— aromatica.

Monœcia.

Euphorbia propinqua.
Dalechampia tripartita.
Croton acuminatum.

Diœcia.

Cissampelos nympheæfolia.

Polygamia.

Acacia læta.
— fasciculata.

Cryptogamia.

Cheilanthes leptophylla.

OBSERVATIONS,

SYSTEMATICAL AND GEOGRAPHICAL,

ON

THE HERBARIUM

COLLECTED BY

PROFESSOR CHRISTIAN SMITH,

IN THE

VICINITY OF THE CONGO,

DURING THE EXPEDITION TO EXPLORE THAT RIVER,

UNDER THE

COMMAND OF CAPTAIN TUCKEY,

IN THE YEAR 1816.

BY

ROBERT BROWN, F.R.S.,

CORRESPONDING MEMBER OF THE ROYAL INSTITUTE OF FRANCE, AND OF THE ROYAL ACADEMIES OF SCIENCES OF BERLIN AND MUNICH: HONORARY MEMBER OF THE LITERARY AND PHILOSOPHICAL SOCIETY OF NEW YORK, MEMBER OF THE WERNERIAN SOCIETY OF EDINBURGH. LIBRARIAN TO THE LINNEAN SOCIETY.

[*Reprinted from a "Narrative of an Expedition to explore the River Zaïre," pp.* 420—485.]

LONDON:

1818.

OBSERVATIONS, &c. [420

THE Herbarium formed by the late Professor Smith and his assistant, Mr. David Lockhart, on the banks of the Congo, was, on its arrival in England, placed at the disposal of Sir Joseph Banks; under whose inspection it has been arranged; the more remarkable species have been determined; and the whole collection has been so far examined as the very limited time which could be devoted to this object allowed.

In the following pages will be found the more general results only of this examination; descriptions of the new genera and species being reserved for a future publication.

In communicating these results I shall follow nearly the same plan as that adopted in the Botanical Appendix to Captain Flinders's Voyage to Terra Australis:

1st. Stating what relates to the three Primary Divisions of Plants.

2dly. Proceeding to notice whatever appears most remarkable in the several Natural Orders of which the collection consists; and

3dly. Concluding with a general comparison of the vegetation on the line of the river Congo, with that of other equinoctial countries.

I. The number of species in the herbarium somewhat exceeds 600; the specimens of several of which are, indeed, imperfect; but they are all referable with certainty to the primary divisions, and, with very few exceptions, to the natural orders to which they belong.

Of the Primary Divisions, the Dicotyledonous plants amount to 460.

The Monocotyledonous to 113

And of the Acotyledonous, in which Ferns are included, there are only 33 species.

It is a necessary preliminary, with reference especially to the first part of my subject, to determine whether this herbarium, which was collected in a period not exceeding two months, and in a season somewhat unfavourable, can
421] warrant any conclusions concerning the proportional numbers of the three primary divisions, or of the principal natural orders in the country in which it was formed.

Its value in this respect must depend on the relation it may be supposed to have to the whole vegetation of the tract examined, and of the probability of the circumstances under which it was formed, not materially affecting the proportions in question.

Its probable relation to the complete Flora of the country examined, can at present be judged of only by comparing it with collections from different parts of the same coast of equinoctial Africa.

The first considerable herbarium from this coast, of which we have any account, is that formed by Adanson, on the banks of the Senegal, during a residence of nearly four years. Adanson himself has not given the extent of his collection, but as he has stated the new species contained in it to be 300,[1] it may, I think, be inferred, that altogether it did not exceed 600, which is hardly equal to that from Congo. Limited as this supposed extent of Adanson's herbarium may appear, it is estimated on the most moderate calculation of the proportion that new species were likely to bear to the whole vegetation of that part of equinoctial Africa, which he was the first botanist to examine; allowance being at the same time made for the disposition manifested in the account of his travels, to reduce the plants which he observed to the nearly related species of other countries.

From the herbarium and manuscripts in the library of

[1] *Fam. des Plant.* 1, *p. cxvi.*

Sir Joseph Banks, it appears that the species of plants collected by Mr. Smeathman at Sierra Leone, during a residence of more than two years, amounted to 450.

On the same authority I find that the herbarium formed in the neighbourhood of Cape Coast by Mr. William Brass, an intelligent collector, consisted of only 250 species.

And I have some reason to believe, that the most extensive and valuable collection ever brought from the west coast of equinoctial Africa, namely, that formed by Professor Afzelius, during his residence of several years at Sierra Leone, does not exceed 1200 species; although that eminent naturalist, in the course of his researches, must have examined a much greater extent of country than was seen in the expedition to Congo.

From these, which are the only facts I have been able to meet with respecting the number of species collected [422
on different parts of this line of coast, I am inclined to regard the herbarium from Congo as containing so considerable a part of the whole vegetation, that it may be employed, though certainly not with complete confidence, in determining the proportional numbers both of the primary divisions and principal natural orders of the tract examined; especially as I find a remarkable coincidence between these proportions in this herbarium and in that of Smeathman from Sierra Leone.

I may remark here, that from the very limited extent of the collections of plants above enumerated, as well as from what we know of the north coast of New Holland, and I believe I may add of the Flora of India, it would seem that the comparative number of species in equal areas within the tropics and in the lower latitudes beyond them, has not been correctly estimated; and that the great superiority of the intratropical ratio given by Baron Humboldt, deduced probably from his own observations in America, can hardly be extended to other equinoctial countries. In Africa and New Holland, at least, the greatest number of species in a given extent of surface does not appear to exist within the tropics, but nearly in the parallel of the Cape of Good Hope.

In the sketch which I have given of the botany of New

Holland, I first suggested the inquiry respecting the proportions of the primary divisions of plants as connected with climate; and I then ventured to state that "from the equator to 30° lat. in the northern hemisphere at least, the species of Dicotyledonous plants are to the Monocotyledonous as about 5 to 1, in some cases considerably exceeding, and in a very few falling somewhat short of this proportion, and that in the higher latitudes a gradual diminution of Dicotyledones takes place until in about 60° N., and 55° S. lat. they scarcely equal half their intratropical proportion."[1]

Since the publication of the Essay from which this quotation is taken, the illustrious traveller Baron Humboldt, to whom every part of botany, and especially botanical geography, is so greatly indebted, has prosecuted this subject further, by extending the inquiry to the natural orders of plants; and in the valuable dissertation prefixed to his great botanical work,[2] has adopted the same equinoctial proportion of Monocotyledones to Dicotyledones as that
423] given in the Paper above quoted; a ratio which seems to be confirmed by his own extensive herbarium.

I had remarked, however, in the Essay referred to, that the relative number of these two primary divisions in the equinoctial parts of New Holland appeared to differ considerably from those which I had regarded as general within the tropics; Dicotyledones being to Monocotyledones only as 4 to 1. But this proportion of New Holland very nearly agrees with that of the Congo and Sierra Leone collections. And from an examination of the materials composing Dr. Roxburgh's unpublished Flora Indica, which I had formerly judged of merely by the index of genera and species, I am inclined to think that nearly the same proportion exists on the shores of India.

Though this may be the general proportion of the coasts, and in tracts of but little varied surface within the tropics,

[1] *Flinders' Voyage to Terra Australis*, 2, *p*. 538. (*Antè*, *p*. 8.)

[2] Nova Genera et Species Plantarum, quas in perigrinatione orbis novi collegerunt, &c. *Amat. Bonpland* et *Alex. de Humboldt.* ex. sched. autogr. in ord. dig. *C. S. Kunth*, 1815, *Parisiis.*

it seems at the same time probable from Baron Humboldt's extensive collections, and from what we know of the vegetation of the West India islands, that in equinoctial America, in tracts including a considerable portion of high land, the ratio of Dicotyledones to Monocotyledones is at least that of 11 to 2, or perhaps nearly 6 to 1. Whether this or a somewhat diminished proportion of Dicotyledones exists also in similar regions of other equinoctial countries, we have not yet sufficient materials for determining.

Upon the whole, however, it would seem from the facts of which we are already in possession, that the proportions of the two primary divisions of phænogamous plants vary considerably even within the tropics, from circumstances connected certainly in some degree with temperature. But there are facts also which render it probable, that these proportions are not solely dependent on climate. Thus the proportion of the Congo collection, which is also that of the equinoctial part of New Holland, is found to exist both in North and South Africa, as well as in Van Diemen's Island, and in the south of Europe.

It is true indeed that from about 45° as far as to 60°, or perhaps even to 65° N. lat. there appears to be a gradual diminution in the relative number of Dicotyledones; but it by no means follows that in still higher latitudes a further reduction of this primary division takes place. On the contrary, it seems probable from Chevalier Giesecke's list of the plants of the west coast of Greenland,[1] on different parts of which, from lat. 60° to 72°, he resided several years, that the relative numbers of the two primary divi- [424
sions of phænogamous plants are inverted on the more northern parts of the coast;[2] Dicotyledones being to Monocotyledones, in the list referred to, as about 4 to 1,

1 Article "Greenland," in Brewster's 'Edinburgh Encyclopædia.'

2 That some change of this kind takes place on that coast might perhaps have been conjectured from a passage in Hans Egede's 'Description of Greenland,' where it is stated, that although from lat. 60° to 65° there is a considerable proportion of good meadow land, yet in the more northern parts, "the inhabitants cannot gather grass enough to put in their shoes, to keep their feet warm, but are obliged to buy it from the southern parts." (English Translation, pp. 44 and 47.)

or nearly as on the shores of equinoctial countries. And analogous to this inversion it appears, that at corresponding Alpine heights, both in the temperate and frigid zones, the proportion of Dictyledones is still further increased.

The ACOTYLEDONOUS or cryptogamous plants of the herbarium from Congo, are to the phænogamous as about 1 to 18. Some allowance is here to be made for the season, peculiarly unfavourable, no doubt, for the investigation of this class of plants. But it is not likely that Professor Smith, who had particularly studied most of the cryptogamous tribes, should have neglected them in this expedition; and the circumstance of the very few imperfect specimens of Mosses in the collection being carefully preserved and separately enveloped in paper, seems to prove the attention paid to, and consequently the great rarity of, this order at least; which, however, is not more striking than what I have formerly noticed with respect to some parts of the north coast of New Holland.[1]

I have in the same place considered the Acotyledones of equinoctial New Holland, as probably forming but one thirteenth of the whole number of plants, while the general equinoctial proportion was conjectured to be one sixth. This general ratio, however, is certainly over-rated, though it is probably an approximation to that of countries containing a considerable portion of high land. Within the tropics, therefore, it would seem that the ratio of acotyledonous to phænogamous plants, varies from that of 1 : 15 to 1 : 5; the former being considered as an approximation to the proportion of the shores, the latter to that of mountainous countries.

425] II. The NATURAL ORDERS of which the herbarium from Congo consists are 87 in number; besides a very few genera not referable to any families yet established. More than half the species, however, belong to nine orders, namely, to Filices, Gramineæ, Cyperaceæ, Convolvulaceæ, Rubiaceæ, Compositæ, Malvaceæ, Leguminosæ, and Euphorbiaceæ; all of which have their greatest

[1] *Flinders' Voyage*, 2, *p*. 539. (*Antè*, *pp*. 9, 10.)

number of species in the lower latitudes, and several within the tropics.

I now proceed to make some observations on the orders above enumerated, and on such of the other families, included in the collection, as present anything remarkable, either in their geographical distribution, or in their structure; more especially where the latter establishes or suggests new affinities; and I shall take them nearly in the same order as that followed in the botanical appendix to Captain Flinders's Voyage.

ANONACEÆ. Only three species of this family are contained in the collection. One of these is *Anona Senegalensis,* of which the genus has been considered doubtful, even by M. Dunal in his late valuable Monograph of the order.[1] That it really belongs to Anona, however, appears from the specimen with ripe fruit preserved in the collection. It is remarkable therefore as the only species of this genus yet known which is not a native of equinoctial America; for *Anona Asiatica,* of which Linnæus had no specimen in his herbarium when he first proposed it under this name, according to the original synonym, is nothing more than *Anona muricata:* and *A. obtusiflora,* supposed by M. Tussac[2] to have been introduced into the American Islands from Asia, does not appear to differ from *A. mucosa* of Jacquin, which is known to be a native of Martinica.

The second plant of this order in the collection is very nearly related to *Piper Æthiopicum* of the shops, the *Unona Æthiopica,* and perhaps also *Unona aromatica* of Dunal:[3] these with several plants already published, form a genus, which, like Anona, is common to America and Africa, but of which no species has yet been observed in Asia.

Of MALPIGHIACEÆ, an order chiefly belonging to equinoctial America, there are also three species from Congo.

One of these is *Banisteria Leona,* first described, from [426

[1] *Monogr. de la famille des Anonacées, p.* 76.

[2] *Flore des Antilles,* 1, *p.* 193.

[3] *Anonac*, *p.* 113 *et* 112.

Smeathman's specimens, by Cavanilles,[1] who has added the fruit of a very different plant to his figure, and quotes the herbarium of M. de Jussieu as authority for this species being likewise a native of America, which is, I believe, equally a mistake.

The two remaining plants of Malpighiaceæ, in the collection, with some additional species from different parts of the coast, form a new genus, having the fruit of Banisteria, but with sufficient distinguishing characters in the parts of the flower, and remarkable in having alternate leaves. From this disposition of leaves, in which the genus here noticed differs from all others decidedly belonging to the order, an additional argument is afforded, for referring *Vitmannia* to Malpighiaceæ, as proposed by M. du Petit Thouars;[2] and the approximation, though perhaps not the absolute union of Erythroxylon to the same family is confirmed.

It may not be improper here to notice a very remarkable deviation from the usual structure of leaves in Malpighiaceæ, which is supposed to occur in a plant of equinoctial Africa, namely *Flabellaria pinnata* of Cavanilles (the *Hiræa pinnata* of Willdenow). It is certain, however, that the figure given by Cavanilles of this species is made up from two very different genera; the pinnated leaf belonging to an unpublished Pterocarpus; the fructification to a species of Hiræa, having simple opposite leaves. The evidence respecting this blunder, which was detected by Mr. Dryander, is to be found in the herbarium of Sir Joseph Banks.

In Malpighiaceæ the insertion of the ovulum is towards its apex, or considerably above its middle; and the radicle of the embryo is uniformly superior. In these points Banisteria presents no exception to the general structure, though Gærtner has described its radicle as inferior, and M. de Jussieu does not appear to have satisfied himself respecting the fact.[3] It appears, however, that M. Richard

[1] *Dissert.* 424, *t.* 247.
[2] *In Nov. gen. Madagasc.* n. 46 (Biporeia).
[3] *Annal. du Mus. d Hist. Nat.* 18, *p.* 480.

is aware of the constancy in the direction of the embryo in this order.[1]

HIPPOCRATICEÆ. M. de Jussieu has lately proposed this as a distinct family,[2] of which there are two plants in the collection. The first is a species of Hippocratea; the second is referable to Salacia.

In Hippocraticeæ, the insertion of the ovula is either [427
towards the base, or is central; the direction of the radicle is always inferior. In these points of structure, which are left undetermined by M. de Jussieu, they differ from Malpighiaceæ, but agree with Celastrinæ, to which, notwithstanding the difference in insertion and number of stamina, and in the want of albumen, they appear to me to have a considerable degree of affinity; especially to Elæodendrum, where the albumen is hardly visible, and to Ptelidium, as suggested by M. du Petit Thouars,[3] in which it is reduced to a thin membrane.

SAPINDACEÆ. Only four plants of this natural family, which is almost entirely equinoctial, occur in the herbarium. Two of these are new species of Sapindus. The third is probably not specifically different from *Cardiospermum grandiflorum* of the West India Islands. And the fourth is so nearly related to *Paullinia pinnata*, of the opposite coast of America, as to be with difficulty distinguished from it. M. de Jussieu,[4] who probably intends the same plant, when he states *P. pinnata* to be a native of equinoctial Africa, has also described a second species from Senegal.[5] No other species of this genus has hitherto been found, except in equinoctial America; for *Paullinia Japonica* of Thunberg, probably belongs even to a different natural order. The species from Congo, however, seems to be a very general plant on this line of coast; having been found by Brass near Cape Coast, and by Park on the banks of the Gambia.

[1] *Mem. du Mus. d'Hist. Nat.* 2, *p.* 400.
[2] *Annal. du Mus. d'Hist. Nat.* 18, *p.* 183.
[3] *Hist. des Véget. des Isles de l'Afrique, p.* 34.
[4] *In Annal. du Mus. d'Hist. Nat.* 4, *p.* 347. [5] *Loc. cit., p.* 348.

In Sapindaceæ there is not the same constancy in the insertion of the ovulum and consequent direction of embryo, as in the two preceding orders. For although, in the far greater part of this family, the ovulum is erect and the radicle of the embryo inferior, yet it includes more than one genus in which both the seeds and the embryo are inverted. With this fact it would seem M. de Jussieu is unacquainted;[1] and he is surely not aware that in his late Memoir on Melicocca[2] he has referred plants to that genus differing from each other in this important point of structure.

TILIACEÆ. It is remarkable that of only nine
428] species belonging to this family in Professor Smith's herbarium, three should form genera hitherto unnoticed.

The *first* of these new genera is a shrub, in several of its characters related to Sparmannia, like which, it has the greater part of its outer stamina destitute of antheræ; in the structure of its fruit, however, it approaches more nearly to Corchorus.

The *second* genus also agrees with Corchorus in its fruit; but differs from it sufficiently in the form and dehiscence of the antheræ; as well as in the short pedicellus, like that of Grewia, elevating its stamina and pistillum.

The *third*, of which the specimens are in fruit only, fortunately, however, accompanied by the persistent flower, is remarkable in having a calyx of three lobes, while its corolla consists of five petals; the stamina are in indefinite number; and the fruit is composed of five single-seeded capsules, connected only at the base. In the want of symmetry or proportion between the divisions of its calyx and corolla it resembles the *Chlenaceæ* of M. Du Petit Thouars,[3] as well as *Oncoba* of Forskael and *Ventenatia* of M. de Beauvois.[4] The existence of this new genus decidedly belonging to Tiliaceæ, and having a considerable resemblance to Vente-

[1] *Annal. du Mus. d'Hist. Nat.* 18, *p.* 476.
[2] *Mém. du Mus. d'Hist. Nat.* 3, *p.* 179.
[3] *Hist. des Végét. des Isles de l'Afrique, p.* 46.
[4] *Flore d'Oware,* 1, *p.* 29, *t.* 17.

natia, whose place in the system is, indeed, not yet determined, but of which the habit is nearly that of Rhodolæna, seems in some degree to confirm M. du Petit Thouars's opinion of the near relation of Chlenaceæ to Tiliaceæ ; though M. de Jussieu, in placing it between Ebenaceæ and Rhodoraceæ,[1] appears to take a very different view of its affinities.

MALVACEÆ. Of this family 18 species were observed on the banks of the Congo. It forms, therefore, about one thirty-fourth part of the phænogamous plants of the collection; which is somewhat greater than the equinoctial proportion of the order, as stated in Baron Humboldt's dissertation,[2] but nearly agrees with that of India, according to Dr. Roxburgh's unpublished Flora Indica.

The greater part of the Malvaceæ of the collection belong to Sida and Hibiscus; and certain species of both these genera are common to India and America. *Urena Americana* and *Malachra radiata,* hitherto supposed to be natives of America only, are also contained in the collection ; and [429
the loftiest tree seen on the banks of the Congo, is a species of Bombax, which, as far as can be determined from the very imperfect specimens preserved in the herbarium, does not differ from *Bombax pentandrum* of America and India. I have formerly remarked[3] that Malvaceæ, Tiliaceæ, Hermanniaceæ, Buttneriaceæ, and Sterculiaceæ, constitute one natural class ; of which the orders appear to me as nearly related as the different sections of Rosaceæ are to each other. In both these, as well as in several other cases that might be mentioned, there seems to be a necessity for the establishment of natural classes, to which proper names, derived from the orders best known, and differing perhaps in termination, might be given.

It is remarkable that the most general character connecting the different orders of the class now proposed, and which may be named from its principal order Malvaceæ, should

[1] *Mirbel, Elem. de Physiol. Veg. et de Bot.* 2, *p.* 855.
[2] *Prolegomena, p. xviii. De Distrib. Geogr. Plant., p.* 43.
[3] *Flinders's Voy.* 2, *p.* 540. (*Antè, p.* 11.)

be that of the valvular æstivation of the calyx; for several, at least, of the genera at present referred to Tiliaceæ, in which this character is not found, ought probably, for other reasons likewise, to be excluded from that order: and hence perhaps also the Chlenaceæ, though nearly related, are not strictly referable to the class Malvaceæ, from all of whose orders, it must be admitted, they differ considerably in habit.

LEGUMINOSÆ. According to Baron Humboldt,[1] this family, or class, as I am rather disposed to consider it, constitutes one twelfth of the phænogamous plants within the tropics. Its proportion, however, is much greater in Professor Smith's herbarium, in which there are 96 species belonging to it, or nearly one sixth of the whole collection. And ample allowance being made for the lateness of the season when the collection was formed, which might be supposed to reduce the number of this family less than many of the others, Leguminosæ may be stated as forming one eighth of the Phænogamous plants on the banks of the Congo. In India, it probably forms about one ninth, which is also nearly the proportion it bears to Phænogamous plants in the equinoctial part of New Holland.

I have formerly proposed to subdivide Leguminosæ into three orders.[2]

Of the first of these orders, MIMOSEÆ, there are only
430] eight species from Congo, seven of which belong to Acacia, as it is at present constituted; the eighth is a sensitive aculeated Mimosa very nearly allied to *M. aspera* of the West Indies, as well as to *M. canescens* of Willdenow, found by Isert in Guinea; and perhaps is not different from the species mentioned by Adanson as being common on the banks of the Senegal.

Of the second order, CÆSALPINEÆ, the collection contains 19 species, among which there are four unpub-

[1] *Op. citat.* [2] *Flinders' Voy.* 2, *p.* 551. (*Antè, p.* 22.)

lished genera. One of these is Erythrophleum of Afzelius, the Red Water Tree of Sierra Leone; another species of which genus is the ordeal plant, or *Cassa* of the natives of Congo. *Guilandina Bonduc* and *Cassia occidentalis,* are also in the herbarium; the former, I believe, is unquestionably common to India and America; whether *Cassia occidentalis* be really a native of India and equinoctial Africa, in both of which it is now at least naturalized, is perhaps doubtful.

Among PAPILIONACEÆ, which constitute the principal part of Leguminosæ in the collection, there is only one plant with stamina entirely distinct. This decandrous plant forms a genus very different from any yet established, but to which *Podalyria bracteata* of Roxburgh[1] belongs.

The genera composing Papilionaceæ on the banks of the Congo have, upon the whole, a much nearer relation to those of India than of equinoctial America. To this, however, there is one remarkable exception. For of the only two species of *Pterocarpus* in the collection, one is hardly to be distinguished from *P. Ecastaphyllum,* unless by the want of the short acumen existing in the plant of Jamaica. The second agrees entirely with Linnæus's original specimen of *P. lunatus* from Surinam, and seems to be not uncommon on the west coast of equinoctial Africa; having been observed by Professor Afzelius at Sierra Leone, and probably by Isert in Guinea;[2] while no species of Pterocarpus related to either of these has hitherto been observed in India. On the other hand *Abrus precatorius* and *Hedysarum triflorum,* both of which occur in the collection, are common to equinoctial Asia and America.

TEREBINTACEÆ, as given by M. de Jussieu, appears to be made up of several orders nearly related to each other, and of certain genera having but little affinity to any of them. Of this, indeed, the illustrous author of the Genera Plantarum seems to have been aware. He pro- [431

[1] *Coromand. Plants,* 3 *tab.*

[2] *Reise nach Guinea, p.* 116.

bably, however, had not the means of ascertaining all their distinguishing characters, and therefore preferred leaving the order nearly as it was originally proposed by Bernard de Jussieu in 1759.

One of the orders included in Terebintaceæ, and which is proposed by M. de Jussieu himself, under the name of CASSUVIÆ, consists of Anacardium, Semecarpus, Mangifera, Rhus, and Buchanania, with some other unpublished genera.

The perigynous insertion of stamina in *Cassuviæ* (or *Anacardeæ*) may be admitted in doubtful cases from analogy, there being an unpublished genus belonging to it even with ovarium inferum. And the ovarium, though in all cases of one cell, with a single ovulum, may, at least in those genera in which the style is divided, be supposed to unite in its substance the imperfect ovaria indicated by the branches of the style, and which in Buchanania are actually distinct from the complete organ. The only plant belonging to this order in the herbarium, is a species of *Rhus*, with simple verticillate leaves, and very nearly approaching in habit to two unpublished species of the genus from the Cape of Good Hope.

AMYRIDEÆ, another family included in Terebintaceæ, and to which the greater part of Jussieu's second section belongs, may, like the former order, be considered as having in all cases perigynous insertion of stamina; this structure being manifest in some of its genera. Of Amyrideæ, there are two plants in the collection. The first of these is a male plant, probably of a species of Sorindeia;[1] the second, which is the *Safu* of the natives, by whom it is cultivated on account of its fruit, cannot be determined from the imperfect state of the specimens; it is, however, probably related to Poupartia or Bursera.

CONNARACEÆ, is a third family which I propose to separate from Terebintaceæ: it consists of Connarus *Linn.* Cnestis *Juss.* and Rourea of Aublet or Robergia of Schre-

[1] *Aubert du Petit Thouars, nov. gen. Madagas. n.* 80.

ber. The insertion of stamina, in this family, is ambiguous; but as in a species of Cnestis from Congo, they originate from, or at least firmly cohere with, the pedicellus of the ovaria, they may be considered perhaps in all the [432
genera rather as hypogynous than perigynous. The most important distinguishing characters of Connaraceæ consist in the insertion of the two collateral ovula of each of its pistilla being near the base; while the radicle of the embryo is situated at the upper or opposite extremity of the seed, which is always solitary. In *Connarus* there is but one ovarium, and the seed (figured by Gærtner under the name of Omphalobium) is destitute of albumen. *Rourea* or Robergia has always five ovaria, though in general one only comes to maturity. Its seed, like that of Connarus, is without albumen, and the æstivation of the calyx is imbricate.

Of *Cnestis* there are several new species in Professor Smith's herbarium. This genus has also five ovaria, all of which frequently ripen; the albumen forms a considerable part of the mass of the seed; and the æstivation of the calyx is valvular. The genera of this group, therefore, differ from each other, in having one or more ovaria; in the existence or absence of albumen; and in the imbricate or valvular æstivation of calyx. Any one of these characters singly is frequently of more than generic importance, though here even when all are taken together, they appear insufficient to separate Cnestis from Connarus.

In considering the place of the Connaraceæ in the system, they appear evidently connected on the one hand with Leguminosæ, from which Connarus can only be distinguished by the relation the parts of its embryo have to the umbilicus of the seed. On the other hand, *Cnestis* seems to me to approach to *Averrhoa*, which agrees with it in habit, and in many respects in the structure of its flower and seed; differing from it, however, in its five ovaria being united, in the greater number of ovula in each cell, in the very different texture of its fruit, and in some degree in the situation of the umbilicus of the seed.

But *Averrhoa* agrees with *Oxalis* in every important

point of structure of its flower, and in most respects in that of its seed.

Oxalis, indeed, differs from Averrhoa in the texture of its fruits, in some respects in the structure of its seed; and very widely in habit, in the greater part of its species. The difference in habit, however, is not so great in some species of Oxalis; as for example, in those with pinnated and even ternate leaves from equinoctial America; and in that natural division of the genus including *O. sensitiva,* of which there are two species in the Congo herbarium.
433] This latter section of Oxalis[1] agrees also with *Averrhoa Carambola*[2] in the foliola, when irritated, being reflected or dependent, which is likewise their position in the state of collapsion or sleep, in all the species of both genera.

To the natural order formed by Oxalis and Averrhoa, the name of OXALIDEÆ may be given, in preference to that of *Sensitivæ,* under which, however, Batsch[3] was the first to propose the association of these two genera, and to point out their agreement in sensible qualities and irritability of leaves.

M. de Jussieu, in a memoir recently published,[4] has proposed to remove Oxalis from Geraniaceæ, to which he had formerly annexed it, and to unite it with Diosmeæ.

It appears to me to have a much nearer affinity to *Zygophylleæ,*[5] though it is surely less intimately connected with that order than with Averrhoa.

I am aware that M. Correa de Serra, one of the most profound and philosophical botanists of the present age, has considered Averrhoa as nearly related to Rhamneæ[6] or rather to Celastrinæ; from which, however, it differs in the number and insertion of stamina and especially in the direction of the embryo, with respect to the pericarpium.

In all these characters Averrhoa agrees with Oxalis; its relation to which is further confirmed on considering the appendage of the seed or arillus, whose modifications in

1 Herba sentiens, *Rumph. Amboin.* 5, *p.* 301.

2 *Bruce in Philos. Transact.* 75, *p.* 356.

3 *Tab. affin. p.* 23.

4 *Mém. du Mus. d'Hist. Nat.* 3, *p.* 448.

5 *Flinders's Voy.* 2, *p.* 545. (*Antè, p.* 16.)

6 *Annal. du Mus. d'Hist. Nat.* 8, *p.* 72.

these two genera seem to correspond with those of their pericarpia.

CHRYSOBALANEÆ. The genera forming this order are Chrysobalanus, Moquilea, Grangeria, Coupea, Acioa, Licania, Hirtella, Thelira, and Parinarium, all of which are at present referred by M. de Jussieu to Rosaceæ, and the greater part to his seventh section of that family, namely, Amygdaleæ. If Rosaceæ be considered as an order merely, these genera will form a separate section, connecting it with Leguminosæ. But if, as I have formerly proposed, both these extensive families are to be regarded as natural classes, then they will form an order sufficiently distinct from Amygdaleæ, both in fructification and habit, as well as in geographical distribution.

The principal distinguishing characters in the fructification of *Chrysobalaneæ* are the style proceeding from the base of the ovarium; and the ovula (which, as in Amyg- [434
daleæ, are two in number) as well as the embryo being erect. The greater part of Chrysobalaneæ have their flowers more or less irregular; the irregularity consisting in the cohesion of the foot-stalk of the ovarium with one side of the tube of the calyx, and a greater number, or greater perfection of stamina on the same side of the flower.

Professor Smith's herbarium contains only two genera of this order, namely, *Chrysobalanus* and *Parinarium.*[1] One species of the former is hardly distinguishable from *Chrysobalanus Icaco* of America, and is probably a very common plant on the west coast of Africa; *Icaco* being mentioned by Isert[2] as a native of Guinea, and by Adanson[3] in his account of Senegal.

Of *Parinarium,* there is only one species from Congo, which agrees, in the number and disposition of stamina, with the character given of the genus. In these respects M. de Jussieu[4] has observed a difference in the two species

[1] *Juss. Gen.* 342. Parinari, *Aublet Guian.* 514. Petrocarya, *Schreb. Gen.* 629. [2] *Reise nach Guinea, p.* 54. [3] *Voyage au Senegal,* 175.
[4] *Gen. Plant.* 342.

found by Adanson at Senegal, and has moreover remarked that their ovarium coheres with the tube of the calyx. In that species most common at Sierra Leone, and which is probably one of those examined by M. de Jussieu, the ovarium itself is certainly free, its pedicellus, however, as in the greater part of the genera of this order and several of Cæsalpineæ, firmly cohering with the calyx, may account for the statement referred to. I am not, indeed, acquainted with any instance among Dicotyledonous plants of cohesion between a simple ovarium, which I consider that of Chrysobalaneæ to be, and the tube of the calyx.

The complete septum between the two ovula of Parinarium, existing before fecundation, is a peculiar structure in a simple ovarium; though in some degree analogous to the moveable dessepiment of Banksia and Dryandra, and to the complete, but less regular, division of the cavity that takes place after fecundation in some species of Persoonia.[1]

MELASTOMACEÆ. Four plants only of this order occur in the collection.

The first is a species of *Tristemma*, very nearly related to *T. hirtum* of M. de Beauvois.[2]

435] The second is perhaps not distinct from *Melastoma decumbens*, of the same author.[3]

The third and fourth are new species referable to *Rhexia*, as characterised by Ventenat,[4] though not to that genus as established by Linnæus; and in some respects differing from the species that have been since added to it, all of which are natives of America.

In the original species of *Tristemma*[5] there are, in the upper part of the tube of the calyx, two circular ciliated membranous processes, from which the name of the genus is derived; the limb of the calyx itself being considered as constituting the third circle. The two circular membranes are also represented as complete in *T. hirtum*.

But in the species from Congo, which may be named *T.*

[1] *Linn. Soc. Transact.* 10, *p.* 35. [2] *Flore d'Oware,* 1, *p.* 94, *t.* 57.

[3] *Op. citat.* 1, *p.* 69, *t.* 49.

[4] *Mém. de l'Institut. sc. phys.* 1807, *prem. semest. p.* 11.

[5] Tristemma virusana, *Vent. Choix de Plantes,* 35.

incompletum, only one circular membrane exists, with the unilateral rudiment of the second.

The rudiment of the inferior membrane in this species points out the relation between the apparently anomalous appendage of the calyx in Tristemma, and the ciliated scales irregularly scattered over its whole surface in Osbeckia; the analogy being established by the intermediate structure of an unpublished plant of this order from Sierra Leone, in Sir Joseph Banks's herbarium, in which the nearly similar squamæ, though distinct, are disposed in a single complete circle; and by *Melastoma octandra* of Linnæus, in which they are only four in number, and alternate with the proper divisions of the calyx.

The two species here referred, though improperly, to Rhexia, agree with a considerable part of the species published in the monograph of that genus by M. Bonpland, and with some other genera of the order, in the peculiar manner in which the ovarium is connected with the tube of the calyx. This cohesion, instead of extending uniformly over the whole surface, is limited to ten longitudinal equidistant lines or membranous processes, apparently originating from the surface of the ovarium; the interstices, which are tubular, and gradually narrowing towards the base, being entirely free.

The function of these tubular interstices is as remarkable as their existence.

In Melastomaceæ, before the expansion of the corolla, the tops of the filaments are inflected, and the antheræ are pendulous and parallel to the lower or erect portion of the filament; their tips reaching, either to the line of complete cohesion between the calyx and ovarium, where that exists; or, where this cohesion is partial, and such as I have now [436
described, being lodged in the tubular interstices; their points extending to the base of the ovarium. From these sheaths, to which they are exactly adapted, the antheræ seem to be disengaged in consequence of the unequal growth of the different parts of the filament; the inflected portion ceasing to increase in length at an early period, while that below the curvature continues to elongate con-

siderably until the extrication is complete, when expansion takes place.

It is singular that this mode of cohesion between the ovarium and calyx in certain genera of Melastomaceæ, and the equally remarkable æstivation of antheræ accompanying it, should have been universally overlooked, especially in the late monograph of M. Bonpland; as both the structure and economy certainly exist in some, and probably in the greater part, of the plants which that author has figured and described as belonging to Rhexia.

On the limits, structure, and generic division of Melastomaceæ, I may remark—

1st. That *Memecylon*, as M. du Petit Thouars has already suggested,[1] and *Petaloma* of Swartz[2] both belong to this order, and connect it with *Myrtaceæ*, from which they are to be distinguished only by the absence of the pellucid glands of the leaves and other parts, existing in all the genera really belonging to that extensive family.

2ndly. There are very few Melastomaceæ in which the ovarium does not in some degree cohere with the tube of the calyx; *Meriana*, properly so called, being, perhaps, the only exception.

And in the greater number of instances where, though the ovarium is coherent, the fruit is distinct, it becomes so from the laceration of the connecting processes already described.

3rdly. That the generic divisions of the whole order remain to be established. On examination, I believe, it will be found that the original species of the Linnean genera, *Melastoma* and *Rhexia*, possess generic characters sufficiently distinguishing them from the greater part of the plants that have been since added to them by various authors. In consequence of these additions, however, their botanical history has been so far neglected, that probably no genuine species of Melastoma, and certainly none of Rhexia, has yet been published in M. Bonpland's splendid and valuable monographs of these two genera.

[1] *Mélanges de Botanique; Observ. address. à M. Lamarck, p.* 57.
[2] *Flor, Ind. Occid.* 2, *p.* 831, *tab.* 14.

Of RHIZOPHOREÆ,[1] as I have formerly proposed to [437
limit it, namely, to Rhizophora, Bruguiera, and Carallia, the collection contains only one plant, which is a species of Rhizophora, the Mangrove of the lower part of the river, and probably of the whole line of coast, but very different both from that of America, and from those either of India or of other equinoctial countries that have been described. There is, however, a plant in the collection which, though not strictly belonging to this order, suggests a few remarks on its affinities.

I referred *Carallia*[2] to Rhizophoreæ, from its agreement with them in habit, and in the structure of its flower. It is still uncertain whether its reniform seed is destitute of albumen; the absence of which, however, does not seem necessary to establish its affinity with the other genera of this order; for plants having the same remarkable economy in the germination of the embryo as that of Rhizophora, may belong to families which either have or are destitute of albumen.

The plant referred to from Congo may be considered as a new species of *Legnotis* having its petals less divided than those of the original species of that genus, and each cell of its ovarium containing only two pendulous ovula. The genus *Legnotis* agrees with Carallia in habit, especially in having opposite leaves with intermediate stipules; in the valvular æstivation of its calyx, and in several other points of structure of its flower. It differs in its divided petals; in its greater number of stamina, disposed, however, in a simple series; and in its ovarium not cohering with the calyx. It is therefore still more nearly related to *Richæia* of M. du Petit Thouars,[3] from which perhaps it may not be generically distinct. The propriety of associating Carallia[4] with Rhizophoreæ is not perhaps likely to be disputed; and its affinity to Legnotis, especially to the species from Congo, appears very probable. It would seem, therefore, that we have already a series of structures

[1] *Flinders's Voy.* 2, *p.* 549. (*Antè, p.* 20.) [2] *Roxburgh. Coromand.* 3, *p.* 8, *t.* 211.
[3] *Nov. Gen. Madagasc. n.* 84.
[4] Or Barraldeia, *Du Petit Thouars, Nov. Gen. Madagasc. n.* 82.

connecting Rhizophora on the one hand with certain genera of *Salicariæ*, particularly with *Antherylium*, though that genus wants the intermediate stipules; and on the other with *Cunoniaceæ*,[1] especially with the simple leaved species of
438] *Ceratopelatum*. While Loranthus and Viscum, associated with Rhizophora by M. de Jussieu, appear to form a very distinct family, and which, as it seems to me, should even occupy a distant place in the system.

HOMALINÆ. In the collection from Congo a plant occurs evidently allied, and perhaps referable, to *Homalium*, from which it differs only in the greater number of glands alternating with the stamina, whose fasciculi are in consequence decomposed: the inner stamen of each fasciculus being separated from the two outer by one of the additional glands. This plant was first found on the banks of the Gambia, by Mr. Park, from whose specimens I have ascertained that the embryo is enclosed in a fleshy albumen.

The same structure of seed may be supposed, from very obvious affinity, to exist in *Astranthus* of Loureiro, to which *Blackwellia* of Commerson ought perhaps to be referred; in *Napimoga* of Aublet, probably not different from Homalium; and in *Nisa*,[2] a genus admitting of subdivision, and which M. du Petit Thouars has referred to Rhamneæ. All these genera appear to me sufficiently different from Rosaceæ, where M. de Jussieu has placed them, and from every other family of plants at present established.

Their distinguishing characters as a separate order are, the segments of the perianthium disposed in a double series, or an equal number of segments nearly in the same series; the want of petals; the stamina being definite and opposite to the inner series of the perianthium, or to the alternate segments where they are disposed apparently in a simple series; the unilocular ovarium (generally in some degree coherent with the calyx) having three parietal placentæ, with one, two, or even an indefinite number of ovula; and the seeds having albumen, as inferred from its existence in the genus from Congo. The cohesion of the ovarium with

[1] *Flinders's Voy.* 2, *p.* 548. (*Antè, p.* 20.)

[2] *Nov. Gen. Madagasc. n.* 81.

the tube of the perianthium, though existing in various degrees in all the genera above enumerated, is probably a character of only secondary importance in Homalinæ. For an unpublished genus found by Commerson in Madagascar, which in every other respect agrees with this family, has ovarium superum. This genus at the same time seems to establish a considerable affinity between Homalinæ and certain genera, either absolutely belonging to *Passifloreæ*, especially *Paropsia* of M. du Petit Thouars,[1] or nearly related to them as *Erythrospermum*, well de- [439
scribed and figured by the same excellent botanist.[2]

The increased number of stamina in Homalium, and particularly in the genus from Congo, instead of presenting an objection to this affinity, appears to me to confirm it. It may be observed also that there are two genera referable to Passifloreæ, though they will form a separate section of the order, which have a much greater, and even an indefinite, number of perfect stamina, namely, *Smeathmania*, an unpublished genus of equinoctial Africa, agreeing in habit, in perianthium, and in fruit, with Paropsia; and *Ryania* of Vahl,[3] which appears to me to belong to the same family.

In Passifloreæ the stamina, when their number is definite, which is the case in all the genera hitherto considered as belonging to them, are opposite to the outer series of the perianthium; a character which, though of general importance, and here of practical utility in distinguishing them from Homalinæ, is not expressed in any of the numerous figures or descriptions that have been published of the plants of this order.

Passifloreæ and Cucurbitaceæ, though now admitted as distinct families, are still placed together by M. de Jussieu; and he considers the floral envelope in both orders as a perianthium or calyx, whose segments are disposed in a double series.[4]

These views of affinity and structure are in some degree confirmed by Homalinæ, in which both ovarium inferum

[1] *Hist. des Végét. des Isles de l'Afrique*, 59. [2] *Op. citat.* 65.
[3] *Eclog.* 1, *p.* 51, *t.* 9. [4] *Annal. du Mus. d'Hist. Nat.* 6, *p.* 102.

and superum occur; and in one genus of which, namely, *Blackwellia*, the segments of the perianthium, though the complete number, in relation to the other genera of the order, be present, are all of similar texture and form, and are disposed nearly in a simple series. If the approximation of these three families be admitted, they may be considered as forming a class intermediate between Polypetalæ and Apetalæ, whose principal characters would consist in the segments of the calyx being disposed in a double series, and in the absence of petals; the different orders nearly agreeing with each other in the structure of their seeds, and to a considerable degree in that of the ovarium.

The formation of this class, however, connected on the
440] one hand with Apetalæ by Samydeæ,[1] and on the other, though as it seems to me less intimately, with Polypetalæ by Violeæ, would not accord with any arrangement of natural orders that has yet been given. While the admission of the floral envelope being entirely calyx; and of the affinity of the class with Violeæ, would certainly be unfavorable to M. de Candolle's ingenious hypothesis of petals in all cases being modified stamina.

VIOLEÆ.[2] This order does not appear to me so nearly related to Passifloreæ as M. du Petit Thouars is disposed to consider it; for it not only has a genuine polypetalous corolla, which is hypogynous, but its antheræ differ materially in structure, and its simple calyx is divided to the base. The irregularity both of petals and stamina in the original genera of the order, namely, Viola, Pombalia,[3] and Hybanthus, though characters of considerable importance, are not in all cases connected with such a difference in habit as to prevent their union with certain regular flowered genera, which it has lately been proposed to associate with them.

The collection from Congo contains two plants belonging to the section of Violeæ with regular flowers. One of

[1] *Ventenat in Mém. de l'Instit. Sc. Phys.* 1807, 2 *sem. p.* 142.
[2] *Juss. Gen. Pl.* 295. *Ventenat Malmais,* 27.
[3] *Vandelli Fasc. Pl. p.* 7, *t.* 1. Ionidium, *Venten. Malmais.* 27.

these evidently belongs to *Passalia,* an unpublished genus in Sir Joseph Banks's herbarium, and described in the manuscripts of Solander from a plant found by Smeathman at Sierra Leone, which is perhaps not specifically distinct from that of Congo, or from *Ceranthera dentata* of the Flore d'Oware. But *Ceranthera,*[1] which M. de Beauvois, being unacquainted with its fruit, has placed in the order Meliaceæ, is not different from *Alsodeia,* a genus published somewhat earlier, and from more perfect materials, by M. du Petit Thouars,[2] who refers it to Violeæ. The latter generic name ought of course to be adopted, and with a change in the termination (*Alsodinæ*) it may also denote the section of this order with regular flowers.

Physiphora of Sir Joseph Banks's herbarium, discovered by himself in Brazil, differs from Alsodeia only in its filaments being very slightly connected at base, and in the form and texture of its capsule, which is membranaceous, and, as the name imports, inflated.

Five species belonging to this section of Violeæ occur in Aublet's History of the Plants of Guiana, where each of [441
them is considered as forming a separate genus. Of three of these genera, namely, *Conohoria, Rinorea,* and *Riana* the flowers alone are described; the two others, *Passura* and *Piparea,* were seen in fruit only.

From the examination of flowers of Aublet's original specimens of the three former genera, in Sir Joseph Banks's herbarium, and of the fruit of *Conohoria,* which entirely agrees with that of *Passura,* and essentially with that of *Piparea,* I have hardly a doubt of these five plants, notwithstanding some differences in the disposition of their leaves, actually belonging to one and the same genus; and as they agree with *Physiphora* in every respect, except in the texture and form of the capsule, and with the *Passalia* of Sierra Leone and Congo, except in having their stamina nearly or entirely distinct, it appears that all these genera may be referred to Alsodeia.

I have also examined, in Sir Joseph Banks's herbarium, a specimen of *Pentaloba sessilis* of the Flora Cochinchi-

[1] *Flore d'Oware,* 2, *p.* 10. [2] *Hist. des Végét. des Isles de l'Afrique,* 55.

nensis, which was sent so named, by Loureiro himself, and have found it to agree in every important point with Alsodeia, even as to the number of parietal placentæ. Loureiro, however, describes the fruit of Pentaloba as a five-lobed, five-seeded berry, and if this account be correct, the genus ought to be considered as distinct; but if, which is not very improbable, the fruit be really capsular, it is evidently referable to Alsodeia; with the species of which, from Madagascar and the west coast of equinoctial Africa, it agrees in the manifest union of its filaments.

It appears therefore that the ten genera now enumerated, and perhaps also *Lauradia* of Vandelli, may very properly be reduced to one; and they all at least manifestly belong to the same section of Violeæ, though at present they are to be found in various, and some rather distant, natural orders.

M. de Jussieu, in adopting Aublet's erroneous description of the stamina of Rinorea and Conohoria, has referred both these genera to Berberides,[1] to which he has also annexed Riana, adding a query whether Passura may not
442] belong to the same genus. With M. de Beauvois, he refers Ceranthera to Meliaceæ; and Pentaloba of Loureiro he reduces also to the same order.[2] Piparea is, together with Viola, annexed to Cistinæ in his Genera Plantarum, and is therefore the most correctly placed, though its structure is the least known, of all these supposed genera.

[1] The genera belonging to BERBERIDEÆ are *Berberis* (to which Ilex Japonica of Thunberg belongs); *Leontice* (including *Caulophyllum*, respecting which see *Linn. Soc. Transac.* 12, p. 145) *Epimedium*; and *Diphylleia* of Michaux. *Jeffersonia* may perhaps differ in the internal structure of its seeds, as it does in their arillus, from true Berberideæ, but it agrees with them in the three principal characters of their flower, namely, in their stamina being equal in number and opposite to the petals; in the remarkable dehiscence of antheræ; and in the structure of the ovarium. *Podophyllum* agrees with Diphylleia in habit, and in the fasciculi of vessels of the stem being irregularly scattered; essentially in the floral envelope, and in the structure of the ovarium; its stamina, also, though numerous, are not altogether indefinite, but appear to have a certain relation both in number and insertion to the petals: in the dehiscence of antheræ, and perhaps also in the structure of seeds, it differs from this order, to which, however, it may be appended. *Nandina* ought to be included in Berberideæ, differing only in its more numerous and densely imbricate bracteæ, from which to the calyx and even to the petals, the transition is nearly imperceptible; and in the dehiscence of its antheræ.

[2] *Mém. du Mus. d'Hist. Nat.* 3, *p.* 440.

An unpublished genus of New Holland, which I have named *Hymenanthera*, in Sir Joseph Banks's herbarium, agrees with Alsodeia in its calyx, in the insertion, expansion, and obliquely imbricate æstivation of its petals, and especially in the structure of its antheræ, which approach more nearly to those of Violeæ properly so called. It differs, however, from this order in having five squamæ alternating with the petals; and especially in its fruit, which is a bilocular berry, having in each cell a single pendulous seed, whose internal structure resembles that both of Violeæ and Polygaleæ, between which I am inclined to think this genus should be placed.

CHAILLETEÆ. The genus *Chailletia* was established by M. de Candolle[1] from a plant found by Martin in French Guiana, and which, as appears by specimens in Sir Joseph Banks's herbarium, had been many years before named *Patrisia* by Von Rohr, who discovered it in the same country. At a still earlier period, Solander, in his manuscripts, preserved in the library of Sir Joseph Banks, described this genus under the name of *Mestotes*, from several species found by Smeathman at Sierra Leone. Both *Dichapetalum* and *Leucosia* of M. du Petit Thouars[2] appear to me, from the examination of authentic specimens, to belong to the same genus; and in Professor Smith's herbarium there is at least one additional species of Chailletia different from those of Sierra Leone.

Of the two generic names given by M. du Petit [443
Thouars, and published somewhat earlier than M. de Candolle's Memoir, Leucosia will probably be considered inadmissible, having been previously applied by Fabricius to a genus of Crustacea; and Dichapetalum is perhaps objectionable, as derived from a character not existing in the whole genus, even allowing it to be really polypetalous. It seems expedient, therefore, to adopt the name proposed by M. de Candolle, who has well illustrated the genus in the memoir referred to. It appears to me that Chailletia,

[1] *Annal. du Mus. d'Hist. Nat.* 17, *p.* 153.
[2] *Nov. Gen. Madagasc. n.* 78 *et* 79.

a genus nearly related to it from India with capsular fruit, and *Tapura* of Aublet (which is *Rohria* of Schreber), form a natural order, very different from any yet established. The principal characters of this order may be gathered from M. de Candolle's figure and description of Chailletia, to which, however, must be added that the cells of the ovarium, either two or three in number, constantly contain two collateral pendulous ovula; and that in the regular flowered genera there exist within, and opposite to, the petal-like bodies an equal number of glands, which are described by M. du Petit Thouars in Dichapetalum, but are unnoticed by him in Leucosia, where, however, they are equally present.

It may seem paradoxical to associate with these genera *Tapura*, whose flower is irregular, triandrous, and apparently monopetalous. But it will somewhat lessen their apparent differences of structure to consider the petal-like bodies, which, in all the genera of this order, are inserted nearly or absolutely in the same series with the filaments, as being barren stamina; a view which M. de Candolle has taken of those of Chailletia, and which M. Richard had long before published respecting Tapura.[1] It is probable also that M. de Candolle at least will admit the association here proposed, as his *Chailletia sessiliflora* seems to be merely an imperfect specimen of *Tapura guianensis*.

The genera to which Chailleteæ most nearly approach appear to me to be *Aquilaria* of Lamarck[2] and *Gyrinops* of Gærtner. But these two genera themselves, which are not referable to any order yet established, may either be regarded as a distinct family, or perhaps, to avoid the too great multiplication of families, as a section of that at present
444] under consideration, and to which I should then propose to apply the name of AQUILARINÆ in preference to Chailleteæ.

The genus Aquilaria itself has been referred by Ventenat to *Samydeæ*. From this order, however, it is sufficiently

[1] *Dict. Elem. de Botanique par Bulliard, revu par L. C. Richard, ed.* 1802, *p.* 34.

[2] Or *Ophiospermum* of the Flora Cochinchinensis, as I have proved by comparison with a specimen from Loureiro himself.

distinct, not only in the structure of its ovarium and seeds, but in its leaves being altogether destitute of glands, which are not only numerous in Samydeæ, but consisting of a mixture of round and linear pellucid dots, distinguish them from all the other families[1] with which there is any probability of their being confounded.

Sir James Smith[2] has lately suggested the near affinity of Aquilaria to Euphorbiaceæ. But I confess it appears to me at least as distinct from that order as from Samydeæ; and I am inclined to think, paradoxical as it may seem, that it would be less difficult to prove its affinity to Thymeleæ than to either of them; a point, however, which, requiring considerable details, I do not mean to attempt in the present essay.

Of EUPHORBIACEÆ there are twenty species in the collection, or one twenty-eighth part of its Phænogamous plants. This is somewhat greater than the intratropical proportion of the order as stated by Baron Humboldt, but rather smaller than that of India or of the northern parts of New Holland.

The most remarkable plants of Euphorbiaceæ in the Congo herbarium are: a new species of the American genus *Alchornea*; a plant differing from *Ægopricon*, a genus also belonging to America, chiefly in its capsular fruit; two new species of *Bridelia,* which has hitherto been observed only in India; and an unpublished genus that I have formerly alluded to,[3] as in some degree explaining the real structure of Euphorbia, and from the consideration of which also it seems probable that what was formerly described as the hermaphrodite flower of that genus, is in reality a compound fasciculus of flowers.[4] From the same species of this unpublished genus a substance resembling caoutchouc is said to be obtained at Sierra Leone.

[1] The only other genus in which I have observed an analogous variety of form in the glands of the leaves, is *Myroxylon* (to which both *Myrospermum* and *Toluifera* belong), in all of whose species this character is very remarkable, the pellucid lines being much longer than in Samydeæ.

[2] *Linn. Soc. Transact.* 11, *p.* 230. [3] *Flinders's Voy.* 2, *p.* 557. (*Antè*, *p.* 29.)

[4] *Linn. Soc. Transact.* 12, *p.* 99.

According to Mr. Lockhart a frutescent species of
445] Euphorbia, about eight feet in height, with cylindrical stem and branches, was observed, planted on the graves of the natives near several of the villages; but of this, which may be what Captain Tuckey has called *Cactus quadrangularis* in his Narrative (p. 115), there is no specimen in the herbarium.

COMPOSITÆ. It is unnecessary here to enter into the question whether this family of plants, of which upwards of 3000 species are already known, ought to be considered as a class or as an order merely; the expediency of subdividing it, and affixing proper names to the divisions, being generally admitted. The divisions or tribes proposed by M. Cassini, in his valuable dissertations on this family, appear to be the most natural, though as yet they have not been very satisfactorily defined.

The number of Compositæ in the collection is only twenty-four, more than half of which are referable to *Heliantheæ* and *Vernoniaceæ* of M. Cassini. The greater part of these are unpublished species, and among them are five new genera. The published species belong to other divisions, and are chiefly Indian: but one of them, *Ageratum conyzoides*, is common to America and India; the *Struchium* (or Sparganophorus) of the collection does not appear to me different from that of the West Indies; and *Mikania chenopodifolia*, a plant very general on this line of coast, though perhaps confined to it, belongs to a genus of which all the other species are found only in America.

Baron Humboldt has stated[1] that Compositæ form one sixth of the Phænogamous plants within the tropics, and that their proportion gradually decreases in the higher latitudes until in the frigid zones it is reduced to one thirteenth. But in the herbarium from Congo Compositæ form only one twenty-third, and both in Smeathman's collection from Sierra Leone and in Dr. Roxburgh's Flora Indica, a still smaller part, of the Phænogamous plants. In the northern part of New Holland they form about one

[1] In *op. citat.*

sixteenth; and in a manuscript catalogue of plants of equinoctial America, in the library of Sir Joseph Banks, they are nearly in the same proportion.

In estimating the comparative value of these different materials, I may, in the first place, observe that though the herbarium from Congo was collected in the dry season of the country, there is no reason to suppose on that account that the proportion of this family of plants, in particular, is materially or even in any degree diminished, nor can [446
this objection be stated to the Sierra Leone collection, in which its relative number is still smaller.

To the Compositæ in Dr. Roxburgh's Flora Indica, however, a considerable addition ought, no doubt, to be made; partly on the ground of his having apparently paid less attention to them himself, and still more because his correspondents, whose contributions form a considerable part of the Flora, have evidently in a great measure neglected them. This addition being made, the proportion of Compositæ in India would not differ very materially from that of the north coast of New Holland, according to my own collection, which I consider as having been formed in more favorable circumstances, and as probably giving an approximation of the true proportions in the country examined. Baron Humboldt's herbarium, though absolutely greater than any of the others referred to on this subject, is yet, with relation to the vast regions whose vegetation it represents, less extensive than either that of the north coast of New Holland, or even of the line of the Congo. And as it is in fact as much the Flora of the Andes as of the coasts of intratropical America, containing families nearly or wholly unknown on the shores of equinoctial countries, it may be supposed to have several of those families which are common to all such countries, and among them Compositæ, in very different proportion. At the same time it is not improbable that the relative number of this family in equinoctial America, may be greater than in the similar regions of other intratropical countries; while there seems some reason to suppose it considerably smaller on the west coast of Africa. This diminished proportion, however, in equinoctial Africa would be the more remarkable, as there is probably no part of the world in which Compositæ form so great a portion of the vegetation as at the Cape of Good Hope.

RUBIACEÆ. Of this family there are forty-three species in the collection, or about one fourteenth of its Phænogamous plants. I have no reason to suppose that this proportion is greater than that existing in other parts of equinoctial Africa; on the contrary, it is exactly that of Smeathman's collection from Sierra Leone.

Baron Humboldt, however, states the equinoctial proportion of Rubiaceæ to phænogamous plants to be one to twenty-nine, and that the order gradually diminishes in relative number towards the poles.

447] But it is to be observed that this family is composed of two divisions, having very different relations to climate; the *first*, with opposite, or more rarely verticillate, leaves and intermediate stipules, to which, though constituting the great mass of the order, the name Rubiaceæ cannot be applied, being chiefly equinoctial; while the *second*, or *Stellatæ*, having verticillate or very rarely opposite leaves, but in no case intermediate stipules, has its maximum in the temperate zones, and is hardly found within the tropics, unless at great heights.

Hence perhaps we are to look for the minimum in number of species of the whole order, not in the frigid zone, but, at least in certain situations, a few degrees only beyond the tropics.

In conformity to this statement, M. Delile's valuable catalogue of the plants of Egypt[1] includes no indigenous species of the equinoctial division of the order, and only five of *Stellatæ*, or hardly the one hundred and sixtieth part of the Phænogamous plants. In M. Desfontaines' Flora Atlantica, Rubiaceæ, consisting of fifteen Stellatæ and only one species of the equinoctial division, form less than one ninetieth part of the Phænogamous plants, a proportion somewhat inferior to that existing in Lapland.

[1] *Flor. Egypt. Illustr. in Descript. de l'Egypte, Hist. Nat. v.* 2, *p.* 49.

In Professor Thunberg's Flora of the Cape of Good Hope, where Rubiaceæ are to Phænogamous plants as about one to one hundred and fifty, the order is differently constituted; the equinoctial division, by the addition of *Anthospermum*, a genus peculiar to southern Africa, somewhat exceeding Stellatæ in number. And in New Holland, in the same parallel of latitude, the relative number of Stellatæ is still smaller, from the existence of *Opercularia*, a genus found only in that part of the world, and by the addition of which the proportion of the whole order to the Phænogamous plants is there considerably increased.

More than half the Rubiaceæ from Congo belong to well known genera, chiefly to Gardenia, Psychotria, Morinda, Hedyotis, and Spermacoce.

Of the remaining part of the order, several form new genera.

The *first* of these is nearly related to Gardenia, which itself seems to require subdivision.

The *second* is intermediate between Rondeletia and Danais, and probably includes Rondeletia febrifuga of Afzelius.[1]

The *third* has the inflorescence and flowers of *Nauclea*, [448
but its ovaria and pericarpia are confluent, the whole head forming a compound spherical fleshy fruit, which is, I suppose, the country-fig of Sierra Leone, mentioned by Professor Afzelius.[2]

The *fourth* is a second species of *Neurocarpæa*, a genus which I have named, but not described, in the catalogue of Abyssinian plants appended to Mr. Salt's Travels.[3]

The *fifth* genus is intermediate between Rubiaceæ and Apocineæ. With the former it agrees in habit, especially in its interpetiolary stipules; and in the insertion and structure of its seeds, which are erect, and have the embryo lodged in a horny albumen forming the mass of the nucleus; while it resembles Apocineæ in having its

[1] *In Herb. Banks.* This is the "New sort of Peruvian Bark" mentioned in his Report, p. 174; which is probably not different from the Bellenda or African Bark of Winterbottom's Account of Sierra Leone, vol. 2, p. 243.

[2] *Sierra Leone Report for* 1794, *p.* 171, *n.* 32.

[3] *Voyage to Abyssinia, append. p. lxiv.* (*Antè, p.* 94.)

ovarium entirely distinct from the calyx; its capsule in appearance and dehiscence is exactly like that of Bursaria.

The existence of this genus tends to comfirm what I have formerly asserted respecting the want of satisfactory distinguishing characters between these two orders, and to prove that they belong to one natural class; the ovarium superum approximating it to Apocineæ; the interpetiolary stipules and structure of seeds connecting it, as it appears to me, still more intimately with Rubiaceæ.

The arguments adduced by M. de Jussieu[1] for excluding *Usteria* from Rubiaceæ and referring it to Apocineæ, are, its having ovarium superum, an irregular corolla, fleshy albumen, and only one stamen; there being no example of any reduction in the number of stamina in Rubiaceæ, (in which Opercularia and Pomax are not included by M. de Jussieu) while one occurs in the male flowers of Ophioxylum, a genus belonging to Apocineæ. From analogous reasoning he at the same time decides in referring *Gærtnera* of Lamarck[2] to Rubiaceæ, though he admits it to have ovarium superum; its flowers being regular, its albumen more copious and horny, and its embryo erect. But all these characters exist in the new genus from Congo. These two genera therefore, together with *Pagamea* of Aublet, *Usteria*, *Geniostoma* of Forster (which is *Anasser* of Jussieu) and *Logania*,[3] might, from their mere agreement in the situation of ovarium, form a tribe inter-
449] mediate between Rubiaceæ and Apocineæ. This tribe, however, would not be strictly natural, and from analogy with the primary divisions admitted in Rubiaceæ, as well as from habit, would require subdivision into at least four sections: but hence it may be concluded that the only combining character of these sections, namely, ovarium superum, is here of not more than generic value; and it must be admitted also that the existence or absence of stipules is in Logania[4] of still less importance.

[1] *Annal. du Mus. d'Hist. Nat.* 10, *p.* 323. [2] *Illustr. Gen. tab.* 167.
[3] *Prodr. Flor. Nov. Holl.* 1, *p.* 455. [4] *Prodr. Flor. Nov. Holl.* 1, *p.* 455.

APOCINEÆ. There are only six plants in the collection belonging to this order.

The *first* of these, together with some other species from Sierra Leone, constitutes an unpublished genus, the fruit of which externally resembles that of *Cerbera*, but essentially differs from it in its internal structure being polyspermous. The Cream fruit of Sierra Leone, mentioned by Professor Afzelius,[1] probably belongs to this genus, of which an idea may be formed by stating its flower to resemble that of Vahea, figured, but not described by M. Lamarck,[2] and its fruit, that of Voacanga[3] of M. du Petit Thouars, from which birdlime is obtained in Madagascar, or of Urceola[4] of Dr. Roxburgh, the genus that produces the caoutchouc of Sumatra.

The *second* belongs to a genus discovered in Sierra Leone by Professor Afzelius, who has not yet described it, but has named it *Anthocleista*. This genus, however, differs from *Potalia* of Aublet (the Nicandra of Schreber) solely in having a four-celled berry; that of Potalia being described both by Aublet and Schreber as trilocular, though according to my own observations it is bilocular. M. de Jussieu has appended *Potalia* to his Gentianeæ, partly determined, perhaps, from its being described as herbaceous. The species of *Anthocleista* from Congo, however, according to the account given me by Mr. Lockhart, the gardener of the expedition, is a tree of considerable size, and its place in the natural method is evidently near *Fagræa*.

Whether these genera should be united with Apocineæ or only placed near them, forming a fifth section of the intermediate tribe already proposed, is somewhat doubtful.

In the perfect hermaphrodite flowers of Apocineæ, no
exception occurs either to the quinary division of the [450
floral envelopes and corresponding number of stamina, or
to the bilocular or double ovarium; and in *Asclepiadeæ*,
which are generally referred by authors to the same order,
something like a necessary connection may be perceived

[1] *Sierra Leone Report*, 1794, *p.* 173, *n.* 47.
[2] *Illustr. Gen. tab.* 169.
[3] *Nov. Gen. Madagasc*, *n.* 32.
[4] *Asiat. Resear.* 5, *p.* 169.

between these relative numbers of stamina and pistilla, and the singular mode of fecundation in this tribe. But in Potalia and Anthocleista, there is a remarkable increase in the number of stamina and segments of the corolla, and at the same time a reduction in the divisions of the calyx. The pistillum in Potalia, however, if my account of it be correct, agrees in division with that of Apocineæ; and the deviation from this division in Anthocleista is only apparent; the ovarium, according to the view I have elsewhere given of this organ,[1] being composed of two united ovaria, again indeed subdivided by processes of the placenta, but each of the subdivisions or partial cells containing only one half of an ordinary placenta, and that not originating from its inner angle, as would be the case were the ovarium composed of four confluent organs.

Of ASCLEPIADEÆ there are very few species in the collection, and none of very remarkable structure. The *Periploca* of Equinoctial Africa alluded to in my essay on this family,[2] was one of the first plants observed by Professor Smith at the mouth of the river; and a species of *Oxystelma*, hardly different from *O. esculentum* of India,[3] was found, apparently indigenous, on several parts of its banks.

The ACANTHACEÆ of the collection, consisting of sixteen species, the far greater part of which are new, have a much nearer relation to those of India than to the American portion of the order. Among these there are several species of *Nelsonia*[4] and *Hypoestes*;[5] a new species of *Ætheilema*,[6] a genus from which perhaps *Phaylopsis* of Willdenow is not different, though its fruit is described by Wendland[7] as a legumen, and by Willdenow, with almost equal impropriety, as a siliqua; a plant belonging to a

1 *Linn. Soc. Transact.* 12, *p.* 89.
2 *Wernerian Nat. Hist. Soc. Trans.* 1, *p.* 40.
3 Periploca esculenta, *Roxb. Coromand.* 1, *p.* 13, *t.* 11.
4 *Prodr. Flor. Nov. Holl.* 1, *p*, 480. 5 *Op. citat.* 1, *p.* 474.
6 *Prodr. Flor. Nov. Holl.* 1, *p.* 478.
7 Micranthus, *Wend. Botan. Beobacht*, 38.

genus I have formerly alluded to as consisting of *Ruellia*
uliginosa and *R. balsamea*;[1] and a new species of *Ble-* [451
pharis. All these genera exist in India, and none of them
have yet been found in America.

CONVOLVULACEÆ. The herbarium of Professor Smith contains twenty-two species of this order, among which, however, there is no plant that presents anything remarkable in its structure; the far greater part belonging to Ipomœa, the rest to Convolvulus.

In the herbarium there is a single species of *Hydrolea*, nearly related to Sagonea palustris of Aublet, which would also be referred to this order by M. de Jussieu. But Hydrolea[2] appears to me to constitute, together with Nama, a distinct family (*Hydroleæ*) more nearly approaching to Polemoniaceæ than to Convolvulaceæ.

SCROPHULARINÆ. The collection contains only ten plants of this family, of which two form new genera, whose characters depend chiefly on the structure of antheræ and form of corolla.

The LABIATÆ of the herbarium consist of seven species, three of which belong to *Ocymum*, a genus common to equinoctial Asia and Africa, but not extending to America; an equal number to *Hyptis*, which is chiefly American, and has not been observed in India; the seventh is a species of *Hoslundia*, a genus hitherto found only on the west coast of Africa, and which, in its inflorescence and in the verticillate leaves of one of its species, approaches to the following order.

VERBENACEÆ, together with Labiatæ form one natural class,[3] for the two orders of which it has already become difficult to find distinguishing characters.

In the Congo herbarium there are seven Verbenaceæ, consisting of three beautiful species of Clerodendron; two

[1] *Prodr. Flor. Nov. Holl.* 1, *p.* 478. [2] *Vid. op. citat, p.* 482.
[3] *Flinders' Voy.* 2, *p.* 565. (*Antè, p.* 38)

new species of Vitex; Stachytarpheta indica of Vahl; and a new species of *Lippia*, which, from its habit and structure, confirms the union of Zapania with that genus, suggested by M. Richard.[1] This species from the Congo has its leaves in threes, and has nearly the same fragrance as
452] Verbena triphylla, whose affinity to Lippia, notwithstanding the difference in calyx and inflorescence, is further confirmed by a peculiarity in the æstivation of its corolla, which extends only to Lippia and Lantana.

OLACINÆ. The herbarium contains a species of Olax differing from all the plants at present referred to that genus, in its calyx not being enlarged after fecundation, but in its original annular form surrounding the base only of the ripe fruit. The existence of this species, which agrees with those of New Holland and with *Fissilia* of Commerson in having only five petals, and in its barren stamina being undivided, while in habit it approaches rather more nearly to the original species *O. Zeylanica* and to *O. scandens* of Roxburgh, both of which I have examined, seems to confirm the union I have formerly proposed,[2] of all these plants into one genus. When I first referred *Fissilia* to this genus, I only presumed from the many other points of agreement that it had also the same structure of ovarium, on which, not only the generic character of Olax, but its affinities, seemed to me in a great measure to depend. M. Mirbel, however, has described the ovarium of Fissilia as trilocular.[3] I can only reconcile this statement with my own observations, by supposing him to have formed his opinion from a view of its transverse section; for on examining one of Commerson's specimens of *Fissilia disparilis*, communicated by M. de Jussieu, I have found its ovarium, like that of all the species of Olax, to be really unilocular; the central columnar placenta, at the top of which the three pendulous ovula are inserted, having no connection whatever with the sides of the cavity.

It was chiefly the agreement of Olax and Santalaceæ in

[1] In *Mich. Flor Bor. Amer.* 2, *p.* 15.
[2] *Prodr. Flor. Nov. Holl.* 1, *p.* 357. [3] *Nouv. Bullet.* 3, *p.* 378.

this remarkable, and I believe, peculiar structure of ovarium, that induced me to propose, not their absolute union into one family, but their approximation in the natural series. I at the same time,[1] however, pointed out all the objections that M. de Jussieu has since stated to this affinity.[2]

Of these objections the two principal are the double floral envelope and ovarium superum of Olax, opposed to the simple perianthium and ovarium inferum in Santalaceæ.

The first objection loses much of its importance, both on considering that *Quinchamalium*, a genus in every other [453
respect resembling Thesium, has an outer floral envelope surrounding its ovarium, and having more the usual appearance of calyx than that of Olax; and also in adverting to the generally admitted association of Loranthus and Viscum, of which the former is provided with both calyx and corolla, the latter, in its male flowers at least, with only a single envelope, and that analagous to the corolla of Loranthus.[3]

The second objection seems to be equally weakened by the obvious affinity of Santalaceæ to *Exocarpus*, which has not only ovarium superum, but the fleshy receptacle of whose fruit, similar to that of Taxus, perfectly resembles, and may be supposed in some degree analogous to, the enlarged calyx of certain species of Olax.

To these objections M. de Jussieu has added a third, which, were it well founded, would be more formidable than either of them, namely, that the ovarium of Santalaceæ is monospermous;[4] a statement, however, which I conclude must have proceeded from mere inadvertency.

URTICEÆ. In the collection the plants of this family, taking it in the most extensive sense, and considering it as a class rather than an order, belong chiefly to *Ficus*, of which there are seven species. One of these is very nearly related to Ficus religiosa; and like that species in India, is regarded as a sacred tree on the banks of the Congo.

[1] *Prodr. Flor. Nov. Holl.* 1, *p.* 351. *Flinders' Voy.* 2, *p.* 571-2. (*Antè, p.* 44.)
[2] *Mém. du Mus. d'Hist. Nat.* 2, *p.* 439.
[3] *Prodr. Flor. Nov. Holl.* 1, *p.* 352.
[4] *Mem. du Mus. d'Hist. Nat.* 2, *p.* 439.

A remarkable tree, called by the natives *Musanga*, under which name it is repeatedly mentioned in Professor Smith's Journal, forms a genus intermediate between Coussapoa of Aublet and Cecropia; agreeing with the latter in habit, and differing from it chiefly in the structure and disposition of its monandrous male flowers, and in the form of its female amenta.

In the inflorescence, and even in the structure of its male flowers, *Musanga* approaches very nearly to *Myrianthus* of M. de Beauvois,[1] which it also resembles in habit. But the fruit of Myrianthus, as given in the 'Flore d'Oware,' is totally different, and, with relation to its male flowers, so remarkable, that a knowledge of the female flowers is wanting to fix our ideas both of the structure and affinities of the genus. This desideratum the expedition to Congo has not supplied, the male plant only of Myrianthus having been observed by Professor Smith.

454] In *Artocarpeæ*, to which *Musanga* belongs, and in *Urticeæ* strictly so called, the ovulum, which is always solitary, is erect, while the embryo is inverted or pendulous. By these characters, as well as by the separation of sexes, they are readily distinguished from those genera of *Chenopodeæ* and of monospermous *Illecebreæ*,[2] in which the albumen is either entirely wanting or bears but a small proportion to the mass of the seed. And hence also *Celtis* and *Mertensia*,[3] in both of which the ovulum is pendulous, are to be excluded from Urticeæ, where they have been lately referred by M. Kunth. The same characters, of the erect ovulum and inverted embryo, characterise Polygoneæ,[4] as I have long since remarked, and exist in *Piperaceæ* and even in *Coniferæ*, if my notions of that remarkable family be correct. But from all those orders Urticeæ are easily distinguished by other obvious and important differences in structure.

PHYTOLACEÆ. In describing Chenopodeæ, in the

1 *Flore d'Oware* 1, *p*. 16, *tabb*. 11 *et* 12.

2 *Prodr. Flor. Nov. Holl.* 1, *pp*. 405, 413, *et p*. 416. Paronychiearum sect. ii. *Jussieu in Mém. du Mus. d'Hist. Nat.* 2, *p*. 388.

3 *Nov. Gen. et Sp. Pl. Orb. Nov*, 2, *p*. 30.

4 *Prodr. Flor. Nov. Holl.* 1, *p*. 419.

Prodromus Floræ Novæ Hollandiæ, I had it particularly in view to exclude Phytolacca, Rivina, Microtea, and Petiveria, which I even then considered as forming the separate family now for the first time proposed.

In *Chenopodeæ* the stamina never exceed in number the divisions of the perianthium, to which they are opposite. In *Phytolaoeæ* they are either indefinite, or when equal in number to the divisions of the perianthium, alternate with them. This disposition of stamina in Phytolaceæ, however, uniting genera with fruits so different as those of Phytolacca and Petiveria, it would be satisfactory to find in the same order a structure intermediate between the multilocular ovarium of the former and the monospermous ovarium, with lateral stigma, of the latter.

Two plants in the herbarium from Congo assist in establishing this connection.

The *first* is a species of *Phytolacca*, related to P. abyssinica, whose quinquelocular fruit is so deeply divided, that its lobes cohere merely by their inner angles, and I believe ultimately separate.

The *second* is a species of *Gisekia*, a genus in which the five ovaria are entirely distinct. This genus is placed by [455
M. de Jussieu in Portulacaceæ; but the alternation of its stamina with the segments of the perianthium, a part of its structure never before adverted to, as well as their insertion, seem to prove its nearer affinity to Phytolacca.[1]

Still, however, the lateral stigma, the spiral cotyledons, and want of albumen in Petiveria, remove it to some distance from the other genera of Phytolaceæ, and at the same time connect it with *Seguieria*, with which also it agrees in the alliaceous odour of the whole plant.

The affinity of *Seguieria* has hitherto remained undetermined, and is here proposed from the examination of three species lately discovered in Brazil, one of which has

[1] *Ancistrocarpus* of M. Kunth (Nov. Gen. et Sp. Pl. Orb. Nov. 2, p. 186) belongs to Phytolaceæ, though its stamina are described to be opposite to the segments of the calyx: and it is not improbable that *Miltus* of Loureiro (Flor. Cochin. p. 302) whose habit, according to the description, is that of Gisekia, from which it differs nearly as Ancistrocarpus does from Microtea, or Rivina octandra from the other species of its genus, may also belong to this order.

exactly the habit of Rivina octandra, and all of which agree with that plant, as well as with several others belonging to the order, in the very minute pellucid dots of their leaves.

Petiveria and Seguieria may therefore form a sub-division of Phytolaceæ And another section of this order exists in New Holland, of which the two genera differ from each other in number of stamina as remarkably as Petiveria and Seguieria.

Of the Monocotyledonous orders, the first on which I have any remarks to offer, is that of

PALMÆ. The collection, however, contains no satisfactory specimens of any plant of this family except of *Elæis guineensis*, the *Maba* of the natives, or Oil Palm, which appears to be common along the whole of this line of coast. In Professor Smith's journal it is stated that a single plant of the Maba Palm[1] was cut down, from which Mr. Lockhart informs me that both the male and female spadices preserved in the collection were obtained. This fact seems to decide that Elæis is monœcious, which, in-
456] deed, Jacquin, by whom the genus was established, concluded it to be, though from less satisfactory evidence.[2] It was first described as diœcious by Gærtner, whose account has been adopted, probably without examination, by Schreber, Willdenow, and Persoon.

In Sir Joseph Banks's collection, however, from which Gærtner received the fruits he has described and figured, and where he may be supposed to have likewise obtained all the original information he had on the subject, there is no proof of the male and female spadices of Elæis guineensis belonging to different individuals.

Gærtner has fallen into a still more important mistake respecting the structure of the fruit of Elæis, the foramina of whose putamen, which are analogous to those of the

[1] *Maba* is, perhaps, rather applied to the fruit than to the tree: *Emba* being, according to Merolla, the name of the single nut, and *Cachio* that of the entire cluster: for the Palm itself, he has no name. *Vide Piccardo Relaz. p.* 122.

[2] *Hist. Stirp. Amer. p.* 281.

cocoa nut, being, according to his description, at the base, as in that genus, whereas they are actually at the apex. It is probable that *Alfonsia oleifera* of Humboldt Bonpland and Kunth, belongs to Elæis, and possibly it may not even differ from the African species.

It is a remarkable fact respecting the geographical distribution of Palmæ, that *Elæis guineensis,* which is universally, and I believe justly, considered as having been imported into the West India colonies from the west coast of Africa, and *Cocos indica,* which there is no reason to doubt is indigenous to the shores of equinoctial Asia and its islands, should be the only two species of an extensive and very natural section of the order, that are not confined to America.

To this section, whose principal character consists in the originally trilocular putamen having its cells when fertile perforated opposite to the seat of the embryo, and when abortive indicated by foramina cæca, as in the Cocoa nut, the name Cocoinæ may be given; though it has been applied by M. Kunth[1] to a more extensive and less natural group, which includes all palms having trilocular ovaria, and the surface of whose fruit is not covered with imbricate scales. I may also remark that from the fruits of *Cocoinæ* only, as I have here proposed to limit the section, the oil afforded by plants of this family, is obtained.

Professor Smith in his journal frequently mentions a species of *Hyphæne,* by which he evidently intended the palm first seen abundantly at the mouth of the river, and afterwards occasionally in the greater part of its course, especially near the Banzas, where it is probably planted for the sake of the wine obtained from it.

According to the gardener's information, this is a palm [457
of moderate height with fan-shaped fronds and an undivided caudex. It therefore more probably belongs to Corypha than to Gærtner's Hyphæne, one species of which is the Cucifera of Delile, the Doom of Upper Egypt; the second, *Hyphæne coriacea,* is a native of Melinda, and

[1] *Nova Gen. et Sp. Orb. Nov.* 1, *p.* 241.

probably of Madagascar, and both are remarkable in having the caudex dichotomous, or repeatedly divided.

As the Palm on the banks of the Congo was seen in fruit only, it is not difficult to account for Professor Smith's referring it rather to Hyphæne than to Corypha; Gærtner having described the embryo of the latter as at the base of the fruit, probably, however, from having inverted it, as he appears to have done in Elæis. It is at least certain that in *Corypha Taliera*[1] of the continent of India, which is very nearly allied to C. umbraculifera, the embryo is situated at the apex, as in Hyphæne.

The journal also notices a species of Raphia, which is probably *Raphia vinifera* of M. de Beauvois,[2] the *Sagus Palma-pinus* of Gærtner.

The collection contains fronds similar to those of *Calamus secundiflorus* of M. de Beauvois,[3] which was also found at Sierra Leone by Professor Afzelius; and a male spadix very nearly resembling that of *Elate sylvestris* of India.

The Cocoa Nut was not observed in any part of the course of the river.

Only five species of Palms appear therefore to have been seen on the banks of the Congo. On the whole continent of Africa thirteen species, including those from Congo, have been found; which belong to genera either confined to this continent and its islands, or existing also in India, but none of which have yet been observed in America, unless perhaps Elæis, if Alfonsia oleifera of Humboldt should prove to be a distinct species of that genus.

CYPERACEÆ. In the collection there are thirty-two species belonging to this order, which forms therefore about one eighteenth of the Phænogamous plants. This is very different from what has been considered its equinoctial proportion, but is intermediate to that of the northern part of New Holland, where, from my own materials, it seems to be as 1 : 14; and of India, in which according to Dr. Roxburgh's Flora it is about 1 : 25.

[1] *Roxb. Coromand.* 3, *tabb.* 255 *et* 256.
[2] *Flore d'Oware* 1, *p.* 75, *tabb.* 44, 45, *et* 46.
[3] *Op citat.* 1, *p.* 15, *tabb.* 9, *et* 10.

In other intratropical countries the proportion may be still smaller; but I can neither adopt the general equinoctial [458
ratio given by Baron Humboldt, namely, that of 1 : 60, nor suppose with him that the minimum of the order is within the tropics. For Cyperaceæ, like Rubiaceæ, and indeed several other families, is composed of tribes or extensive genera, having very different relations to climate. The mass of its equinoctial portion being formed of Cyperus and Fimbristylis, genera very sparingly found beyond the torrid zone; while that of the frigid and part of the temperate zones consists of the still more extensive genus Carex, which hardly exists within the tropics, unless at great heights. Hence a few degrees beyond the northern tropic, on the old continent at least, the proportion of Cyperaceæ is evidently diminished, as in Egypt, according to M. Delile's valuable catalogue;[1] and the minimum will, I believe, be found in the Flora Atlantica of M. Desfontaines and in Dr. Russel's catalogue of the plants of Aleppo.[2] It is not certain, however, that the smallest American proportion of the order exists in the same latitude. And it appears that in the corresponding parallel of the southern hemisphere, at the Cape of Good Hope and Port Jackson, the proportion is considerably increased by the addition of genera either entirely different from, or there more extensive than, those of other countries.

Among the Cyperaceæ of the Congo herbarium there are fifteen species of Cyperus, of which *C. Papyrus* appears to be one. The abundance of this remarkable species, especially near the mouth of the river, is repeatedly noticed in Professor Smith's journal, but from the single specimen with fructification in the collection, its identity with the plant of Egypt and Sicily, though very probable, cannot be absolutely determined. I perceive a very slight difference in the sheaths of the radii of the common umbel, which in the plant from Congo are less angular and less exactly truncated, than in that of Egypt; in other respects the two plants seem to agree. I have not seen C. laxiflorus, a

[1] *Flor. Ægypt. Illustr. in Descrip. de l'Egypte, Hist. Nat.* 2, *p.* 49.
[2] *Nat. Hist. of Aleppo, 2nd ed. vol.* 2, *p.* 242.

species discovered in Madagascar by M. du Petit Thouars, and said to resemble C. Papyrus except in the vaginæ of the partial umbels.[1]

Among the species of Cyperaceæ in the collection, having the most extensive range, are *Cyperus articulatus*, which is
459] common to America, India, and Egypt; *Fuirena umbellata* and *Eleocharis capitata*,[2] both of which have been found in America, India, and New Holland; and *Cyperus ligularis* indigenous to other parts of Africa and to America.

Hypælyptum argenteum, a species established by Vahl from specimens of India and Senegal, and since observed in equinoctial America by Baron Humboldt, is also in the collection.

The name *Hypælyptum*, under which I have formerly described the genus that includes *H. argenteum*,[3] was adopted from Vahl, without inquiry into its origin. It is probably, however, a corruption of *Hypælytrum*,[4] by which M. Richard, as he himself assures me, chiefly intended another genus, with apparently similar characters, though a very different habit, and one of whose species is described by Vahl in Hypælyptum; his character being so constructed as to include both genera. M. Kunth has lately published *H. argenteum* under the name of Hypælytrúm;[5] but in adopting the generic character given in the 'Prodromus Floræ Novæ Hollandiæ,' he has, in fact, excluded the plants that M. Richard more particularly meant to refer to that genus. It is therefore necessary, in order to avoid further confusion, to give a new name to Hypælyptum as I have proposed to limit it, which may be *Lipocarpha*, derived from the whole of its squamæ being deciduous.

In describing *Lipocarpha* (under the name of Hypælyptum) in the work referred to, I have endeavoured to establish the analogy of its structure to that of *Kyllinga*; the inner or upper squamæ being in both genera opposite to the inferior squama, or anterior and posterior, with relation to the axis of the spikelet: while the squamæ of

[1] *Encyc. Method. Botan. vol.* 7, *p.* 270.

[2] *Prodr. Flor. Nov. Holl.* 1, *p.* 225. Scirpus capitatus *Willd. sp. pl.* 1, *p.* 294, exclus. syn. Gronovii.

[3] *Prodr. Flor. Nov. Holl.* 1, *p.* 219.

[4] *Persoon Syn. Plant* 1, *p.* 70.

[5] *Nov. Gen. et Sp. Plant* 1, *p.* 218.

Richard's Hypælytrum being lateral, or right and left with respect to the axis of the spikelet,[1] were compared to those of the female flowers of *Diplacrum,* to the utriculus or nectarium of *Carex,* and to the lateral bracteæ of *Lepyrodia,* a genus belonging to the nearly related order Restiaceæ.[2] But as in *Hypælytrum,* according to M. Richard's description, and I believe also in his *Diplasia,*[3] there are sometimes more than two inner squamæ, which are then imbricate, they may in both these genera be considered as a spikelet reduced to a single flower, as in several other genera of Cyperaceæ, and in Lipocarpha itself, from which, [460 however, they are still sufficiently different in their relation to the including squamæ and to the axis of the spike.

This view of the structure of Hypælytrum, of which there is one species in the Congo herbarium, appears to me in some degree confirmed by a comparison with that of *Chondrachne* and *Chorizandra*;[4] for in both of these genera the lower squamæ of the ultimate spikelet are not barren, but monandrous, the central or terminating flower only being hermaphrodite.

GRAMINEÆ. Of this extensive family there are forty-five species from the Congo, or one twelfth of the Phænogamous plants of the collection. This is very nearly the equinoctial proportion of the order as given by Baron Humboldt, namely, one to fifteen, with which that of India seems to agree. On the north coast of New Holland, the proportion is still greater than that of Congo.

The two principal tribes which form the far greater part of Gramineæ, namely, *Poaceæ* and *Paniceæ* have, as I have formerly stated,[5] very different relations to climate, the maximum both in the absolute and relative number of species of Paniceæ being evidently within the tropics, that of Poaceæ beyond them.

I have hitherto found this superiority of Paniceæ to Poaceæ, at or near the level of the sea within the tropics,

[1] *Prodr. Flor. Nov. Holl.* 1, *p.* 219. [2] *Flinders's Voy.* 2, *p.* 579. (*Antè, p.* 53.)
[3] *Persoon Syn. Pl.* 1, *p.* 70. [4] *Prodr. Flor. Nov. Holl.* 1, *p.* 220.
[5] *Prodr. Flor. Nov. Holl.* 1, *p.* 169. *Obs. II. Flinders's Voy.* 2, *p.* 583. (*Antè, p.* 58.)

10

so constant, that I am inclined to consult the relative numbers of these two tribes, in determining whether the greater part of any intratropical Flora belongs to level tracts, or to regions of such elevation as would materially affect the proportions of the principal natural families: and in applying this test to Baron Humboldt's collection, it is found to partake somewhat of an extratropical character, Poaceæ being rather more numerous than Paniceæ. While in conformity to the usual equinoctial proportions, considerably more than half the grasses in the Congo herbarium consist of Paniceæ.

Among the Paniceæ of the collection, there are two unpublished genera. The *first* is intermediate, in character, to Andropogon and Saccharum, but with a habit very different from both. The *second*, which is common to
461] other parts of the coast and to India, appears to connect in some respects Saccharum with Panicum.

The remarks I have to make on the *Acotyledonous Plants* from Congo, relate entirely to

FILICES, of which there are twenty-two species in the collection. The far greater part of these are new, but all of them are referable to well established genera, particularly to Nephrodium, Asplenium, Pteris, and Polypodium. There are also among them two new species of *Adiantum*, a genus of which no species had been before observed on this line of coast. *Trichomanes* and *Hymenophyllum* are wanting in the collection, and these genera, which seem to require constant shade and humidity, are very rare in equinoctial Africa. Of *Osmundaceæ*, the herbarium contains only one plant, which is a new species of *Lygodium*, and the first of that genus that has been noticed from the continent of Africa.

Among the few species common to other countries, the most remarkable is Gleichenia Hermanni,[1] which I have compared and found to agree with specimens from the con-

[1] *Prodr. Flor. Nov. Holl.* 1, *p.* 161. Mertensia dichotoma *Willd. Sp. Pl.* 5, *p.* 71.

tinent of India, from Ceylon, New Holland, and even from the Island of St. Vincent.

Acrostichum stemaria of M. de Beauvois,[1] which hardly differs from A. alcicorne of New Holland, and of several of the islands of the Malayan Archipelago, was also observed; and *Acrostichum aureum*, which agrees with specimens from equinoctial America, was found growing in plenty among the mangroves near the mouth of the river.

I have formerly observed that the number of Filices, unlike that of the other Cryptogamous orders, (Lycopodineæ excepted,) is greatest in the lower latitudes; and, as I then supposed, near or somewhat beyond the tropics. The latter part of this statement, however, is not altogether correct; the maximum of the order, both in absolute and relative number of species, being more probably within the tropics, though at considerable heights.

The degree of latitude alone being given, no judgment can be formed respecting the proportion of Filices: for besides a temperature somewhat inferior, perhaps, to [462
that of equinoctial countries of moderate elevation, a humid atmosphere and protection from the direct rays of the sun, seem to be requisite for their most abundant production.

When all these conditions co-exist, their equinoctial proportion to Phænogamous plants is probably about one to twenty, even on continents where the tracts most favourable to their production form only a small part, their number being increased according as such tracts constitute a more considerable portion of the surface.

Hence their maximum appears to exist in the high, and especially the well wooded, intratropical islands. Thus in Jamaica, where nearly two hundred species of Ferns have been found, their proportion to Phænogamous plants is probably about one to ten. In the Isles of France and Bourbon, from the facts stated by M. du Petit Thouars,[2] they appear to be about one to eight.

In Otaheite, according to Sir Joseph Banks's observations,

[1] *Flore d'Oware* 1, *p.* 2, *t.* 2.

[2] *Mélanges de Bot. Observ. add. à M. de Lamarck, p.* 6, *et* 38.

they are as one to four. And in St. Helena, from Dr. Roxburgh's Catalogue,[1] they exceed one to two.

This high proportion extends to the islands considerably beyond the southern tropic. Thus in the collection formed by Sir Joseph Banks in New Zealand, they are about one to six: in Norfolk Island, from my friend Mr. Ferdinand Bauer's observations, they exceed one to three: and in Tristan Da Cunha, both from the Catalogue published by M. du Petit Thouars,[2] and the still more complete Flora of that Island, for which I am indebted to Captain Dugald Carmichael, they are to the Phænogamous plants as two to three.

The equinoctial proportion of Ferns in level and open tracts, is extremely different from those already given; and it is not improbable that as the maximum of this order is equinoctial, so its minimum will also be found either within or a few degrees beyond the tropics. Thus in several of the low Islands in the Gulf of Carpentaria, having a Flora of upwards of two hundred Phænogamous plants, not more than three species of Ferns were found, and those very sparingly. In Egypt it appear, both by Forskål's catalogue and the more extensive Flora of M. Delile, that only one Fern[3] has been observed.

463] In Russel's catalogue of the plants of Aleppo two only are noticed: and even in M. Desfontaines' Flora Atlantica not more than eighteen species occur, or with relation to the Phænogamous plants, about one to one hundred.

The Ferns in the herbarium from Congo, are to the Phænogamous plants as about one to twenty-six, which agrees nearly with their proportion in Forskål's catalogue of the plants of Arabia, with that of the north coast of New

[1] *Beatson's Tracts relative to St. Helena, p.* 295. [2] *Mélanges de Botanique.*

[3] Named *Adiantum capillus veneris* by both these authors; but possibly a nearly related species that has often been confounded with it. Of the species I allude to, which may be called *Adiantum Africanum,* I have collected specimens in Madeira, and have seen others from Teneriffe, St. Jago, Mauritius or Isle de Bourbon, and Abyssinia. Adiantum Africanum has also been confounded with *A. tenerum* of Jamaica, and other West India islands, and the latter with *A. capillus veneris,* which has in consequence been supposed common to both hemispheres, to the old and new continent, and to the torrid and temperate zones.

Holland, according to my own observations, and which is probably not very different from their proportion in India.

In concluding here the subject of the proportional numbers of the Natural Orders of plants contained in the herbarium from Congo, I may observe, that the ratios I have stated, do not always agree with those given in Baron Humboldt's learned dissertation, so often referred to. I have ventured, however, to differ from that eminent naturalist with less hesitation, as he has expressed himself dissatisfied with the materials from which his equinoctial proportions are deduced. Whatever may be the comparative value of the facts on which my own conclusions depend, I certainly do not look upon them as completely satisfactory in any case. And it appears to me evident, that with respect to several of the more extensive natural orders, other circumstances besides merely the degrees of latitude and even the mean temperature must be taken into account in determining their relative numbers. To arrive at satisfactory conclusions in such cases, it is necessary to begin by ascertaining the geographical distribution of genera, a subject, the careful investigation of which may likewise often lead to important improvements in the establishment or sub-divisions of these groups themselves, and assist in deciding from what regions certain species, now generally diffused, may have originally proceeded.

To the foregoing observations on the principal Natural Orders of Plants from the banks of the Congo, a few remarks may be added on such families as are general in equinoctial countries, but which are not contained in the collection.

These are Cycadeæ, Piperaceæ, Begoniaceæ, Laurinæ [464
(Cassytha excepted,) Passifloreæ, Myrsineæ, Magnoliaceæ, Guttiferæ, Hesperideæ, Cedreleæ, and Meliaceæ.

Cycadeæ, although not found in equinoctial Africa, exist at the Cape of Good Hope and in Madagascar.

Piperaceæ, as has been already remarked by Baron Humboldt,[1] are very rare in equinoctial Africa; and indeed

[1] *Nov. Gen. et Sp. Pl. Orb. Nov.* 1, *p.* 60.

only two species have hitherto been published as belonging to the west coast: the first, supposed to be *Piper Cubeba*, and certainly very nearly related to it, is noticed by Clusius;[1] the second is imperfectly described by Adanson in his account of Senegal. A third species of Piper, however, occurs in Sir Joseph Banks's herbarium, from Sierra Leone: and we know that at least one species of this genus and several of Peperomia, exist at the Cape of Good Hope.

The extensive genus *Begonia*, which it is perhaps expedient to divide, may be considered as forming a natural order, whose place, however, among the Dicotyledonous families, is not satisfactorily determined. Of *Begoniaceæ*,[2] no species has yet been observed on the continent of Africa, though several have been found in Madagascar and the Isles of France and Bourbon, and one in the Island of Johanna.

No genus of *Laurinæ*, is known to exist in any part of the continent of Africa, except the paradoxical Cassytha, of which the only species in the Congo collection can hardly be distinguished from that of the West Indies, or from *C. pubescens* of New Holland. The absence of Laurinæ on the continent of Africa is more remarkable, as several species of Laurus have been found both in Teneriffe and Madeira, and certain other genera belonging to this family exist in Madagascar and in the Isles of France and Bourbon.

Passifloreæ. A few remarkable plants of this order have been observed on the different parts of the west coast of Africa, especially Modecca of the Hortus Malabaricus and Smeathmania, an unpublished genus already mentioned in treating of Homalinæ.

Myrsineæ. No species of any division of this order, has been met with in equinoctial Africa, though several of the
465] first section, or Myrsineæ, properly so called, exist both at the Cape of Good Hope and in the Canary Islands.[3]

[1] Piper ex Guinea, *Clus. exot. p.* 184, who considers it as not different from the Piper caudatum, figured on the same page, and which is no doubt Piper Cubeba of the Malayan Archipelago. [2] *Bonpland Malmais*, 151.

[3] To the first section belong *Myrsine*, *Ardisia*, and *Bladhia*. The second, including *Embelia*, and perhaps also *Othera* of Thunberg, differs from the first merely in its corolla being polypetalous. *Ægiceras* may be considered as

Magnoliaceæ and *Cedreleæ*, which are common to America and India, have not been found on the continent of Africa, nor on any of the adjoining Islands.

Guttiferæ and *Hesperideæ* exist, though sparingly, on other parts of the coast.

A few plants really belonging to *Meliaceæ* have been found on other parts of western equinoctial Africa, and a species of *Leea* (or *Aquilicia*, for these are only different names for the same genus) which was formerly referred to this order, occurs in the herbarium from Congo.

M. de Jussieu, who has lately had occasion to treat of the affinity of Aquilicia,[1] does not venture to fix its place in the system. Its resemblance to Viniferæ in the singular structure of seeds, in the valvular æstivation of the corolla, in the division of its leaves, the presence of stipules, and even in inflorescence, appears to me to determine, if not its absolute union, at least its near affinity to that order. Of *Viniferæ*, Vitis is at present the only certain genus; for *Cissus* and *Ampelopsis* having, as Richard has already observed, exactly the same structure of ovarium, namely, two cells with two erect collateral ovula in each, should surely be referred to it; nor is there any part of the character or description of *Botria* of Loureiro, which prevents its being also included in the same genus.

Lasianthera of M. de Beauvois,[2] referred by its author to Apocineæ, but which M. de Jussieu has lately sug- [466
gested may belong to Viniferæ,[3] is too imperfectly known to admit of its place being determined.

forming a third section, from the remarkable evolution of its embryo and consequent want of albumen. In the æstivation of calyx and corolla it agrees with *Jacquinia*, which together with *Theophrasta*, (or *Clavija* of the Flora Peruviana), forms the fourth section; characterised by the squamæ, more or less distinct, of the faux of the corolla, and by generally ripening more than one seed. The fifth, includes only *Bæobotrys* of Forster (the *Mæsa* of Forskål) which, having ovarium inferum and five barren filaments alternating with the segments of the corolla, bears the same relation to the other genera of this order, that Samolus does to Primulaceæ. On the near affinity, and slight differences in fructification, between this family and Myrsineæ, I have formerly made a few remarks in the Prodr. Flor. Nov. Holl. 1, p. 533.

[1] *Mém. du Mus. d'Hist. Nat.* 3, *p.* 437 *et* 441. [2] *Flore d'Oware*, 1, *p.* 85. [3] *Loc. cit.*

III. In the third part of my subject I am to compare the vegetation of the line of the river Congo with that of other equinoctial countries, and with the various parts of the continent of Africa and its adjoining Islands.

The first comparison to be made is obviously with the other parts of the *West coast of equinoctial Africa.*

The most important materials from this coast to which I have had access are contained in the herbarium of Sir Joseph Banks, and consist chiefly of the collections of Smeathman from Sierra Leone, of Brass from Cape Coast (Cabo Corso), and the greater part of the much more numerous discoveries of Professor Afzelius already referred to. Besides these, there are a few less extensive collections in the same herbarium, especially one from the banks of the Gambia, made by Mr. Park in returning from his first journey into the interior; and a few remarkable species brought from Suconda and other points in the vicinity of Cape Coast, by Mr. Hove. The published plants from the west coast of Africa are to be found in the splendid and interesting *Flore d'Oware et Benin* of the Baron de Beauvois; in the earlier volumes of the Botanical Dictionary of the Encyclopèdie Méthodique by M. Lamarck, chiefly from Sierra Leone and Senegal; in the different volumes of Willdenow's Species Plantarum from Isert; in Vahl's Enumeratio Plantarum from Thonning; a few from Senegal in the Genera Plantarum of M. de Jussieu; and from Sierra Leone in a memoir on certain genera of Rubiaceæ by M. de Candolle, in the Annales du Museum d'Histoire Naturelle. Many remarkable plants are also mentioned in Adanson's Account of Senegal, and in Isert's Travels in Guinea.

On comparing Professor Smith's herbarium with these materials, it appears that from the river Senegal in about 16° N. lat. to the Congo, which is in upwards of 6° S. lat., there is a remarkable uniformity in the vegetation, not only as to the principal natural orders and genera, but even to a considerable extent in the species of which it consists. Upwards of one third part of the plants in the collection from Congo had been previously observed on other parts of the coast, though of these the greater part are yet unpublished.

Many of the Trees, the Palms, and several other remark- [467 able plants, which characterise the landscape, as *Adansonia, Bombax pentandrum, Anthocleista, Musanga* of the natives (the genus related to Cecropia,) *Elæis Guineensis, Raphia vinifera* and *Pandanus Candelabrum,* appear to be very general along the whole extent of coast.

Sterculia acuminata,[1] the seed of which is the *Cola,* mentioned in the earliest accounts of Congo, exists, and is equally valued, in Guinea and Sierra Leone, and, what is remarkable, has the same name in every part of the west coast.

The *Ordeal Tree* noticed in Professor Smith's journal under the name of Cassa, and in Captain Tuckey's narrative erroneously called a species of Cassia, if not absolutely the same plant as the *Red Water Tree* of Sierra Leone,[2] and as it is said also of the Gold Coast, belongs at least to the same genus.

A species of the *Cream Fruit,* mentioned by Professor Afzelius,[3] remarkable in affording a wholesome and pleasant saccharine fluid, used by the natives of Sierra Leone even to quench their thirst, though the plant belongs to Apocineæ, a family so generally deleterious, was also met with.

The *Sarcocephalus* of the same author,[4] which is probably what he has noticed under the name of the country-fig of Sierra Leone,[5] was found, and seems to be not uncommon, on the banks of the Congo.

Anona Senegalensis, whose fruit, though smaller than that of the cultivated species of the genus, has, according to Mr. Lockhart, a flavour superior to any of them, was everywhere observed, especially above Embomma, and appears to be a very general plant along the whole extent of coast.

And *Chrysobalanus Icaco,* or a species very nearly related to it, which is equally common from Senegal to

[1] *De Beauvois. Flore d'Oware,* 1, *p.* 41, *t.* 24.
[2] *Winterbottom's Sierra Leone,* 1, *p.* 129.
[3] *Sierra Leone Report for* 1794, *p.* 173, *n.* 47.
[4] *In Herb. Banks.* [5] *Op. cit. p.* 171, *n.* 32.

Congo, was found abundantly near the mouth of the river.

The remarks I have to make on *Esculent Plants,* my knowledge of which is chiefly derived from the journals of Captain Tuckey and Professor Smith, and the communi-
468] cations of Mr. Lockhart, may be here introduced; the cultivated as well as the indigenous species being very similar along the whole of the west coast.

On the banks of the Congo, as far as the expedition proceeded, the principal articles of vegetable food were found to be Indian Corn or Maize (*Zea Mays*); Cassava, both sweet and bitter, (*Iatropha Manihot L.*); two kinds of Pulse, extensively cultivated, one of which is *Cytisus Cajan* of Linnæus, the other not determined, but believed to be a species of *Phaseolus;* and Ground Nuts (*Arachis hypogæa L.*)

The most valuable fruits seen were Plantains (*Musa sapientum*); the Papaw (*Carica papaya*); Pumpkins (*Cucurbita Pepo*); Limes and Oranges (*Citrus medica et aurantium*); Pine Apples (*Bromelia Ananas*); the common Tamarind (*Tamarindus indica*); and *Safu,* a fruit the size of a small plum, which was not seen ripe.

One of the most important plants not only of Congo, but of the whole extent of coast, is *Elæis Guineensis* or the *Oil Palm,* from which also the best kind of Palm Wine is procured. Wine is likewise obtained from two other species of Palms, which are probably *Raphia vinifera,* and the supposed *Corypha,* considered as an Hyphæne by Professor Smith.

Among the other Alimentary Plants which are either of less importance or imperfectly known, may be mentioned the "*Shrubby Holcus,*" noticed by Captain Tuckey (p. 138); the common *Yam,* which Mr. Lockhart informs me he saw only near Cooloo; and another species of *Dioscorea* found wild only, and very inferior to the Yam, requiring, according to the narrative, "four days boiling to free it from its pernicious qualities." On the same authority, "Sugar Canes of two kinds" were seen at Embomma, and Cabbages at Banza Noki: a kind of Capsicum or Bird Pepper, and Tobacco, were both observed to be generally cultivated: and I find in the herbarium, a specimen of the *Malaguetta Pepper*, or one of the species of Amomum, confounded under the name of *A. Granum-Paradisi*.

Mr. Lockhart believes there was also a second kind of Ground Nut or Pea, which may be that mentioned by Merolla, under the name of *Incumba*,[1] and the second sort perhaps noticed in Proyart's account of Loango,[2] which is
probably *Glycine subterranea* of Linnæus, the Voandzeia [469
of M. du Petit Thouars,[3] or Voandzou of Madagascar, where it is generally cultivated.[4]

Of the indigenous fruits, Anona Senegalensis, Sarcocephalus, a species of Cream fruit, and Chrysobalanus Icaco, have been already mentioned, as trees common to the whole line of coast.

A species of *Ximenia* was also found by Professor Smith, who was inclined to consider it as not different from *X. Americana:* its fruit, which, according to his account, is yellow, the size of a plum, and of an acid but not disagreeable taste, is in the higher parts of the river called Gangi, it may therefore probably be the *Ogheghe* of Lopez,[5] by whom it is compared to a yellow plum, and the tree producing it said to be very generally planted.

An *Antidesma*, probably like that mentioned by Afzelius, as having a fruit in size and taste resembling the currant, is also in the herbarium.

It is particularly deserving of attention, that the greater part of the plants now enumerated, as cultivated on the banks of the Congo, and among them nearly the whole of the most important species, have probably been introduced from other parts of the world, and do not originally belong even to the continent of Africa. Thus it may be stated with confidence that the Maize, the Manioc or Cassava, and the Pine Apple, have been brought from America, and probably the Papaw, the Capsicum, and Tobacco; while the

[1] *Piccardo Relaz. del Viag. nel Reg. di Congo*, *p.* 119. [2] *P.* 18.
[3] *Nov. Gen. Madagasc. n.* 77. [4] *Flacourt Madagasc. pp.* 114 *et* 118.
[5] *Pigafetta, Hartwell's Translat. p.* 115.

Banana or Plantain, the Lime, the Orange, the Tamarind, and the Sugar Cane, may be considered as of Asiatic origin.

In a former part of this essay, I have suggested that a careful investigation of the geographical distribution of genera might in some cases lead to the determination of the native country of plants at present generally dispersed. The value of the assistance to be derived from the source referred to, would amount to this; that, in doubtful cases, where other arguments were equal, it would appear more probable that the plant in question should belong to that country in which all the other species of the same genus were found decidedly indigenous, than to that where it was the only species of the genus known to exist. It seems to me that this reason-
470] ing may be applied with advantage towards determining the original country of several of the plants here enumerated, especially of the Banana, the Papaw, the Capsicum, and Tobacco.

The *Banana* is generally considered to be of Indian origin: Baron Humboldt, however, has lately suggested[1] that several species of *Musa* may possibly be confounded under the names of Plantain and Banana; and that part of these species may be supposed to be indigenous to America. How far the general tradition said to obtain both in Mexico, and Terra Firma, as well as the assertion of Garcilasso de la Vega respecting Peru, may establish the fact of the Musa having been cultivated in the new continent before the arrival of the Spaniards,[2] I do not mean at present to inquire. But in opposition to the conjecture referred to, it may be advanced that there is no circumstance in the structure of any of the states of the Banana or Plantain cultivated in India, or the islands of equinoctial Asia, to prevent their being all considered as merely varieties of one and the same species, namely, *Musa sapientum*; that their

[1] *Nouv. Espag. vol.* 2, *p.* 360.

[2] *Op. cit.*, *p.* 361. It may be observed, however, that this is not the opinion in every part of the continent of South America, for with respect to Brazil, Marcgraf and Piso assert that both the Banana and Plantain are considered as introduced plants, and the latter apparently from Congo. (*Marcg. p.* 137, *et Piso Hist. Nat. Bras. p.* 154.)

reduction to a single species is even confirmed by the multitude of varieties that exist;[1] by nearly the whole of these varieties being destitute of seeds; and by the existence of a plant indigenous to the continent of India, producing perfect seeds; from which, therefore, all of them may be supposed to have sprung.

To these objections to the hypothesis of the plurality of species of the Banana, may be added the argument referred to as contributing to establish its Asiatic origin; for we are already acquainted with at least five distinct species of Musa in equinoctial Asia, while no other species has been found in America; nor does it appear that the varieties of Banana, cultivated in that continent, may not equally be reduced to Musa sapientum as those of India: and lastly, it is not even asserted that the types of any of those supposed species of American Banana, growing without cultivation, and producing perfect seeds, have any where been found.[2]

That the Bananas now cultivated in equinoctial Africa, [471
come originally from India, appears to me equally probable, though it may be allowed that the *Ensete* of Bruce[3] is perhaps a distinct species of this genus, and indigenous only to Africa.

[1] Musa sapientum, *Rox. Corom. tab.* 275.

[2] M. Desvaux, in a dissertation on the genus Musa (*in Journ. de Botanique appl.* vol. 4, *p.* 1), has come to the same conclusion respecting the original country of the cultivated Banana, and also that its numerous varieties are reducible to one species. In this dissertation he takes a view of the floral envelope of Musa peculiar to himself. The perianthium in this genus is generally described as consisting of two unequal divisions or lips. Of these, one is divided at top into five, or more rarely into three segments, and envelopes the other, which is entire, of a different form and more petal-like texture. The enveloping division M. Desvaux regards as the calyx, the inner as the corolla. It seems very evident to me, however, that the deviation in Musa from the regular form of a Monocotyledonous flower, consists in the confluence of the three divisions of the outer series of the perianthium, and in the cohesion, more or less intimate, with these of the two lateral divisions of the inner series; the third division of this series, analogous to the labellum in the Orchideæ, being the inner lip of the flower. This view seems to be established by the several modifications observable in the different species of Musa itself, especially in *M. superba* of Roxburgh, (*Plants of Coromand.* 3, *tab.* 223), and in the flower of Musa figured by Plumier, (*Nov. Gen. t.* 34), but still more by the irregularity confined to the inner series in Strelitzia, and by the near approach to regularity, even in this series, in Ravenala (or Urania), both of which belong to the same natural order.

[3] *Travels, vol.* 5, *p.* 36.

The *Papaw* (Carica papaya), from analogous reasoning, may be regarded as of American origin; there being several other decidedly distinct species natives of that continent, while no species except the cultivated Papaw, nor any plant nearly related to this singular genus, is known to exist either in Asia or Africa. But in the present case, the assistance derived from the argument adduced, may perhaps be considered as unnecessary; for the circumstance of there being no Sanscrit name for so remarkable a plant as the Papaw,[1] is nearly decisive of its not being indigenous to India. And in the Malay Islands, the opinion of the inhabitants, according to Rumphius,[2] is that it was there introduced by the Portuguese.

The same argument may be extended to *Capsicum*, of which all the known species probably belong to the new continent; for the only important exception stated to this genus being wholly of American origin, namely, *C. frutescens*,
472] seems to be set aside merely by the appellations of *Tchilli* and *Lada Tchilli*, as given to it in the Malay Islands; *Chilli*, either simply, or in composition, being the Mexican name for all the species and varieties of this genus.[3]

All the species of *Nicotiana* appear to be American, except *N. Australasiæ* (the *N. undulata* of Ventenat and Prod. Flor. Nov. Holl. but not of Flora Peruviana,) which is certainly a native of New Holland. The exception here, however, does not materially invalidate the reasoning, *N. Australasiæ* differing so much from the other species as to form a separate section of the genus.

The same argument might perhaps be applied to other plants of doubtful origin, as to *Canna indica*, which it would derive from America.

It is certainly not meant, however, to employ this reasoning in every case, and in opposition to all other evidence; and instances may be found, even among the alimentary plants, where it is very far from being satisfactory. Thus the Cocoa Nut, though it will probably be considered as

[1] *Fleming in Asiat. Resear. ii, p.* 161. [2] *Herb. Amboin. i, p.* 147.

[3] *Hernandez, Rer. Medic. Nov. Hispan. Thesaur. p.* 134, *et Nieremb. Hist. Nat. p.* 363.

indigenous to the shores and islands of equinoctial Asia, is yet the only species of its genus that does not belong exclusively to America.

Cytisus Cajan, may be supposed to have been introduced from India. This plant, which is very generally cultivated in the vicinity of the Congo, I conclude is the *Voando*, mentioned by Captain Tuckey as being ripe in October; and as Mr. Lockhart understood from the natives, that Cytisus Cajan continues to bear for three years, it is probably Merolla's *Ovvando*, of which he gives a similar account.[1]

Whether *Arachis hypogæa* be indigenous or introduced, cannot now perhaps be satisfactorily determined. This remarkable plant, whose singular structure and economy were first correctly described by M. Poiteau,[2] and which was every where seen in abundance, as far as the river was examined, appears to form an important article of cultivation along the whole of the west coast of Africa, and probably also on the east coast, on several parts of which it was found by Loureiro.[3]

According to the same author, it is also universally cultivated in China and Cochinchina.

From China it has probably been introduced into the [473
continent of India, Ceylon, and the Malayan Archipelago, where, though now generally cultivated, there is reason to believe, particularly from the names given to it, that it is not indigenous. I think it not very improbable that it may have been carried from Africa to various parts of equinoctial America, though it is noticed in some of the early accounts of that continent, especially of Peru and Brazil.

According to Professor Sprengel,[4] it is mentioned by Theophrastus as cultivated in Egypt: but it is by no means evident that Arachis is the plant intended in the passage of Theophrastus referred to; and it is probable that had it been formerly cultivated in Egypt, it would still be found in that country; it is not, however, included either in

[1] *Piccardo Relaz.* p. 120.
[2] *Mem. de l'Instit. Sc. Phys. Sav. Etrang.* 1, p. 455.
[3] *Flor. Cochin.* 430.
[4] *Hist. Rei Herb.* 1, p. 98.

Forskål's Catalogue, or in the more extensive Flora Egyptiaca of M. Delile.

There is nothing very improbable in the supposition of Arachis hypogæa being indigenous to Asia, Africa, and even America; but if it be considered as originally belonging to one of those continents only, it is more likely to have been brought from China through India to Africa, than to have been carried in the opposite direction.

Glycine subterranea, however, which is extensively cultivated in Africa, Madagascar, and several parts of equinoctial America, is probably of African origin; it is stated, at least both by Marcgraf and Piso, to have been introduced into Brazil from Angola or Congo.[1]

The *Holcus* noticed by Captain Tuckey, of which the specimens in the herbarium do not enable me to determine whether it be a distinct species, or a variety only of *H. sorghum* or *saccharatus*, may be considered as indigenous, or at least as belonging to Africa. According to Mr. Lockhart, it is very generally found wild, and it is only once mentioned as cultivated: it may, however, have been formerly cultivated, along with other species of Millet, to a much greater extent; its place being now supplied by the Maize, which gives probably both a more productive and a more certain crop.

The *Dioscorea* or bitter Yam, which was observed only in a wild state, may be presumed to be a native species; and
474] if ever it has been cultivated, it may in like manner be supposed to have been superseded by the Manioc or Cassava.

The *Safu*,[2] which Mr. Lockhart understood from the natives was one of their most esteemed fruits, he observed to be very generally planted round the villages, especially from Embomma upwards, and to be carefully preserved from birds: its importance is perhaps increased from its ripening in October, a season when the general supply of vegetable food may be supposed to be scanty.

[1] Mandubi d'Angola. *Marcg. Hist. Nat. Brasil.* 43. Mandobi, *Piso, Hist. Nat. Brasil.* p. 256.

[2] Probably the *Zaffo* of some of the earlier accounts of Congo, vide *Malte-Brun Précis de la Geogr.* 5, *p.* 9.

There seems no reason to doubt that this tree, whose probable place in the system I have stated in my remarks on Amyrideæ, belongs originally to the west coast of Africa.

Elæis Guineensis, of which the oil is distinctly described in the beginning of the sixteenth century by *Da Ca da Mosto,* in his account of Senegal,[1] is without doubt indigenous to the whole extent of this coast; as is *Raphia vinifera,* of which the remarkable fruit also very early attracted attention;[2] and the supposed species of *Corypha.*

Of Alimentary Plants, whether cultivated or indigenous, that are known or supposed to belong to the west coast of equinoctial Africa, but which were not seen on the banks of the Congo, a few of the more important may be mentioned.

Among these are the Cocoa Nut and Rice, the former, according to the natives, not being found in the country. The absence of these two valuable plants is the more remarkable, as the Cocoa Nut is said to exist in the neighbouring kingdom of Loango; and, according to Captain Tuckey, a certain portion of land was seen on the banks of the river well adapted to the production of Rice, which is mentioned as cultivated in some of the earlier accounts of Congo.

The Sweet Potatoe (*Convolvulus Batatas*), also noticed by the Portuguese Missionaries, was not met with.

The Butter and Tallow Tree of Afzelius, which forms a new genus belonging to Guttiferæ; the Velvet-Tamarind of Sierra Leone (Codarium acutifolium;[3]) and the Monkey Pepper, or Piper Æthiopicum of the shops (*Unona Æthiopica* of Dunal), which is common on many parts of the coast, were not observed.

Two remarkable plants, the *Akee*[4] and the *Jamaica* [475

[1] *Ramusio* 1, *p.* 104. *Gryn. Nov. Orb.* 28.

[2] Palma-Pinus, *Lobel. advers. p.* 450.

[3] *Afzel. Gen. Plant. Guineen. par. prim. p.* 23. Codarium nitidum *Vahl, enum.* 1, *p.* 302.

[4] Blighia sapida, *König in Annals of Bot.* 2, *p.* 571. *Hort. Kew. ed.* 2*da. vol.* 2, *p.* 350.

At the moment that this sheet was about to have been sent to the press, Sir Joseph Banks received a small collection of specimens and figures of plants,

or *American Nutmeg*,[1] now cultivated in the West India colonies; and the former undoubtedly, the latter probably, introduced from Africa by the Negroes, were neither met with on the banks of the Congo, nor have they been yet traced to any part of the west coast.

The relation which the vegetation of the *Eastern shores of equinoctial Africa* has to that of the west coast, we have at present no means of determining; for the few plants, chiefly from the neighbourhood of Mozambique, included in Loureiro's Flora Cochinchinensis, and a very small number collected by Mr. Salt on the same part of the coast, do not afford materials for comparison.

The character of the collections of *Abyssinian Plants* made by Mr. Salt in his two journeys, forming part of Sir Joseph Banks's herbarium, and amounting to about 260 species, is somewhat extratropical, and has but little affinity to that of the vegetation of the west coast of Africa.

To the Flora of *Egypt*, that of Congo has still less relation, either in the number or proportions of its natural
476] families: the herbarium, however, includes several species which also belong to Egypt, as Nymphæa Lotus, Cyperus Papyrus and articulatus, Sphenoclea zeylanica, Glinus lotoides, Ethulia conyzoides, and Grangea maderaspatana.

observed in the late Mission to Cummazee, the capital of Ashantee; and among them a drawing of the fruit and leaf of a plant, there called *Attueah* or *Attuah*, which is no doubt the *Akee*, whose native country is therefore now ascertained.

[1] Monodora myristica, *Dunal Annonac. p.* 80. *Decand. Syst. Nat. Reg. Veget.* 1, *p.* 477. Anona myristica, *Gært. Sem.* 2, *p.* 194, *t.* 125, *p.* 1. *Lunan Hort. Jamaic.* 2, *p.* 10. This remarkable plant is very properly separated from Anona, and considered as a distinct genus by M. Dunal in his monograph of Anonaceæ. The character given of this new genus, however, is not altogether satisfactory, M. de Candolle's description, from which it is derived, having probably been taken from specimens which he had it not in his power to examine completely. Both these authors have added to this genus Annona microcarpa of Jacquin (*Fragm. Bot. p.* 40, *t.* 44, *f.* 7), established by that author from the fruit of my *Cargillia Australis* (*Prodr. Flor. Nov. Holl.* 1, *p.* 527), which belongs to the very different family of Ebenaceæ.

Long, in his History of Jamaica (*vol.* 3, *p.* 735), has given the earliest account of *Monodora Myristica*, under the name of the *American Nutmeg*, and considers it to have been probably introduced from South America: according to other accounts, it comes from the Mosquito shore: but there is more reason to suppose that it has been brought by the Negroes from some part of the west coast of Africa.

Of the many remarkable genera and orders characterising the vegetation of *South Africa*, no traces are to be found in the herbarium from Congo. This fact is the more worthy of notice, because even in Abyssinia a few remains, if I may so speak, of these characteristic tribes, have been met with; as the *Protea Abyssinica*,[1] observed by Bruce, and *Pelargonium Abyssinicum* and *Geissorrhiza Abyssinica*[2] found by Mr. Salt.

Between the plants collected by Professor Smith in the island of *St. Jago* and those of the Congo herbarium, there is very little affinity; great part of the orders and genera being different, and not more than three species, of which Cassia occidentalis is one, being common to both. To judge from this collection of St. Jago, it would seem that the vegetation of the Cape Verd Islands is of a character intermediate between that of the adjoining continent and of the Canary Islands, of which the Flora has, of course, still less connection with that of Congo.

It might perhaps have been expected that the examination of the vicinity of the Congo would have thrown some light on the origin, if I may so express myself, of the Flora of *St. Helena*. This, however, has not proved to be the case; for neither has a single indigenous species, nor have any of the principal genera, characterising the vegetation of that Island, been found either on the banks of the Congo, or on any other part of this coast of Africa.

There appears to be some affinity between the vegetation of the banks of the Congo and that of *Madagascar* and the *Isles of France* and *Bourbon*. This affinity, however, consists more in a certain degree of resemblance in several natural families and extensive or remarkable genera, than in identity of species, of which there seems to be very few in common.

The Flora of Congo may be compared with those of equinoctial countries still more remote.

With that of *India*, it agrees not only in the proportions of many of its principal families, or in what may be termed

1 Gaguedi *Bruce's Travels* 5, *p.* 52.

2 *Salt's Travels in Abyssinia, append. pp. lxiii. and lxv.* (*Antè, pp.* 93 *and* 95.)

the equinoctial relation, but also, to a certain degree, in the 477] more extensive genera of which several of these families consist: and there are even about forty species common to these distant regions.

To the vegetation of *Equinoctial America* it has certainly much less affinity. Several genera, however, which have not yet been observed in India or New Holland, are common to this part of Africa and America:[1] and there are upwards of thirty species in the Congo herbarium, which are also natives of the opposite coasts of Brazil and Guiana.

As the identity of species, especially of the Dicotyledonous division, common to equinoctial America and other intratropical countries, has often been questioned, I have subjoined two lists of plants included in the Congo herbarium, of which the first consists of such species as are common to America and India: and the second, of such as are found in America only.

I have given also a third list, of species common to Congo and India, or its Islands, but which have not been observed in America.

And a fourth is added, consisting of doubtful plants, to which I have, in the mean time, applied the names of those species they most nearly resemble, and to which they may really belong, without, however, considering their identity as determined.

I. *List of Plants common to Equinoctial Africa, America, and Asia.*

Gleichenia Hermanni, *Prodr. Flor. Nov. Holl.*
Mertensia dichotoma, *Willd.* } (same species)
Agrostis Virginica, L.
Cyperus articulatus, L.
— niloticus, Vahl. } *ead. sp.*
Lipocarpha argentea, *Nob.*
Hypælyptum argenteum, *Vahl.* }
Eleocharis capitata, *Prodr. Fl. N. Holl.*
Fuirena umbellata, *L. fil.*
Pistia Stratiotes, L.
Boerhaavia mutabilis, *Prodr. Flor. Nov. Holl.*
Ipomœa pes-capræ, *Nob.*
Convolvoulus pes-capræ, L.,
— brasiliensis, L. } *ead. sp.*
Ipomœa pentaphylla, *Jacqu.*
Scoparia dulcis, L.

[1] Namely, Elæis, *Jacqu.* Rivina, *L.* Telanthera, *Nob.* (Alternantheræ pentandræ). Alchornea, *Sw.* Blechum, *Prodr. Flor. Nov. Holl.* (Blechi *sp. Juss.*) Schwenckia, *L.* Hyptis, *Jacqu.* Vandellia, *L.* Annona, *L.* Banisteria, *Nob.* (Banisteriæ *sp. L.*) Paullinia, *Juss.* (Paulliniæ *sp. L.*) Vismia, *Ruiz. et Pav.* Conocarpus, *L.* Legnotis, *Sw.* (Cassipourea, *Aubl.*) Chailletia, *Decand.*

Heliotropium indicum, L.
Sphenoclea zeylanica, *Gært.*
Ageratum conyzoides, L.
Waltheria indica, L. } *ead. sp.*
— americana, L.
Hibiscus tiliaceus, L.

Sida periplocifolia, L.
Cassia occidentalis, L.
Guilandina Bonduc, L. } *ead. sp.* [478
— Bonducella, L.
Abrus precatorius, L.
Hedysarum triflorum, L.

II. *Plants common to Equinoctial Africa and America: but not found in India.*

Octoblepharum albidum, *Hedw.*
Acrostichum aureum, L.
Eragrostis ciliaris. }
Poa ciliaris, L.
Cyperus ligularis, L.
Schwenckia americana, L.
Hyptis obtusifolia, *Nob.*
Struchium (Americanum), *Br. jam.* 312.

Sida juncea, *Banks et Soland. MSS. Brasil.*
Urena americana, L. } *ead. sp.*
— reticulata, *Cavan.*
Malachra radiata, L.
Jussiæa erecta, L.
Crotalaria axillaris, *Hort. Kew. & Willd.*
Pterocarpus lunatus, L.

III. *Plants common to equinoctial Africa and India: but not found in America.*

Roccella fuciformis, *Achar. Lichenog.* 440.
Perotis latifolia, *Soland. in Hort. Kew.*
Centotheca lappacea, *Beauv.*
Eleusine indica, *Gært.*
Flagellaria indica, L.
Gloriosa superba, L.
Celosia argentea, L. } *ead. sp.*
— margaritacea, L.
— albida? *Willd.*
Desmochæta lappacea, *Decand.*

Grangea (maderaspatana), *Adans.*
Lavenia erecta, *Sw.*
Oxystelma esculentum, *Nob.*
Periploca esculenta, *Roxb.*
Nymphæa Lotus, L. } *ead. sp.*
— pubescens, *Willd.*
Hibiscus surattensis, L.
Leea sambucina, L.
Hedysarum pictum, L.
Indigofera lateritia, *Willd.*
Glinus lotoides, L.

IV. *List of Species which have not been satisfactorily ascertained.*

Acrostichum alcicorne, *Sw.* }
— stemaria, *Beauv.*
Imperata cylindrica, *Prodr. Flor. Nov. Holl.*
Panicum crus-galli, L.
Typha angustifolia, L.
Gisekia pharnaceoides, L.
Cassytha pubescens, *Prodr. Flor. Nov. Holl.*

Celtis orientalis, L.
Cardiospermum grandiflorum, *Sw.*
Paullinia pinnata, L.
Hydrocotyle asiatica, L. [479
Hedysarum adscendens, *Sw.*
Hedysarum vaginale, L.
Pterocarpus Ecastophyllum, L.

On these lists it is necessary to make some observations.

1st. The number of species in the first three lists taken

together is equal at least to one-twelfth of the whole collection. The proportion, indeed, which these species bear to the entire mass of vegetation on the banks of the Congo is probably considerably smaller, for there is no reason to believe that any of them are very abundant except Cyperus Papyrus and Bombax pentandrum, and most of them appear to have been seen only on the lower part of the river.

2nd. The relative numbers of the species belonging to the primary divisions in the lists, are analogous to, and not very materially different from, those of the whole herbarium; Dicotyledones being to Monocotyledones nearly as 3 to 1; and Acotyledones being to both these divisions united as hardly 1 to 16: hence the Phænogamous plants of the lists alone form about one-thirteenth of the entire collection.

The proportions now stated are very different from those existing in the catalogue I have given of plants common to New Holland and Europe;[1] in which the Acotyledones form one-twentieth, and the Phænogamous plants only one-sixtieth part of the extra-tropical portion of the Flora; while the Monocotyledones are to the Dicotyledones as 2 to 1.

The great proportion of Dicotyledonous plants in the lists now given, and especially in the first two, which are altogether composed of American species, is singularly at variance with an opinion very generally received, that no well established instance can be produced of a Dicotyledonous plant, common to the equinoctial regions of the old and new continent.

3rd. The far greater part of the species in the lists are strictly equinoctial; a few, however, have also been observed in the temperate zones, namely, Agrostis Virginica, belonging, as its name implies, to Virginia, and found also on the shores of Van Diemen's Island, in a still higher latitude; Cyperus Papyrus and articulatus, Nymphæa Lotus, and
480] Pistia Stratiotes, which are natives of Egypt; Glinus lotoides of Egypt and Barbary; and Flagellaria indica,

[1] *Flinders' Voy.* 2, *p.* 592. (*Antè, p.* 68.)

existing on the east coast of New Holland, in as high a latitude as 32° S.

4th. It may perhaps be suggested with respect to these lists, that they contain, or even chiefly consist of, plants that during the constant intercourse which has now subsisted for upwards of three centuries between Africa, America, and India, may have, either from design or accidentally, been carried from one of these regions to another, and therefore are to be regarded as truly natives of that continent only from which they originally proceeded.

It appears to me, however, that there is no plant included in any of the lists which can well be supposed to have been *purposely* carried from one continent to another, unless perhaps *Chrysobalanus Icaco*, and *Cassia occidentalis;* both of which may possibly have been introduced into America by the Negroes, from the west coast of Africa; the former as an eatable fruit, the latter as an article of medicine. It seems at least more likely that they should have travelled in this than in the opposite direction. But I confess the mode of introduction now stated, does not appear to me very probable, even with respect to these two plants; both of them being very general in Africa, as well as in America; though Crysobalanus Icaco is considered of but little value as a fruit in either continent; and for Cassia occidentalis, which exists also in India, another mode of conveyance must likewise be sought.

Several species in the lists, however, may be supposed to have been *accidentally* carried, from adhering to, or being mixed with, articles of food or commerce; either from the nature of the surface of their pericarpial covering, as Desmochæta lappacea, Lavenia erecta, Ageratum conyzoides, Grangea maderaspatana, Boerhaavia mutabilis, and Hyptis obtusifolia; or from the minuteness of their seeds, as Schwenckia americana, Scoparia dulcis, Jussiæa erecta, and Sphenoclea zeylanica. That the plants here enumerated have actually been carried in the manner now stated is, however, entirely conjectural, and the supposition is by no means necessary; several of them, as Lavenia erecta, Scoparia dulcis, and Boerhaavia mutabilis, being also natives of the intratropical part of New Holland; their transportation to or from which cannot be supposed to have been affected in any of the ways suggested.

The probability, however, of these modes of transportation, with respect to the plants referred to, and others of similar structure, being even admitted, the greater part of the lists would still remain; and to account for the disper-
481] sion of these, recourse must be had to natural causes, or such as are unconnected with human agency. But the necessity of calling in the operation of these causes implies the adoption of that theory according to which each species of plants is originally produced in one spot only, from which it is gradually propagated. Whether this be the only, or the most probable opinion that can be held, it is not my intention to inquire: it may however be stated as not unfavorable to it, that, of the Dicotyledonous plants of the lists, a considerable number have the embryo of the seed highly developed, and at the same time well protected by the texture of its integuments.

This is the case in Malvaceæ, Convolvulaceæ, and particularly in Leguminosæ, which is also the most numerous family in the lists, and in several of whose species, as *Guilandina Bonduc*, and *Abrus precatorius*, the two conditions of development and protection of the embryo coexist in so remarkable a degree, that I have no doubt the seeds of these plants would retain their vitality for a great length of time either in the currents of the ocean,[1] or in the digestive organs of birds and other animals; the only means apparently by which their transportation from one continent to another can be effected: and it is deserving of notice that these seem to be the two most general plants on the shores of all equinoctial countries.

The Dicotyledonous plants in the lists which belong to other families have the embryo of the seed apparently less

[1] Sir Joseph Banks informs me, that he received some years ago the drawing of a plant, which his correspondent assured him was raised from a seed found on the west coast of Ireland, and that the plant was indisputably *Guilandina Bonduc*. Linnæus also seems to have been acquainted with other instances of germination having taken place in seeds thrown on shore on the coast of Norway. *Vide Coloniæ Plantarum, p.* 3, *in Amœn. Acad. vol.* 8.

advanced, but yet in a state of considerable development, indicated either by the entire want or scanty remains of albumen: the only exception to this being *Leea*, in which the embryo is many times exceeded in size by the albumen.

In the Monocotyledonous plants, on the other hand, consisting of Gramineæ, Cyperaceæ, Gloriosa, Flagellaria and Pistia, the embryo bears a very small proportion to the mass of the seed, which is formed of albumen, generally farinaceous. But it may here be observed that the existence of a copious albumen in Monocotyledones does not equally imply an inferior degree of vitality in the embryo, but [482
may be considered as the natural structure of that primary division; seeds without albumen occurring only in certain genera of the paradoxical Aroideæ, and in some other Monocotyledonous orders which are chiefly aquatic.

5th. Doubts may be entertained of the identity of particular species. On this subject I may observe, that for whatever errors may be detected in these lists, I must be considered as solely responsible; the insertion of every plant contained in them being founded on a comparison of specimens from the various regions of which their existence in the particular lists implies them to be natives. The only exception to this being Lipocarpha argentea, of which I have not seen American specimens; as a native of that continent therefore it rests on the very sufficient authority of Baron Humboldt and M. Kunth.

In my remarks on the natural orders, I have already suggested doubts with respect to certain species included in the lists, and shall here add a few observations on such of the others as seem to require it.

Acrostichum aureum, L. was compared, and judged to agree, with American specimens; and I have therefore placed it in the 2nd list, without, however, meaning to decide whether those plants originally combined with A. aureum, and now separated from it, should be regarded as species or varieties.

Fuirena umbellata, L. fil. from Congo, has its umbels somewhat less divided than either the American plant or that from the continent of India; but from specimens collected in the Nicobar Islands, this would appear to be a variable circumstance.

Gloriosa superba, L. which seems to be very general along the whole of the west coast of Africa, is considered as a variety of the Indian plant by M. Lamarck. This African variety has no doubt given rise to the establishment of the second species of the genus, namely *G. simplex,* which Linnæus adopted from Miller;[1] and which Miller founded on the account sent to him by M. Richard, of the Trianon Garden, along with the seeds of what he called a new Gloriosa, brought from Senegal by Adanson, and having blue flowers. Miller had no opportunity of determining the correctness of this account; for though the seeds vegetated, the plant died without flowering; but he added a character not unlikely to belong to the seedling plants of G. superba, namely the want of tendrils. Adanson himself,
483] indeed, notices what he considers a new species of Gloriosa in Senegal,[2] but he says nothing of the colour of its flowers, which he would hardly have omitted, had they been blue: that his plant, however, was not without tendrils may be inferred from their entering into the character he afterwards gave of the genus,[3] as well as from M. Lamarck's account of his variety β of G. superba,[4] which he seems to have described from Adanson's specimens. And as no one has since pretended to have seen a species of this genus, either with blue flowers, or leaves without tendrils, *G. simplex,* which has long been considered as doubtful, may be safely left out of all future editions of the Species Plantarum. As the supposed G. superba of this coast, however, seems to differ from the Indian plant in the greater length and more equal diameter of its capsule, it may possibly be a distinct species, though at present I am inclined to consider it as only a variety.

Sphenoclea zeylanica, Gært. I have compared this plant from Congo with specimens from India, Java, China,

[1] Gloriosa 2, *Mill. Dict. ed.* 7.
[2] Nouvelle espèce de Methonica, *Hist. Nat. du Senegal, p.* 137.
[3] Mendoni, *Fam. des Plant.* 2, *p.* 48.
[4] *Encyc. Method. Botan.* 4, *p.* 134.

Cochinchina, Gambia, Demerary, and the island of Trinidad.

I was at one time inclined to believe that Sphenoclea[1] might be considered as an attendant on Rice, which it very generally accompanies, and with which I supposed it to have been originally imported from India into the various countries where it is found. This hypothesis may still account for its existence in the rice fields of Egypt;[2] but as it now appears to have been observed in countries where there is no reason to believe that rice has ever been cultivated, the conjecture must be abandoned.

Hibiscus tiliaceus, L. agrees with the plant of India, except in a very slight difference in the acumen of the leaf; but the specimens from America have their outer calyx proportionally longer.

Sida periplocifolia, L. corresponds with American specimens; those in Hermann's herbarium, from which the species was established, have a longer acumen to the leaf: in other respects I perceive no difference.

Waltheria indica, L. I consider *W. americana* to [484
be a variety of this sportive species, which seems to be common to all equinoctial countries.

Urena americana, L. and *U. reticulata, Cavan.* appear to me not to differ specifically; and the plant from Congo agrees with West India specimens.

Jussiæa erecta, L. from Congo, agrees with West India specimens in having linear leaves; a specimen, however, from Miller's herbarium, which has been compared, and is said to correspond, with that in the Linnean collection, has elliptical leaves.

Chrysobalanus Icaco, L. has its leaves more deeply retuse than any American specimens I have seen, but in this respect it agrees with Catesby's figure.

Guilandina Bonduc, L. from which *G. Bonducella* does not appear to differ in any respect, is one of the most general plants on the shores of equinoctial countries.

[1] *Rapinia herbacea* of the Flora Cochinchinensis (p. 127) is certainly *Sphenoclea zeylanica*, as appears by a specimen sent to Sir Joseph Banks by Loureiro himself.

[2] *Delile Flor. Egypt. illust. in op. cit.*

Pterocarpus lunatus, L. I have compared the plant from Congo with an authentic specimen from the Linnean herbarium, the examination of which proves that the appearance of ferruginous pubescence in the panicle, noticed in Linné's description, is the consequence of his specimen having been immersed in spirits.

Several of the plants included in the fourth list, I am inclined to consider varieties only of the species to which they are referred; but I have placed them among the more doubtful plants of this list, as their differences seem to be permanent, and are such as admit of being expressed. One of these is

Cardiospermum grandiflorum, Sw. of which the specimens from Congo differ somewhat in inflorescence from the West India plant.

Paullinia pinnata, L. is distinguished rather remarkably from the American plant by the figure of the leaflets, which approach to cuneiform, or widen upwards, but I can perceive no other difference.

Pterocarpus Ecastophyllum, L. differs merely in the want of the very short acumen or narrow apex of the leaf, which I have constantly found in all the West India specimens I have examined.

Gisekia pharnaceoides, L. from Congo, has nearly linear leaves; but I have seen specimens from Kœnig with leaves of an intermediate form.

I shall conclude this essay, already extended considerably beyond my original plan, with a general statement of the proportion of new genera and species contained in Professor Smith's herbarium.

485] The whole number of species in the collection is about 620; but as specimens of about thirty of these are so imperfect as not to be referable to their proper genera, and some of them not even to natural orders, its amount may be stated at 590 species.

Of these about 250 are absolutely new: nearly an equal number exist also in different parts of the west coast of equinoctial Africa, and not in other countries; of which, however, the greatest part are yet unpublished: and about 70 are common to other intratropical regions.

Of unpublished genera there are 32 in the collection; twelve of which are absolutely new, and three, though observed in other parts of this coast of equinoctial Africa, had not been found before in a state sufficiently perfect, to ascertain their structure; ten belong to different parts of the same line of coast; and seven are common to other countries.

No natural order, absolutely new, exists in the herbarium; nor has any family been found peculiar to equinoctial Africa.

The extent of Professor Smith's herbarium proves not only the zeal and activity of my lamented friend, but also his great acquirements in that branch of science, which was his more particular province, and to his excessive exertions in the investigation of which he fell a victim, in the ill-fated expedition to Congo.

Had he returned to Europe, he would assuredly have given a far more complete and generally interesting account of his discoveries than what is here attempted: and the numerous facts which he could no doubt have communicated respecting the habit, the structure, and the uses of the more important and remarkable plants, would probably have determined him to have followed a very different plan from that adopted in the present essay.

It remains only that I should notice the exemplary diligence of the Botanic Gardener, Mr. David Lockhart, the only survivor, I believe, of the party by whom the river above the falls was examined, in that disastrous journey which proved fatal to the expedition.

From Mr. Lockhart I have received valuable information concerning many of the specimens contained in the herbarium, and also respecting the esculent plants observed on the banks of the Congo.

February 2, 1818.

LIST OF PLANTS

COLLECTED BY

THE OFFICERS, &c., IN CAPT. ROSS'S VOYAGE,

ON THE

COASTS OF BAFFIN'S BAY.

BY

ROBERT BROWN, F.R.S.

[*Reprinted from "A Voyage of Discovery for the purpose of exploring Baffin's Bay," by John Ross, K.S., Captain Royal Navy. Appendix, pp.* cxli—cxliv.]

LONDON:

1819.

LIST OF PLANTS [cxli

COLLECTED ON THE

COASTS OF BAFFIN'S BAY,

From Lat. 70° 30′ *to* 76° 12′ *on the East Side;*

AND AT

POSSESSION BAY,

In Lat. 73° *on the West Side.*

The List is formed chiefly from Capt. Ross's collection; a considerable number of additional species to which (S.) is annexed, were collected by Capt. Edward Sabine, and a few marked (F.) were received from Mr. Fisher, the surgeon of the Alexander.

TRIANDRIA.

Eriophorum polystachyon, *Linn.*
Alopecurus alpinus, *Smith, Flor. Brit.* iii, p. 1386.
Agrostis algida, *Phipps's Voy.* p. 200. *Wahlenb. Lapp.* p. 25, t. i. (S.) Gramen sui generis.
——— paradoxa, *nov. sp.* Vix hujus, forsan proprii, generis.
Poa laxa, *Willden. Sp. Pl.* i, p. 386.

HEXANDRIA.

Rumex digynus, *L.* Distinctum genus (DONIA *nob.*[1]) efformat.

DECANDRIA.

Andromeda tetragona, *L.*
Pyrola rotundifolia, *L.*? Absque floribus haud determinanda.
Saxifraga oppositifolia, *L.* [cxlii
——— propinqua, *nov. sp.* S. Hirculo, cui proxima, minor, et diversa præsertim calycibus nudis et petalis inappendiculatis.
——— flagellaris, *Sternberg, Saxifr.* p. 25, t. 6. S. setigera, *Pursh, Amer.* i, p. 312. (F.)
——— tricuspidata, *Willden Sp. Pl.* ii, p. 657. (S.)
——— cæspitosa, *L.* Notis nonnullis differt, forsan distincta.
——— petiolaris, *nov. sp.* proxima S. rivulari. (S.)
——— cernua, *L.*
Silene acaulis, *L.*
Lychnis apetala, *L.*

[1] Corrected by Mr. Brown in the second edition of the voyage to "OXYRIA, *Hill.*"—EDIT.

Lychnis triflora, *nov. sp.* (S.)
Cerastium alpinum, *L.*

ICOSANDRIA.

Potentilla pulchella, *nov. sp.* P. sericeæ affinis. (S.)
——— grœnlandica, *nov. sp.?* nimis affinis P. frigidæ et Braunianæ. (S.)
Dryas integrifolia. *Vahl in Flor. Dan.* 1216.

POLYANDRIA.

Papaver nudicaule, *L.*
Ranunculus —, sulphureus forte vel glacialis; species e fragmentis non determinanda. (F.)

DIDYNAMIA.

Pedicularis hirsuta, *L.*

clxiii] TETRADYNAMIA.

Draba muricella, *Wahlenb. Lapp.* p. 174, t. xi, f. 2? (S.)
—— oblongata, *nov. sp.* (S.)
—— corymbosa, *nov. sp.?* præcedenti valde affinis et ambæ D. rupestri (*Hort. Kew.* iv, p. 91) proximæ. (S.)
Cochlearia fenestrata, *nov. sp.* A C. anglica et danica, quibus valde propinqua, differt valvulis subaveniis et dissepimenti elliptico-lanceolati axi dehiscente.

SYNGENESIA.

Leontodon Taraxacum, *L.?* varietas nana? vix species distincta.

MONŒCIA.

Carex compacta, *nov. sp.* C. pullæ affinis. (F.)

DIŒCIA.

Empetrum nigrum, *L.*
Salix arctica, *nov. sp.*
——— specimen mancum dubiæ speciei, præcedenti proximæ.

POLYGAMIA.

Hierochloe alpina, *Br.* Holcus alpinus, *Wahlenb. Lapp.* p. 51. (S.)

CRYPTOGAMIA.

Lycopodium Selago, *L.* (S.)
Polytrichum juniperinum, *Hooker and Taylor, Musc. Brit.* p. 25.
Orthotrichum cupulatum, *Musc. Brit.* p. 72?
Trichostomum lanuginosum, *Musc. Brit.* p. 60.
Dicranum scoparium, *Musc. Brit.* p. 57. [cxliv
Mnium turgidum, *Wahlenb. Lapp.* p. 351.
Bryum ——, absque capsulis.
Hypnum aduncum, *L.*
Jungermannia ——, fructificatione nulla.
Gyrophora hirsuta, *Achar. Syn.* p. 69. (S.)
——— erosa, *Achar. Syn.* p. 65. (S.)
Cetraria islandica, *Achar. Syn.* p. 229.
——— nivalis, *Achar. Syn.* p. 228.
Cenomyce rangiferina, *Achar. Syn.* p. 277.
——— fimbriata, *Achar. Syn.* p. 254?
Dufurea? rugosa, *nov. sp.*
Cornicularia bicolor, *Achar. Syn.* p. 301.
Usnea?——, *nov. sp.?* absque scutellis.
Ulva crispa. *Lightf. Scot.* 972?

Algarum genus?? Confervis simplicissimis et Tremellæ cruentæ (*Eng. Bot.* 1800) quodammodo affine?? Minute globules, the colouring matter of the Red Snow, of which extensive patches were seen in lat. 76° 25′ N., and long. 65° W.

CATALOGUE OF PLANTS

FOUND IN

SPITZBERGEN

BY

CAPTAIN SCORESBY.

BY

ROBERT BROWN, F.R.S.

[*Reprinted from* " *An Account of the Arctic Regions*," *by W. Scoresby, Jun., F.R.S.E. Vol.* 1, *Appendix*, No. V, pp. 75, 76.]

EDINBURGH.

1820.

CATALOGUE OF PLANTS

FOUND IN

SPITZBERGEN.[1] [75

HEXANDRIA.

Luzula campestris, Juncus campestris, *L.*

DECANDRIA.

Andromeda tetragona, *Linné.*
Saxifraga oppositifolia, *L.*
—— cernua, *L.*
—— var. nivalis, *L.*
—— cæspitosa, β grœnlandica, *Wahlenb. lapp.*, 119.
Cerastium alpinum, α hirsutum, *Wahlenb. lapp.*, 136.

ICOSANDRIA.

Dryas octopetala, *L.*

POLYANDRIA.

Papaver radicatum, *Rottb.* Vix diversum a P. nudicaule, *L.*
Ranunculus sulphureus, *Soland. in Phipps' Voyage.*

DIDYNAMIA.

Pedicularis hirsuta, *L.*

TETRADYNAMIA.

Cochlearia grœnlandica? Vel C. Anglica, *Wahl. lapp.*
Cardamine bellidifolia, *L.*
Draba alpina, *L.*

DIŒCIA.

Salix polaris, *Wahlenb. lapp.*, 261.

CRYPTOGAMIA.

Trichostomum lanuginosum.
Hypnum dendroides.
—— rufescens?
Bryum ventricosum, *Smith brit.*
—— ligulatum?
Dicrani species?
Andreæa alpina.
Ulva ?
Fucus forsan nov. sp. prope [76
alatum, sed absque fructific.
—— plumosus.
—— sinuatus.
Conferva ?
—— nigra?
Cenomyce furcata, *Achar. Syn.*, 276.
—— pocillum, *Id.*, 253.
Solorina crocea, *Id.*, 8.
Alectoria jubata, β chalybeiformis, *Id.* 291.
Lecanora murorum, var. *Id.* 181.
Lecidea atrovirens, *Id.* 24.
Gyrophora hirsuta, *Id.* 69.
—— erosa, *Id.* 65.
—— proboscidea, *Id.* 64.
Endocarpum sinopicum, *Id.* 98.
Sphærophoron coralloides, *Id.* 287.
Parmelia stygia, *Id.*
—— recurva, *Id.* 206?
—— sp. nov.? sed absque fructific.
Peltidea canina?
Cetraria nivalis, *Id.* 228.
Cornicularia aculeata, β spadicea, *Id.* 300.
Usnea? prope U. melaxantham, *Id.* 303.
Stereocaulon paschale, *Id.* 284.

---

[1] This list includes the whole of the plants I met with, excepting some of the larger fuci, in three or four visits to the shore about King's Bay and Mitre Cape. Some of the specimens being imperfect, or without fructification, their species could not always be determined.—W. SCORESBY.

CHLORIS MELVILLIANA.

A

LIST OF PLANTS

COLLECTED IN

MELVILLE ISLAND,

(LATITUDE 74°—75° N. LONGITUDE 110°—112° W.)

IN THE YEAR 1820;

BY THE OFFICERS OF THE VOYAGE OF DISCOVERY

UNDER THE ORDERS OF

CAPTAIN PARRY.

WITH

CHARACTERS AND DESCRIPTIONS

OF THE

NEW GENERA AND SPECIES.

BY

ROBERT BROWN, F.R.S., L.S.,

MEMBER OF THE IMPERIAL ACADEMY NATURÆ CURIOSORUM, OF THE ROYAL ACADEMY OF SCIENCES OF STOCKHOLM, AND OF THE ROYAL SOCIETY OF COPENHAGEN; CORRESPONDING MEMBER OF THE ROYAL ACADEMIES OF SCIENCES OF PARIS, BERLIN, AND MUNICH, &c.

[*Reprinted from ' A Supplement to the Appendix to Captain Parry's Voyage,' pp.* cclxi—ccc.]

LONDON:

1823.

LIST OF PLANTS [cclxi

COLLECTED IN

MELVILLE ISLAND.

THE following list of the Plants observed in Melville Island, chiefly in the vicinity of Winter Harbour, is drawn up from the Herbaria of Captain Sabine, Mr. Edwards, Mr. James Ross, Captain Parry, Mr. Fisher, and Mr. Beverley, whose names are here given in the order of the extent of their collections.

To Captain Parry, Mr. Edwards, Mr. Ross, and Mr. Fisher, I am indebted for complete series of specimens of their respective collections; and I have to offer my acknowledgments to Captain Sabine for having allowed me freely to examine his more extensive herbarium, and to retain it until he was about to leave England, in October, 1821, when the whole, in compliance with his request, was returned to him.

The delay that has taken place in the publication of the present account has been in part owing to the state of my health during a considerable portion of the time that has elapsed since the collections were placed in my hands. I have also experienced much greater difficulty than I had anticipated in determining many of the species; arising either from their extremely variable nature, from the incomplete state of the specimens contained in the collections, or from the want of authentic specimens of other countries, [cclxii
with which it was necessary to compare them. I may

notice, likewise, as a third cause of the delay, the greater extent of my original plan, which included remarks on the state and relative proportions of the primary divisions and natural orders contained in the list; a comparison with the vegetation of regions of nearly similar climates; and observations on the range of those species common to Melville Island and other parts of the world. Towards the completion of this plan I had made considerable progress. But to have satisfactorily treated some of the subjects referred to would have required more time than I have had it in my power to devote to them, and in several cases better materials than I have hitherto been able to obtain.

I have consequently found it necessary to relinquish, for the present, this part of my plan,[1] and to confine myself to a systematic list, adding only characters and descriptions
cclxiii] of the new or imperfectly known genera and species; the only indication left of my intention to treat any of the subjects alluded to being a greater number of references

[1] I shall here offer a single remark on the relative proportions of the two primary divisions of Phænogamous plants.

In my earliest observations on this subject I had come to the conclusion that from 45° as far as 60° or perhaps 65° of north latitude, the proportion of Dicotyledonous to Monocotyledonous plants gradually diminished. (*Flinders' voy.* 2, p. 538. *Antè, p.* 8.) But from a subsequent examination of the list of Greenland plants, given by Professor Giesecke (Art. "Greenland," in Brewster's 'Edinburgh Encyclopædia'), as well as from what I had been able to collect respecting the vegetation of alpine regions, I had supposed it not improbable that in still higher latitudes, and at corresponding heights above the level of the sea, the relative numbers of these two divisions were again inverted (*Tuckey's Congo, p.* 423. *Antè, p.* 103); in the list of Greenland plants referred to, Dicotyledones being to Monocotyledones as four to one, or in nearly the equinoctial ratio; and in the vegetation of Spitzbergen, as well as it could be judged of from the materials hitherto collected, the proportion of Dicotyledones appearing to be still further increased.

This inversion in the cases now mentioned was found to depend at least as as much on the reduction of the proportion of Gramineæ, as on the increase of certain Dicotyledonous families, especially Saxifrageæ and Cruciferæ.

The Flora of Melville Island, however, which, as far as relates to the two primary divisions of Phænogamous plants, is probably as much to be depended on as any local catalogue hitherto published, leads to very different conclusions; Dicotyledones being in the present list to Monocotyledones as five to two, or in as low a ratio as has been anywhere yet observed; while the proportion of Grasses, instead of being reduced, is nearly double what has been found in any other part of the world (see Humboldt, in 'Dict. des Sciences Nat.,' tom. 18, table at p. 416); this family forming one fifth of the whole Phænogamous vegetation.

to authors than is absolutely necessary for the present list, though essential to my original design.

With this more limited plan, and with its execution, as far at least as regards the determination of several of the species, I am so little satisfied, that had the publication depended entirely on myself, and related solely to the present essay, I should have deferred it still longer, probably until the return of Captain Parry from the arduous enterprise in which he is now embarked.

I have, however, to express my regret for the delay that has already taken place, as it has prevented the appearance of the valuable memoirs in other departments of Natural History, which have been long ready for publication; and also as it has till now deprived Botanists of the excellent figures so admirably illustrating the structure of the plants selected for engraving, and for which it is hardly necessary to add that I am indebted to the friendship of Mr. Bauer.[1]

[1] It has not been thought necessary to reproduce the engravings illustrative of the plants described in the present memoir; in the Appendix to Captain Franklin's journey; in Mr. Clarke Abel's journey in the interior of China; and in Captain King's survey of the coasts of Australia. For these plates, in all ten in number, the reader is referred to the works in which they originally appeared.—EDIT.

cclxiv]
DICOTYLEDONES.

RANUNCULACEÆ.

1. Ranunculus nivalis, foliis radicalibus elongato-petiolatis dilatatis lobatis: lobis subovatis; caulinis subsessilibus palmatis, caule erecto subunifloro, petalis obovatis integerrimis longioribus calyce hirsutissimo, stylis rectiusculis ovaria glabra æquantibus.

Ranunculus nivalis, *Wahlenb. lapp. p.* 156. *Schlechtend. ranuncul. sect. post. p.* 14.

β. folia radicalia basi cuneata vix ad medium lobata, lobo medio semiovato basi latiore, petala orbiculato-obovata calyce hirsutissimo sesquilongiora.

Ranunculus nivalis β. *Wahlenb. lapp. p.* 157 (exclus. syn. Martens spitzb.)

Ranunculus sulphureus, *Soland. in Phipps' voy. p.* 202, (fide speciminis unici biflori absque foliis radicalibus, in Herb. Banks). *De Cand. syst. nat. p.* 274 (exclus. syn. Martens spitzb., Laxmanni, Willdenovii et Smithii). *Br. spitzb. pl. in Scoresby's arct. reg.* 1, *append. p.* 75. (*Antè, p.* 181) *Richardson in Franklin's journ. p.* 742.

γ. folia radicalia basi subcuneata v. transversa alte lobata, lobo medio cuneato-obovato basi angustiore.

Obs. Varietas γ, cujus exemplaria duo tantum à nobis visa proxime accedit α quæ, in Insula Melville haud observata, sequentibus notis distinguenda.

α. folia radicalia reniformia alte lobata, lobo medio cuneato-obovato basi angustiore.

Ranunculus nivalis, *De Cand. syst. nat.* 1, *p.* 273, exclus. cit. ad *Sw. in act. holm.* 1789, *p.* 47, quæ R. pygmæus, et syn. *Martens spitzb.* ad var. β pertinente.

A R. nivali differt R. frigidus *Willden.* foliis radicalibus minus alte incisis lobulis pluribus, petalis obcordatis venis anastomozantibus, quæ in R. nivali distinctæ, et statura paulo majore.

2. Ranunculus Sabinii, foliis radicalibus elongato-petiolatis tripartitis: lobis ellipticis: lateralibus semibifidis; caulinis sessilibus tripartitis linearibus, calycibus hirsutis petala retusa subæquantibus.

Obs. Planta inter R. nivalem et pygmæam media in Herb. D. Sabine exstat, ulterius examinanda, forsan haud distincta à R. nivali cujus cfr. ic. Flor. Dan. 1699, ubi petala retusa et folium radicale pinnatifidum.

3. Ranunculus hyperboreus, foliis petiolatis trifidis: lobis divaricatis obtusis: lateralibus subbifidis medio integerrimo, caule repente, acheniis lævibus stigmate sessili apiculatis.

Ranunculus hyperboreus, *Rottb. in act. Hafn.* 10, *p.* 458, *t.* 4, *n.* 16. *Flor. Dan.* 331. *Zœg. flor. island. in Olafs. reise* 2, *p.* 237. *Willden. sp. pl.* 2, *p.* 1322. *Pers. syn.* 2, *p.* 104. *Wahlenb. lapp. p.* 158. *De Cand. syst. nat.* 1, *p.* 272. *Schlechtend. ranuncul. sect. post. p.* 12.

Ranunculus foliis subrotundis trilobis integerrimis, [cclxv caule repente. *Gmel. Sib.* 4, *p.* 204, *t.* 83, *b.*

Desc. *Herba* pusilla, glabra. *Folia* elongato-petiolata, alte trifida, lobo medio ovali, sæpissime indiviso, lateralibus sæpius bifidis lobulo exteriore minore, nunc indivisis, rarissime trifidis. *Petioli* filiformes basi vaginantes. *Pedunculi* oppositifolii, petiolum subæquantes, sæpius pilis sparsis adpressis. *Calyx* tetraphyllus nunc triphyllus (an unquam 5-phyllus?), foliolis concavis pilosiusculis. *Petala* 5, calyce manifeste longiora, *lamina* obovata, intus nitenti trinervi, *ungue* lineari, apice foveola angusta marginata. *Stamina* 15—18, petalis breviora, *filamentis* inæqualibus, *antheris* ovalibus. *Achenia* (30 circiter) in capitulum ovatum congesta, stigmate brevi mucronulata.

4. Ranunculus affinis, foliis radicalibus pedato-multifidis petiolatis; caulinis subsessilibus digitatis; lobis omnium linearibus, caule erecto 1-2-floro cum calycibus ovariisque pubescentibus, fructibus oblongo-cylindraceis, acheniis rostro recurvo.

Obs. R. auricomo proxima species.

5. Caltha arctica, caule repente, foliis reniformibus crenato-repandis obtusis, folliculis (12—16) imbricatis, stigmate persistente adnato apice recurvo, antheris linearibus viginti pluribus.

Obs. Affinitate C. radicanti accedit; figura foliorum et caule repente convenit cum C. natante, quæ facile distinguenda pistillis stamina longitudine et numero superantibus, in capitulum sphæricum dense congestis, stigmatibus rectis simplicibus subsessilibus, antheris ovalibus, floribus albis foliisque aliquoties minoribus, et facie diversissima.

PAPAVERACEÆ.

6. Papaver nudicaule, *Linn. sp. pl. ed.* 2, *p.* 725. *Flor. Dan.* 41. *Willden. sp. pl.* 2, *p.* 1145. *Pers. syn.* 2, *p.* 62. *Br. in Ross' voy. ed.* 2, *vol.* 2, *p.* 193. (*Antè, p.* 178). *Hooker in Scoresby's Greenl. p.* 413.

Papaver nudicaule γ radicatum, *De Cand. syst. nat.* 2, *p.* 70.

Papaver radicatum, *Rottb. in act. Hafn.* 10, *p.* 455, *t.* 8, *p.* 24. *Br. spitzb. pl. in Scoresby's arct. reg.* 1, *append. p.* 75. (*Antè, p.* 181.)

CRUCIFERÆ.

7. Draba alpina, *Linn. sp. pl. ed.* 1, *p.* 642, *ed.* 2, *p.* 896. *Willden. sp. pl.* 3, *p.* 425. *Pers. syn.* 2, *p.* 190. *Wahlenb. lapp. p.* 173. *De Cand. syst. nat.* 2, *p.* 338.

α. siliculæ glabræ.

Draba alpina, *Herb. Linn.*

β. siliculæ pilosæ.

Draba alpina, *Br. spitzb. pl. in Scoresby's arct. reg.* 1, *append. p.* 75. (*Anté, p.* 181)

cclxvi] 8. Draba pauciflora, scapis aphyllis pedicellisque pilosis, foliis lanceolatis integerrimis pilis furcatis simplicibusque, petalis (flavis) spathulatis calycem hirsutum vix superantibus, ovariis glabris.

Obs. Dubia species, alpinæ proxima, cujus exemplar unicum in Herb. D. Sabine vidi.

9. Draba lapponica, *De Cand. syst. nat.* 2, *p.* 344.
Draba androsacea, *Wahlenb. lapp. p.* 174, *t.* 11, *f.* 5, exclus. syn.

Desc. *Radix* fusiformis, fibris nonnullis longis simplicibus, multiceps. *Caules* breves, divisi, basi reliquiis petiolorum emarcidis albis squamati, partiales semunciales, dense foliati. *Folia* lanceolata v. oblongo-lanceolata acutiuscula, plana, integerrima, venis alte immersis anastomozantibus, marginibus ciliatis pilis patentibus simplicibus paucissimisque furcatis, paginis adultorum glabris, novellorum pube brevi ramosa substellata conspersis. *Scapi* unciales—sesquiunciales, sæpissime aphylli, nunc folio unico lanceolato-lineari instructi, glaberrimi, læves. *Corymbi* 5-6-flori pedicellis glaberrimis patentibus, inferioribus flore sæpe longioribus. *Calyx*: foliolis concavis, ovalibus, extus vel pilis nonnullis simplicibus conspersis vel sæpius glaberrimis. *Petala* alba, calyce duplo longiora, *ungue* brevi, *lamina* obovata venosa. *Stamina* tetradynama, calyce longiora, petalis breviora, *filamentis* edentulis, *antheris* uniformibus, subrotundis ochroleucis. *Ovarium* sessile ovatum glabrum. *Stylus* brevissimus. *Stigma* capitato-bilobum, stylo manifeste latius. *Siliculæ* racemoso-corymbosæ, lanceolato-ovatæ, glabræ, stigmate subsessili apiculatæ, pedicellis patentibus paulo longiores polyspermæ. *Semina* biseriata, immarginata.

10. Cochlearia fenestrata, siliculis ellipticis ovalibusve, valvis subaveniis, dissepimento elliptico-lanceolato axi sæpius fenestrato, foliis radicalibus cordatis integerrimis; caulinis spathulato-oblongis subdentatis.

Cochlearia fenestrata, *Br. in Ross' voy. ed.* 2, *vol.* 2, *p.* 193. (*Antè, p.* 178.) *De Cand. syst. nat.* 2, *p.* 367.

Desc. Species polymorpha. *Folia* radicalia reniformi-cordata, citò decidua; caulina sessilia, integra vel paucidentata. *Calyx* sæpe purpurascens. *Petala* alba, obovata, calyce longiora. *Antheræ* subrotundæ. *Stylus* brevis.

Stigma capitatum. *Silicula* obtusa, stylo brevi cum stigmate apiculata. *Valvæ* ventricosæ, venis altè immersis. *Dissepimentum* nunc ellipticum, nunc oblongum v. angustato-oblongum, e lamellis duabus tenuissimis facile separandis; loculi polyspermi. *Funiculi umbilicales* basibus connexis ope membranæ angustæ dissepimento parallelæ. *Semina* contraria, h. e. cruribus embryonis invicem septoque parallelis, ovata, reticulata, immarginata.

Obs. In exemplari unico Siliculas passim triloculares trivalves dissepimento pariter fenestrato observavi.

PLATYPETALUM.

CHAR. GEN. *Silicula* ovalis, *valvis* convexiusculis. *Semina* biseriata. *Cotyledones* incumbentes. *Stylus* brevissimus. *Calyx* sub-patens. *Petalorum laminæ* dilatatæ.

cclxvii] Habitus *fere* Brayæ *quacum structura floris cotyledonibusque incumbentibus convenit; satis diversum pericarpii forma. Affine quoque* Subulariæ *esse videtur, quæ ob cotyledones angustas bicrures, in embryone tantum bicruri ab eadem tribu minime removenda. Notis fructificationis pluribus accedit etiam* Stenopetalo *nob. quod calyce clauso, petalis subulatis! glandulis receptaculi et habitu diversissimum, nec revera affine.*

11. PLATYPETALUM PURPURASCENS, stigmate bilobo patenti, stylo manifesto, scapis nudis unifoliisque pubescentibus, siliculis glabriusculis.

DESC. *Radix* perennis, fusiformis sæpe multicaulis. *Caules* breves, indivisi, basi denudati, supra densè foliati. *Folia* lanceolata, obtusiuscula, integerrima, rarius dente uno alterove instructa, crassa, avenia, læte-viridia, apice pilis nonnullis albis acutis simplicibus rariusve furcatis plerumque obsita; *petioli* basi dilatati membranacei pallidi. *Scapi* terminales, sæpius aphylli, vix unciales, basi nunc glabrati. *Corymbus* 4-6-florus, ebracteatus. *Calyx* modice patens, sepalis ovatis concavis subæqualibus, extus fusco-purpureis, limbo angusto albo, apice sæpe pilosiusculis quandoque glaberrimis, tardius deciduis. *Petala* alba, purpureo dilute tincta, unguiculata, laminis dilatatis, latioribus quam longioribus, integris, obtusissimis, ungues lineares superantibus. *Glandulæ* receptaculi quatuor, per paria approximatæ, latera filamentorum breviorum stipantes. *Stamina* tetradynama, *filamentis* edentulis distinctis; *antheris* uniformibus subrotundis ochroleucis. *Ovarium* sessile, ovale, pubescens pilis acutis simplicibus numerosis albis. *Stylus* brevissimus, tamen manifestus. *Stigma:* lobis patentibus, obtusis, papulosis. *Siliculæ* corymbosæ, ovales, stylo brevissimo cum stigmate patenti apiculatæ, biloculares, polyspermæ, valvis modice concavis, dissepimento completo. *Semina* immarginata, fusca.

12. Platypetalum dubium, stigmate indiviso subsessili, siliculis scapisque pubescentibus.

Obs. Floribus ignotis dubiæ generis planta cujus exemplaria tria in Herb. D. Sabine exstant. Cotyledones certè incumbentes et lineares, basibus tamen crus radiculare embryonis vix occupantibus.

EUTREMA.

Siliqua (abbreviata) anceps, valvis carinatis, dissepimento incompleto! *Cotyledones* incumbentes.

Herba *habitu omnino* Brayæ *et* Platypetali, *quibus maxime affine genus, distinguendum tamen facile siliqua ancipiti, dissepimento incompleto, et seminum funiculis.*

13. EUTREMA EDWARDSII.

Desc. *Herba* perennis, glabra, 2-3-uncialis. *Radix* fusiformis, crassa, biuncialis, striis transversis tenuibus sæpe subannulata, fibrillas numerosas exserens, multicaulis. *Caules* simplicissimi, erecti, paucifolii. *Folia* radicalia elongato-petiolata, ovato-lanceolata, integerrima rarissime paucidentata, crassiuscula, plana, uninervia, venis alte immersis crebre anastomozantibus inconspicuis, glaberrima: [cclviii
petiolis folio 4-5-ies longioribus, linearibus membranaceis, albicantibus, adversus lucem trinerviis; *caulina* radicalibus conformia, inferiora brevi petiolata, superiora subsessilia.

Corymbi 7-10-flori, densi, folio florali sessili sæpe subtensi, cæterum ebracteati. *Calyx* glaber, sepalis æqualibus, ovatis, obtusis, modice concavis, trinerviis, extra medium purpurascentibus, insertione parum inæqualibus. *Petala* alba, calyce sesquilongiora, *ungues* breves, *laminæ* obovatæ (vel ex ovali obovatæ), obtusæ, integerrimæ, planæ, obsoletè uninerviæ, vix manifeste venosæ. *Glandulæ* receptaculi quatuor, per paria approximatæ, latera filamentorum breviorum stipantes, parvæ. *Stamina* tetradynama. *Filamenta* subulata, glabra, edentula, duo lateralia paulo breviora basi aversa (acie nec superficie plana ovarium spectanti). *Antheræ* uniformes, ovato-subrotundæ, incumbentes, infra medium affixæ, loculis parallelo-contiguis, longitudinaliter dehiscentibus. *Pollen* flavum, sphæricum, simplex quantum observare potui per lentem centies augentem. *Ovarium* sessile, glabrum, oblongo-ovatum, uniloculare, placentis duabus parietalibus polyspermis. *Stylus* brevissimus vix manifestus. *Stigma* capitatum, indivisum v. simibilobum, stylo vix amplius. *Siliquæ* (siliculosæ) racemosæ, erectæ, lineari-lanceolatæ, ancipites, glaberrimæ, vix trilineares, stigmate obtuso indiviso subsessili apiculatæ. *Valvæ* carinatæ, carina manifesta, venis immersis, cortice demum ad margines solubili, in disco arctius adherenti; *replum* cortice pariter separabili. *Dissepimentum*, præter basin apicemque ubi sæpius completum, plerumque margo perangustus ad utrumque latus cujus processus membranaceus angustior e quo funiculi umbilicales brevissimi obtusi crassi papillæformes orti. *Semina* immarginata, fusca, lævia. *Cotyledones* incumbentes, lineari-oblongæ, plano-convexiusculæ, basi attenuata brevi in crure radiculari sita.

Obs. This species is named in honour of Mr. Edwards, Surgeon of the Hecla, from whose extensive and well-preserved herbarium I have derived great assistance in drawing up the present list, and in which only perfect specimens with ripe siliquæ of Eutrema Edwardsii were found.

EXPLICATIO TABULÆ—A.[1]

Eutrema Edwardsii.—1. Planta florida, et 17. fructifera; utraque magnitudine naturali. Sequentes magnitudine auctæ; 2. flos integer; 3. petalum; 4. flos petalis orbatus; 6. sepalum (foliolum calycis); 6. stamina et pistillum integumentis floralibus avulsis; 7. stamen longius; 8, stamen brevius; 9. pollen ad augmentum 200; 10. pistillum receptaculo insidens à facie visum; 11. idem duplo auctius; 12. ejusdem portio transverse secta; 13. idem valvis avulsis; 14. pistillum à latere visum; 15. idem valvis avulsis; 16. placentæ parietalis portio cum ovulis; 18. siliqua matura dehiscens à facie visa; 19. siliqua matura clausa à latere visa; 20. eadem valvis orbata; 21. eadem duplo auctius; 22. semen; 23. idem transverse sectum; 24. idem longitudinaliter sectum; 25. embryo.

PARRYA.

Char. Gen. *Siliqua* lato-linearis, *valvis* venosis. *Semina* biseriata, *testæ epidermide* laxo, corrugato. *Cotyledones* accumbentes. *Stigmata* approximata basibus connatis [cclxix
in stylum (brevissimum) decurrentibus. *Filamenta* edentula.

Herbæ *perennes, subacaules.* Folia *radicalia integerrima v. dentata, crassiuscula, opaca, venis immersis inconspicuis, petiolorum basibus dilatatis scariosis semivaginantibus.* Scapi *radicales, aphylli, ebracteati.* Flores *purpurei.* Calyx *subpatens.* Glandulæ hypogynæ 4, *filamenta longiora extus stipantes.*

Obs. Affinitate proximum genus Arabidi, diversum siliquarum figura, structura seminum et stigmatis, et denique habitu.

This Genus is named in honour of Captain Parry, the distinguished commander of the Expedition in which it was discovered, and whose herbarium contained very complete specimens of the species here described.

[1] *See Note at p.* 187.

14. PARRYA ARCTICA.

Parrya, siliquis lineari-oblongis, antheris ovalibus, foliis (fere omnibus) integerrimis, pedunculis glaberrimis.

Desc. *Herba* humilis, perennis, glaberrima. *Radix* perpendicularis, crassa, sublignea, striis transversis tenuibus notata, sæpe multiceps. *Caules* brevissimi, dense foliati. *Folia* petiolata, lanceolata passimve spathulato-lanceolata, integerrima, nonnulla rarissime paucidentata, crassiuscula, opaca, immerse uninervia, venis altè immersis inconspicuis. *Petioli* dimidio superiore angusto lineari textura laminæ, inferiore dilatato semivaginanti scarioso albicanti. *Scapus* caulem abbreviatum terminans vel sæpe axillaris, aphyllus, ebracteatus, glaberrimus, florifer foliis sæpe duplo fructifer triplo—quadruplove longior. *Flores* corymbosi, pedunculis patentibus glaberrimis. *Calyx* glaber, modicè patens, deciduus: *sepala* ovalia, obtusa, concava, insertione parum inæqualia, immerse nervosa, nervis passim oblique connexis. *Petala* quatuor, æqualia, unguiculata, purpurea, rarius alba, calyce duplo longiora; *ungues* lineares; *laminæ* obovatæ, uninerviæ, venosæ venis apice dichotomis. *Stamina* 6, tetradynama. *Filamenta* edentula; 4 *longiora* latiora, altero latere extra medium paulo angustiora. *Antheræ* uniformes, infra medium affixæ, oblongo ovales, ochroleucæ, basi cordata lobulis approximato-parallelis, connectivo perangusto. *Pollen* sphæricum, simplex (nec compositum quantum observare licuit per lentum 114-ies augentem). *Glandulæ hypogynæ* quatuor, filamenta longiora extus stipantes. *Ovarium* sessile, glabrum, biloculare, polyspermum, ovulis numerosis. *Stylus* brevissimus. *Stigma* bipartitum, lobis placentis oppositis, obtusis, mutuo sæpius appressis, basibus confluentibus et quasi in latera styli decurrentibus. *Siliquæ* racemosæ, erectæ, nonnullæ quandoque pendulæ, pedicellis patentibus, intra cicatrices floris sessiles, lineari-oblongæ, passim siliculiformes, utrinque obtusæ. *Valvæ* planæ, uninerviæ, venosæ. *Dissepimentum* completum (rarissime fenestratum foramine magno v. parvo) arachnoideo-areolatum, axi quandoque opaciori paulo incrassato, lineisve duabus opacioribus axi approximatis. *Funiculi umbilicales* marginati, latiusculi, dimidio inferiori septo cohærentes. *Semina* 6-8 in singulo loculo, sæpiusque biseriata, epidermis testæ laxus, tenuissimus, albus, ultra ipsam testam in limbum latiusculum extensus, supra nucleum rugosus, testa ipsa, dempto epidermide, crasso-membranacea e duabus lamellis invicem arctè cohærentibus conflata, [cclxx membrana interna nulla nisi lamella interior testæ. *Embryo* curvatus, plumbeus. *Cotyledones* ovali-obovatæ, planiusculæ, accumbentes, aveniæ. *Radicula* teres, acuta.

Obs. Parryæ altera species est *P. macrocarpa,* siliquis lanceolato-linearibus utrinque acutis inter semina sæpe constrictis, antheris linearibus, pedicellis floriferis hispidiusculis, foliis incisis dentatisque; quæ Cardamine nudicaulis, *Linn. sp. pl. ed.* 1, *p.* 654, fide speciminis unici fructiferi absque floribus illius herbarii. Cardamine, &c. *Gmel. sib.* 3, *p.* 273, *n.* 43. Cardamine articulata, *Pursh. am.* 2, *p.* 439. *De Cand. syst.* 2, *p.* 268. Arabis nudicaulis, *De Cand. syst.* 2, *p.* 240.

EXPLICATIO TABULÆ—B.[1]

Parrya arctica.—1, 2, et 3. Plantæ floridæ et 21 planta fructifera; omnes magnitudine naturali. Sequentes magnitudine auctæ, 4. flos integer, 5. petalum, 6. flos petalis orbatus, 7. sepalum, 8. genitalia integumentis floralibus avulsis, 9. stamen longius antice, 10. idem postice visum, 11. stamen brevius, 12. pollen 200-ies auctum, 13. pistillum receptaculo insidens à facie visum, 14. idem duplo auctius, 15. idem valvis avulsis, 16. ejusdem (14) sectio transversalis, 17. pistillum à latere visum, 18. idem valvis avulsis ovula exhibens loculi alterius, 19. ejusdem (17) sectio transversa, 20. placentæ portio cum ovulis et funiculis suis, 22. siliqua matura dehiscens à facie visa, 23. siliqua matura clausa à latere visa, 24. eadem valvis orbata exhibens dissepimentum et semina loculi alterius, 25. placentæ portio cum seminibus duobus epidermide laxo rugoso arilliformi tectis, 26. seminis maturi integumentis ambobus instructi

[1] *See Note at p.* 187.

sectio transversa, 27. semen epidermide arilliformi orbatum, 28. ejusdem sectio longitudinalis, 29. embryo situ naturali, 30. idem cotyledonibus arte expansis, 31. semen abortivum.

15. Cardamine bellidifolia. *Linn. sp. pl. ed.* 2, *p.* 913. *Flor. Dan. t.* 20. *Wahlenb. lapp. p.* 179. *De Cand. syst. nat.* 2, *p.* 249. *Br. in Scoresby's arct. reg.* 1, *append. p.* 75. (*Antè, p.* 181.)

Cardamine foliis simplicibus ovatis petiolis longissimis. *Linn. lapp. p.* 214, *n.* 260 (cum figura respectu habitus bona, quoad flores pessima, *tab.* 9, *f.* 2), exclus. syn. Clusii et Gerardi, ad Arabidem bellidifoliam pertinentibus, monente D. Smith, in Flor. lapp. ed. 2.

CARYOPHYLLEÆ.

16. Lychnis apetala. *Linn. sp. pl. ed.* 2, *p.* 626. *Flor. Dan.* 806. *Willden, sp. pl.* 2, *p.* 810. *Pers. syn.* 1, *p.* 520. *Wahlenb. lapp. p.* 135, *t.* 7. *Br. in Ross' voy. ed.* 2, *vol.* 2, *p.* 192. (*Antè, p.* 178.) *Richardson in Franklin's journ. p.* 738.

Cucubalus caule simplicissimo unifloro corolla inclusa. *Linn. lapp.* 143, *n.* 181, *t.* 12, *f.* 1.

17. Cerastium alpinum. *Linn. sp. pl. ed.* 2, *p.* 628.
cclxxi] *Willden. sp. pl.* 2, *p.* 814. *Pers. syn.* 1, *p.* 521. *Smith brit.* 2, *p.* 500. *Engl. bot.* 472. *Hooker scot. p.* 144 *et* 280. *Soland. in Phipps' voy. p.* 202. *Br. in Ross' voy. ed.* 2, *vol.* 2, *p.* 192. (*Antè, p.* 178.) *Spitz. pl. in Scoresby's arct. reg.* 1, *append. p.* 75. (*Antè, p.* 181.) *Hooker in Scoresby greenl. p.* 413. Cerastium latifolium. *Lightf. scot.* 1, *p.* 242, *t.* 10.

Obs. Species polymorpha cujus tres varietates sequentes in Insula Melville observatæ.

α. folia oblonga rariusve brevè ovalia, pedunculi dichotomi rarius uniflori, pili pedunculorum plerique glanduloso-capitati, capsulæ oblongæ calyce duplo fere longiores.

β. folia late ovata, pedunculi dichotomi pilis plerisque acutis, calycis foliola interiora glabriuscula.

γ. hirsuta, folia elliptica v. lanceolata, pedunculi divisi et solitarii, pilis plerisque acutis, capsulæ calyce paulo longiores.

18. Stellaria Edwardsii, foliis ovato-lanceolatis integerrimis enerviis nitidis, pedunculis terminalibus unifloris trifidisve, petalis bipartitis calyce immerse trinervi longioribus, antheris purpureis.

Obs. Duplex varietas.

In α. (cujus exemplaria plurima in Melville Island, et aliqua anno 1792, ad Chesterfield Inlet lecta vidi) folia ovata acuta v. ovato-lanceolata, pedunculi solitarii v. trifidi, lateralibus sæpissime unifloris altero nunc abortiente, dum solitarii ebracteati, dum divisi bibracteati, bracteis semifoliaceis margine membranaceo ciliato, pedicellis lateralibus pariter bibracteatis. Caulis et folia sæpius glaberrima, caulis nunc villosiusculus et folia basi ciliata villis tenuibus laxis.

β. (cujus exemplaria duo, quorum alterum multicaule), folia ovato-lanceolata apice subattenuata nitidissima, pedunculi sæpius uniflori. Caules et folia glaberrima.

In utraque fructus desideratur, qui exstat in S. Edwardsii, *Richardson in Franklin's Journ. p.* 738. In hac vero, quæ forsan distincta species, antheræ ochroleucæ minimæ et polline destitutæ, styli elongati et stigmata manifestiora, caules et folia glaberrima, capsula erecta calyce fere duplo longior semisexvalvis, semina reniformia lævia fusca.

De Stellaria nitida *Hooker in Scoresby greenl. p.* 411, cui secundum auctorem folia lanceolata siccitate subtrinervia, flores subpaniculati et antheræ flavæ, incertus sum.

19. Arenaria quadrivalvis, foliis subulatis acutis glaberrimis trinerviis, pedunculis unifloris elongatis pubescentibus, calycibus acutissimis trinerviis petala elliptica superantibus capsula quadrivalvi (nunc 3—5-valvi) sæpius brevioribus.

Alsine rubella, *Wahlenb. lapp.* 128, *t.* 6, forsan haud distincta; sed secundum auctorem capsula 3-valvis petala rubella et in icone subspathulata basi valde attenuata.

Desc. *Herba* 1-2-uncialis. *Radix* perennis, descendens. *Caulis* à basi ramosissimus cæspitem densum efformans,

infra vaginis petiolaribus emarcidis nervisque foliorum denudatis obsitus, supra dense foliatus. *Folia* opposita basi connata, subulata, acuta, mutica, super concaviuscula, subter convexa, trinervia, marginibus nudis. *Pedunculi*
cclxxii] terminales solitarii, uniflori, prope basin bibracteati, *bracteis* lanceolatis, semifoliaceis margine membranaceo, pubescentes pilis brevibus, porrectis, glanduloso-capitatis, numerosis. *Calyx* 5-partitus, *sepalis* lanceoloatis, acutissimis, vix acuminatis, concaviusculis, trinerviis, viridibus nunc fusco-purpureo tinctis margine albo membranaceo, extus pilis nonnullis brevissimis minute capitatis conspersis, persistens. *Petala* 5, integerrima, alba, calyce paulo breviora, ovali-oblonga v. elliptica, integerrima, baši parum attenuata, persistentia. *Stamina* decem, margini disci brevissimi subcarnosi, dubiæ originis, perigyni potius quam hypogyni, inserta. *Filamenta* subulato-filiformia, glabra. *Antheræ* ochroleucæ, subrotundæ, loculis approximatis, appositis, longitudinaliter dehiscentibus. *Ovarium* sessile, ovatum, glabrum, uniloculare, polyspermum. *Stigmata* quatuor (passim 3 et 5) filiformia, alba intus longitudinaliter hispidula. *Capsula* calyce persistenti appresso sæpius paulo longior, nunc eundem subæquans, quadrivalvis, passim 3 et 5-valvis, valvis vix omnino ad basin distinctis. *Receptaculum* seminum centrale, longitudine fere capsulæ, cum apice cavitatis primo connexum mox solutum. *Semina* reniformia, lævia, fusca, funiculis umbilicalibus cum receptaculo communi persistentibus.

20. Arenaria Rossii, glaberrima, foliis triquetro subulatis obtusiusculis muticis enerviis florem vix æquantibus, pedunculis unifloris elongatis, petalis oblongis calyces obsolete trinervios paulo superantibus.

Desc. *Herba* pusilla, glaberrima. *Caules* ramosissimi, cæspitosi, densè foliati. *Folia* opposita basibus connatis, carinata. *Pedunculi* foliis aliquoties longiores. *Calyx* 5-partitus, purpurascens; *sepala* æqualia, ovata, acutiuscula, modice concava, obsoletissime trinervia, marginibus membranaceis nudis. *Petala* 5, angusto-oblonga, obtusa, integra, alba, calyce paululum longiora. *Stamina* 10. *Filamenta* disco scutelliformi subcarnoso potius perigyno quam hypogyno inserta, è latiore basi filiformia, glabra, alba. *Antheræ* ovales, ochroleucæ. *Ovarium* ovatum, sessile, uniloculare, glabrum, polyspermum. *Stigmata* 3, filiformia.

Obs. Arenaria Rossii, *Richardson in Franklin's journ. p.* 738, paulo diversa est statura majore, foliis calycem longitudine superantibus, minus crassis nec adeo obtusis, internodio sæpius brevioribus, calycis foliolis duobus exterioribus parum brevioribus, nervis lateralibus omnium manifestioribus, petalis longitudine calycis. In hac capsula trivalvis calycem æquat.

Alsine stricta *Wahlenb. lapp. p.* 127, ab Arenaria Rossii *Richards. l. c.* differt statura duplo majore, foliis longioribus acutis aliisque notis.

SAXIFRAGEÆ.

SAXIFRAGA. *Linn.*

CHAR. GEN. *Stamina* 10, *antheris* didymis. *Petala* indivisa. *Styli* 2. *Capsula* (v. adhærens v. libera,) bilocularis, birostris v. biloba, foramine inter rostra ipsisve lobis intus longitudinaliter dehiscens, polysperma. *Semina:* testa nucleo subconformi.

Obs. Characterem in paucis mutatum structuram antherarum et seminum respicientem proposui ob genus [cclxxiii maxime affine (LEPTARRHENA *nob.* quæ Saxifraga amplexifolia, *Sternb. saxifr. suppl. p.* 2, *t.* 2. Saxifraga pyrolifolia, *Don in Linn. soc. trans.* 13, *p.* 389) cui antheræ uniloculares bivalves septo incompleto parallelo, et semina (capsulæ altè bilobæ) scobiformia, testa utrinque ultra nucleum ovalem elongata, subulata!

21. SAXIFRAGA OPPOSITIFOLIA. *Linn. sp. pl. ed.* 2, *p.* 575. *Willden. sp. pl.* 2, *p.* 648. *a. Smith Brit,* 2, *p.* 450. *Engl. bot. t.* 19. *Wahlenb. lapp. p.* 113. *Carpat. p.* 118. *Soland. in Phipps' voy. p.* 202. *Br. in Ross' voy. ed.* 2, *vol.* 2, *p.* 192. (*Antè, p.* 177.) *Spitz. pl. in Scoresby's arct. reg.* 1, *append. p.* 75. (*Antè, p.* 181.) *Don in Linn. soc. trans. v.* 10, *p.* 400.

22. Saxifraga hirculus. *Linn. sp. pl. ed.* 2, *p.* 576.
β. Petala obovata, ungue nudo: caulis uniflorus.
Saxifraga propinqua. *Br. in Ross' voy. ed.* 2, *vol.* 2, *p.* 192. (*Antè, p.* 177.)
Hirculus propinquus. *Haw. Saxif. enum. p.* 41.
Obs. Petala quandoque, sæpius forsan, appendiculata et calyces ciliati; ideoque à S. Hirculo vix differt nisi petalis plerumque obovatis ungue nudo nec ciliato, caule fere semper unifloro et statura minore. Hæc varietas solum in Insula Melville observata fuit.

23. Saxifraga flagellaris, flagellis filiformibus, caule erecto simplici 1-3-floro calycibusque glanduloso-pubescentibus, foliis radicalibus caulinisque inferioribus obovato-spathulatis ciliatis; superioribus villosiusculis, petalis persistentibus capsula semisupera longioribus.
Saxifraga flagellaris. *Sternb. saxifr. p.* 25 *et* 58, *t.* 6. *Steven in Mem. soc. nat. cur. mosq.* 4, *p.* 79. *Marschall flor. taur-caucas.* 3. *p.* 291. *Br. in Ross' voy. ed.* 2, *vol.* 2, *p.* 192. (*Antè, p.* 177.) *Don in Linn. soc. trans.* 13, *p.* 373.
Saxifraga setigera. *Pursh. am.* 1, *p.* 312.
Desc. *Radix* perpendicularis, fibras longas subsimplices dimittens, elevans *Caulem* unicum, simplicissimum, 2-4-uncialem, foliatum, pubescentem, pilis brevibus strictis purpureo-capitatis, sursum crebrioribus, basi demum glabratum. *Folia* indivisa, radicalia et caulina inferiora confertissima, patentia, superiora sparsa: *radicalia* cuneato-obovata et subspathulata, acutiuscula, plana, basi angustata in petiolum brevem latiusculum, immerse nervosa, nervis lateralibus dichotomis ramis interioribus in extimum margini folii approximatum desinentibus, apicibus mox infra apicem folii confluentibus, ibique callo subovali in pagina superiore parum elevato aucta, marginibus longitudinaliter ciliatis, pilis subulatis strictis rigidulis brevibus albicantibus, capitulo glanduloso purpurascente demum deciduo apiculatis, terminali dilatato pariter apiculato; *caulina inferiora* conferta, radicalibus subsimilia figura, ciliis marginalibus et paginis glabris; *superiora* sparsa, paulo minora, oblonga, acutiuscula, basi vix attenuata, subsessilia utraque pagina margini-

busque pilis brevibus glanduloso-capitatis iisque calycis et caulis similibus. *Flagella* ex alis foliorum radicalium et inferiorum caulis solitaria, filiformia, 3—5-uncias longa, angulata v. anguste marginata, arcuato-deflexa, pilis glandulosis rarissimis conspersa, aphylla, apice sobolifera : [cclxxiv
gemmula parva, turbinata, è foliolis nanis numerosis conniventibus, arcte imbricatis, obovatis acutiusculis, immersè nervosis, glabris, marginibus ciliis nonnullis brevibus ornatis ; et in ipsa basi radiculis 2—3 simplicibus singulis è vagina (coleorhiza) membranacea, primo clausa dein lacerata erumpentibus. *Flores* 1—3, pedunculati, erecti, medius præcocior, ebracteatus, laterales bractea unica nunc duabus alternis lato-linearibus sessilibus foliaceis. *Calyx* basi adhærens, dimidio libero 5-partito, laciniis ovatis, obtusiusculis extus pube glandulosa foliorum instar conspersis, intus glaberrimis, nervis alte immersis. *Petala* 5, aurea, calyce duplo longiora, *ungue* brevissimo, *lamina* obovata, 5—7-nervi, sæpissime inappendiculata, quandoque squamula obsoleta extra nervos extimos. *Stamina* decem, calyce parum longiora, petalis breviora. *Filamenta* subulata, subæqualia. *Antheræ* uniformes, cordatæ, flavæ, loculis contiguis medio (margine) longitudinaliter dehiscentibus. *Pollen* flavum. *Ovarium* basi brevè turbinata adherenti, dimidio libero bifido ; biloculare, polyspermum. *Styli* vix ulli. *Stigmata* suborbiculata, depressa, papulosa, nec omnino glabra. *Capsula* plusquam semisupera, bilocularis, biloba, calycis laciniis erectis cincta, et petalis persistentibus (vix emarcidis) longioribus occultata, lobis brevibus crassiusculis longitudinaliter, fere ad stigmata persistentia usque, dehiscentibus. *Semina* in cavitate biloculari indivisa solum, lobis vacuis, minuta, lævia, cylindraceo-oblonga, castanea : *testa* membranacea. *Albumen* semini conforme, album, farinoso-carnosum. *Embryo* rectus, axilis, teres, longitudine fere dimidii albuminis. *Cotyledones* radicula breviores.

24. Saxifraga tricuspidata. *Rottb. in act. Hafn.* 10, *p.* 446, *t.* 6, *n.* 21. *Gunn. norv.* 2, *p.* 135, *n.* 1046. *Flor. Dan.* 976. *Willden. sp. pl.* 2, *p.* 657. *Pers. syn.* 1, *p.* 490. *Sternb. saxifr. p.* 54. *Pursh. am.* 1, *p.* 312. *Giesecke*

Greenl. in Edin. Encyclop. Br. in Ross' voy. ed. 2, *vol.* 2, *p.* 192. (*Antè, p.* 177.) *Don in Linn. soc. trans.* 13, *p.* 440. *Richardson in Franklin's jour. p.* 737.

Obs. In planta Insulæ Melville, quæ statura humilior, folia radicalia passim et caulina omnia indivisa.

25. Saxifraga hyperborea, foliis glaberrimis; radicalibus palmatis elongato-petiolatis, caule lanato subbifloro, bracteis oblongo-linearibus sessilibus, petalis uninerviis, capsulis semiinferis.

Obs. Proxime accedit S. rivulari *Linn.* et ejusdem forsan varietas. S. rivularis autem differt bracteis ambabus sæpius, inferioribus semper subpetiolatis obovatis, caule inferne minus lanato. Ab utraque distincta est S. petiolaris (*Br. in Ross' voy. ed.* 2, *v.* 2, *p.* 192. (*Antè, p.* 177) foliis omnibus glandulis subsessilibus conspersis: radicalibus scapum æquantibus v. superantibus, folio florali lobato, petalis trinerviis.

26. Saxifraga uniflora, foliis radicalibus aggregatis trifidis; caulinis linearibus indivisis distantibus, caule unifloro ovarioque infero viscido: pube glandulosa brevissima, calycibus obtusis, petalis obovato-oblongis.

Saxifraga cæspitosa, *Br. in Ross' voy. ed.* 2, *vol.* 2, *p.* 192. (*Antè,* p. 177.)

cclxxv] Saxifraga venosa, *Haworth, enum. saxifr. p.* 28?

Obs. Nimis affinis S. cæspitosæ, *Linn.*; vix distincta species.

27. Saxifraga nivalis, *Linn. sp. pl. ed.* 2, *p.* 573. *Willden. sp. pl.* 2, *p.* 645. *Pers. syn.* 1, *p.* 488. *Smith brit.* 2, *p.* 449. *Engl. bot.* 440. *Wahlenb. lapp. p.* 113.

α. corymbus multiflorus thyrsoideus, pedunculis inferioribus trifloris.

β. corymbus simplicissimus pauciflorus. *Linn. lapp. t.* 2, *f.* 5.

Obs. varietas *β.* dimidio minor, pube caulis et pedicellorum parciore breviore stricta (nec, ut in *α.* laxa decumbente lanam brevem referente); in utraque petala persistentia.

Saxifraga longiscapa, *Don in Linn. soc. transact.* 13, *p.* 388, à varietate β. vix differt nisi scapo longiore.

28. Saxifraga foliolosa, foliis radicalibus cuneatis subdentatis, scapis divisis: ramis apice unifloris infra tectis foliolis nanis fasciculatis, calycibus inferis obovatis, petalorum laminis cordato-lanceolatis.

Saxifragæ stellaris var. *Linn.* Saxifraga caule nudo simplici foliis dentatis coma foliolosa *Linn. lapp. p.* 137, γ. *tab.* 2, *f.* 3.

Saxifraga stellaris β. comosa. *Willden. sp. pl.* 2, *p.* 644.

Obs. Distincta videtur à S. stellari, *Linn.* (quæ in Insula Melville haud observata fuit) scapo densè foliolato floribus paucissimis (v. nullis) calycibus obovatis, et præsertim petalorum æqualium laminis basi cordatis.

29. Saxifraga cernua, *Linn. sp. pl. ed.* 2, *p.* 577, *flor. lapp. n.* 172, *t.* 2, *f.* 4. *Willden. sp. pl.* 2, *p.* 652. *Pers. syn.* 1, *p.* 489. *Smith brit.* 2, *p,* 453. *Engl. bot. t.* 664. *Flor. Dan.* 22. *Wahlenb. lapp. p.* 116. *Hooker. scot. p.* 130. *Gmel. sib.* 4, *p.* 162, *n.* 74. *Sternb. saxifr. p.* 18, *t.* 12, *f.* 2. *Soland. in Phipps' voy. p.* 202. *Br. in Ross' voy. ed.* 2, *vol.* 2, *p.* 192. (*Antè, p.* 177.) *Spitz. pl. in Scoresby's arct. reg.* 1, *app. p.* 75. (*Antè, p.* 181.) *Don in Linn. soc. trans.* 13, *p.* 364. *Richardson in Franklin's journ. p.* 737.

Obs. Variat rarius caule ramoso, ramis unifloris, quæ S. cernua, *Gunn. nor. n.* 528, *t.* 8, *f.* 2, et Saxifraga bulbifera? *Flor. Dan.* 390. *Zoega pl. island. in Olafs. reise* 2, *p.* 236.

30. Chrysosplenium alternifolium, *Linn. sp. pl. ed.* 2, *p.* 569. *Willden. sp. pl.* 2, *p.* 637. *Pers. syn.* 1, *p.* 487. *Smith brit.* 2, *p.* 453. *Eng. bot. t.* 54. *Hooker scot. p.* 128. *Wahlenb. lapp. p.* 111. *Carpat p.* 116. *Marschall caucas.* 1, *p.* 313. *Richardson in Franklin's journ. p.* 737.

ROSACEÆ.

31. Dryas integrifolia, foliis integerrimis passimque infra medium inciso-crenatis: venis subtus inconspicuis; novellis semisiccatisque marginibus rovolutis.

Dryas integrifolia, *Vahl in act. soc. hist. nat. hafn. vol.* 4, cclxxvi] *par.* 2, *p.* 171. *Flor. Dan.* 1216. *Pers. syn.* 2, *p.* 57. *Br. in Ross' voy. ed.* 2, *vol.* 2. *p.* 193. (*Antè, p.* 178.) *Richardson in Franklin's journ. p.* 740.

Dryas tenella, *Banks' mss.* (fid. specim. à Terra Nova, ubi primum an. 1766, à D. Banks detecta) *Pursh. am.* 1, *p.* 350.

Obs. Nimis affinis D. octopetalæ, quæ differt statura sæpissime majore, foliis semper longitudinaliter inciso-crenatis, subtus costatis venis prominulis.

SIEVERSIA.

Sieversia, *Willden. in Mag. der gesell. naturf. fr. zu Berlin* 5. *jahrg.* (1811), *p.* 397, charactere emendato.

Char. Gen. *Calyx* decemfidus, laciniis alternis accessoriis. *Petala* 5. *Stamina* indefinite numerosa. *Ovaria* indefinita, ovulo adscendente. *Styli* terminales, continui. *Achenium* stylo toto persistenti aristatum. *Embryo* erectus.

Habitus *fere, nec omnino,* Gei, *quod differt* Stylis *geniculatis articulo superiore dissimili sæpiusque deciduo.*

Ab utroque genere distinguendum Geum potentilloides (Coluria *nob.*) ob Stylum basi cum apice ovarii articulatum deciduum, et Achenia (glandulosa) tubo elongato turbinato calycis inclusa.

Sieversiæ species sunt Geum montanum et reptans *Linn.* radiatum *Michaux,* Peckii et triflorum *Pursh,* glaciale *Adams,* humilis, congesta et dilatata *nob.* et Geum anemonoides *Willden. sp. pl.* quæ Dryas pentapetala *Linn.* cui certè styli terminales nec laterales ut perhibet Willdenow in charactere generis Sieversiæ, pro hac specie solum ab illo instituti.

32. Sieversia Rossii, aristis nudis, foliis radicalibus interruptè pinnatis glabris: pinnis trilobis; accessoriis imisque nanis indivisis, caule unifloro subdiphyllo, petalorum venis omnibus distinctis.

Desc. *Herba* perennis, 2—6-uncialis, glabra. *Caudex* demersus, radiciformis, squamis scariosis fuscis (petiolorum reliquiis) tectus, infra medietatem fibras descendentes simplices fibrillosas crassiusculas proferens. *Folia radicalia* numerosa (4—7,) glabriuscula, petiolata, interrupte pinnata, exstipulata, pinnis circumscriptione ovatis cuneatisve, trifidis vel bifidis (lobo superiore lateralium deficiente) basi inæquali, inferne in rachin decurrenti, superioribus approximatis, nanis indivisis interpositis inter medias; imis ipsis minimis integerrimis. *Petioli* infra medium dilatati ibique scariosi, pallide fusci. *Scapi* ex alis foliorum radicalium vel squamarum superiorum caudicis demersi, infra nudi, extra medium foliis sæpius duobus, alternis, sessilibus, pinnatifidis, exstipulatis, in statu florescentiæ (cum scapi folia radicalia vix æquant) invicem apicique scapi approximatis, in fructiferis folia radicalia aliquoties superantibus, ab invicem at ab apice sæpe distantibus; teretes, pubescentes, pube descendendo sensim parciore. *Flos* solitarius, erectus, ebracteatus. *Calyx* extus pubescens, decemfidus, tubo brevi turbinato, laciniis 5 majoribus interioribus, late semiovatis, acutiusculis, quinque alternis dimidio minoribus, ovalibus, petalis oppositis. *Petala* 5, obovata, integerrima, venosa, aurea, sinubus laciniarum majorum calycis inserta [cclxxvii
iisque sesquilongiora. *Stamina* fauci calycis inserta, indefinita, 30 plura. *Filamenta* subulata, glabra. *Antheræ* ovatæ, flavæ, basi semibifidæ, loculis parallelo-approximatis, longitudinaliter dehiscentibus. *Pollen* globosum, simplex. *Ovaria* indefinite numerosa, receptaculo subcylindraceo imbricato inserta, breve pedicellata, ab apice pedicellorum solubilia, hirsuta, pilis acutis strictis, monosperma, ovulo adscendente. *Styli* terminales, filiformes, subulati, glabri, stricti. *Stigmata* dilatata, obliqua, retusa, papulosa.

Obs. This species is named in honour of Lieutenant James Ross, in whose well-preserved herbarium several plants were found not contained in the other collections.

EXPLICATIO TAB. C.[1]

Sieversia Rossii. 1, 2. Planta florida, magnitudine naturali. Sequentes auctæ. 3 et 4. flos antice et postice visus. 5. flos petalis et staminibus orbatus. 6. petalum. 7. portio calycis cum staminibus respondentibus ejusdem basi insertis. 8, 9. stamen antice et postice visum. 10. pollen 200-ies auctum. 11. pistillum. 12. id. longitudinaliter sectum. 13. pistilla receptaculo insidentia. 14. receptaculum commune pistillorum cum pedicellis. 15. achenium fere maturum. 16. id. longitudinaliter sectum. 17. id. transverse sectum. 18. semen. 19. embryo.

33. Potentilla pulchella, foliis pinnatis bijugis super villosis subter sericeis, foliolis pinnatifidis pari inferiori minore: lobis omnium lanceolato-linearibus, caulibus paucifloris (uniflorisve), stylo basi glanduloso-dilatata.

Potentilla pulchella, *Br. in Ross' voy. ed.* 2, *vol.* 2, *p.* 193. (*Antè*, p, 178.)

Potentilla sericea? *Greville in Mem. Wern.- soc.* 3, *p.* 430; fide speciminis in herb. grœnlandico D. Jameson.

Obs. P. sericea *Linn.* facile distinguitur foliis 3—5-jugis, et lana elongata receptaculi, quod in P. pulchella pube brevi ovaria vix æquante instructum. Nostra planta affinitate propius accedit P. niveæ, haud obstante hujus divisione ternata foliorum, quæ nunc, rarissime quamvis, addito foliolorum pari nano similiter pinnata evadunt.

34. Potentilla nivea. *Linn. sp. pl. ed.* 2, *p.* 715. *Rottb. in act. Hafn.* 10, *p.* 451, *t.* 7, *n.* 22, *optima fig. var.* α. *Willden. sp. pl.* 2, *p.* 1109. *Pers. syn.* 2, *p.* 56. *Wahlenb. lapp. p.* 146. *Nestler potent. p.* 73. *Lehman potent. p.* 184.

α. folia super villosiuscula viridia, subter niveo-tomentosa.

β. folia utrinque villosiuscula, paginis concoloribus.

[1] *See Note at p.* 187.

Potentilla nivea β, *Wahlenb. lapp. p.* 147.

Potentilla Groenlandica, *Br. in Ross' voy. ed.* 2, *vol.* 2, *p.* 193. (*Antè, p.* 178.)

Potentilla frigida? *Greville in Mem. Wern. soc.* 3, *p.* 430, sec. exempl. in herbario D. Jameson.

Potentilla verna, *Hooker in Scoresby's greenl. p.* 413.

Obs. Polymorpha species, cui nimis affinis est [cclxxviii Potentilla Vahliana *Lehm. potent. p.* 172, quæ P. hirsuta *Flor. Dan. t.* 1390, secundum exemplar Groenlandicum à D. Giesecke; et P. Jamesoniana *Greville in Mem. Wern. soc.* 3, *p.* 417, *t.* 20, fide exempl. à D. Jameson; nec diversa videtur P. macrantha *Ledeb.* secundum specimen ex Oonalaska à D. Fischer.

PAPILIONACEÆ.

35. Astragalus alpinus, *Linn. sp. pl. ed.* 2, *p.* 1070. *Flor. lapp. p.* 218, *n.* 267, *t.* 9, *f.* 1. *Flor. Dan.* 51. *Gmel. sib.* 4, *p.* 45, *n.* 59. *Pall. astrag. p.* 41, *t.* 32. *Willden. sp. pl.* 3, *p.* 1297. *Wahlenb. lapp. p.* 190, *t.* 12, *f.* 5 (*fruct.*) *Helv.* 131. *Carpat.* 223. *Pursh. am.* 2, *p.* 472.

Phaca astragalina, *De Cand. Astrag. p.* 52. *Pers. syn.* 2, *p.* 331. *Richardson in Franklin's journ. p.* 745.

36. Oxytropis arctica, subacaulis sericea, stipulis petiolaribus, foliolis oppositis alternisque ovali-oblongis, capitulo subumbellato paucifloro, leguminibus erectis oblongis acuminatis calycibusque nigro-pubescentibus.

Desc. *Radix* lignea, perpendicularis, longissima, crassa, subramosa, multiceps. *Caules* brevissimi, dense foliati et basi stipulis villosissimis persistentibus imbricatis tecti. *Folia* conferta, foliola 11—17, novella utrinque villosa sericea, adulta super glabriuscula, ovalia v. oblonga, sæpius obtusa raro acutiuscula. Stipulæ membranaceæ, infra petiolo adnatæ, apicibus solutis semilanceolatis, acutissimis. *Scapi* foliis longiores, teretes, villosi, villis albo-cinereis, nunc cinereis nigrisque intermixtis, nunc omnino nigris. *Flores* majusculi. *Capitulum* 3—5-florum, pedicellis brevissimis. *Bracteæ* lineares, acutæ, patulæ, calyce breviores, extus pube nigricante. *Calyx* villis nigris subadpressis copiosis tectus, dentibus erectis brevibus. *Corolla* cæruleo-violacea, calyce duplo longior (9—10-lin. æquans). *Vexillum* obcordatum lateribus reflexis, lamina basi attenuata absque callis auriculisve. *Alæ* vexillo breviores, obtusissimæ, apice dilatato oblique retuso, prope basin lateris auriculati intus plica saliente, hinc auriculo mediocri. *Carina* alis paulo brevior, obtusa cum mucrone brevi acutiusculo. *Stamina* inclusa 1—9-fid. antheris uniformibus. *Legumen* erectum, calyce hinc longitudinaliter fisso infra auctum, oblongum, acuminatum, sutura superiore intrusa intusque septifera, septo incompleto bipartibili, funiculis adnatis parallelo-striato. *Semina* reniformia, in singulo loculo 7—9, funiculis apice solutis è margine dissepimenti quasi ortis.

Obs. Species proxima O. uralensi quæ diversa floribus leguminibusque spicatis, foliolis numerosioribus et semper acutissimis, calycibus leguminibusque cinereis pilis nonnullis atris pluribus albis.

COMPOSITÆ.

37. Leontodon palustre, *Smith brit.* 2, *p.* 823. *Engl. bot.* 553. *Pers. syn.* 2, *p.* 367. *Hooker scot. p.* 227. *Flor. Dan.* 1708. *Richardson in Franklin's journ. p.* 746.

cclxxix] Leontodon lividus, *Waldst. et. Kitaib. pl. rar. hung.* 2, *p.* 120, *t.* 115. *Willden. sp. pl.* 3, *p.* 1545. *Marsch. taur-caucas.* 2, *p.* 246, *vol.* 3, *p.* 531.

Leontodon taraxacum? *Br. in Ross' voy. ed.* 2, *vol.* 2, *p.* 194. (*Antè, p.* 178.)

Leontodon taraxacum β, *Wahlenb. carpat.* 238. *Upsal. p.* 257.

Obs. Nimis affinis L. Taraxaco L. videtur.

38. Arnica montana β, *Linn. sp. pl. ed.* 2, *p.* 1245. *Willden. sp. pl.* 3, *p.* 2106. *Pers. syn.* 2, *p.* 453. *Wahlenb. lapp.* 210.

Arnica angustifolia, *Vahl in Flor. Dan.* 1524, fide exempl. Groenland. à D. Giesecke.

Doronicum foliis lanceolatis, *Linn. lapp.* 241, *n.* 305.

Obs. Planta nostra Groenlandicâ sæpius humilior (2—4-uncialis) cum exemplaribus nonnullis à D. Richardson prope littora maris arctici quadrans vix specie distinguenda ab Arnica montana *α*, cujus insuper varietates sunt Arnica plantaginea et fulgens, *Pursh. am.*

39. Cineraria congesta, capitulo lanato, foliis lineari-lingulatis undulatis, caule simplicissimo.

Desc. *Herba* 3—4-uncialis lanata. *Radix* fasiculato-fibrosa. *Folia* radicalia et ima caulina numerosa indivisa, lingulata, obtusa, undulata, demum glabriuscula, viridia; caulina superiora 2—3, alterna, lana decumbente. *Caulis* erectus, simplicissimus, lana implexa tardius decidua tectus. *Anthodia* in capitulum terminale subsphæricum ebracteatum dense congesta, lana copiosa semi-involuta, radiata. *Involucrum* (calyx communis) simplici serie polyphyllum, lana decumbenti copiosa, è villis longis implexis articulatis, dense tectum. *Ligulæ* numerosæ, femineæ, lamina oblongo-lineari, integra, 2—3-nervi. *Flosculi* hermaphroditi perfecti. *Tubus* gracilis. *Limbus* infundibuliformis semiquinquefidus, decemnervis, laciniis semilanceolatis trinerviis nervis axilibus tenuioribus. *Antheræ* semi-exsertæ basibus muticis, appendicibus apicis linearibus acutis. *Ovaria* glabra, subcylindracea. *Stigmata* intus canaliculata apice subtruncata. *Pappus* sessilis, filiformis, albus, radiis numerosis longitudinaliter denticulatis.

Obs. Distincta species videtur, attamen non longe distat à C. palustri statura et inflorescentia insigniter variabili.

40. Tussilago corymbosa, corymbo femineo laxo pauciflоro: corollulis ligularibus nervosis; masculo congesto, foliis cordatis sinuatis inæqualiter dentatis subtus tomentosis.

Desc. *Radix* repens. *Folia* radicalia longius petiolata, cordata, nunc sagittato-cordata, sinuata, sæpius ad $\frac{1}{3}$ nunc ad $\frac{1}{2}$ fere radii, lobis inæqualiter dentatis, dentibus mucronulo eglanduloso terminatis, adulta super glabra cum tomento aliquo in nervis venisque primariis, subter lana

brevi alba implexa, diametro sesquiunciali usque $2\frac{1}{2}$ uncias æquanti. *Scapi* 4—8-unciales, adulti tomento parco obsiti, bracteis (petiolis dilatatis) amplexicaulibus, sæpius foliolo nano dentato terminatis. *Anthodia* polygamo-dioica. Mas. *Corymbus* coarctatus pauciflorus: *anthodiis* radiatis: *ligulis* femineis, lamina oblonga: *flosculis* hermaphrodito-masculis,
cclxxx] infundibuliformibus, stigmatibus hispidis, incrassatis, exsertis. Fem. *Corymbus* simplex, 5—8-florus: *pedunculi* involucro longiores, bracteis nonnullis linearibus acuminatis pilis articulatis pubescentes. *Involucrum* (calyx communis) simplici serie polyphyllum, foliolis acutis, extus pubescentibus, pilis articulatis brevibus. *Corollulæ* omnes ligulatæ, femineæ, præter 2—3 centrales, hermaphrodito-masculas. *Femineæ* involucro longiores, ligula 2—3-nervi indivisæ, stigmatibus patulis, stylis extra tubum hispidulis.

Obs. Proxima species T. frigidæ, quæ differt præsertim thyrso femineo multifloro congesto demum fastigiato, masculo laxiore, foliis minus altè sinuatis.

41. Antennaria alpina, *Br. in Linn. soc. transact.* 12, *p.* 123.

Gnaphalium alpinum. *Linn. sp. pl. ed.* 2, *p.* 1199, *lapp. n.* 301. *Willden. sp. pl.* 3, *p.* 1883. *Pers. syn.* 2, *p.* 421. *Wahlenb. lapp.* 202, *Helv. p.* 149. *Carpat. in obs. ad. p.* 258. *Pursh. am.* 2, *p.* 525. *Richardson in Franklin's journ. p.* 747.

Obs. Planta feminea tantum in Melville Island lecta; mascula à nobis nondum visa (nisi hujus forsan varietas pusilla ab Oonalaska), et nullibi, quantum scio, observata!

CAMPANULACEÆ.

42. Campanula uniflora, *Linn. sp. pl. ed.* 2, *p.* 231, *flor. lapp. n.* 85, *t.* 9, *f.* 5, 6. *Rottb. in act. hafn.* 10, *p.* 432, *t.* 6, *n.* 19. *Willden. sp. pl.* 1, *p.* 890. *Pers. syn.* 1, *p.* 188. *Wahlenb. lapp. p.* 63. *Flor. Dan.* 1512. *Svensk bot.* 526. *Richardson in Franklin's journ. p.* 733.

ERICINÆ.

43. Andromeda tetragona, *Linn. sp. pl. ed.* 2, *p.* 563, *lapp. n.* 166, *t.* 1, *f.* 4. *Willden. sp. pl.* 2, *p.* 607. *Pers. syn.* 1, *p.* 480. *Flor. Dan.* 1030. *Pall. ross.* 2, *p.* 56, *t.* 73, *f.* 4. *Wahlenb. lapp. p.* 200. *Br. spitzb. pl. in Scoresby's arct. reg.* 1, *append. p.* 75. (*Antè, p.* 181.) *Ross' voy. ed.* 2, *v.* 2, *p.* 192. (*Antè, p.* 177.) *Richardson in Franklin's journ. p.* 737.

SCROPHULARINÆ.

44. Pedicularis arctica, caule simplici lanato, foliis pinnatifidis lobis sub-ovatis dentato-incisis: adultis glabris; caulinis petiolo dilatato, calycibus quinquefidis lanatis, galea obtusa truncata bidentata, filamentis longioribus hirsutis.

Desc. *Radix* fasciculata, fibris crassis carnosis. *Caulis* simplex, foliatus, 2-3-uncialis, lana alba implexa tardius nec omnino decidua. *Folia* circumscriptione linearia, pinnatifida; lobis sæpius approximatis, dentatis, primò lanata, adulta glabriuscula; *petioli* omnium, radicalium præcipue, lanati. *Spica* multiflora, densa, florida sesquiuncialis, fructifera 2-3-uncialis: *bracteæ* foliaceæ, pinnatifidæ. *Calyx* lanatus, lana copiosa, alba, implexa, persistenti, semiquinquefidus, laciniis inæqualibus, semilanceolatis, inte- [cclxxxi
gerrimis, vel obsoletissime dentatis. *Corolla* purpurea, glaberrima: *galea* leviter falcata, obtusa, antice apice oblique truncata et ad truncaturæ basin utrinque dente unico acuto brevi quandoque brevissimo. *Stamina* inclusa: *Filamenta* duo longiora extra medium hirsuta, duo breviora longitudinaliter glabra: *Antheræ* uniformes, imberbes, basi bifidæ. *Stigma* subcapitatum, sæpius exsertum. *Capsula* calyce persistenti duplo longior, ovata, acuminata, inæquilatera, margine inferiore rectiusculo superiore modice arcuato, bilocularis, bivalvis, valvis medio septigeris, septi dimidio inferiori placentifero. *Semina* oblonga, teretius-

cula, altero latere margine perangusto aucta, utraque extremitate areola nigricanti notata.

Obs. Species proxima P. sudeticæ *Willden. sp. pl.* 3, *p.* 209, quæ differt statura majore, caule glabro, foliorum lobis linearibus inciso-pinnatifidis; caulinis petiolo haud dilatato, corollæ labio inferiore manifestè dentato. P. sudetica *Richardson in Franklin's jour. p.* 742, à sudetica verá vix diversa est nisi corollæ labio superiore breviore, denticulo longiore, caule subunifolio, nec species distincta videtur.

POLYGONEÆ.

45. Polygonum viviparum, *Linn. sp. pl. ed.* 2. *p.* 516, *fl. lapp. n.* 152. *Gmel. sib.* 2, *p.* 44, *n.* 34, *t.* 7, *f.* 2. *Willden. sp. pl.* 2, *p.* 441. *Pers. syn.* 1, *p.* 439. *Smith brit.* 1, *p.* 428. *Engl. bot.* 669. *Fl. Lond. new ser.* 1, *t.* 81. *Wahlenb. lapp.* 99. *Flor. Dan.* 13. *Svensk. bot.* 336. *Marsch. taur-caucas.* 1, *p.* 301. *Pursh. am.* 1, *p.* 271. *Giesecke Greenl. in Edin. encyclop. Hooker in Scoresby's greenl. p.* 410. *Richardson in Franklin's journ. p.* 737.

Natter Wurtz, *Marten's Spitzb. lib.* 3, *cap.* 7, *t.* I, *a.*

OXYRIA.

Oxyria, *Hill, veg. syst.* 10, *p.* 24 (genus omnino artificiale, *Hill l. c.*). *De Cand. fl. franc.* 3, *p.* 379 (Rumicis subgenus). *Br. in Ross' voy. ed.* 2, *vol.* 2, *p.* 192 (*Antè, p.* 177) (genus distinctum). *Campdera rumex, p.* 153. *Hooker Scot. p.* 99.

Char. Gen. *Perianthium* tetraphyllum (duplici serie). *Stamina* 6. *Styli* 2. *Stigmata* penicillata. *Achenium* lenticulare, membranaceum, utrinque alatum, perianthio infra cinctum. *Embryo* centralis.

Obs. Genus propius accedens Rheo quam Rumici, ab utroque satis distinctum.

A Rheo differt numero binario perianthii et stylorum,

stigmatibus penicillatis (quæ in Rheo capitata sublobata), et textura achenii: convenit numero proportionali et situ staminum (quæ geminatim nempe foliolis exterioribus et solitarie interioribus perianthii opposita) pericarpio semidenudato alato, et embryone centrali.

Rumex ab Oxyria diversus est numero ternario omnium partium floris, situ staminum, quæ sex tantum et geminatim foliolis exterioribus perianthii opposita, fructu nucamentaceo aptero, foliolis interioribus mutatis perianthii tecto, embryone laterali: convenit fere stigmatum divisione.

Ovuli insertionem et Radiculæ embryonis situm inter notas genericas haud introduxi: Semen enim erectum [cclxxxii
cum Embryone inverso uti character totius ordinis (incluso certe Calligono contra assertionem Campderæ l. c.) eundem à Chenopodeis optime distinguens in *prodr. flor. nov. holl. p.* 419, primus proposui. Inter ordines apetalos similem structuram seminis in Urticeis et Piperaceis, aliis notis distinguendis, obtinet: dum Embryo inversus cum ovulo pendulo characterem essentialem CHLORANTHEARUM (*Br. in Bot. magaz.* 2190, *nov.* 1820) efformat.

46. OXYRIA RENIFORMIS, *Hooker scot. p.* 111. *Scoresby's greenl. p.* 410. Oxyria digyna, *Campd. rumex, p.* 155.

Rheum digynum. *Wahlenb. lapp.* 101, *tab.* 9, *fructus. Helv. p.* 74, *Carpat.* 114.

Rumex digynus, *Linn. sp. pl. ed.* 2, *p.* 480, *fl. lapp. n.* 132, *obs.* β. *Willden. sp. pl.* 2, *p.* 258. *Pers. syn.* 1, *p.* 395. *Smith brit.* 1, *p.* 395. *Eng. bot.* 910. *Flor. Dan.* 14.

47. SALIX ARCTICA, ovariis subsessilibus tomentosis, stigmate quadrifido stylum subæquante, squamis orbiculato-obovatis, foliis integerrimis ovalibus obovatisve: adultis super glabris subter villosiusculis.

Salix arctica, *Br. in Ross' voy. ed.* 2, *v.* 2, *p.* 194. (*Antè, p.* 178.) *Richardson in Franklin's journ. p.* 752.

Salix n. 37. *Hooker in Scoresby's greenl. p.* 414*, secundum specim. à D. Scoresby.

Salix, *Greville in Mem. Wern. soc.* 3, *p.* 432, fide specim. in herb. grœnl. D. Jameson.

DESC. *Frutex* depressus; *radice* lignea crassa longa. *Rami* decumbentes, floriferi omnes et sterilium nonnulli adscendentes, adulti glabri. *Folia* sparsa, petiolata, elliptico-obovata, v. obovata, integerrima, obtusa, quandoque retusa, novella super glabra, subter villis longis laxis decumbentibus, adulta utrinque glabra, venis subter paulo eminentibus venulis anastomozantibus. *Amenta* utriusque sexus ramos breves villosos foliatos terminantes. *Squamæ* orbiculato-obovatæ sæpe retusæ, fusco-nigricantes, villosæ. *Masc.* 8-10-lin. longa, densa. *Stamina* 2-3, forsan sæpius 3, filamentis distinctis. *Squamulæ* (Nect.) duæ, interiore paulo majore, utraque apice incrassato. *Fem.* *Squamula* unica, interior. *Ovarium* brevissime pedicellatum, pedicello diametrum transversum capsulæ vix æquante, dense tomentosum, cinereum. *Stylus* longitudine varians nunc stigmata æquans, nunc fere dimidio brevior.

MONOCOTYLEDONES.

JUNCEÆ.

48. JUNCUS BIGLUMIS. *Linn. sp. pl. ed.* 2, *p.* 467. *Montin in Amæn. acad.* 2, *p.* 266, *t.* 3, *f.* 3. *Flor. Dan.* 120. *Zoëg. pl. island. in Olafs. reise* 2, *p.* 235. *Vahl in act. soc. hist. nat. hafn.* 2, *par.* 1, *p.* 38. *Willden. sp. pl.* 2, *p.* 216. *Pers. syn.* 1, *p.* 385. *Smith brit.* 1, *p.* 382. *Engl. bot.* 898. *Bicheno in Linn. soc. transact.* 12, *p.* 320. *Hooker scot. p.* 106.

cclxxxiii] 49. LUZULA HYPERBOREA, spicis multifloris subumbellatis pedunculatis sessilibusque (nunc omnibus sessilibus), bractea umbellæ fol acea; partialibus omnibus fimbriatis, capsulis obtusis perianthia acuta subæquantibus, caruncula basilari seminis obsoleta, foliis planis.

Luzula campestris, *Br. spitzb. pl. in Scoresby's arct. reg.* 1, *append. p.* 75. (*Antè, p.* 181.)
Juncus arcuatus, *Hooker in Scoresby's greenl. p.* 410, secund. exempl. à D. Scoresby.
Juncus campestris, *Soland. in Phipps' voy. p.* 201, fide exempl. in Herb. Banks.

Obs. Vix distincta species, et potius ad *L. campestrem* mire variantem, quam ad *L. arcuatam* referenda; præsertim ob bracteam umbellæ sæpissime, non vero semper, foliaceam, et folia plana. L. arcuatæ β. (*Wahlenb. lapp. p.* 88, *cujus fig. in Flor. Dan.* 1386, sed excl. syn. Villars), tamen accedit, inflorescentia, spicis multifloris, longius-pedunculatis, quandoque etiam arcuato-recurvis, bractea umbellæ nunc, rarissime quamvis, squamacea, partialibus omnibus fimbriatis, et caruncula seminis obsoleta. Hæc autem forsan distincta à L. arcuata *a*, *Wahlenb. lapp. p.* 87, *t.* 4. *Hooker flor. lond. n. ser. t.* 151, cui spicæ longius pedunculatæ paucifloræ, et semina ni fallor absque caruncula.

In Luzulis omnibus, quas examini subjeci, excepta L. pilosa, observavi funiculum umbilicalem è filis spiralibus (decompositione partiali funiculi denudatis?) compositum.

CYPERACEÆ.

50. Carex misandra, spicis (4—6) pedunculatis ovalibus pendulis: terminali basi mascula; reliquis femineis, fructibus lanceolatis acuminatis bidentatis margine denticulatis squama ovali longioribus, stigmatibus 2-3.

Desc. (exemplarium quatuor incompletorum cum spicis fructiferis et portione culmi, in herbario D. Ross). *Folium supremum* breve, lineare, marginibus longitudinaliter denticulatis. *Spicæ* v. umbellatæ, v. alternæ, fructiferæ ovales v. oblongæ pendulæ, pedunculis viridibus, laxis, angulatis, spica longioribus. *Bractea* umbellæ communis vaginans, basi atro-fusca, supra viridis, in folium breve subulato-lineare, planum, marginibus denticulato-asperis producta, includens nonnullas partiales, quarum ima communi subsimilis, foliolo breviore terminata, nunc exserta. *Squamæ* ovales, obtusiusculæ, læves, glabræ, nigro-fuscæ, apice limbo angusto albo. *Fructus* circumscriptione lanceolatus, acuminatus, basi attenuata, fusco-ater, ore ipso albicanti emarginato, marginibus acuminis et dimidii superioris denticulatis, cæterum lævis. *Achenium* intra cupulam brevè pedicellatum, obovatum, ventre plano, dorso dum stigmata duo modice convexo dum tria angulato.

Obs. Nimis affinis C. fuliginosæ *Sternb. et Hoppe in act. soc. bot. Ratisb.* 1, *p.* 159, *t.* 3, vix distincta species.

51. Carex concolor, spicis sexu distinctis: mascula unica; femineis 2-3 erectis subsessilibus, squamis omnibus obtusis axi subconcolori, bracteis basi auriculatis, capsulis lævibus ovalibus mucronulo brevissimo integerrimo, stigmatibus 2, culmis lævibus.

cdxxxiv] *Obs.* C. cæspitosæ proxima et vix differt nisi statura minori (3-4 unciali) squamis (nigro-spadiceis) axi sæpius marginibus semper concoloribus, foliis utrinque viridibus et culmis lævibus. An revera distincta species?

52. Eriophorum capitatum, *Host gram. austr.* 1, *p.* 30, *t.* 38. *Schrad. germ.* 1. *p.* 151. *Wahlenb. lapp. p.* 18. *Smith comp. ed.* 2, *p.* 11. *Engl. bot.* 2387. *Hooker scot. p.* 20.

53. Eriophorum angustifolium, *Willden. sp. pl.* 1, *p.* 313. *Smith brit.* 1, *p.* 59. *Engl. bot.* 564. *Schrad. germ.* 1, *p.* 153. *Hooker scot. p.* 21.

Eriophorum polystachion, *Wahlenb. lapp. p.* 18.

Obs. Plantæ nostræ, quasi mediæ inter E. angustifolium et polystachyon forsan ab utroque distinctæ, duæ varietates adsunt.

α, pedunculis lævibus.

β, pedunculis scabris, denticulis crebris minutis.

Hæc ab E. gracile, *Roth catalect.* 2, *add. et Wahlenb. lapp. p.* 19, fid. exempl. ab ipsis auctoribus in Herb. Banks., certe diversa, statura humiliori, foliis latioribus, squamis enerviis omnino nigricantibus, et acheniis oblongo-obovatis.

GRAMINEÆ.

54. Alopecurus alpinus, spica ovata, arista perianthii glumam sericeam lateribus villosissimis subæquante, vagina suprema ventricosa folio suo plano lanceolato triplo longiore.

Alopecurus alpinus, *Smith brit.* 3, *p.* 1386. *Engl. bot.* 1126. *Hooker scot. p.* 22. *Roem. et Schul. syst.* 2, *p.* 272. *Br. in Ross' voy. ed.* 2, *v.* 2, *p.* 191. (*Antè, p.* 177.) *Hooker in Scoresby's greenl.* 410. *Richardson in Franklin's journ. p.* 731.

Alopecurus ovatus, *Knapp gram. brit.* 15. *Hornem. in Flor. Dan.* 1565.

Alopecurus antarcticus, *Giesecke greenl. in Brewster's edin. encyclop.*

Obs. Species (quam primus in Scotiæ monte Loch ny Gaar anno 1794 legi) variat culmo, qui sæpius adscendens, erecto, spica oblongo-cylindracea, arista nunc gluma duplo longiore, rarius nulla.

A. antarcticus, *Vahl symb.* 2, *p.* 18. *Willden. sp. pl.* 1, *p.* 357, ab. A. alpino differt spica sæpius cylindracea, arista glumam bis superante, folio supremo lineari apice attenuato vaginam suam superante v. æquante.

A. pratensis *L.* distinguitur spica cylindracea, glumis acutis latere tantum villosiusculis, arista glumis duplo longiore, vagina suprema laxiusculè cylindracea folium suum lineare multoties superante.

PHIPPSIA. [cclxxxv

Phippsia (subgenus Vilfæ) *Trinius in Spreng. neue entdeck.* 2, *p.* 37.

Char. Gen. *Gluma* uniflora, abbreviata, inæquivalvis. *Perianthium* muticum, obtusum, imberbe; *valvula superiore* nervis sursum divergentibus. *Lodiculæ* 2. *Stam.* 1-3. *Stigmata* 2, sessilia. *Caryopsis* libera, teres, exsulca.

Gramen *pusillum, aquaticum v. in inundatis nascens.* Culmi *basi divisi.* Folia *plana;* vagina *integra, ipso apice tantum fisso.* Panicula *coarctata, ramis semiverticillatis.* Glumæ *enerves, inferiore minori.* Stamina 1-3. Stigmata *persistentia.*

Obs. E graminibus unifloris proxime accedit Vilfæ et Colpodio, affinitatem habet etiam quandam cum Schmidtia *Trattin.* (Coleanthus *Roem. et. Sch. syst.* 2, *p.* 11), cui certe gluma nulla, et perianthium bivalve, probante valvula superiore dinervi.

Inter genera locustis bifloris Phippsia affinis est Catabrosæ, conveniens glumis abbreviatis, perianthiis obtusis concavis et foliorum vaginis apice tantum fisso: differt locustis unifloris, caryopside tereti, nec lateraliter compressa.

55. PHIPPSIA ALGIDA.

Agrostis algida, *Soland. in Phipps' voy. p.* 200, cum descriptione accurata. *Wahlenb. lapp. p.* 25, *t.* 1, ubi perianthium pro gluma, omnino prætervisa, depictum, et lodicula, perperam indivisa et aucta, pro perianthio univalvi. *Flor. Dan.* 1505, structuram eandem exhibens ac in Wahlenb. l. c. *Br. in Ross' voy. ed.* 2, *v.* 2, *p.* 191 (Gramen sui generis.) (*Antè, p.* 177.)

Trichodium algidum, *Svensk bot.* 545, f. 2, ab *ic.* Wahlenb. mutuata. *Roem. et. Sch. syst.* 2, *p.* 283.

Desc. *Gramen* biunciale, glaberrimum, cæspitosum. *Culmi* ipsa basi divisi ibique vaginis scariosis tecti. *Folia* linearia, obtusiuscula, lævia: *ligula* brevis, obtusissima, indivisa: *vagina* laxiuscula, integra, ipso apice tantum fisso. *Panicula* coarctata, ramis semiverticillatis, paucifloris, lævibus. *Locustæ* unifloræ. *Gluma* nana, bivalvis, inæqualis, valvulæ muticæ, obtusiusculæ, concavæ, haud carinatæ, integræ, membranaceæ; *inferior* minor, enervis; *superior* plus duplo major, ipso perianthio triplo circiter brevior, obsolete uninervis; ambæ sæpissime post lapsum perianthii cum rachide persistentes, inferiore quandoque decidua. *Perianthium* intra glumam brevissime pedicellatum: *valvula inferior* concava, ovato-lanceolata, trinervis, nervorum dimidio inferiore hispidulo; *superior* ejusdem fere longitudinis et latitudinis sed diversæ figuræ, obtusa, 3-4-dentata, dinervis, nervis hispidulis à basi sursum paulo divergentibus, ipsa basi sub-approximatis. *Lodiculæ* 2, subovatæ, membranaceæ, indivisæ, glabræ. *Stamina* 1-3. *Stigmata* 2, sessilia, longa, hyalina, ramulis simplicibus. *Caryopsis* ovali-oblonga, teres, exsulca, stigmatibus emarcidis diu coronata. *Embryo* caryopside quadruplo brevior.

Obs. Hæc è speciminibus à Melville Island ; species [cclxxxvi
autem variat perianthii nervis lævibus, staminibus 2, et quandoque unico, nervo alteri valvulæ superioris perianthii opposito.

In Terra Tschutski à Dav. Nelson, in tertio it. Cook, lecta fuit varietas (?) insignis, duplo major, culmis ramosis foliis laxioribus aliisque notis diversa : vix species distincta.

COLPODIUM.

Colpodium. *Trin. agrost. p.* 119, *f.* 7. Subgenus Vilfæ *Trin. in Spreng. neue entdeck.* 2, *p.* 37.

CHAR. GEN. *Gluma* uniflora, subæquivalvis, mutica. *Perianthium* gluma longius, submuticum, obtusum, apice scarioso; *valvulis* subæqualibus, integerrimis, *superiore* exserta, dinervi, lateribus parallelis. *Lodiculæ* 2. *Styli* 2. *Stigmata* plumosa. *Caryopsis.* - - - -

Gramen *glabrum.* Culmi *erecti v. adscendentes.* Folia *plana,* ligula *indivisa imberbi folio latiore,* vagina *longitudinaliter fissa.* Panicula *coarctata, ramis semiverticillatis.* Locustæ *oblongæ, glabriusculæ cum v. absque rudimento, sæpius setuliformi, flosculi secundi.*

Obs. Gramen hocce habitu fere peculiari, primo intuitu Poæ propius accedit quam Agrostidi s. Vilfæ, relationem quodammodo etiam cum Dupontia et Deschampsia habere videtur. Caryopside ignota autem genus haud stabilitum, et de ejusdem affinitate cum Colpodii speciebus Trinii, præsertim *C. Steveni* et *compresso,* incertus sum.

56. COLPODIUM LATIFOLIUM, panicula coarctata lanceolata, foliis planis lato-linearibus.

Agrostis paradoxa, *B. in Ross' voy. ed.* 2, v. 2, *p.* 192. (*Antè, p.* 177.)

Desc. *Gramen* robustum, spithameum—pedale, glabrum. *Culmus* è basi decumbenti v. radicanti adscendens, nunc erectus, teres, lævis, foliatus, basi vaginis scariosis tectus. *Folia* plana, lineari-lanceata, acuta, stricta, utrinque marginibusque retrorsum scabris: *vaginæ* scabriusculæ, ad basin usque fissæ, suprema folio proprio longior: *ligula* obtusa, imberbis, erosa, denticulata, folio latior. *Panicula* coarctata, angusto-lanceolata, fusco-purpurea, perianthiorum apicibus albis, sesquiuncialis—biuncialis, ramis brevibus, semiverticillatis, appressis, inferioribus demum modice patentibus, pedunculis pedicellisque pauci-denticulatis, strictis, apice vix dilatato cum locusta continuo. *Gluma* uniflora, bivalvis, mutica, herbaceo-membranacea, glabra, valvulis suboppositis, concavis vix carinatis, obtusiusculis v. acutis, integris, semitrinerviis, *inferiore* paulo breviore, nervis lateralibus brevissimis, *superiore* acutiore, nervis lateralibus magis manifestis sed longe infra apicem evanescentibus. *Perianthium* intra glumam, qua haud duplo longius, brevissime pedicellatum, cum pedicello crasso articulatum, basi obliqua, herbaceo-membranaceum, textura fere glumæ, muticum, per lentem pube brevissima conspersum, intra glumam è majore parte viride, supra eandem fusco-purpureum, apice scarioso albicanti. *Valvulæ* concavæ, textura omnino similes, longitudine subæquales, *inferior* nervo centrali manifesto sæpius apicem muticum attingenti, nunc
cclxxxvii] in setulam dorsalem brevissimam altitudinem valvulæ subæquantem desinente, lateralibus utrinque duobus obsoletis, infra apicem prorsus evanescentibus; *superior* obtusior, integerrima, dorso angusto planiusculo vel leviter convexo, lineari, dinervi, nervis parallelis, tenuibus, nudis, lateribus dorso aliquoties latioribus, parallelis, marginibus nudis. *Lodiculæ* duæ, subcollaterales, membranaceæ, semibifidæ, dentibus acutis, imberbes, longitudine ovarii. *Stamina* 3, filamentis capillaribus, antheris fusco-stramineis, utrinque bifidis. *Ovarium* ovatum, acutum, glabrum, exsulcum. *Styli* brevissimi, approximati, vix manifesti. *Stigmata* hyalina, dense plumosa, apicibus acutis.

Obs. In exemplaribus plerisque nullum certe rudimentum flosculi secundi, quod tamen in nonnullis à Melville Island atque in exemplari à Possession Bay adest, setuliforme, hispidulum; et in specimine unico à Melville Island locustas nonnullas bifloras flosculo secundo pedicellato perfecto observavi.

57. Poa angustata, panicula simplici coarctata lineari-lanceolata, locustis 4-5-floris, gluma inferiore dimidio minore, perianthiis apice erosis: valvula inferiore basi elanata lateribus glabriusculis, foliis angusto-linearibus.

Desc. *Gramen* 4-6-unciale, glabrum, erectum; *radice* fibrosa. *Culmi* foliati, basi quandoque divisi, læves. *Folia* angusto-linearia, plana, acuta, glabra, lævia; *vaginæ* subcylindraceæ, læves, suprema folio proprio longior, omnes ipsa basi integra; *ligula* subquadrata tam lata quam longa, apice dentato dente medio paulo longiore. *Panicula* erecta, angustata, circumscriptione lineari-lanceolata, ramis paucifloris, pedicellis denticulatis, strictis, viridibus, apice paulo dilatato, cum locusta haud omnino continuo. *Locustæ* oblongæ, coloratæ, sæpius quadrifloræ. *Glumæ* hyalinæ, glaberrimæ, uninerviæ, cum pedicellis persistentes, valvula inferiore fere dimidio minore; superiore duplo latiore et fere duplo longiore, obtusiore, perianthio dimidio circiter breviore, nervis lateralibus obsoletis. *Perianthia* separatim decidentia, rachi locustæ glabra; *valvula inferior* oblonga, concava, acutiuscula, apice scarioso eroso-denticulato, quinquenervis, lateribus infra medium pube rara in nervis extimis crebriore instructis, ipsa basi absque lana implexa; *superior* paulo brevior, dinervis, nervis viridibus, denticulatis, lateribus complicatis. *Lodiculæ* 2, hyalinæ, imberbes, semibifidæ. *Stamina* 3.

58. Poa abbreviata, panicula simplicissima coarctata subovata, locustis 4-5-floris, glumæ valvulis subæqualibus acutissimis perianthia basi lanata lateribus pubescentia æquantibus, foliis involuto-setaceis.

Desc. *Gramen* 3-4-unciale. *Culmi* foliati, basi sæpe divisi, læves. *Folia* involuta, subsetacea, retrorsum scabra, *vaginæ* fere ad basin usque fissæ, cylindraceæ. *Panicula*

vix semuncialis, ramis alternis, subbifloris, strictis, lævibus, vix denticulatis. *Locustæ* oblongæ, coloratæ. *Glumæ* acutissimæ, valvulis longitudine subæqualibus, carinatis, glaberrimis, *inferiore* manifeste angustiore, paululum breviore, uninervi; *superiore* basi trinervi. *Perianthia* glumas paulo superantia; *valvula inferior* ipsa basi lana implexa parca instructa, carina à basi ad duas tertias partes longitudinis sericea, linea pariter sericea utrique margini approxi-
cclxxxviii] mata, à basi ad eandem fere altitudinem attingenti, intersticiis pubescentibus subsericeis; *superior* dinervis, nervis pectinatim denticulatis, lateribus induplicatis latiusculis. *Lodiculæ* 2. *Stamina* 3, antheris stramineis. *Ovarium* imberbe. *Stigmata* 2, subsessilia, plumosa, hyalina.

59. Poa arctica, panicula effusa: ramis paucifloris capillaribus lævibus locustisque coloratis ovatis 3-4-floris, glumis subæqualibus, perianthii valvula inferiore basi lanata carina lineaque submarginali sericeis: intersticiis pubescentibus, foliis linearibus: ligula subquadrata erosa.

Poa laxa, *Br. in Ross' voy. ed.* 2, *v.* 2, *p.* 192. (*Antè*, *p.* 177.) *Hooker in Scoresby's greenl. p.* 410, non Willdenovii.

Desc. *Gramen* 5-8-pollicare. *Culmi* erecti v. adscendentes, basi quandoque divisi, graciles, læves, foliati. *Folia radicalia* angusto-linearia, canaliculata, culmo aliquoties breviora; *culmea* paulo latiora, plana, marginibus lævibus, denticulis obsoletissimis: *vaginæ* strictæ, striatæ, læves, ipsa basi integra; *ligula* subquadrata, nunc paulo longior quam lata, apice eroso-inciso. *Panicula* sæpius effusa, nunc rara, nunc minus effusa, rarissime subcoarctata, rachi ramisque fuscis, ramis 3-4, semiverticillatis, 1-2-floris, capillaribus, lævibus. *Locustæ* ovatæ v. oblongo-ovatæ, fusco-purpureæ, apicibus valvularum stramineo-fuscis ipsoque margine albo, 3-4-floræ, cum rudimento minuto scarioso longius pedicellato quarti v. quinti; rachi articulatim solubili, per lentem scabriuscula. *Glumæ* subæquivalves, carinatæ, acutæ, fusco-purpureæ, glaberrimæ, carina extra medium obsoletissime denticulata; *inferiore* angustiore, nervis lateralibus obsoletioribus, altero obsoletissimo; *supe-*

riore vix longiore, nervis lateralibus manifestioribus. *Perianthii valvula inferior* oblonga, subcarinata, ipsa basi v. potius ex apice articuli racheos lana longa contortuplicata flosculos subnectenti, carina à basi ad duas tertias partes longitudinis sericea, villis brevibus, supra obsoletissime denticulata, lateralibus à basi ad eandem circiter altitudinem ac portio sericea carinæ pubescentibus, linea intramarginali sericea : *valvula superior* inferiore paulo brevior, dinervis, nervis viridibus, pectinato-ciliatis pilis brevibus, lateribus induplicatis axin fere attingentibus. *Lodiculæ* 2, cuneiformes, semibifidæ dentibus acutis, hyalinæ, imberbes, ovario breviores. *Stamina* 3, antheris stramineis. *Ovarium* oblongum, imberbe. *Styli* 2, brevissimi. *Stigmata* hyalina, laxe plumosa ramis denticulatis.

Obs. Exemplaria nonnulla statura majore, locustis acutioribus, glumis acuminatis perianthia inferiora subæquantibus, foliis latioribus.

Poa laxa, *Willden. sp. pl.* 1, *p.* 386, quam ex eodem monte Silesiæ ubi à b. Haenke detecta fuit habeo à D. Trevirano communicatam, differt statura minore, panicula coarctata, rachi ramisque paniculæ et glumis infra medium viridibus, perianthiis acutioribus lana baseos parciore; locustæ rachi lævi.

Poa flexuosa, *Host gram. austr.* 4, *p.* 15, *t.* 26, quæ similis videtur P. arcticæ panicula effusa et locustæ colore figura et pubescentia, differt paniculæ rachi ramisque viridibus magis divisis scabris, glumarum carinis longitudinaliter denticulatis.

60. Festuca brevifolia, racemo subsimplici erecto, [cclxxxix
flosculis teretibus supra scabriusculis arista duplo longioribus, foliis setaceis vaginisque lævibus : culmeo supremo multoties breviore vagina sua laxiuscula.

Obs. Facies et statura fere F. ovinæ inter quam et F. Halleri media ; priori forsan nimis affinis.

61. Festuca vivipara.

Obs. Nullam observationem habeo de exemplari unico Festucæ cujusdam viviparæ olim viso in herbario D. Sabinè, ulterius examinando.

PLEUROPOGON.

CHAR. GEN. *Locustæ* multifloræ, cylindraceæ. *Gluma* abbreviata, inæquivalvis, mutica. *Perianthii valvula inferior* mutica, obtusa, concava, nervosa, apice scarioso: *superior* nervo utroque lateraliter biseto! *Lodiculæ* distinctæ. *Styli* 2. *Stigmata* plumosa. *Caryopsis* libera, lateribus compressis.

Gramen *elegans*. Folia *plana*, *angusta*, vagina *integra*, *ipso apice tantum fisso*. Racemus *simplex*, *locustis cernuis*, *purpureis*, *nitentibus*. Gluma *valvula inferiore acuta*, *superiore latiore obtusa*. Perianthia *distincta*, *valvula inferiore* 5-7-*nervi*, *superiore lanceolata emarginata*, *pari superiore setarum brevissimo*.

Obs. Genus Glyceriæ proximum, quacum locustis teretibus, perianthiis obtusissimis et vaginis foliorum integris convenit; differt præsertim setis lateralibus nervorum valvulæ superioris perianthii, lodiculis distinctis, stigmatibus haud decompositis, caryopside lateraliter compressa et inflorescentia.

Character fere essentialis in nervis valvulæ superioris perianthii latere setigeris; analoga structura enim vix, quantum scio, in ullo alio gramine obtinet nisi in Uniola latifolia *Mich. am.*, ubi equidem nullis aliis differentiis comitata pro charactere specifico tantum habenda.

62. PLEUROPOGON SABINII.

DESC. *Gramen* 3-unciale usque spithameum, glabrum. *Culmi* erecti, foliati, striati, læves, simplices. *Folia* radicalia angustiora, longiora; culmea linearia, plana, brevia, lævia: *vaginæ* paulo compressæ, striatæ, glabræ, læves, fere ad apicem integræ, ipso apice fisso, marginibus scariosis, suprema folio proprio longior: *ligula* brevissima, rotundata, emarginata. *Spica* racemosa, simplicissima, rachi striato-angulata, lævi, viridi, pedunculis lateralibus glumam vix

superantibus, recurvis, lævibus, indivisis, alternis, distantibus. *Locustæ* subcylindraceæ, cernuæ v. pendulæ, semunciales, purpureæ, nitidæ, per lentem tenuissime pubescentes. *Gluma* bivalvis, nana, inæqualis, membranacea, purpurea, mutica; *valvula inferiore* ovata, acuta; *superiore* obovata, [ccxc
obtusissima, inferiore duplo latiore, paulo longiore. *Perianthia* alterna, distincta. *Valvula inferior* obovato-oblonga, obtusissima, concava, quinquenervis, extus pube brevissima appressa conspersa, apice marginibusque ab apice ad medium albis, scariosis, nervis omnibus infra apicem desinentibus, medio in mucronulum brevissimum, marginem valvulæ vix attingentem producto. *Valvula superior* longitudine fere inferioris, manifeste angustior, elliptico-lanceolata, apice profunde emarginato, lateribus induplicatis, dinervis, nervis brevissime ciliatis, singulis bisetis, *setis* lateralibus, per paria oppositis, *duæ inferiores* infra medium valvulæ ortæ, subulato-filiformes, strictæ, modice patentes, denticulatæ, longitudine circiter dimidii totius valvulæ; *duæ superiores* paulo supra medium valvulæ ortum ducentes, brevissimæ, denticulatæ, mucroniformes, altera quandoque obsoleta. *Lodiculæ* 2, collaterales, approximatæ, brevissimæ, truncatæ, basi leviter cohærentes, sed absque læsione separandæ. *Stamina* 3, filamentis capillaribus, antheris linearibus utrinque semibifidis. *Ovarium* ovatum, imberbe. *Styli* 2, glabri. *Stigmata* laxe plumosa, hyalina, ramis denticulatis, superioribus vix brevioribus. *Caryopsis* libera, lateraliter compressa, ventre angusto-lineari, leviter canaliculato, axi longitudinaliter saturatiore. *Embryo* caryopside triplo brevior.

Obs. Duplex varietas.

α, elatior, subspithamea, antheris stramineis. Tab. D, f. 1—7.

β, 3-4-uncialis, antheris purpureis. Tab. D, f. 8—10.

The specific name is given in honour of Captain Edward Sabine, in whose herbarium, the most extensive formed in the voyage, numerous specimens were found of both varieties of this remarkable grass.

EXPLICATIO Tabulæ D.[1]

Pleuropogon Sabinii. 1. Varietatis α, planta magnitudine naturali. 2. ejusd. locusta cum pedunculo et portione racheos magis aucta. 3. perianthium clausum articulo racheos insidens, auctius. 4. id. expansum, pariter auctum. 5. valvula superior perianthii facie visa ad id. augment. 6. pollen. 7. flosculus perianthio orbatus exhibens stamina pistillum et lodiculas auct. uti 4 et 5. 8. Var. β, planta mag. natur. 9. ejusd. locusta cum pedunculo ad augm. id. ac. 2. 10. perianthium expansum genitalia et lodicularum alteram exhibens ad augm. n. 4.

DUPONTIA.

Char. Gen. *Gluma* subæquivalvis, scariosa, concava, mutica, locustam 2-3-floram subæquans. *Perianthia* mutica, scariosa, (basi barbata,) altero pedicellato; valvulis integris, inferiore concava. *Lodiculæ* 2. *Ovarium* imberbe. *Stigmata* subsessilia. *Caryopsis* - - - -

Gramen *glabrum, erectum.* Folia *linearia, plana, vaginis semifissis, basi integra.* Panicula *simplex, coarctata, fusco et purpurascenti varia, pedicellis cum locustis continuis, perianthiis separatim solubilibus.*

ccxci] *Obs.* Ad Deschampsiam proxime accedit hocce genus; distinguitur perianthiis muticis, valvulis integris nec dentatis. Cum Catabrosa, facie diversissima, convenit pluribus notis, differt glumis locustam subæquantibus, perianthiis basi brevè barbatis. A Poa diversum locustis haud compressis, glumis perianthiisque concavis nec carinatis. Ad confirmandum genus caryopsis desideratur.

This genus is named in honour of Monsieur Dupont, of Paris, author of a valuable essay on the Sheath of the leaves of Grasses, and of observations on the genus Atriplex.

[1] *See Note at p.* 187.

63. Dupontia Fisheri.

Desc. *Gramen* 6-10-unciale, erectum. *Culmi* simplices, foliati, læves, glaberrimi. *Folia radicalia* et inferiora culmi canaliculata, angusto-linearia, acuta, lævia, 2-3-uncialia, *vaginis* strictis, scariosis, vix ad medium fissis; *culmea* 1-2 superiora breviora, plana, lævia, vaginis propriis laxiusculis foliaceis ultra medium fissis longiora: *ligula* mediocris, obtusa, subtruncata, imberbis. *Panicula* coarctata, spiciformis, basi quandoque interrupta, purpureo-fusca, nitens, sesquiuncialis—biuncialis, ramis subgeminatis, paucifloris, pedicellisque lævibus cum locusta continuis. *Locustæ* ovatæ, bifloræ, cum rudimento clavato setuliformi tertii flosculi, nunc trifloræ flore tertio completo, nunc bifloræ absque tertii rudimento. *Gluma* bivalvis, subæqualis, mutica, glaberrima, purpurascens, subnitens, margine pallido scarioso, longitudine locustæ. *Valvulæ* concavæ nec carinatæ, oblongo-lanceolatæ, *inferior* paulo angustior, acuminata v. acutissima, uninervis; *superior* semi-trinervis, medio paulo infra apicem lateralibus longe intra marginem evanescentibus. *Perianthia* subconformia; *inferius* intra glumam subsessile, à pedicello brevissimo separabile; *superius* cum apice paulo dilatato pedicelli brevis articulatum, facile solubile; utriusque *valvula inferior* ovata mutica obtusa, vix unquam acuta, integra, concava, ipsa basi pilis brevibus strictis albis barbata, et à basi fere ad medium pilis brevioribus strictis subadpressis subsericea, trinervis, nervis lateralibus intra marginem evanescentibus, medio paulo infra apicem desinente: *superior* longitudine inferioris, manifeste angustior, lineari-oblonga, glaberrima, dinervis, nervis brevibus, intersticio lineari concaviusculo. *Lodiculæ* duæ, distinctæ, collaterales, membranaceæ, hyalinæ, subovatæ, v. cuneatæ, apice eroso-dentato, ovario longiores. *Stamina* 3, filamentis distinctis, capillaribus, antheris fusco-purpureis, linearibus utrinque bifidis. *Ovarium* ovale, glabrum. *Stigmata* 2, subsessilia, hyalina, dense plumosa, ramis apicem versus brevioribus.

Obs. The specific name is that of Mr. Fisher, whose herbarium contained the most complete series of specimens of this grass.

64. Deschampsia brevifolia, panicula coarctata lanceolata: pedicellis lævibus, locustis 2-3-floris, arista stricta valvulam subæquante, foliis involutis: caulinis abbreviatis.

Desc. *Gramen* 3-5-unciale, glabrum. *Culmi* simplices, erecti, foliati. *Folia* inferiora involuto-subulata, stricta, uncialia—sesquiuncialia; *vaginis* strictis, folio brevioribus, ipsa basi integra: *ligula* oblonga, lacinulata; *supremum* brevissimum, vagina elongata, laxiuscula, ligula breviore. *Panicula* coarctata, lanceolata v. oblonga, fusco-purpurascens, scariosa, ramis semiverticillatis. *Locustæ* bifloræ, raro trifloræ, semper cum rudimento pedicelliformi flos-
ccxcii] culi alterius. *Gluma* subæquivalvis, mutica, acuta, valvulis lanceolatis, concavis, acutissimis, scariosis, disco purpurascenti, limbo pallido, uninerviis, locusta paulo brevioribus. *Perianthia* subuniformia, scarioso-membranacea, separatim solubilia, inferius sessile; *valvula inferior* ipsa basi barbata, pilis brevibus, strictis, albis, cæterum glabra, concava, subquinquenervis, nervis omnibus lævibus, lateralibus obsoletis, apice eroso-multidentato, dorso sæpius infra medium aristata, arista setacea, recta, denticulata, valvulam ipsam vix vel paulo superanti: *superior* longitudine inferioris, angustior, dinervis, apice bidentato, quandoque semibifido. *Lodiculæ* 2, collaterales, hyalinæ, imberbes, acutæ, ovario longiores. *Stamina* 3, antheris purpureis, utrinque bifidis. *Ovarium* glabrum. *Stigmata* 2, sessilia, hyalina, dense et breve plumosa. *Flosculus superior* pedicello barbato quocum articulatus insidens, paulo minor, arista valvulæ inferioris medio vel supra medium dorsi inserta. *Rudimentum* flosculi tertii *setula* est extus longitudinaliter barbata, clavula scariosa minutissima terminata.

β. Perianthia mutica.

Hujus quatuor exemplaria tantum visa à varietati *α*. facie paulo diversa folio supremo longiori.

TRISETUM.

Triseti species *Palis. agrost. p.* 88, charactere reformato.

Char. Gen. *Locustæ* 2-5-floræ, ancipites. *Gluma* carinata, membranacea, subæquivalvis. *Perianthii valvula inferior* carinata apice bidentata v. biseta, dorso (supra medium) aristata. *Caryopsis* libera, exsulca, lateraliter compressa.

Gramina *cæspitosa; vaginis longitudinaliter fissis.* Panicula *sæpe coarctata, aristis arcuato-patulis.*

Obs. A Deschampsia differt locustis ancipitibus, glumis carinatis, perianthii valvula inferiore carinata apice attenuato bidentato v. biseto, caryopside lateraliter compressa. Ab Avenis plerisque glumis perianthiisque carinatis; ab omnibus caryopside exsulca et lateribus compressis.

65. Trisetum subspicatum, *Palis. agrost. p.* 88. Trisetum airoides, *Roem. et Sch. syst.* 2, *p.* 666, exclus. syn. Wulfen et Host. *Richardson in Franklin's journ. p.* 731.

Aira spicata, *Linn. sp. pl. ed.* 2, *p.* 95, *fl. lapp. n.* 47. *Flor. Dan. t.* 228, mala. *Gunn. norv. n.* 422. *Wahlenb. lapp. p.* 33.

Aira subspicata, *Linn. syst. nat. ed.* 12, *v.* 2, *p.* 91. *Willden. sp. pl.* 1, *p.* 377. *Pers. syn.* 1, *p* 77. *Zoëg. pl. island. in Olafs. reise* 2, *p.* 234. *Giesecke greenl. in Brewster's Edin. encyclop.*

HIEROCHLOE.

Hierochloe *Gmel. sib.* 1, *p.* 100. *Br. prodr. flor. nov. holl. p.* 208. *Trin. agrost. p.* 130.

Hierochloa et Toresia, *Palis. agrost. p.* 62 *et* 63.

Char. Gen. *Gluma* subæquivalvis, locustam trifloram [ccxcii æquans. *Perianthia* bivalvia, *lateralia* mascula, triandra; *terminale* hermaphroditum, diandrum.

Obs. Relationem veram Anthoxanthi ad Hierochloem, in prodr. flor. nov. holl. p. 209, primum indicatam, optime confirmat planta Javanica intermediæ structuræ à D. Horsfield detecta; in hac enim perianthium *lateralium inferius* masculum bivalve, *superius* univalve, neutrum: *terminale* hermaphroditum. Hujus novi generis (Ataxia) habitus potius est Anthoxanthi, quocum etiam gluma inæquivalvi quadrat.

Cum Hierochloe characteribus nonnullis convenit Arthrochloa *nob.* (Holcus *Palis. Trinii,* et *Wahlenb.* non *Linnæi gen. et sp. pl. ed. prima,* nec *Schreberi* nec *Gærtneri ;*) quæ tamen facile distinguitur ab hoc genere uti et ab Aira et Arrhenathero, gluma cum apice pedicelli articulata et unà cum locusta decidua.

66. HIEROCHLOE ALPINA, *Roem. et Sch. syst.* 2, *p.* 515. *Br. in Ross' voy. ed.* 2, *vol.* 2, *p.* 194. (*Antè, p.* 178.) *Richardson in Franklin's journ. p.* 731.

Holcus alpinus, *Swartz in Schrad. neue journ.* 2, *st.* 2, *p.* 45, *t.* 3, *Wahlenb. lapp. p.* 31, *t.* 2. *Svensk bot.* 438. *Flor. Dan.* 1508. *Giesecke greenl. in Brewster's Edin. encyclop.*

67. HIEROCHLOE PAUCIFLORA, racemo simplici, flosculo masculo superiore brevissime setigero, foliis culmi brevissimis ; radicalibus involutis.

DESC. *Gramen* 3-5-unciale. *Radix* repens. *Culmi* erecti, infra foliati supra nudi, striati. *Folia radicalia* subulata, marginibus involutis, culmo aliquoties breviora ; *culmea* abbreviata latè subulata, marginibus inflexis, vaginis suis laxiusculis multoties breviora. *Racemus* erectus, simplex vel subsimplex, pauciflorus, pedicellis lævibus. *Locustæ* ovatæ, acutæ, trifloræ. *Glumæ* bivalves, scariosæ, ovatæ, concavæ, acutiusculæ, glaberrimæ, locustam subæquantes, valvula inferiore manifeste minore. *Flosculi laterales* masculi, triandri, bivalves, chartacei, *valvula inferior* ovata, concava, marginibus infra medium nudiusculis supra omnino nudis, flosculi superioris mox sub apice emarginato setigera, seta brevissima stricta valvulam vix superante ; flosculi inferioris mutica v. per-obselete setigera ; utriusque valvula superior angustior, linearis, dinervis, semibifida. *Flosculus terminalis* hermaphroditus, diander, muticus : *valvula inferior* concava, quinquenervis, extra medium dorso lateribusque pilosiusculis, chartacea, fusca, apice scarioso ; *superior* linearis, hyalina, glabra, acuta, indivisa, uninervis. *Lodiculæ* 2, collaterales, lanceolatæ, acuminatæ, hyalinæ, ovario longiores. *Ovarium* glabrum. *Styli* 2. *Stigmata* alba, dense plumosa.

ACOTYLEDONES. [ccxciv

MUSCI.

68. Polytrichum propinquum, caule simplici elongato, foliis margine serrulatis dorso lævibus.

Obs. Species, absque fructificatione haud determinanda, à Polytricho communi satis diversa videtur.

69. Polytrichum hyperboreum, caule ramoso, foliis piliferis marginibus induplicatis discum (totum lamelliferum) operientibus, capsula tetragona apophysata.

Desc. *Caules* sæpius ramosi ramis fastigiatis, nunc simpliciores innovatione una alterave divisi. *Folia* è dilatata semivaginanti basi subulata, madore patula, siccitate appressa, disco toto lamellifero; marginibus latis, induplicatis, integerrimis, membranaceis, à basi dilatata usque ad apicem altero alterum equitante; *pilus* apicis hyalinus folio aliquoties (2-3-plo) brevior, per lentum denticulatus, strictus. *Masculi flores* disciformes, in distincto individuo sæpe minore. *Seta* nitens caulibus procerioribus (biuncialibus) subsimplicibus brevior, fastigiato-ramosos superans v. æquans. *Capsula* erecta v. inclinans tetragona, angulis in aciem attenuatis; apophysis angulata angustior. *Operculum* hemisphæricum cum mucronulo brevi. *Peristomium* dentibus 64. *Epiphragma* demum separabile. *Calyptra exterior* è villis dense implexis.

Obs. Duplex varietas.

a, caulibus fastigiato-ramosis setam vix æquantibus.

β, caulibus innovando subramosis seta longioribus.

Hæc P. pilifero proxima ab eodem differt caulibus elongatis innovando ramosis, pilis folio aliquoties brevioribus.

70. Polytrichum brevifolium, caule ramoso, foliis serrulatis muticis madore erectis siccitate appressis, capsula inclinata obovata exapophysata.

Desc. *Muscus* sesquiuncialis. *Caules* divisi, ramis fastigiatis. *Folia* è basi dilatata semimembranacea subulata, extra medium serrulata, acuta, mutica, disco toto lamellifero, dorso lævi. *Seta* lævis, pallida. *Capsula* lævis, cernua, inæquilatera. *Operculum* conico-hemisphæricum, rostro subulato recurvo diametrum baseos vix æquante. *Peristomii* dentes 40, æquidistantes, intersticiis angustiores. *Epiphragma* crassiusculum. *Calyptra exterior* è villis arcte implexis.

Obs. Muscus, cujus tria tantum exemplaria à nobis visa in herbario D. Ross, nimis forsan affinis P. alpino *L.*

ccxcv] 71. Polytrichum septentrionale, *Sw. in act. holm.* 1795, *p.* 270. *Musc. suec. p.* 107, *t.* 9, *f.* 18. *Menzies in Linn. soc. transact.* 4, *p.* 82, *t.* 7, *f.* 5.

Obs. In herbario D. Fisher absque fructificatione visum, ideoque dubium.

72. Polytrichum lævigatum, *Wahlenb. lapp. p.* 349, *t.* 22. *Hooker musc. exot. t.* 81.

Catharinea lævigata, *Bridel mant. p.* 202.

Catharinea glabrata, *Hooker isl.* 2, *p.* 340, *et* 1, *p.* 24.

Obs. Peristomii dentes sæpius quantum determinare potui 16, quandoque 32, lineares, acutiusculi, hyalini, per lentem longitudinaliter striati, striis sæpius paulo flexuosis, in hemisphærium conniventes ; dum 16 approximati intersticiis angustissimis, parum inæquales, latioribus nunc bidentatis ; dum 32 æquales, intersticiis manifestis. *Epiphragma* hyalinum, diametro longitudinem dentis vix æquante. *Membrana interior capsulæ* exteriori approximata, intus lævis absque processubus plicisve. *Columella* libera, angulata, longitudine fero capsulæ. *Capsula* per lentem modice augentem manifeste areolata.

73. Hypnum nitens, *Hedw. sp. musc. p.* 255. *Smith brit.* 3, *p.* 1316. *Engl. bot.* 1646. *Musc. brit. p.* 100. *Wahlenb. lapp. p.* 381.

74. Hypnum cordifolium, *Hedw. stirp. crypt.* 4, *p.* 97,

t. 37. *Sp. musc. p.* 254. *Smith brit.* 3, *p.* 1318. *Engl. bot.* 1447. *Musc. brit. p.* 107.

75. Hypnum aduncum, *Linn. sp. pl. ed.* 2, *p.* 1592. *Smith brit.* 3, *p.* 1327. *Hedw. stirp. crypt.* 4, *p.* 62, *t.* 24. *Sp. musc. p.* 295.

76. Leskia rufescens, *Schwaegr. suppl.* 1, *sect. post. p.* 178, *t.* 86.

Hypnum rufescens, *Dicks. crypt. fasc.* 3, *p.* 9, *t.* 8, *f.* 4. *Smith brit.* 3, *p.* 1316. *Engl. bot.* 2296. *Musc. brit. p.* 99.

77. Mnium turgidum, *Wahlenb. lapp. p.* 351, *t.* 23. *Schwaegr. suppl.* 1, *sect. post. p.* 123, *t.* 77. *Br. in Ross' voy. ed.* 2, *vol.* 2, *p.* 194. (*Antè, p.* 178.) *Richardson in Franklin's journ. p.* 756.

78. Timmia megapolitana, *Hedw. stirp. crypt.* 1, *p.* 83, *t.* 31. *Sp. musc. p.* 176. *Schwaegr. suppl.* 1, *sect. post. p.* 84. *Richardson in Franklin's journ.* 756. Timmia cucullata, *Michaux am.* 2, *p.* 304.

79. Bryum rostratum, *Schrad. spicil. p.* 72. *Smith brit.* 3, *p.* 1369. *Engl. bot.* 1745. *Musc. brit. p.* 126, *t.* 30. Mnium rostratum, *Schwaegr. suppl.* 1, *sect. post. p.* 136, *t.* 79.

Obs. Muscus hicce, necnon sex proxime præcedentes absque fructificatione tantum visi.

80. Bryum calophyllum, foliis ovatis obtusis con- [ccxcvi
cavis: marginibus simplicibus integerrimis, capsulis obovatis pendulis.

Desc. *Cæspites* densi. *Caules* innovationibus continuis divisi, 2-5 unciales, vetusti tomento radicali copioso et foliis emarcidis tecti. *Rami annotini* fastigiati, basi tantum tomento radicali parciore instructi, supra glabri. *Folia* uniformia, sparsa, approximata, ovata v. subovalia, modice concava, obtusa, mutica, marginibus simplicibus nec recurvis nec incrassatis, areolis subrotundis, uniformibus, nervo valido, apicem folii attingenti absque mucronulo excurrenti, sæpius purpurea, quandoque viridia, madore patenti-erecta, siccitate appressa et paulo undulata. *Seta* terminalis, ramos annotinos superans, castanea, lævis, apice arcuato-recurvo. *Capsula* obovata, basi acutiuscula, vix attenuata, lævis. *Operculum* concolor, hemisphæricum, papilla minuta. *Peristomium* duplex, *exterius* dentibus 16, rufescentibus acumine pallidiore, tranversim striatis ; *interius* album, è membrana lata leviter carinata, terminata ciliis 16, imperforatis, cum dentibus exterioris alternantibus, intersticiis subdenticulatis.

Obs. Peristomii structura Pohliæ accedit.

81. Pohlia bryoides, foliis ovato-lanceolatis acuminatis integerrimis margine recurvis, capsulis pyriformi-oblongis, operculo conico, floribus masculis capitato-discoideis.

Desc. *Cæspites* densi. *Caules* innovatione continuo ramosi, infra tomento radicali castaneo-rufo reliquiisque foliorum tecti. *Folia* læte viridia, ovato-lanceolata, acuminata, nervo valido, in acumen excurrenti, marginibus integerrimis angustissime recurvis, areolis parvis oblongo-trapezoideis. *Masculi Flores* monoici, ramos annotinos terminantes, gemmaceo discoidei, foliis perigonialibus exterioribus erectis, intimis nanis. *Antheræ* numerosæ, cylindraceæ, brevissime pedicellatæ. *Paraphyses* filiformes, articulatæ. *Feminei Flores* terminales ; *vaginula* capsulæ maturæ pistillis abortivis numerosis paraphysibusque fere ad apicem truncatum stipata. *Seta* mediocris, lævis, fusca, apice arcuato. *Capsula* pendula, fusca, lævis, oblongo-pyriformis, basi attenuata in apophysim obconicam ipsa theca breviorem. *Operculum* hemisphærico-conicum, capsula quandoque paulo saturatius. *Annulus* latiusculus, striatus. *Peristomium* duplex : *exterius* dentibus 16, acuminatis, integerrimis, transversim striatis, fusco-rufescentibus, acumine pallido ; *interioris* membrana vix carinata, ciliis 16 cum dentibus exterioris alternantibus, absque intermediis minoribus, cum exteriore diu cohærens sed demum liberum.

82. Pohlia arctica, foliis (viridibus) ovato-lanceolatis

acuminatis: marginibus integerrimis recurvis, capsulis pyriformi-oblongis, operculo hemisphærico, floribus hermaphroditis.

Obs. *Muscus* per singula fere puncta præcedenti simillimus, præter flores hermaphroditos et operculum hemisphæricum; ambo forsan ad unam eandemque speciem polygamam pertinentes. *Flores* gemmacei, terminales, foliis perichætialibus interioribus nanis. *Antheræ* numerosæ, cum pistillis vix paucioribus intermistæ, et cum horum abortientibus paraphysibusque filiformibus vaginulâ capsulæ maturæ fere ad ejusdem apicem insidentes. *Peristomium interius* structura præcedentis pariterque cum exteriori [ccxcvii
diu cohærens, demum vero liberum et in omni statu separabile. Huic et præcedenti valde affinis videtur Ptychostomum compactum *Hornschuch*, et *Schwaegr. suppl.* 2, *sect.* 1, *p.* 56, *t.* 115, cui peristomium interius cum exteriore arctius cohæret. Hujus generis? alteram speciem arcticam habeo, *Ptychostomum pulchellum*, capsula sphærico-obovata, operculo hemisphærico mutico, dentibus peristomii exterioris apice liberis basi mediante membrana (peristomio interiore) cohærentibus, foliis ovato-lanceolatis acuminatis integerrimis.

83. Pohlia purpurascens, foliis (purpurascentibus) ovato-lanceolatis acutissimis: marginibus integerrimis recurvis, capsulis pyriformi-oblongis, operculo hemisphærico obtuso, floribus hermaphroditis.

Obs. Præcedentis forsan varietas, vix distinguenda nisi notis supra datis.

Propter peristomii interni structuram hanc cum duabus præcedentibus ad Pohliam retuli, facies tamen potius Bryi est, et omnes B. cæspiticio quam maxime affines.

84. Trichostomum lanuginosum, *Hedw. stirp. crypt.* 3, *p.* 3, *t.* 2. *Sp. musc. p.* 109. *Schwaegr. suppl.* 1, *sect.* 1, *p.* 149. *Smith brit.* 3, *p.* 1240. *Engl. bot.* 1348. *Turner musc. hibern. p.* 38. *Musc. brit. p.* 60, *t.* 19. *Hooker scot. par.* 2, *p.* 134. *Wahlenb. lapp. p.* 329. *Richardson in Franklin's journ. p.* 755.

Racomitrium lanuginosum, *Brid. mant. p.* 79.
Obs. Specimina pauca et absque fructificatione.

85. Didymodon capillaceum, *Schrad. spicil. p.* 64. *Sw. in act. holm.* 1795, *p.* 237. *Musc. suec. p.* 28. *Roth. germ.* 3, *p.* 199. *Web. et Mohr. tasch. p.* 155. *Schkuhr deut. moos. p.* 66, *t.* 29. *Wahlenb. lapp. p.* 314. *Carpat. p.* 336. *Voit musc. herbip. p.* 34. *Musc. brit. p.* 67, *t.* 20. *Brid. mant. p.* 100. *Hooker scot. par.* 2, *p.* 136. *Richardson in Franklin's journ. p.* 755.

Swartzia capillacea, *Hedw. stirp. crypt.* 2, *p.* 72, *t.* 26.

Cynontodium capillaceum, *Hedw. sp. musc. p.* 57. *Schumach. sælland.* 2, *p.* 40.

Cynodontium capillaceum, *Schwaegr. suppl.* 1, *sect.* 1, *p.* 114.

Trichostomum capillaceum, *Smith brit.* 3, *p.* 1236. *Engl. bot.* 1152. *Turner musc. hibern. p.* 35.

Bryum capillaceum, *Dicks. crypt. fasc.* 1, *p.* 4, *t.* 1, *f.* 6.

Bryum tenuifolium, *Villars dauph.* 4, *p.* 868.

Bryum n. 1806. *Hall. hist.* 3, *p.* 44, *t.* 45, *p.* 1.

Obs. Duas varietates à Melville Island habeo, quarum.

α. Statura et foliis laxiusculis cum D. capillaceo europæo convenit, paululum differt capsulis ovalibus nec oblongis.

β. Statura humiliore, foliis strictioribus et brevioribus; media quasi inter D. capillaceum vulgare et D. subulatum *Schkuhr deut. moos. p.* 65, *t.* 28, quod ad eandem speciem pertinere videtur.

ccxcviii] In utraque varietate atque in D. capillaceo *Richardson, l. c.* flores monoicos, masculis gemmiformibus alaribus prope apicem ejusdem rami cum femineo gemmiformi, necnon annulum manifestum, in D. capillaceo, jamjam à Voitio l. c. notatum, et dentes peristomii 16 bipartitos cruribus transversim connexis observavi.

86. Barbula leucostoma, caule subsimplici, foliis ovato-lanceolatis mucronulatis integerrimis, capsula cylindracea erecta, operculo conico, peristomii dentibus obliquis apice tortis.

Desc. *Muscus* cæspitosus, semuncialis. *Caules* breves, dense foliati, sæpius indivisi, quandoque parum ramosi. *Folia* mucrone brevissimo, minute areolata, marginibus anguste revolutis, nervo valido, siccitate adpressa et parum torta. *Seta* caule longior, lævis, fusca. *Capsula* lævis, æquilatera. *Operculum* conicum, acutum, paulo inclinans, capsula dimidio brevius, tenuissimè spiraliter striatum. *Peristomium* album, dentibus 32, filiformibus, per paria approximatis, dimidioque inferiore trabeculis connexis, supra distinctis, apicibus parum tortis. *Calyptra* lævis.

Obs. Inter Barbulam et Didymodon media.

87. Syntrichia ruralis, *Web. et Mohr tasch. p.* 215. *Voit mus. herbip. p.* 52. *Brid. mant. p.* 98.

Tortula ruralis, *Smith brit.* 3, *p.* 1254. *Engl. bot.* 2070. *Turner musc. hibern. p.* 50. *Sw. musc. suec. p.* 39. *Schwaegr. suppl.* 1, *sect.* 1, *p.* 137. *Wahlenb. carpat. p.* 338. *Musc. brit. p.* 31, *t.* 12. *Hooker scot. par.* 2, *p.* 127. *Richardson in Franklin's journ. p.* 755.

Barbula ruralis, *Hedw. sp. musc. p.* 121. *Wahlenb. lapp.* 318.

Obs. Specimina duo tantum et sine fructificatione.

88. Syntrichia mucronifolia, caule ramoso, foliis ovato-oblongis siccitate adpressis: pilo integerrimo latitudine folii breviore, capsula cylindracea inæquilatera erecta duplo longiore; operculo subulato-conico.

Tortula mucronifolia, *Schwaegr. suppl.* 1, *sect.* 1, *p.* 136, *t.* 35? *Wahlenb. lapp. p.* 317?

Desc. *Muscus* v. cæspitosus v. aliis intermistus. *Caules* erecti, breves, semper ramosi, ramis fastigiatis, dense foliati. *Folia* concava, marginibus integerrimis, infra medium leviter recurvis, minute areolatis, areolis baseos paulo laxioribus, nervo valido in pilum integerrimum excurrente, madore erecto-patentibus, siccitate imbricatis adpressis nec contortis, pilo parum flexo. *Seta* capsula haud duplo longior, concolor, siccitate tortilis. *Capsula* saturate castanea, lævis. *Operculum* badium, per lentem pluries augentem spiraliter striatum, dimidium capsulæ vix æquans.

Peristomii membrana alba, pulchre reticulata, longior ciliis contortis. *Calyptra* novella tantum visa, lævis.

Obs. Syntrichia subgenus tantum esse videtur Barbulæ (s. Tortulæ), cujus dentes è membrana angusta ortum ducunt; et in speciebus omnibus utriusque quas investigavi operculum spiraliter striatum est.

ccxcix] De synonymis supra citatis *S. mucronifoliæ* haud omnino certus sum, figura tamen Schwaegrichenii bene respondet, et descripto Wahlenbergii in omnibus convenit nisi longitudine cuspidis foliorum inferiorum.

89. Encalypta ciliata, *Hedw. sp. musc. p.* 61? *Schwaegr. suppl.* 1, *sect.* 1, *p.* 59? *Smith brit.* 3, *p.* 1181? *Engl. bot.* 1418? *Wahlenb. lapp.* 311? *Musc. brit. a, p.* 35, *t.* 13?

Leersia ciliata, *Hedw. stirp. crypt.* 1, *p.* 49, *t.* 19?

Obs. Exemplaria nonnulla Encalyptæ speciei in herb. D. Sabine olim visa ad hanc, ni fallor, pertinent; posthac determinanda.

90. Gymnostomum obtusifolium, foliis oblongo-ovatis obtusis integerrimis, capsula oblonga duplo longiore operculo conico columellæ adnato.

Desc. *Caules* ramosi, dense foliati. *Folia* concava, infra laxiusculè supra medium minutè reticulata, marginibus planis, nervo vix apicem attingenti, madore erecto-patula, siccitate appressa et parum flexa. *Seta* fusca, lævis, caule longior. *Capsula* erecta, lævis, fusca, reticulata. *Operculum* brevè conicum, cum columella cylindracea diu cohærens.

APLODON.

Char. Gen. *Peristomium* simplex: *dentibus* 16, æquidistantibus, indivisis, reflexilibus. *Capsula* apophysata, erecta. *Calyptra* lævis. *Flores* terminales: *masculi* discoideo-capituliformes.

Obs. Subgenus Splachni, à quo differt solummodo dentibus 16 æquidistantibus, et forsan columella capsulæ maturæ inclusa. Sed quoniam axis pellucidus dentis cujusvis compositionem ejusdem indicat, ad Systylium (quod Splachni alterum subgenus), dentibus 16, æquidistantibus, bipartitis, planè accedit; in hoc enim cohærentia operculi cum columella, ex analogia cum Gymnostomis quibusdam, pro charactere specifici tantum valoris habenda sit; et ad eandem structuram approximatio indicata est in *Splachno tenue* et *longicollo*, in quibus columella tota apice subulato persistit, quamvis ab operculo cito soluta est. Transitus ab Aplodonte ad Splachnum facilis est per S. longicollum (*Dicks. crypt. fasc.* 4, *p.* 4, *t.* 10, *f.* 9, Americæ occidentali nec Scotiæ indigenum), cui dentes vix manifeste per paria approximati, qua nota differt à S. tenue valde affine sed dentibus geminatis reflexilibus instructo. Ad Aplodontem proxime accedit Weissia Splachnoides *Schwaegr.* (Cyrtodon *nob.*, alterum subgenus Splachni quasi constituens), diversa præsertim dentibus erectis apicibus incurvis, ideoque S. Frœlichiano dentibus erectis sed geminatis affinis.

91. Aplodon Wormskioldii.

Splachnum Wormskioldii, *Hornem. in Flor. Dan.* 1659. *Schwaegr. suppl.* 2, *sect.* 1, *p.* 27, *t.* 108.

a. Folia acuminata. [ccc

Desc. *Muscus* læte virens, dense cæspitosus. *Caules* 1-3-unciales, innovationibus repetitis ramosi, infra tomento radicali castaneo foliisque emarcidis tecti; ramis annotinis herbaceis, viridibus, foliatis. *Folia* alterna, descendendo remotiora, læte viridia, ovato-lanceolata, acuminata, integerrima, laxè reticulata, nervo tenui, ad ortum acuminis concoloris, diametrum transversum folii vix æquantis, desinenti. *Masculus Flos* discoideo-capituliformis, ramum paucifolium ejusdem cum femineo vel distincti caulis terminans; *foliis perigonialibus* caulinis subconformibus, infra conniventibus coloratis, apicibus patulis viridibus. *Antheræ* numerosæ, brevissimè pedicellatæ, cylindraceæ. *Paraphyses* plures, lutescentes, articulis sursum crassioribus brevioribusque, ultimo obtuso. *Pistilla* nulla. *Femineus Flos* terminalis, masculo angustior, *foliis perichætialibus* rameis conformibus et concoloribus. *Pistilla* 3-5; *paraphysibus* paucissimis; *antheris* nullis. *Seta* ramum fructiferum subæquans, herbacea, sæpissime viridis, etiam post lapsum operculi, quandoque demum pallide fusca. *Vaginula* laxiuscula, dilute fusca, ore nigro-castaneo, quandoque inæquali, basi pistillis abortientibus stipata. *Calyptra* glabra, lævis, subcampanulata, sed altero latere fere ad apicem usque fissa, capsulâ adultâ brevior. *Apophysis* obovata, basi vix attenuata, capsulam crassitie subæquans, nunc paulo amplior, concolor, demum pallida et alte corrugata. *Capsula* erecta, cylindraceo-obovata, lævis, castanea, stomate haud coarctato et quandoque dentibus deciduis nudo, deoperculata apophysi brevior. *Peristomium* simplex, dentibus 16, æquidistantibus, lato-subulatis, indivisis, axi longitudinali semipellucido, transversim striatis, siccitate arcte reflexis, madore conniventibus, semisiccatis patulis. *Columella* capsula matura brevior, apice simplici. *Operculum* depresso-hemisphæricum, obtusissimum, altero latere stomati diutius adhærens.

β. Folia acutiuscula.

Obs. Ab α differt, præter folia absque acumine et quandoque obtusiuscula, caulibus brevioribus vix uncialibus, stomate patentiore.

Planta groenlandica inter has duas varietates quasi media, cum α. foliis acuminatis conveniens; ad β. habitu propius accedens.

SPLACHNUM.

Linn. Hedw.

Char. Gen. *Peristomium* simplex: *dentibus* (reflexilibus) v. 8, geminatis (coalitione nunc indivisis): v. 4, quaternatis. *Capsula* erecta, apophysata. *Calyptra* glabra, lævis. *Flores* terminales: *masculi* (cum v. absque pistillis sterilibus), discoideo-capituliformes.

Obs. In *S. octoblepharo* Insulæ Diemeni et *magellanico* peristomium octodentatum, sed dentium striæ longitudinales

semipellucidæ eorundem compositionem indicant. In *S. angustato*, *arctico* et *propinquo* peristomii dentes quaternatim approximati et basi coadunati. Dum *S. Frœlichianum*, [ccci
et forsan *Wulfenianum*, capsula inclinata et dentibus erectis à Splachnis genuinis distinguitur et subgenus efformat.

92. Splachnum vasculosum, *Linn. sp. pl. ed.* 2, *p.* 1572, exclus. syn. Buxb. *Hedw. stirp. crypt.* 2, *p.* 44, *t.* 15, optime, *Sp. musc. p.* 53. *Schkuhr deut. moos. p.* 41, *t.* 17, icone à supra citata Hedwigii mutuata. *Schwaegr. suppl.* 1, *sect.* 1, *p.* 51. *Wahlenb. lapp. p.* 308. *Musc. brit. p.* 21, *t.* 31, bene. *Hooker scot. par.* 1, *p.* 125.

Desc. *Caules* innovando subramosi, unciales, laxe foliati, inferne fibras purpureas ramosas supra-axillares nonnullas exserentes. *Folia* alterna, orbiculato-obovata, obtusissima, parum concava, basi angustata, semiamplexicaulia, marginibus integerrimis planis, nervo mox infra apicem evanescenti; *perichætialia* similia, intimis 2-3 exceptis minoribus ovatis acutiusculis. *Seta* caulem subæquans, castanea, lævis. *Vaginula* basi stipata pistillis pluribus abortivis. *Apophysis* subsphærica vel obovata, capsula duplo amplior, semisiccata rugosa, nigro-fusca. *Capsula* cylindracea, lævis, minute reticulata, fusca. *Peristomium* dentibus 16, per paria approximatis, sæpiusque ad medium, quandoque fere ad apicem, connatis, singuli axi pellucentiori tenuissimo, omnes è basi angusta annulari orti, arcte reflexiles dorso capsulæ appressi. *Columella* cylindracea, longitudine thecæ, apice dilatato, plano-depresso. *Masculi Flores* caulem distinctum paucifolium ejusdem cæspitis terminantes, capitato-discoidei; *foliis perigonialibus* extimis obtusiusculis, interioribus longioribus, è basi latiore lutescenti conniventi patulis, lanceolatis apice angustatis, integerrimis. *Antheræ* numerosæ, viginti plures. *Paraphyses* numerosissimæ, antheris longiores, subclavatæ, articulis superioribus crassioribus brevioribusque. *Pistilla* nulla.

Obs. Ab exemplaribus in Scotiæ montibus à D. Hooker lectis hoc paulo tantum differt foliis remotioribus et seta longiore.

93. Splachnum arcticum, peristomii dentibus quaternatim approximatis, apophysi obconica capsula clausa angustiore deoperculata latiore, operculo conico-hemisphærico, floribus masculis sessilibus, seta perichætium bis superante, foliis ovato-lanceolatis concavis cuspidatis integerrimis.

Desc. *Muscus* dense cæspitosus. *Caules* innovationibus ramosi, sesquiunciales, infra foliis vetustis emarcidis tomentoque radiculoso copioso tecti. *Rami* annotini læte virides, foliati, basin versus foliis rarioribus et brevioribus. *Folia* lanceolata-ovata, concava, integerrima, cuspidata, cuspide concolori fere ½ longitudine laminæ, laxe reticulata, læte viridia. *Femineus Flos* gemmiformis, angustus. *Pistilla* 3-5, filis succulentis, paucis, hyalinis; *staminibus* nullis. *Seta* longitudine fere rami annotini, parum angulata, lævis, castanea, capsula tota, apophysi simul sumpta, duplo longior. *Capsula* vera cylindracea, lævis, nigro-castanea, ore dilatato, patulo. *Apophysis* obconica basi attenuata, capsula paulo longior. *Operculum* madore conicum, siccitate conico-hæmisphæricum mucronulo manifesto. *Peristomium* intra marginem membranæ exterioris, ubi desinet interior, ortum: *dentibus* 16, quaternatim ad medium usque connatis, singulis absque stria longitudinali manifesta. *Masculus Flos* cum femineo collateralis, ramum terminans, discoideo-capituliformis, semper sessilis, etiam dum femineus, primo pariter sessilis,
cccii] florescentia peracta ramulo suo proprio elongato insidet. *Folia perigonialia* è basi lanceolatâ erectâ in cuspidem basi longiorem, subulatam producta. *Antheræ* numerosæ viginti circiter, levissime arcuatæ, brevissime pedicellatæ. *Paraphyses* stramineæ, sursum incrassatæ articulis brevioribus crassioribusque. *Pistilla* nulla.

Obs. Facies omnino S. mnioidis, quocum pluribus notis convenit, satis diversum dentium dispositione.

94. Splachnum propinquum, peristomii dentibus basi quaternatim cohærentibus, apophysi obconica capsula operculata paulo latiore, operculo siccitate depresso mutico, floribus masculis brevè pedunculatis, seta perichætium vix superante, foliis ovatis concavis cuspidatis integerrimis.

Desc. *Cæspites* densi. *Caules* innovando divisi, unciales.

Folia viridissima, acumine subulato-setaceo, concolori, longitudine $\frac{1}{3}$ folii. *Seta* foliis floralibus paulo longior, angulata, lævis, capsulam cum apophysi sumptam vix superans. *Capsula* cylindracea, brevis, ore dilatato. *Apophysis* primo viridis, mox fusca, capsulâ ante lapsum operculi paulo tantum crassior, demum nigricans, pyriformis, capsula deoperculata concolori fere duplo amplior. *Operculum* conico-hemisphæricum, muticum, siccitate planiusculo-depressum. *Peristomii dentes* 16, quaternatim approximati et ad medium usque cohærentes, singuli absque stria longitudinali manifesta. *Columella* crasso-cylindracea, pulposa, apice hemisphærico cavitatem operculi replenti. *Masculus Flos* capitato-discoideus, ramulum brevem, femineo collateralem, terminans, antheris paraphysibusque numerosis, pistillis certe nullis.

Obs. Proximum S. arctico, an ejusdem varietas ?

95. Splachnum exsertum, capsula interiore soluta siccitate semiexserta; exteriore ore dilatato, apophysi obconica capsula (concolori) angustiore, foliis lanceolato-ovatis acuminatis integerrimis.

Desc. *Caules* annotino-ramosi; *Folia* omnino S. arctici et propinqui. *Masculus Flos* capitato-discoideus, ramulum distinctum, femineo breviorem, foliatum, ejusdem caulis terminans, foliis perigonialibus basi lutescentibus, acumine brevi viridi. *Antheræ* paucæ, cylindraceæ, leviter arcuatæ: *paraphysibus* numerosis sursum crassioribus: *pistillis* nullis. *Seta* terminalis, perichætium vix superans, dilute fusca, lævis. *Capsula* cum apophysi sumpta turbinata; *theca exterior* obovata; *interior* pedicello insidens libera, demum exsiccatione exterioris exserta. *Peristomium: dentes* 16, mox intra marginem capsulæ exterioris orti, primo quaternatim basi cohærentes, demum quaternatim vel quandoque geminatim reflexi.

Obs. Muscus valde affinis hinc *S. arctico* et *propinquo* inde *paradoxo;* et hi omnes adeo approximati præsertim figura et textura foliorum ut varietates unius ejusdemque speciei forsan considerari possunt.

96. Splachnum paradoxum, capsula adulta absque sutura operculi (demum separabilis ?); interiore pedicellata, apophysi attenuata capsula angustiore, foliis lanceolato-ovatis acuminatis integerrimis.

ccciii] Desc. *Caules* vix semunciales, innovationibus ramosi. *Folia* ovato-lanceolata, concaviuscula, carinata, laxe reticulata, integerrima, acumine subulato diametrum transversum folii subæquanti, demum decolori pilum referenti. *Masculi Flores* discoideo-capituliformes, terminantes ramos proprios pedunculiformes, paucifolios, foliolis nanis alternis: *folia perigonialia* lanceolata, basi conniventia, apicibus patulis acuminatis. *Antheræ* numerosæ, cylindraceæ, levissime incurvæ. *Paraphyses* numerosæ, subclavatæ. *Flos femineus* terminalis. *Seta* fusca, lævis, caule longior. *Capsula* erecta, oblongo-obovata, basi in apophysin obconicam seipsa angustiorem et breviorem attenuata, lævis, per lentem pluries augentem punctis minutis longitudinaliter seriatis, depressis, adversus lucem semipellucidis tenuissime quasi striata, absque operculo ejusve ulla indicatione, apiculo obtuso paulo constricto. *Theca vera* dimidiam superiorem tantum capsulæ exterioris occupans, pedicello cylindraceo, ex apice apophysis derivato, insidens, libera, ad ortum dentium desinens ibique cum capsula exteriore confluens. *Dentes* 16, quaternatim ad medium cohærentes, subulati, pallide fusci, apicem cavitatis capsulæ attingentes. *Semina* minutissima, in cumulo olivaceo-viridia, seorsim hyalina, lævia.

Obs. Hæc omnia è specimine unico cum capsulis 8 maturis plenis et duabus vetustis vacuis pariter clausis, varietatem nanam S. arctici referente, desumpta sunt. Exemplaria dein plura varietatis, ut videtur, ejusdem Musci, in herbario D. Richardson, inter Fort Enterprise et mare arcticum lecta, et cum S. mnioidi *Schwaegr. in Franklin's journ. p.* 755 (non Hedwigii), intermista inveni: horum capsulæ adultæ numerosæ cinnamomeo-fuscæ, clausæ et absque sutura vel ulla alia operculi indicatione. In hac varietate β., quæ statura major et calyptra dimidiata donata, seta longior quam in α. dentesque 16 subæquidistantes et fere ad basin distincti.

E duplicis varietatis hisce speciminibus diu in animo fuit proponere novum genus sub nomine Cryptodontis, ob capsulam operculo destitutam dentibus verò inclusis instructam: sed omnibus iterum examini subjectis capsulam unam alteramve vetustam operculo delapso et peristomio dentato, in eodem cæspite cum clausis, et quantum determinare licuit ad eandem speciem pertinentem, observavi, ideoque ad Splachnum, haud tamen absque dubitatione, muscum paradoxum demum retuli. In Splachneis autem, præter annuli defectum in tota tribu, approximationes nonnullæ ad capsulam clausam occurrunt, scilicet in Aplodonte ubi operculum cum altero latere stomatis diutius cohæret, et in Systylio in quo cum columella cohærens persistit: nec transitus difficilis à Splachneis ad Voitiam habitu et statione iisdem bene convenientem.

VOITIA.

Hornschuch comment. de voit. et syst. p. 5. *Hooker musc. exot.* 97. *Nees. v. Esenb. et Hornsch. bryol. germ.* 1, *p.* 79. *Schwaegr. suppl.* 2, *sect.* 1, *p.* 2. *Greville et Arnott in Wern. soc. transact.* 4.

Char. Gen. *Capsula* clausa (absque operculo dentibusve inclusis), rostrata. *Calyptra* dimidiata, capsula adulta longior, tardius decidua. *Flores* terminales: *masculus* femineo collaterali subconformis.

Obs. Genus à Phasco ægre distinguendum, habitu [ccciv
quamvis necnon statione valde diversum, et ad Splachneas mediante S. paradoxo (s. Cryptodonti) accedens. *Calyptra* multo amplior equidem et diutius remanens quam in Phasco, sed demum decidua, nec persistens. *Vaginula* in V. hyperborea certe indivisa, nec eandem bivalvem neque fissam in exemplaribus paucis V. nivalis à nobis investigatis observare licuit. In utraque specie ejusdem margo manifeste inæqualis et sublacera, sed eandem fere structuram in Phascis quibusdam, præsertim in P. bryoidi et curvicollo, observavimus. *Capsula* cum seta sua elongata sæpe decidua sed quandoque nec raro vel cum eadem persistens, vel à seta

persistenti decidens: et seta minime post lapsum capsulæ in Phascis omnibus persistit. *Membrana interna* libera, cum processu subulato, rostrum capsulæ penetranti, in P. bryoidi et curvicollo pariter exstat: et florum dispositio subsimilis in Phascis nonnullis obtinet. *Semina* minutissima affinitatem Voitiæ cum Cryptodonti potius quam Phasco indicant.

Voitia vogesiana *Nestl.* dubia hujus generis species mihi videtur, et habitu Phascis nonnullis, præsertim P. flexuoso *Schwaegr. suppl.* 2, *sect.* 1, *p.* 1, *t.* 101, convenit: à Voitia diversa floribus sæpe dioicis, masculorum forma, capsulis basi in apophysin angustiorem attenuatis, seminibus majusculis, et forsan magnitudine proportionali calyptræ à me nondum visæ.

97. Voitia hyperborea, capsula globoso-ovata basi subtruncata, foliis dilatato-ovatis acuminatis.

Voitia hyperborea, *Greville et Arnott in Wern. soc. mem.* 4, *tab.* 7, *f.* 19, *capsula, et* 21, *folium.*

Desc. *Muscus* cæspites densos sæpius efformat, raro aliis, Splachnis præsertim, intermixtus. *Caules* 6-9-lineas longi, tomento radicali inferne arcte cohærentes, innovationibus subramosi, basi foliis-vetustis tomentoque radicali rufo-castaneo copioso tecti; ramis annotinis dense foliatis eradiculosis. *Folia* late ovata, modice concava, integerrima, acumine è nervo valido producto formato, ¼ longitudinis folii æquante, sed concolori nec nisi vetustate canescenti pilumque referenti, laxiuscule reticulata, areolis rectangulis, invicem inæqualibus sed per totam folii longitudinem uniformibus, marginalibus vix majoribus, madore erecta, siccitate subappressa. *Perichætialia* paulo majora, acumine proportionatim longiore. *Vaginula* cylindracea, basi pistillis paucis abortivis stipata, indivisa, nec fissa nec bivalvis, apice membranaceo inæquali lacero. *Seta* elongata, caulem totum æquans v. parum superans, lævis, castanea, siccitate tortuosa. *Capsula* erecta, castanea, lævis, dilatato-ovata, basi transversa subtruncata, rostro apicis inclinato longitudine dimidii capsulæ, absque operculo ejusve omni vestigio: *exterior* coriacea, minute reticulata, areolis quadratis;

interior ab exteriore libera, centro baseos umbilicatæ affixa, apice clauso processu subulato longitudine rostri exterioris, pallida, tenuè membranacea, utrinque lævis nec intus septis processubusve inæqualis. *Columella* angulata subtetragona, longitudine capsulæ interioris. *Semina* minutissima, Phasci bryoidis decies fere minora, in cumulo viridia, separatim hyalina, subglobosa, per lentem centies augentem striis nonnullis insignita, sed simplicia nec divisibilia. *Masculus Flos* ramulum proprium, brevissimum, femineo collateralem hoc vero post fæcundationem elongato demum quasi lateralem terminans, discoideo-gemmiformis, femineo sub- [cccv similis, foliis perigonialibus perichætialibus conformibus. *Antheræ* numerosæ, cylindraceæ, leviter arcuatæ. *Paraphyses* copiosæ, articulis superioribus sensim crassioribus et brevioribus.

Obs. Valde affinis Voitiæ nivali quæ differt capsula oblongo-ovata basi acuta, foliis elongato-ovatis laxioribus, statura majori.

HEPATICÆ.

98. JUNGERMANNIA MINUTA, *Schreb. in Crantz grönl. forts. p.* 285. *Dicks. fasc.* 2, *p.* 13. *Wahlenb. lapp. p.* 393. *Hooker brit. junger. t.* 44. *Engl. bot.* 2231.

Jungermannia bicornis, *Flor. Dan.* 888, *f.* α. *Schwaegr. prodr. hepat. p.* 27. *Richardson in Franklin's journ. p.* 757.

Obs. Planta nostra, cujus exemplaria perpauca et fructificatione destituta tantum visa, media quasi inter *J. minutam* et *ventricosam,* ab utraque foliis explanatis, nec margine inferiore induplicatis, differt.

99. MARCHANTIA POLYMORPHA, *Linn. sp. pl. ed.* 2, *p.* 1603, *Flor. lapp. n.* 422. *Wahlenb. lapp. p.* 397. *Schmid. ic. p.* 106, *t.* 29. *Engl. bot.* 210. *Hooker scot. par.* 2, *p.* 119. *Mich. am.* 2, *p.* 277. *Br. in Flind. voy.* 2, *p.* 593. (*Antè, p.* 69.) *Richardson in Franklin's iourn. p.* 757.

LICHENOSÆ.

100. Gyrophora proboscidea, *Achar. syn. p.* 64. *Engl. bot.* 2484. *Hooker scot. par.* 2, *p.* 41.

Gyrophora proboscidea β, *Richardson in Franklin's journ. p.* 758, *tab.* 30, *f.* 4.

Gyromium proboscideum, *Wahlenb. lapp. p.* 483.

Obs. In nostra planta pagina inferior, quæ semper lævis fibrillisque destituta, sæpius cinerea, nunc tota atra; quandoque thallo ad ambitum cribroso G. erosæ accedit.

101. Lecanora elegans, *Achar. syn. p.* 182. *Hooker scot. par.* 2, *p.* 50. *Richardson in Franklin's journ. p.* 760.

Lichen elegans, *Wahlenb. lapp. p.* 417, *Carpat. p.* 373. *Engl. bot.* 2181.

102. Borrera? aurantiaca, thallo adscendenti aurantiaco tereti-compresso nudo subdichotomo basi pallido: ramulis ultimis brevissimis obtusis.

Obs. Affinis B. flavicanti *Achar. l. c.*, utraque thallo teretiusculo fruticuloso à reliquis diversa. In hac Apothecia ignota ideoque dubii generis est.

103. Cetraria juniperina, *Achar. syn. p.* 226.

cccvi] *Obs.* Vix omnino cum C. juniperinâ quadrant specimina nostra quibus laciniæ crenatæ nec erosæ, margines pulvere destitutæ, discus lævis vix manifeste lacunosus, et paginæ, quæ citrinæ, concolores.

104. Cetraria nivalis, *Achar. syn. p.* 228. *Hooker scot. par.* 2, *p.* 57. *Br. in Ross' voy.* 2 *ed. v.* 2, *p.* 195. (*Antè, p.* 178.) *Spitz. pl. in Scoresby's arct. append. p.* 76. (*Antè, p.* 181.) *Richardson in Franklin's journ. p.* 761.

Lichen nivalis, *Linn. lapp. n.* 446, *t.* 11. *f.* 1. *Soland. in Phipps' voy. p.* 203. *Wahlenb. lapp. p.* 433, *Carpat. p.* 379. *Engl. bot.* 1994. *Svensk bot.* 384.

105. Cetraria cucullata, *Achar. syn. p.* 228. *Richardson in Franklin's journ. p.* 761.
Lichen cucullatus, *Smith in Linn. Soc. transact.* 1, *p.* 84, *t.* 4, *f.* 7. *Wahlenb. lapp. p.* 433, *Upsal. p.* 413, *Carpat. p.* 379.

106. Cetraria islandica, *Achar. syn. p.* 229. *Hooker scot. par.* 2, *p.* 58. *Br. in Ross' voy. ed.* 2, *vol.* 2, *p.* 195. (*Antè, p.* 178.) *Richardson in Franklin's journ. p.* 761.
Lichen islandicus, *Linn. sp. pl. ed.* 2, *p.* 1611. *Flor. Dan.* 155. *Engl. bot.* 1330. *Svensk. bot.* 34. *Wahlenb. lapp. p.* 434, *Carpat. p.* 379, *Upsal. p.* 413. *Soland. in Phipps' voy. p.* 203.
Physcia islandica, *Mich. am.* 2, *p.* 326.

107. Cetraria odontella, *Achar. syn. p.* 230.
Lichen odontellus, *Wahlenb. lapp. p.* 434.

108. Peltidea aphthosa, *Achar. syn. p.* 238. *Wahlenb. lapp. p.* 446, *Carpat. p.* 380. *Svensk bot.* 318. *Hooker scot. par.* 2, *p.* 60. *Richardson in Franklin's journ. p.* 761.
Lichen aphthosus, *Linn. sp. pl. ed.* 2, *p.* 1616. *Eng. bot.* 1119. *Wulfen. in Jacqu. coll.* 4, *p.* 266, *t.* 17.

109. Cornicularia ochroleuca, *Achar. syn. p.* 301. *Hooker scot. par.* 2, *p.* 69. *Richardson in Franklin's journ. p.* 762.
Usnea ochroleuca, *Hoffm. pl. lichen.* 2, *p.* 7, *t.* 26, *f.* 2.
Lichen ochroleucus, *Wahlenb. lapp. p.* 438, *Carpat.* 382. *Engl. bot.* 2374.

110. Cornicularia lanata, *Achar. syn. p.* 302. *Hooker scot. par.* 2, *p.* 69.
Lichen lanatus, *Linn. sp. pl. ed.* 2, *p.* 1623. *Engl. bot.* 846. *Wahlenb. lapp. p.* 440, *Carpat. p.* 383.
Lichen normöricus, *Gunn. norv. par.* 2, *p.* 123, *t.* 2, *f.* 9—14.

111. Cerania vermicularis, *Achar. syn. p.* 278. [cccvii

Cenomyce? vermicularis, *Hooker scot. par.* 2, *p.* 65. *Richardson in Franklin's journ. p.* 762. *Br. in Flinders' voy.* 2, *p.* 594. (*Antè, p.* 69.)

Bœomyces vermicularis, *Wahlenb. lapp. p.* 458.

Cladonia subuliformis, *Hoffm. pl. lichen.* 2, *p.* 15, *t.* 29, *f.* 1—3.

Lichen vermicularis, *Dicks. crypt. fasc.* 2, *p.* 23, *t.* 6, *f.* 10. *Engl. bot.* 2029.

Obs. Apothecia (?) lateralia, sparsa, atra, thallo innata eoque submarginata, apotheciis Roccellæ aliquo modo accedentia, in exemplaribus nonnullis à D. Fisher lectis, observavi.

112. Cenomyce pyxidata, *Achar. syn. p.* 252.

113. Stereocaulon paschale, *Achar. syn. p.* 284. *Mich. am.* 2, *p.* 331. *Br. in Flinders' voy.* 2, *p.* 594. (*Antè, p.* 70.) *Spitzb. pl. in Scoresby's arct.* 1, *append. p.* 76. (*Antè, p.* 181.) *Giesecke Greenl. in Edin. encyclop. Hooker scot. par.* 2, *p.* 66. *Richardson in Franklin's journ. p.* 762.

Bœomyces paschalis, *Wahlenb. lapp. p.* 450, *Carpat. p.* 386.

Lichen paschalis, *Linn. sp. pl. ed.* 2, *v.* 2, *p.* 1621. *Soland. in Phipps' voy. p.* 204.

Lichen ramulosus, *Sw. fl. ind. occid.* 3, *p.* 1917.

114. Usnea sphacelata, thallo erectiusculo fruticuliformi, ramis primariis ochroleucis nigro-vittatis lævibus: ultimis attenuatis nigris, sorediis confertis concoloribus ochroleucisve.

Usnea? prope melaxantham, *Br. spitzb. pl. in Scoresby's arct.* 1, *append. p.* 76. (*Antè, p.* 181.)

Obs. Proxima U. melaxanthæ *Ach. syn. p.* 303, differt statura aliquoties minore, ramis primariis lævibus, sorediorum præsentia. Apothecia nondum visa. Eandem speciem, sorediis pariter instructam apotheciisque destitutam, in summitate Montis Tabularis Insulæ Van Diemen, anno 1804, legi.

FUNGI.

115. Cantharellus lobatus, *Fries. syst. mycolog.* 1, *p.* 323.
Helvella membrancea, *Flor. Dan.* 1077, *f.* 1.

116. Lycoperdon pratense, *Pers. syn. fung. p.* 142.

Præter plantas supra enumeratas, species nonnullæ in herbariis citatis exstant, scilicet Muscorum quinque, Lecideæ v. Leprariæ unica, et Agarici tres: has vero è speciminibus vel fructificatione destitutis vel male exsiccatis haud determinare potui.

Algæ submersæ prorsus nullæ reportatæ fuere.

Species quæ Florulæ Melvillianæ adhucdum propriæ [cccviii
remanent sequentes sunt.

Ranunculus Sabinii, qui *nivali* nimis affinis.
Ranunculus affinis, proximus *duricomo.*
Draba pauciflora, valde dubia species.
Platypetalum dubium, cujus flores ignoti.
Sieversia Rossii, proxima *S. humili* Oonalashkæ indigena.
Tussilago corymbosa, valde affinis *T. frigidæ.*
Pedicularis arctica, prope *P. sudeticam* et *Langsdorfii.*
Dupontia Fisheri, gramini nulli cognito affinis.
Barbula leucostoma, quæ species distincta videtur.
Gymnostomum obtusifolium, species insignis, sed non satis cognita.
Splachnum arcticum, proximum *S. mnioidi.*
Borrera aurantiaca, Lichenosa distincta, sed dubii generis.

Genus itaque Insulæ Melville peculiare nullum restat nisi *Dupontia*, si hoc equidem servari meretur.

Aliquas observationes, species nonullas Florulæ Melvillianæ illustrantes, derivatas ex herbarii inspectione ad litora orientalia Americæ arcticæ, inter grad. 66 et 70 lat., in novissima navigatione duce D. Parry, formati à D. Ross, cujus amicitiæ specimina totius collectionis debeo, hic subjungere licet; ordine Florulæ servato numerisque specierum præfixis.

11. Platypetalum purpurascens.

Siliculæ v. ovali-oblongæ v. oblongæ, glabræ v. pilis raris brevibus simplicibus bifidisque conspersæ, *stigmate* quandoque capitato emarginato, nec semper bilobo lobis patentibus, coronatæ; *valvulis* aveniis, ecarinatis, planiusculis; *dissepimento* rarius fenestrato. *Semina* biseriata. *Cotyledones* incumbentes, angusto-oblongæ, rectæ nec basibus crus radiculare embryonis occupantibus.

Platypetalum itaque hinc *Subulariæ* affine inde *Eudemæ*, hæc vero differt stylo elongato, dissepimento semper fenestrato, et forsan aliis notis è floris examini accuratiore derivandis.

13. Eutrema Edwardsii.

Herba quandoque 4-6-uncialis.

18. Stellaria Edwardsii.

Exemplaria omnia ad var. α pertinent, foliis ovatis acutis caulibusque glaberrimis, pedunculis unifloris, antheris purpureis, capsulis erectis semisexvalvibus calycem vix superantibus, seminibus lævibus fuscis.

Species forsan polygama, ad quam referenda S. Edwardsii *Richardson l. c.*? et S. nitida *Hooker?*

cccix] 26. Saxifraga uniflora.

Exemplaria omnia staturæ majoris sunt, et pleraque caulibus 2-3-floris donata; ideoque hæc, quæ potius pro varietate insigni quam distincta specie supra proposita fuit, ad S. cæspitosam absque dubio reducenda.

36. Oxytropis arctica.

Hujus varietas notabilis, vix enim distincta videtur species, statura minore, scapo sæpe unifloro passimque umbella biflora, dentibus calycis respectu tubi paulo longioribus, foliolis sæpius 7, quandoque 9, villis persistentibus utrinque argenteo-sericeis.

Cineraria congesta.

Herba quandoque spithamea, folia sæpius sinuato-dentata, nunc alte sinuata, passim indivisa; hæc exemplaria itaque propius *C. palustri* accedunt, inflorescentia densiore, lana magis copiosa et longiore præsertim distinguenda.

50. Carex misandra.

Hujus exemplaria aliqua completa, 6-9-uncialia, foliis linearibus apice attenuato, marginibus deorsim scabris, culmo lævi, spicis 3-4, alternis, raro subumbellatis, terminali basi solum rariusve tota mascula, stigmatibus sæpissime tribus. Hinc ad C. fuliginosam *Sternb. l. c.* procul dubio referenda.

51. Carex concolor.

Specimina proceriora, spicis femineis longioribus, axi squamarum pallido, ad C. cæspitosam propius accedunt, et culmo lævi præcipue distingui possunt.

56. Colpodium latifolium.

In exemplaribus plerisque rudimentum breve setuliforme flosculi secundi adest; necnon valvulæ inferioris perianthii setula denticulata dorsalis, $\frac{1}{5}$ circiter ab ejusdem apice, nervum centralem terminans, altitudinem valvulæ vix æquans. Aliqua autem omnino mutica sicut pleraque ab Insula Melville.

57. Poa angustata.

Hujus, ni fallor, varietas nana (2-3-uncialis), perianthiis glaberrimis, locustis viridibus apicibus purpureis valvulæ inferioris perianthiorum solum exceptis.

58. Poa abbreviata.

Specimina pleraque vix triuncialia.

60. Festuca brevifolia.

Triviale nomen his exemplaribus vix convenit, quibus folia radicalia dimidium et ultra culmi æquant, et culmea vaginis suis proportionatim longiora sunt.

cccx] 62. Pleuropogon Sabinii.

Exemplar unicum cæspitosum, in palude à D. Ross lectum ad var. β pertinet, culmis partialibus quadriuncialibus, antheris purpureis.

91. Aplodon wormskioldii.

Exemplaria nonnulla varietatis à supra enumeratis diversæ, cujus folia acutiuscula absque acumine, apophysis ovato-globosa, nec basi attenuata, pallidè straminea, cava, axi solido, capsulâ castaneâ amplior.

97. Voitia hyperborea.

Sæpius aliis Muscis, Splachnis præsertim, intermista crescit.

OBSERVATIONS

ON THE

STRUCTURE AND AFFINITIES

OF THE

MORE REMARKABLE PLANTS

COLLECTED BY

THE LATE WALTER OUDNEY, M.D.,

AND

MAJOR DENHAM, AND CAPTAIN CLAPPERTON,

IN THE YEARS 1822, 1823, AND 1824,

DURING THEIR

EXPEDITION TO EXPLORE CENTRAL AFRICA.

BY

ROBERT BROWN,

HON. M.R.S.E., F.R.S., AND F.L.S.; MEMBER OF THE ROYAL SWEDISH ACADEMY OF SCIENCES, OF THE ROYAL SOCIETY OF DENMARK, AND OF THE IMPERIAL ACADEMY NATURÆ CURIOSORUM; CORRESPONDING MEMBER OF THE ROYAL INSTITUTE OF FRANCE, AND OF THE ROYAL ACADEMIES OF SCIENCES OF PRUSSIA AND BAVARIA, ETC. ETC.

[*Reprinted from the 'Narrative of Travels and Discoveries in Northern and Central Africa,' by Major Dixon Denham and Captain Hugh Clapperton. Appendix, pp.* 208—246.]

LONDON:

1826.

OBSERVATIONS, &c.

The Herbarium formed, chiefly by the late [Appcnd. p. 208 Dr. Oudney, during the expedition, contains specimens, more or less perfect, of about three hundred species. Of these one hundred belong to the vicinity of Tripoli; fifty were collected in the route from Tripoli to Mourzuk, thirty-two in Fezzan, thirty-three on the journey from Mourzuk to Kouka, seventy-seven in Bornou, and sixteen in Haussa or Soudan.

These materials are too inconsiderable to enable us to judge correctly of the vegetable productions of any of the countries visited by the mission, and especially of the more interesting regions, Bornou and Soudan.

For the limited extent of the herbarium, the imperfect state of many of the specimens, and the very scanty information to be found respecting them, either in the herbarium itself or in the journal of the collector, it is unfortunately not difficult to account.

Dr. Oudney was sufficiently versed in botany to have formed collections much more extensive and instructive, had the advancement of natural history been the principal purpose of his mission. His time and attention, however, were chiefly occupied by the more important objects of the expedition; as a botanist he had no assistant; and the state of his health during his residence in Bornou must, in a great degree, have rendered him unable to collect or observe the natural productions of that country.

For the few specimens belonging to Soudan, we are indebted to Captain Clapperton, who, after the death of Dr.

Oudney, endeavoured to preserve the more striking and useful plants which he met with. His collection was originally more considerable; but before it reached England many of the specimens were entirely destroyed. It still includes several of the medicinal plants of the natives; but these being without either flowers or fruit, cannot be determined.

209] In the whole herbarium, the number of undescribed species hardly equals twenty; and among these not one new genus is found.

The plants belonging to the vicinity of Tripoli were sent to me by Dr. Oudney, before his departure for Fezzan. This part of the collection, amounting to one hundred species, was merely divided into those of the immediate neighbourhood of Tripoli, and those from the mountains of Tarhona and Imsalata.

It exceeds in extent the herbarium formed by Mr. Ritchie near Tripoli, and on the Gharian hills, which, however, though containing only fifty-nine species, includes twenty-seven not in Dr. Oudney's herbarium.

The specimens in Mr. Ritchie's collection are carefully preserved, the particular places of growth in most cases given, and observations added on the structure of a few; sufficient at least to prove, that much information on the vegetation of the countries he visited might have been expected from that ill-fated traveller.

In these two collections united, hardly more than five species are contained not already published in the works that have appeared on the botany of North Africa; particularly in the 'Flora Atlantica' of M. Desfontaines, in the 'Flore d'Egypte' of M. Delile, and in the 'Floræ Libycæ Specimen' of Professor Viviani, formed from the herbarium of the traveller Della Cella.

The plants collected in the Great Desert and its oases, between Tripoli and the northern confines of Bornou, and which somewhat exceed a hundred, are, with about eight or ten exceptions, also to be found in the works now mentioned. And, among those of Bornou and Soudan, which fall short of one hundred, very few species occur not already known as natives of other parts of equinoctial Africa.

A complete catalogue of the herbarium, such as I have now described it, even if the number and condition of the specimens admitted of its being satisfactorily given, would be of but little importance, with reference to the geography of plants. Catalogues of such collections, if drawn up hastily, and from imperfect materials, as must here have been the case, are indeed calculated rather to injure than advance this department of the science, which is still in its infancy, and whose progress entirely depends on the scrupulous accuracy of its statements. To produce confidence in these statements, and in the deductions founded on them, it should in every case distinctly appear that, in establishing the identity of the species enumerated, due attention has been paid to the original authorities on which they depend, and, [210
wherever it is possible, a comparison actually made with authentic specimens.

In the account which I am now to give of the present collection, I shall confine myself to a slight notice of the remarkable known plants it contains, to characters or short descriptions of the more interesting new species, and to some observations on such of the plants as, though already published, have either been referred to genera to which they appear to me not to belong, or whose characters require essential alteration.

In proceeding on this plan, I shall adopt the order followed in the botanical appendix to Captain Tuckey's 'Expedition to the River Congo.' And as there will seldom be room for remarks on the geographical distribution of the species I have to notice, I shall chiefly endeavour to make my observations respecting them of some interest to systematic botanists.

Cruciferæ. Fifteen species belonging to this family exist in the collection, one of which only appears to be undescribed, and of this the specimens are so imperfect that its genus cannot with certainty be determined. Of those already published, however, the generic characters of several require material alterations, some of which suggest observations relative to the structure and arrangement of the natural order.

Savignya Ægyptiaca (*De Cand. Syst.* 2, p. 283) is the first of these. It was observed near Bonjem by Dr. Oudney, whose specimens slightly differ from those which I have received from M. Delile, by whom this plant was discovered near the pyramid of Saqqârah, and who has well figured and described it in his 'Flore d'Egypte,' under the name of Lunaria parviflora. By this name it is also published by M. Desvaux. Professor Viviani, in giving an account of his Lunaria libyca, a plant which I shall presently have occasion to notice more particularly, has remarked,[1] that Savignya of De Candolle possesses no characters sufficient to distinguish it as a genus from Lunaria; and still more recently, Professor Sprengel has referred our plant to Farsetia.[2] The genus Savignya, however, will no doubt be ultimately established, though not on the grounds on which it was originally constituted; for the umbilical cords certainly adhere to the partition, the silicule, which is never
[112] absolutely sessile, is distinctly pedicellated in Dr. Oudney's specimens, the valves are not flat, and the cotyledons are decidedly conduplicate. In describing the cotyledons of his plant as accumbent, M. De Candolle has probably relied on the external characters of the seed, principally on its great compression, its broad margin or wing, and on the whole of the radicle being visible through the integuments. It would appear, therefore, that the true character of the cotyledons of Savignya has been overlooked, chiefly from its existing in the greatest possible degree. To include this degree of folding, in which the margins are closely approximated, and the radicle consequently entirely exposed, a definition of conduplicate cotyledons, somewhat different from that proposed in the 'Systema Naturale' becomes necessary. I may here also observe, that the terms Pleurorhizæ and Notorhizæ, employed by M. De Candolle to express the two principal modifications of cotyledons in Cruciferæ, appear to me so far objectionable, as they may

[1] *Floræ Libycæ Specim.* p. 35.

[2] *Syst. Vegetab.* p. 871.

seem to imply that in the embryo of this family the position of the radicle is variable, and that of the cotyledons fixed. It is at least deserving of notice that the reverse of this is the fact; though it is certainly not necessary to change these terms, which are now generally received.

On the subject of Savignya, two questions naturally present themselves. In the first place—Is this genus, solely on account of its conduplicate cotyledons, to be removed from Alyssineæ, where it has hitherto been placed, to Velleæ, its affinity with which has never been suspected, and to whose genera it bears very little external resemblance? Secondly—In dividing Cruciferæ into natural sections, are we, with M. De Candolle, to expect in each of these subdivisions an absolute uniformity in the state of the cotyledons? As far as relates to the accumbent and flatly incumbent states, at least, I have no hesitation in answering the latter question in the negative; and I believe that in one case, namely, Hutchinsia, these modifications are not even of generic importance, for it will hardly be proposed to separate H. alpina from petræa, solely on that ground. I carried this opinion further than I am at present disposed to do, in the second edition of Mr. Aiton's 'Hortus Kewensis,' where I united in the genus Cakile plants, which I then knew to differ from each other, in having accumbent and conduplicate cotyledons; and I included Capsella bursa-pastoris in the genus Thlaspi, although I was aware, both from my own observations and from Schkuhr's excellent figure,[1] that its cotyledons were incumbent. I am at present, [218
however, inclined to adopt the subdivision of both these genera, as proposed by several authors and received by M. De Candolle; but to this subdivision the author of the 'Systema Naturale' must have been determined on other grounds than those referred to; for in these four genera, in which the three principal modifications of cotyledons occur, he has taken their uniformity for granted.

As to the place of Savignya in the natural family, I believe, on considering the whole of its structure and habit, that it ought to be removed from Alyssineæ to a subdivision

[1] *Handb. tab.* 180.

of the order that may be called *Brassiceæ,* but which is much more extensive than the tribe so named by M. De Candolle; including all the genera at present known with conduplicate cotyledons, as well as some others, in which these parts are differently modified.

There are two points in the structure of Savignya, that deserve particular notice. I have described the æstivation of the calyx as valvular; a mode not before remarked in this family, though existing also in Ricotia. In the latter genus, however, the apices of the sepals are perhaps slightly imbricate, which I cannot perceive them to be in Savignya.

The radicle is described by M. De Candolle as superior with relation to the cotyledons. I am not sure that this is the best manner of expressing the fact of its being horizontal, or exactly centrifugal, the cotyledons having the same direction. This position of the seed is acquired only after fecundation; for at an earlier period the foramen of the testa, the point infallibly indicating the place of the future radicle, is ascendent. From the horizontal position of the radicle in this and some other genera, especially Farsetia, we may readily pass to its direction in Biscutella, where I have termed it descendent, a character which I introduced to distinguish that genus from Cremolobus. But in Biscutella the embryo, with reference to its usual direction in the family, is not really inverted, the radicle being still placed above the umbilicus. On the contrary, in *Cremolobeæ,* a natural tribe belonging to South America, and consisting of Cremolobus and Menonvillea, though the embryo at first sight seems to agree in direction with the order generally, both radicle and cotyledons being ascendent, it is, in the same sense, not only inverted, but the seed must also be considered as resupinate; for the radicle is seated below the umbilicus, and also occupies the inner side of the seed, or that next the placenta—peculiarities which, taken together, constitute the character of the
213] tribe here proposed. It appears to me singular that M. De Candolle, while he describes the embryo of these two genera as having the usual structure of the order, should

consider that of Iberis, in which I can find no peculiarity, as deviating from that structure.[1]

Lunaria libyca of Viviani[2] is the second plant of Cruciferæ on which I have some observations to offer. This species was described and figured, by the author here quoted, in 1824, from specimens collected in 1817 by Della Cella. The specimens in the herbarium were found near Tripoli, where the plant had also been observed in 1819 by Mr. Ritchie, who referred it to Lunaria, and remarked that the calyx was persistent. Professor Sprengel, in his 'Systema Vegetabilium,' considers it a species of Farsetia.

That this plant ought not to be associated either with the original species of Lunaria, or with Savignya, as now constituted, is sufficiently evident. And if it is to be included in Farsetia, it can only be on the grounds of its having a sessile silicule, with compressed valves, an indefinite number of seeds in each cell, and accumbent cotyledons. But in these respects it accords equally with Meniocus, a genus proposed by M. Desvaux, and with some hesitation received by M. De Candolle, and with Schivereckia of Andrzejowski, which he has also adopted.

[1] SAVIGNYA.

Savignya. *De Cand. Syst.* 2, p. 283. Lunariæ sp. *Delile. Desvaux. Viviani.*

Char. Gen. *Calyx* basi æqualis; æstivatione valvata. *Silicula* oblonga, septo conformi, valvis convexiusculis. *Semina* biseriata imbricata marginata. *Cotyledones* conduplicatæ.

Herba *annua, glabra (quandoque pube rara simplici).* Folia *crassiuscula, inferiora obovata in petiolum attenuata grosse dentata, media sæpe incisa, superiora linearia.* Racemi *oppositifolii, ebracteati.* Flores *parvi erecti, petalis violaceis venis saturatioribus.* Siliculæ *racemosæ, divaricatæ, inferiores sæpius deflexæ.*

Calyx erectus, æstivatione valvata, ipsis apicibus vix imbricatis. *Petala* unguiculata, laminis obovatis sub æstivatione mutuo imbricatis. *Stamina* distincta, edentula, singulum par longiorum *glandula* subquadrata extus stipatum; breviora, quantum e speciminibus observare licuit, eglandulosa. *Ovarium* brevissime pedicellatum, ovulis adscendentibus nec horizontalibus. *Stylus* brevis. *Stigma* capitatum vix bilobum. *Silicula* breviter manifeste tamen stipitata, oblonga nunc oblongo-elliptica. *Valvulæ* uninerviæ reticulato-venosæ. *Dissepimentum* e lamellis duabus separabilibus uninerviis venis anastomozantibus obsoletis: areolis subtransversim angustato-linearibus, parietibus (tubulis) rectis subparallelis. *Funiculi* horizontales, dimidio inferiore septo arcte adnato superiore libero.

[2] *Flor. Lib. Specim. p.* 34, *tab.* 16, *f.* 1.

It does not, however, agree with either of those genera in habit, and it is easily distinguished from both by its simple
214] filaments and other characters, which I shall notice hereafter. Is this plant, then, *sui generis*? ought it to be united with Alyssum, the character of that genus being modified to receive it? or does not Alyssum require subdivision, and may not our plant be referred to one of the genera so formed? A brief result of the examination of these questions, so far as they are connected with the subject under consideration, will be found annexed to the charaater which is given of the genus formed by the union of Lunaria libyca with Alyssum maritimum, a plant also in the collection, from the neighbourhood of Tripoli.

Alyssum maritimum, which is described both as an Alyssum and as a Clypeola by Linnæus, is the *Konig* of Adanson, who founded his generic distinction on the monospermous cells and supposed want of glands of the receptacle, and M. Desvaux, admitting Adanson's genus, has named it Lobularia. In the second edition of 'Hortus Kewensis' I included this plant in Alyssum, which M. De Candolle has also done in his great work.

For the genus here proposed I shall adopt Adanson's name, altering only the termination, and wishing it to be considered as commemorating the important services rendered to Botany by my friend Mr. Konig, of the British
215] Museum.[1] In comparing these two species of Koniga,

[1] KONIGA.

Konig. *Adans. fam.* 2, *p.* 420. Lobularia. *Desvaux in Journ. de Botan. appl.* 3, *p.* 172. Alyssi sp. *Hort. Kew. ed.* 2, vol. 4, p. 95. *De Cand. Syst. Nat.* 2, *p.* 318. Lunariæ sp. *Viv. Libyc. p.* 34. Farsetiæ sp. *Spreng. Syst. Veg.* 2, *p.* 871.

CHAR. GEN. *Calyx* patens. *Petala* integerrima. *Glandulæ* hypogynæ 8! *Filamenta* omnia edentula. *Silicula* subovata, valvis planiusculis, loculis 1-polyspermis, funiculis basi septo (venoso, nervo deliquescenti) adnatis. *Semina* (sæpissime) marginata. *Cotyledones* accumbentes.

Herbæ (*annuæ v. perennes*) *pube bipartita appressa incanæ*. Folia *integerrima sublinearia*. Racemi *terminales, nunc basi foliati*. Flores *albi*.

Calyx basi subæqualis. *Petalorum* laminæ dilatatæ. *Antheræ* ovatæ. *Glandularum* quatuor per paria filamenta longiora lateraliter adstantes; reliquæ quatuor abbreviatæ geminatim filamenta breviora stipantes. *Dissepimentum*, præter *areolas* ultimas (laminæ duplicis) transversim lineares parietibus (tubulis) rectis subparallelis, *venis* crebre anastomozantibus a *nervo* descendenti e duobus

their agreement is very striking in habit, in leaves, in the closely pressed bipartite pubescence, in the calyx, petals, stamina, and stigma. They correspond, also, in some other points, less obvious but equally important, which I shall separately notice. The first of these is in having eight glands on the receptacle; a character peculiar, I believe, to these plants, and which first suggested the generic name Octadenia. The glands in Alyssum maritimum were entirely overlooked by Adanson, are not noticed by M. Desvaux, and M. De Candolle has described only the four that subtend the longer stamina. These certainly are much more conspicuous than the remaining four, which, however, occupy the place of the only glands existing in several of the most nearly related genera.

The number and position of the glands in this genus give some support, perhaps, to the hypothesis which I have formerly advanced, of the divisions of an hypogynous disk being in most cases formed of abortive filaments; an opinion more strikingly confirmed, however, in this family of plants, by their form and texture in Alyssum calycinum and minimum.

The second point in which the two species of Koniga agree is in the structure of the septum. On this, which I consider as a new source of character in Cruciferæ, I shall offer some remarks in speaking of Farsetia.

arcte approximatis formato supra basin evanescenti in monospermis obsoleto ortis descendentibus. *Funiculi* in dispermis polyspermisque in diversis loculis alterni.

Obs. Koniga ad Alyssinearum tribum *De Cand.* pertinens, hinc Alysso auctorum inde Farsetiæ accedit. Sed Alyssum, uti in Hort. Kew. et De Cand. Syst. Nat. constitutum est, certe divisione eget.

Alyssum *nob.* facile distinguendum sequentibus notis: Silicula subrotunda, disco convexo, limbo compresso, apice retuso, loculis dispermis, funiculis basi septo adnatis et post lapsum seminum persistentibus, supra liberis et cum iisdem deciduis, in diversis loculis oppositis, in eodem a styli basi equidistantibus: Petalis emarginatis: Filamentis omnibus nonnullisve appendiculatis in speciebus omnibus præter A. calycinum in quo filamenta filiformia simplicia sunt et glandularum loco setulæ quatuor filamenta nana æmulantes exstant.

Ad Alyssum sic constitutum et herbas plerumque annuas pube stellari foliisque integerrimis complectens pertinent A. campestre et calycinum, *Linn.*, strigosum *Russell*, minimum *Willd.* et strictum *ejusd.* a quo densiflorum *Desfont.* vix differt; fulvescens *Smith*, umbellatum *Desv.* rostratum *Stev.* micropetalum *Fisch.* hirsutum *Bieb.* aliasque species ineditas.

The third point of agreement is the adhesion of the funiculi umbilicales to the septum. This adhesion, though really existing, is not very obvious in the monospermous cells of Koniga maritima; but in the supposed variety of this species from Teneriffe, in which the cells are occasionally dispermous, it is manifest, and is very remarkable in all states of Koniga libyca.

I first introduced this adhesion of the funiculi to the septum as a generic character in distinguishing Petrocallis from Draba. It has since been advantageously employed in the character of Lunaria by M. De Candolle, who, however, supposes this structure of much rarer occurrence in
216] Cruciferæ than it really is. According to my observations, it is neither unfrequent, nor always of generic importance. Thus, I find it to exist in some species only of Arabis, namely A. Turrita, pendula, and canadensis, and hence I did not introduce it into my generic character of Parrya, though I have noticed it in my description of the species.

The principal difference existing between these two species of Koniga is that the cells of the ovarium and silicula of *K. maritima* are monospermous, while those of *libyca* are polyspermous, the number being variable, apparently indefinite, but not exceeding six. There are, however, other instances in this family, in which the mere difference between definite and indefinite number of seeds is of specific importance only, as in Draba and Meniocus, in each of which a species exists with dispermous cells; and the objection arising from the apparently still greater difference between unity and indefinite number in the two species of Koniga is removed by a supposed third species or variety of K. maritima, in which two seeds are occasionally produced in each cell. It may even be observed that from unity to the indefinite number in this case, where the ovula in the different cells are alternate, the transition is perhaps more easy than from the binary to the indefinite, in cases where, as in Alyssum properly so called, the ovula are placed opposite in the different cells, and are in the same cell equidistant from its apex; this symmetry, probably, admitting of addition only by fours.

The next genus of Cruciferæ to be noticed is FARSETIA, a fragment of the original species of which is in the collection. There are also several specimens of a plant, found in the desert, supposed to be new, and which, though without flowers, and considerably different in the form of its stigma, I am inclined, from the resemblance in habit, in pubescence, in silicula, in seeds, and especially from the exact similarity in the structure of the septum, to refer to the same genus.[1]

As the introduction of the structure of the dissepiment [217
into the generic characters of Cruciferæ is now proposed for the first time, and as I believe that its texture and appearance should always be attended to in constituting genera in this family of plants, I shall here offer a few remarks respecting it.

According to the particular view which I briefly but distinctly published in 1818, and which M. De Candolle first adopted in 1821, of the composition of the pistillum in Cruciferæ,[2] the dissepiment in this family is necessarily

[1] FARSETIA.

Farsetia. *Turra, Farsetia,* p. 5. Farsetiæ sp. *Hort. Kew.* ed. 2, vol. 4, p. 69.. *De Cand. Syst.* 2, p. 286.

CHAR. GEN. *Calyx* clausus, basi vix bisaccatus. *Filamenta* omnia edentula. *Antheræ* lineares. *Silicula* ovalis v. oblonga, sessilis, valvis planiusculis, loculis polyspermis (raro 1-2-spermis), funiculis liberis. *Dissepimentum* uninerve, venosum. *Semina* marginata. *Cotyledones* accumbentes.

Herbæ *suffruticosæ ramosæ, pube bipartita appressa incanæ.* Folia *integerrima.* Racemi *subspicati.*

OBS. Dissepimentum in omnibus exemplaribus utriusque speciei a nobis visis completum, sed in F. ægyptiaca quandoque basi fenestratum, fide D. Desfontaines. (*Flor. Atlant.* 2, *tab.* 160.)

F. ægyptiaca species unica certa est, nam F. stylosa, cujus flores ignoti, ob stigmatis lobos patentes non absque hæsitatione ad hoc genus retuli.

FARSETIA? *stylosa,* ramosissima, siliculis oblongis polyspermis passimque brevè ovalibus 1-2-spermis, stylo diametrum transversum siliculæ subæquante, stigmatis lobis patentibus.

Obs. Exemplaria omnia foliis destituta, sed illorum cicatrices ni fallor obviæ.

[2] In a work published in 1810, the following passage, which has some relation to this subject, occurs:—"Capsulas omnes pluriloculares e totidem thecis conferruminatas esse, diversas solum modis gradibusque variis cohæsionis et solubilitatis partium judico." (*Prodr. Flor. Nov. Holl.* 1, p. 558.) This opinion, however, respecting the formation of multilocular ovaria, might be held, without necessarily leading to the theory in question of the composition of the fruit in Cruciferæ, which I first distinctly stated in an essay on Compositæ, read

218] formed of two lamellæ, derived from the parietes of the fruit. These lamellæ are in many cases easily separable,

before the Linnean Society in February, 1816, and printed in the twelfth volume of their 'Transactions,' published in 1818. In this volume (p. 89), I observe that "I consider the pistillum of all phænogamous plants to be formed on the same plan, of which a polyspermous legumen, or folliculus, whose seeds are disposed in a double series, may be taken as the type. A circular series of these pistilla disposed round an imaginary axis, and whose number corresponds with that of the calyx or corolla, enters into my notion of a flower complete in all its parts. But from this type, and number of pistilla, many deviations take place, arising either from the abstraction of part of the complete series of organs, from their confluence, or from both these causes united, with consequent abortions and obliterations of parts in almost every degree. According to this hypothesis, the ovarium of a syngenesious plant is composed of two confluent ovaria, a structure in some degree indicated externally by the division of the style, and internally by the two cords (previously described), which I consider as occupying the place of two parietal placentæ, each of these being made up of two confluent chordulæ, belonging to different parts of the compound organ."

In endeavouring to support this hypothesis by referring to certain natural families, in which degradations, as I have termed them, are found, from the assumed perfect pistillum to a structure equally simple with that of Compositæ, and after noticing those occurring in Goodenoviæ, I add, "The natural order Cruciferæ exhibits also obliterations more obviously analogous to those assumed as taking place in syngenesious plants; namely, from a bilocular ovarium with two polyspermous parietal placentæ, which is the usual structure of the order, to that of Isatis, where a single ovulum is pendulous from the apex of the unilocular ovarium; and, lastly, in the genus Bocconia, in the original species of which (*B. frutescens*), the insertion of the single erect ovulum has the same relation to its parietal placentæ, as that of Compositæ has to its filiform cords, a second species (*B. cordata*) exists, in which these placentæ are polyspermous."

From this quotation it is, I think, evident, that in 1818 I had published, in my essay on Compositæ, the same opinion, relative to the structure of the pistillum of Cruciferæ, which has since been proposed, but without reference to that essay, by M. De Candolle, in the second volume of his 'Systema Naturale;' and I am not aware that when the essay referred to appeared, a similar opinion had been advanced by M. De Candolle himself, or by any other author; either directly stated of this family in particular, or deducible from any general theory of the type or formation of the pistillum. I am persuaded, however, that neither M. De Candolle, when he published his 'Systema,' nor M. Mirbel, who has very recently adverted to this subject, could have been acquainted with the passage above quoted. This, indeed, admits of a kind of proof; for if they had been aware of the concluding part of the quotation, the former author would probably not have supposed that all the species referred to Bocconia were monospermous (*Syst. Nat.* 2, p. 89); nor the latter that they were all polyspermous. (*Mirbel in Ann. des Scien. Nat.* 6, p. 267). Respecting *Bocconia cordata*, though it is so closely allied to Bocconia as to afford an excellent argument in favour of the hypothesis in question, it is still sufficiently different, especially in its polyspermous ovarium, to constitute a distinct genus, to which I have given the name (MACLEAYA *cordata*) of my much valued friend Alexander Macleay, Esq., Secretary to the Colony of New South Wales, whose merits as a general naturalist, a profound entomologist, and a practical botanist, are well known.

and where their union is more intimate, their existence is still evident from the want of correspondence, and consequent decussation of their areolæ. The lamellæ, which are usually very thin and transparent, have their surface divided into areolæ, in different genera of very different forms, some of which may, with sufficient clearness, be described. In many cases, no other appearance exists; in some, however, the axis of the septum resembles either a single nerve, or two distinct parallel nerves; and from this axis, whether formed of one or two nerves, tubes having the appearance and ramification of the veins of a leaf, and which generally terminate within the margin, not unfrequently proceed. This is remarkably the case in Farsetia, as I here propose to limit that genus; the central vessels in both its species being closely approximated, so as to form a single cord, extending from the apex to the base of the septum, and the veins being numerous and uncommonly distinct. Approaches more or less manifest to this structure of Farsetia exist in several other genera, as in Parrya, Savignya, and Koniga. But in this last-mentioned genus [219 the nerve, which originates, as in all cases, at the apex, hardly extends, even in the polyspermous species, beyond the middle of the septum, and the veins which are much less distinct, are descendent.

As far as my observations on this subject at present extend, I expect, with great confidence, uniformity in the structure of the septum of strictly natural genera, and in many cases, though certainly not in all, I have found a resemblance in this respect in more extensive groups. Thus Draba, Arabis, and Aubrietia, agree in having amorphous areolæ, bounded by flexuose tubes or lines; while Alyssum, Berteroa, and Fibigia, have narrow linear areolæ, bounded by parallel or slightly arched lines. Capsella bursa differs from Thlaspi and Æthionema, as Draba from Alyssum, and agrees with Lepidium procumbens, *Linn.*, improperly referred to Hutchinsia, and which equally has incumbent cotyledones. Cochlearia differs in like manner from Kernera. And numerous other examples of the same agreement in nearly related plants, and of differences where the usual sources of distinction are less available, might be noticed.

HESPERIS NITENS of Viviani is sparingly in the herbarium, both in flower and fruit. The seeds, though not ripe, are sufficiently advanced to show that the direction of the cotyledons is in this stage accumbent; and, as I have found in Cruciferæ generally that the ultimate agrees with the early state of cotyledons, I conclude they are likewise accumbent in the ripe seed. The plant is also abundantly different from Hesperis in other respects, and does not appear to be referable to any genus yet published. This new genus* I have dedicated to the memory of Dr. Oudney, who found the present species in many of the wadeys between Tripoli and Mourzuk, and remarks that camels and mules eat it.

220] HESPERIS RAMOSISSIMA, which is also in the herbarium, was found in Fezzan. This plant differs in aspect from most of the other species of Hesperis, approaching in some points to Malcomia, in others to Mathiola; and as its cotyledons are very obliquely incumbent, it may form a section or subgenus, with a name, Hesperis (Plagiloba) ramosissima, indicating that character.

CAPPARIDEÆ, of which eight species occur in the collection, is the family next to be noticed. I consider this order as belonging to the same natural class with Cruciferæ; and that this class includes also Resedaceæ, Papaveraceæ, and Fumariaceæ.

M. De Candolle, in defining Capparideæ, appears to

[1] OUDNEYA.

CHAR. GEN. *Calyx* clausus, basi bisaccatus. *Filamenta* distincta, edentula. *Stigmata* connata apicibus distinctis. *Siliqua* sessilis linearis rostrata, valvis planis uninerviis, funiculis adnatis, septo avenio areolarum parietibus subparallelis. *Semina* uniseriatia. *Cotyledones* accumbentes.

Suffrutex (O. Africana *nob.* Hesperis nitens, *Viv. lib. p.* 38, *tab.* 5, *f.* 3), *glaberrimus, ramosus.* Folia *integerrima sessilia avenia, inferiora obovata, superiora sublinearia.* Racemi *terminales, ebracteati.* Flores *mediocris magnitudinis, petalorum laminis obovatis venosis.*

Obs. Oudneya ab Arabidi differt stigmatis forma, siliquæ rostro, et dissepimenti areolarum figura. Parrya ad quam genus nostrum accedit diversa est dissepimento binervi venoso! calyce haud clauso, siliquæ forma, et seminibus biseriatis testa corrugata.

regard the ovarium as having in all cases only two placentæ, and therefore formed of two pistilla or carpella. But to this, which is certainly the more usual number, there are many exceptions. These exceptions occur chiefly in the genus Capparis, which, as it is at present constituted, includes species differing from each other in having an ovarium with from two to eight placentæ, and consequently composed of an equal number of pistilla. Capparis spinosa is the most decided instance of the increased number of placentæ, and this, as well as some other nearly related species, are also remarkable in having septa subdividing the placentæ, and uniting in the centre of the compound ovarium.

In the herbarium there are three species of the genus Cleome. Two of these, C. pentaphylla and arabica, are in many respects well-known plants; the third I believe to be an undescribed species, but nearly related to monophylla.

If the very natural group, formed by the Linnean genus Cleome, is not to be preserved entire, its subdivision must be carried much further, and established on other grounds, than has been done by M. De Candolle, whose genera and sections appear to me to have been equally founded on partial considerations. Thus, his *Polanisia*, uniting all the Cleomes whose stamina exceed six, contains in its first section, in addition to the species from which the genus was formed, at least two sets of plants, having very little affinity either with each other or with the original species, whose only congener is placed in a second section.

Gynandropsis also consists of two groups not very intimately connected; the first is composed of species belonging to South America, and having the usual æstivation of the family: the second, of which *C. pentaphylla* may be taken as the type, is chiefly African, and is readily distinguished by its very different æstivation,—the great peculiarity of which consists in the petals not covering the stamina at any period. To this mode of æstivation [22] of petals, which has never before been noticed, though it equally exists in Crateva and in Resedaceæ, I shall apply the term *aperta*. It is constantly conjoined, and, perhaps, necessarily connected, with the early opening of the calyx, whose segments are originally connivent and slightly imbricate: for it may be here remarked, that in all the modifications of what I have termed imbricate æstivation of petals, they are, I believe, in the very early stage in like manner erect, and the sexual organs equally exposed.

If the expediency of preserving the genus Cleome entire were admitted, a question which I do not pretend at present to decide, it would still be of the greatest importance to arrange its numerous species according to their affinities, and carefully to distinguish the subordinate groups that compose it. To such inferior groups, whether termed subgenera or sections, names, in fact, have been of late years very generally assigned, both by zoologists and botanists.

It has not yet been proposed, however, that these subgeneric names should form an essential part of the name of the species; although by employing them in this manner, while the principal groups would be kept in view, their subdivision would be carried to the same extent, and the subordinate groups as well expressed, as if they had been actually separated into distinct genera.

The adoption of this method, which would not materially disturb names already existing, would probably lead to a greater consistency in the formation of genera, with reference to the natural orders of which they are subdivisions. In this way also the co-operation of two classes of naturalists, at present opposed to each other on the question of the construction of genera, might to a certain extent be expected, and greater uniformity in nomenclature consequently secured.

These advantages appear to me so important, that some expedient for obtaining them will, I am persuaded, at no distant period, be generally adopted.

In favour of the present plan it may be remarked, that it is analogous to the method followed by the Romans in the construction of the names of persons, by which not only the original family, but the particular branch of that family to which the individual belonged was expressed. Thus, the generic name corresponds with the nomen (Cornelius), the name of the section with the cognomen (Scipio), and that of the species with the prænomen (Publius).

Without attempting at present to obviate the objections to which the proposed innovation is no doubt liable, I shall proceed to apply it to Cleome pentaphylla. Accord- [222
ing to my view, the genus Cleome would include Gynandropsis, a name which, as that of a section, may be continued to those species of M. De Candolle's genus belonging to equinoctial America, and having the common æstivation of the family: while *Gymnogonia,* derived from its remarkable æstivation, may be employed for the section that includes C. pentaphylla, of which the name might be given in the following manner:

Cleome (Gymnogonia) pentaphylla. This plant, the earliest known species of Cleome, and that on which the genus was chiefly constituted, was found in Bornou. The species is regarded by M. De Candolle as a native of the West India Islands, and he doubts whether it may not also belong to Egypt and India. On the other hand, I consider it a native of Africa and India, and am not satisfied with the evidence of its being also indigenous to the American Islands, where, though now very common, it has probably been introduced by the negroes, who use it both as a potherb and in medicine. It is not unlikely that M. De Candolle, in forming his opinion of the original country of this plant, has been in part determined by finding several species of his Gynandropsis decidedly and exclusively natives of the new continent. But if I am correct in separating these species from the section to which Cleome (Gymnogonia) pentaphylla belongs, this argument, which I have formerly applied to analogous cases,[1] would be clearly in favour of the opinion I have here advanced; those species of the section with which I am acquainted being undoubtedly natives of Africa or of India.

Cleome (Siliquaria) Arabica (*Linn. sp. pl. ed.* 2, *p.*

[1] *Tuckey's Congo, p.* 469. (*Antè, p.* 156.)

939, *De Cand. prodr.* 1, p. 240), a supposed variety of which was found both in the neighbourhood of Tripoli and in Soudan, belongs to another subdivision of the genus, equally natural, and readily distinguishable. The species of this subdivision are included in M. De Candolle's second section of Cleome, but are there associated with many other plants, to which they have very little affinity.

All the species of *Cleome Siliquaria* are indigenous to North Africa and Middle Asia, except *violacea*, which is a native of Portugal. *Cleome deflexa* of M. De Candolle (*prodr.* 1, p. 240), founded on specimens in Mr. Lambert's herbarium, which were sent by Don Joseph Pavon as belonging to Peru, seems to present a remarkable exception to this geographical distribution of the section. But on examining these specimens I find them absolutely iden-
228] tical with some states of *violacea*. I think it probable, therefore, either that they are erroneously stated to have come from Peru, or that this species may have been there introduced from European seeds.

Cadaba farinosa (*Forsk. Arab.* p. 68, *De Cand. prodr.* 1, p. 244) is in the herbarium from Bornou. The specimen is pentandrous, and in other respects agrees with all those which I have seen from Senegal, and with Strœmia farinosa (*Antè, p.* 94) of my catalogue of Abyssinian plants, collected by Mr. Salt, and published in his travels. M. De Candolle, who had an opportunity of examining this Abyssinian plant, refers it to his *C. dubia*, a species established on specimens found in Senegal, and said to differ from *farinosa*, slightly in the form of the leaves and in being tetrandous. Of the plant from Abyssinia I have seen only two expanded flowers, one of which is decidedly pentandrous, the other apparently tetrandrous. Mr. Salt, however, from an examination of recent specimens, states it to be pentandrous. It is probably, therefore, not different from C. farinosa of Forskal, whose specimens M. De Candolle has not seen. And as the form of the leaves is variable in the specimens from Senegal, and not elliptical, but between oval and oblong, in those of Abyssinia, C. dubia is probably identical with, or a variety merely of, farinosa, as M. De Candolle himself seems to suspect.

Crateva Adansonii (*De Cand. prodr.* 1, *p.* 243) is in the collection from Bornou. This species is established by M. De Candolle upon a specimen in M. de Jussieu's herbarium, found in Senegal by Adanson, and is supposed to differ from all the other species in having its foliola equal at the base. I have examined the specimen in M. de Jussieu's herbarium, in which, however, the leaves not being fully developed, I was unable to satisfy myself respecting their form, but in a specimen, also from Senegal, which I received from M. Desfontaines, the lateral foliola, though having manifestly unequal sides, are but slightly unequal at the base, and the inequality consists in a somewhat greater decurrence of the lamina on the anterior or inner margin of the footstalk. As well as can be determined, in very young leaves, this is also the case in the specimen from Bornou; and it is manifestly so in my specimen of *C. læta*, which appears to belong to the same species.

Crateva læta was founded by M. De Candolle on a plant from Senegal, communicated by M. Gay, from whom I also received a specimen in 1824, with the remark that it was not different from C. Adansonii. In that specimen [224
the flowers are male with an imperfect pistillum; in the plant from Bornou they are hermaphrodite, with elongated filaments; and in the specimen received from M. Desfontaines they are also hermaphrodite, but the stamina, though apparently perfect, are fewer in number and shorter than the stipes of the ovarium. I have observed, however, the flowers to be in like manner polygamous in some other species of Crateva, belonging both to India and America, a fact which materially lessens the dependence to be placed on characters taken from the number and length of the stamina in this genus.

Crateva Adansonii, it would appear, then, is the only known species of the African continent, for C. fragrans does not belong to the genus, and it will be difficult to distinguish this African Crateva from a plant which seems to be the most general species of India; except that in the latter, as in all the other species of the genus, the inequality of the lateral foliola, which is also more marked, consists in the

greater decurrence of the lamina being on the outer or posterior margin of the footstalk. This Indian species, which may be named *C. Roxburghii*, is the Capparis trifoliata of Dr. Roxburgh's manuscripts, but not Nürvala of Hortus Malabaricus (*vol.* 3, *p.* 49, *t.* 42), as he considers it. I have little doubt of its being also the plant described as C. Tapia, by Vahl (*symb.* 3, *p.* 61), his specific character well according with it, and not applying, as far as relates to the petals, to any known species of America. But as this character is adopted by Sir James Smith (*in Rees's Cyclop.*), it may likewise be C. Tapia of the Linnean herbarium; a conjecture the more probable as Linnæus has distinguished his Tapia by its ovate petals from gynandra, in which they are said to be lanceolate (*Sp. pl. ed.* 2, *p.* 637). This celebrated herbarium, however, is here of no authority, for Linnæus was never in possession of sufficient materials to enable him to understand either the structure and limits of the genus Crateva, or the distinctions of its species; and the specific name in question, under which he originally included all the species of the genus, ought surely to be applied to an American plant, at least, and if possible, to that of Piso, with whom it originated. It is hardly to be supposed that the plant intended by Piso can now with certainty be determined; the only species from Brazil, however, with which I am acquainted, well accords with his figure and short description. This Brazilian species is readily distinguishable both from C. Adansonii and Roxburghii, by the form of its petals, which, as in all the other
225] American species, are narrow-oblong or lanceolate; and from C. gynandra by the shortness of its stipes genitalium, or torus.

Crateva Tapia, so constituted, is, on the authority of a fragment communicated by Professor Schrader, the *Cleome arborea* of that author (*in Gœtt. Anzeig.* 1821, *p.* 707, *De Cand. Prodr.* 1, *p.* 242); nor is there anything in the character of *C. acuminata* of De Candolle (*Prodr.* 1, *p.* 243), which does not well apply to our plant.

C. Tapia, as given by M. De Candolle (*op. cit.*), is characterised chiefly on the authority of Plumier's figure, in the accuracy of which, either as to the number or length of stamina, it is difficult to believe, especially when we find it also representing the petals inserted by pairs on the two upper sinuses of the calyx.

The genus Crateva agrees, as I have already stated, in the remarkable æstivation of its flower with Cleome Gymnogonia, by which character, along with that of its fruit, it is readily distinguished from every other genus of the order. Although this character of its æstivation has never before been remarked, yet all the species referred to Crateva by M. De Candolle really belong to it, except *C. fragrans*, which, with some other plants from the same continent, forms a very distinct genus, which I shall name RITCHIEA, in memory of the African traveller, whose botanical merits have been already noticed.

CAPPARIS SODADA *nob.* Sodada decidua, *Forsk. Arab. p.* 81. *Delile, Flore d'Egypte, p.* 74, *tab.* 26. *De Cand. Prodr.* 1, *p.* 245.

The specimen in the herbarium is marked by Dr. Oudney as belonging to a tree common on the boundaries of Bornou. It is probably the *Suag*, mentioned in his journal, observed first at Aghedem, and said to be "a tetrandrous plant, having a small drupa, which is in great request in Bornou and Soudan, for removing sterility in females: it is sweetish and hot to the taste, approaching Sisymbrium Nasturtium;" and that "in passing the plant a heavy narcotic smell is always perceived."

I have here united Sodada with Capparis, not being able to find differences sufficient to authorise its separation even from the first section of that genus, as given by De Candolle.

Forskal describes his plant as octandrous, and M. De Candolle has adopted this number in his generic character. M. Delile (*op. cit.*), however, admits that the stamina vary from eight to fifteen; and, in the specimen which I received [226
from M. Jomard, I have found from fourteen to sixteen. But were the number of stamina even constantly eight, this alone would not justify its separation from Capparis,

several octandrous species of which, belonging to the same section, are already known.

Another species of Capparis, also from Bornou, exists in the herbarium. It appears to be undescribed, and to belong to M. De Candolle's first section of the genus; but the specimen is too imperfect to be satisfactorily determined.

Both these species have aculei stipulares, and it may here be remarked that all the plants belonging either to Capparis, or to any of the genera of the order whose fruit is a berry, in which these aculei are found, are indigenous either to Asia, Africa, or Europe; while all the aculeated Cleomes, with the exception of perhaps a single African species, are natives of equinoctial America.

Mærua rigida. This plant, of which flowering specimens were collected at Aghedem, certainly belongs to Forskal's genus Mærua, adopted by Vahl and De Candolle; and I believe it to be a species distinct from the three already published. It is very nearly related, however, to a fourth species (M. Senegalensis *nob.*), of which I received a specimen from M. Desfontaines. M. De Candolle has placed the genus Mærua at the end of Cappárideæ, between which and Passifloreæ he considers it intermediate. This view of its relation to these two orders I cannot adopt. To me it appears truly a Capparidea, having very little affinity with Passifloreæ, to which it seems to approach in one point only, namely, the corona of the calyx. But of a similar corona rudiments exist in several other African Cappárideæ, and from some of these the genus Mærua is with difficulty distinguished.[1]

[1] MÆRUA.

Mærua. *Forsk. Arab.* p. 104. *Vahl. Symb.* 1, p. 36. *De Cand. Prodr.* 1, p. 254.

Char. Gen. *Calyx* tubulosus: *limbo* 4-partito, æstivatione simplici serie valvata: *corona* faucis petaloidea. *Petala* nulla. *Stipes genitalium* elongatus. *Stamina* numerosa. *Pericarpium* (siliquiforme ?) baccatum.

Frutices *inermes, pube, dum adsit, simplici.* Folia *simplicia coriacea: petiolo cum denticulo rami articulato:* stipulis *minutissimis setaceis.*

Mærua *rigida,* corymbis terminalibus paucifloris, foliis obovatis crassis rigidis aveniis nervo obsoleto, corona lacero-multipartita.

Desc. Frutex ? *Rami* stricti teretes tenuissime pubescentes. *Folia* sparsa,

RESEDACEÆ. The herbarium contains two species of [227 Reseda. The specimens of one of these are too imperfect to be determined. The other is probably undescribed, though very nearly related to R. suffruticulosa, and undata of Linnæus. This supposed new species (*Reseda propinqua*) was found near Tripoli by Mr. Ritchie, and between Tripoli and Mourzuk by Dr. Oudney. It is remarkable in having the ungues of all the petals simple; that is, neither dilated, thickened, nor having any process or appendage at the point of union with the trifid lamina, into which they gradually pass. We have here, therefore, a species of Reseda with petals not different in any respect from those of many other families of plants; and, although this is an exception to their usual structure in the genus, I shall endeavour to show that all the deviations existing, however complex in appearance, are reducible to this more simple state of the organ.

RESEDACEÆ, consisting of *Reseda*, divisible into sections or subgenera, and *Ochradenus*, which may perhaps

obovata cum mucronulo brevissimo, plana, semiunguicularia, utrinque pube tenuissima brevissima simplici, nervo obsoleto, venis fere inconspicuis. *Petioli* lineam circiter longi. *Stipulæ* laterales, setaceæ, petioli dimidio breviores, ramulo appressæ, post lapsum folii persistentes. *Ramuli floriferi* sæpius laterales abbreviati, e foliis confertis floribusque corymboso-fasciculatis (3—6), quorum exteriores folio subtensi; quandoque corymbus ramum terminat. *Pedunculi* teretes, tenuissime pubescentes, ebracteati excepto foliolo florali dum adsit ejusque stipulis vix conspicuis. *Calyx* infundibuliformis, extus tenuissime pubescens: *tubus* subcylindraceus, 8-striatus striis elevatis æqualibus, intus lineis duabus prominulis subcarnosis, cum limbi laciniis alternantibus, altera crassiore: *limbus* tubo paulo longior, 4-partitus laciniis æqualibus, ovatis acutiusculis, obsolete venosis, 5-nerviis, nervis extimis margini approximatis, e furcatione costarum quatuor tubi cum laciniis alternantium ortis; æstivatione simplici serie valvata marginibus tamen paulo inflexis. *Corona faucis* monophylla, laciniis limbi multoties brevior, lacero-multipartita lacinulis subulatis inæqualibus. *Stipes genitalium* liber, cylindraceus, glaber, altitudine tubi. *Stamina: Filamenta* indeterminatim numerosa, viginti circiter, filiformia, glabra, æstivatione contortuplicata. *Antheræ* incumbentes, ovali-oblongæ obtusæ, basi semibifidæ, loculis parallelo-approximatis, intus longitudinaliter dehiscentibus, æstivatione erectæ. *Ovarium* e centro filamentorum stipitatum, cylindraceum, glabrum, uniloculare placentis duabus parietalibus polyspermis. *Stylus* nullus. *Stigma* depresso-capitatum.

OBS. Species hæcce proxime accedit Mæruæ senegalensi *nob.* quæ vix pubescens et foliis venosis distincta; in multis quoque convenit, fide descriptionis Forskalii, cum Mærua uniflora *Vahl*, a nobis non visa. Mærua angolensis, *De Cand.* (in Museo Parisiensi visa), cui flores pariter corymbosi et corona lacero-multipartita, satis diversa est foliis ovalibus.

be regarded as only one of these subdivisions, I consider very nearly related to Capparideæ, and as forming part of the same natural class. It differs in the variable number
228] of the parts of its floral envelopes, from the other orders of the class, in which the quaternary or binary division is without exception; and it is especially remarkable in having the ovarium open even in its earliest state. From Cruciferæ and Capparideæ, two families of the class to which they most nearly approach, Resedaceæ also differ in the apparent relation of the stigmata to the placentæ. The stigmata in this order terminate the lobes of the pistillum, and as these lobes are open sterile portions of the modified leaves, from the union of which in the undivided part I suppose the compound ovarium to originate, they necessarily alternate with the placentæ. I have generally found, however, the upper part of each placenta covered by a fleshy or fungous process, which is connected with the margins of the lobes, and therefore with the stigmata, and is probably essential to the fecundation of the ovula. The singular apparent transposition of the placentæ in Sesamoides of Tournefort, so well described by M. Tristan in his ingenious 'Memoir on the Affinities of Reseda,'[1] appears to me necessarily connected with the extreme shortness of the undivided base of the ovarium; for in supposing this base to be elongated, the placentæ would become parietal, and the ovula, which are actually resupinate, would assume the direction usual in the order.

M. De Jussieu, in his *Genera Plantarum*, has included Reseda in Capparideæ, and to this determination I believe he still adheres. M. Tristan, in the memoir referred to, is inclined to separate it as a family intermediate between Passifloreæ and Cistineæ, but more nearly approaching to the latter. M. De Candolle, who first distinguished Reseda as an order under the name here adopted, in 1819[2] placed it between Polygaleæ and Droseraceæ, and consequently at no great distance from Capparideæ. He must, since, however, have materially altered his opinion respecting it; for

[1] *Annal. du Mus. d'Hist. Nat.* 18, *p.* 392. [2] *Théor. Elem. ed.* 2, *p.* 244.

the order Resedaceæ is not included in the first or second part of his 'Prodomus,' and I can find no observation respecting it in these two volumes. It is probable, therefore, that he may intend to place it near Passifloreæ, as suggested by M. Tristan, or, which is more likely, that he has adopted the hypothesis lately advanced, and ingeniously supported, by Mr. Lindley, respecting its structure and affinities.[1]

According to this hypothesis, in Reseda the calyx of authors is an involucrum, its petals neutral flowers, and [229
the disk or nectary becomes the calyx of a fertile floret in the centre; and, as a deduction from this view of its structure, the genus has been placed near Euphorbiaceæ.

The points in the structure of Reseda, which appear to have led Mr. Lindley to this hypothesis, are the presence and appearance of the hypogynous disk, the anomalous structure of the petals, and the singular æstivation of the flower; but it is no slight confirmation of the correctness of M. De Jussieu's opinion, that all these anomalies occur in a greater or less degree in Cappårideæ, and have been found united in no other family of plants. The remarkable æstivation of Reseda equally exists in Crateva, and in more than one subdivision of the genus Cleome; the hypogynous disk is developed in as great a degree in several Capparideæ; and an approximation to the same kind of irregularity in the petals occurs in two sections of Cleome.

The analogical argument alone then might, perhaps, be regarded as conclusive against the hypothesis. But the question, as far as relates to the petals, and consequently to the supposed composition of the flower, may be decided still more satisfactorily on other grounds. Both MM. Tristan and Lindley regard the upper divided membranaceous part of the petal as an appendage to the lower, which is generally fleshy. On the other hand, I consider the anomaly to consist in the thickening, dilatation, and inner process of the lower portion, and that all these deviations from ordinary structure are changes which take place after

[1] *Collect. Bot. tab.* 22.

the original formation of the petal. To establish these points, and consequently to prove that the parts in question are simple petals, and neither made up of two cohering envelopes, as M. Tristan supposes, nor of a calyx and abortive stamina, according to Mr. Lindley's hypothesis, I shall describe their gradual development, as I have observed it in the common Mignonette, a plant in which all the anomalies that have led to this hypothesis exist in a very great degree.

The flower-bud of Reseda odorata, when it first becomes visible, has the divisions of its calyx slightly imbricate and entirely enclosing the other parts. In this stage the unguis of each of the two upper petals is extremely short, not broader than the base of the lamina, and is perfectly simple; there being no rudiment of the inner process so remarkable in the fully expanded flower. The lamina at the same period may be termed palmato-pinnatifid, its divisions are all in the same plane, the terminating or middle segment is whitish or opaque, and several times longer than the lateral segments, which are semi-transparent.

230] Of the remaining four petals, the two middle are dimidiato-pinnatifid, their lateral segments existing only on the upper side; and the two lower are undivided, being reduced to the middle segment or simple lamina. All the petals are erect, and do not cover the stamina in the slightest degree, either in this or in any other stage. The disk is hardly visible. The antheræ are longer than their filaments, of a pale-green colour; those on the upper or posterior side of the flower being manifestly larger, and slightly tinged with brown. The pistillum is very minute and open at the top. In the next stage the calyx is no longer imbricate, but open; the petals have their segments in nearly the same relative proportions; the interior margin of the unguis is just visible; but the transition from unguis to lamina is still imperceptible; the apex of the former not being broader than the base of the latter. It is unnecessary to follow the development through the more advanced stages of the flower, the facts already stated being, in my opinion, absolutely conclusive as to the real nature of the parts in question: and I may remark that similar observations on certain genera of Caryophylleæ, especially Dianthus, Lychnis, and Silene, clearly establish the analogy between their petals and those of Reseda.

I am aware that it has lately been proposed to include *Datisca* in Resedaceæ, to which it is nearly similar in the structure of its ovarium, as M. de Jussieu has long since remarked. But this is the only point of resemblance between them; for the calyx of Datisca is certainly adherent, and in most of its other characters it differs widely both from Reseda and from every other genus yet published. Among the numerous discoveries made by Dr. Horsfield in Java, there is a genus (TETRAMELES *nob.*), however, manifestly related to Datisca, and remarkable in the regular quarternary division of every part of its diœcious flowers. These two genera form an order very different from every other yet established, and which may be named DATISCEÆ.

CARYOPHYLLEÆ. Five species only of this family were collected near Tripoli, none of which are new.

Of ZYGOPHYLLEÆ, six species exist in Dr. Oudney's herbarium, namely, Tribulus terrestris, found in Bornou; Fagonia cretica, from Tripoli to Benioleed; Fagonia arabica, at Aghedem; Fagonia Oudneyi *nob.* with Zygophyllum simplex in Fezzan; and Zygophyllum album everywhere in the desert.

This family, so distinct in habit from Diosmeæ or Rutaceæ, with which it was formerly united, is not easily characterised by any very obvious or constant peculiarities in its parts of fructification.

The distinguishing characters in its vegetation or habit are the leaves being constantly opposite, with lateral or [231
intermediate stipulæ, being generally compound, and always destitute of the pellucid glands, which universally exist in true Diosmeæ, though not in all Rutaceæ properly so called.

M. Adrien de Jussieu, in his late very excellent Memoir on the great order or class Rutaceæ, in distinguishing

Zygophylleæ[1] from the other subdivisions of that class in which he has included it, depends chiefly on the endocarp, or inner lamina of the pericarp, not separating from the outer lamina or united epicarp and sarcocarp, and on the texture of the albumen. His first section of Zygophylleæ, however, is characterised by the want of albumen; and in his second section I find exceptions to the remaining character, especially, in Fagonia Mysorensis, in which the two laminæ of the ripe capsule separate as completely as in Diosmeæ. Another plant, in my opinion, referable to the same order, and which, in memory of a very meritorious African traveller, I have named *Seetzenia africana,* has in its ripe capsule the epicarp, or united epicarp and sarcocarp, confined to the dorsal carina of each cell, the endocarp being the only membrane existing on the sides, which are exposed long before the bursting of the fruit. The plant in question has, indeed, many other peculiarities, some of which may, perhaps, be considered sufficient to authorise its separation from the order to which I have referred it; for the æstivation of its calyx is valvular, it has no petals, its five styles are distinct to the base, and the cells of its ovarium appear to me to be monospermous. It completely retains, however, the characters of vegetation, on which I chiefly depend in distinguishing Zygophylleæ; and I have no doubt of its being Zygophyllum lanatum of Willdenow,[2] by whom it is stated to be a native of Sierra Leone; I suppose, however, on insufficient authority, for the specimens in the Banksian herbarium, from which I have made my observations, were found in South Africa near Olifant's River, by Francis Masson.

In all the species of Fagonia, and in the two species of Zygophyllum in Dr. Oudney's collection, a character in the fructification still remains which is not found in Diosmeæ or Rutaceæ, and which, were it general in Zygophylleæ, would satisfactorily distinguish this order from all the families it has usually been compared with. This character consists in the direction of the embryo with relation to the

[1] *Mém. du Mus. d'Hist. Nat.* 12, *p.* 450.

[2] *Sp. Plant.* 2, *p.* 564.

insertion of the funiculus, its radicle being seated at the opposite extremity of the seed, or to express, in the unimpregnated ovarium, the infallible indication of this [232 position, the direction of the inner membrane and nucleus of the ovulum corresponds with that of its testa.

But this character, in general very uniform in natural families, and which, equally existing in Cistineæ, so well defines the limits of that order, as I have long since remarked,[1] would seem to be of less importance in Zygophylleæ.

M. Adrien de Jussieu, who, in his memoir already cited, admits its existence in Fagonia, and in both our species of Zygophyllum, considers it as an exception to the general structure of the latter genus, in the definition of which he retains the character of "radicula hilo proxima." I believe, however, that in all the species of Zygophyllum, except Fabago, which, possesses, also, other distinguishing characters, this opposition of the radicle to the external hilum will be found; for in addition to the two species contained in the herbarium, in both of which it is very manifest, I have observed it in Z. coccineum, and in all the species of South Africa that I have had an opportunity of examining. In some of these species, indeed, it is much less obvious, partly from the greater breadth of the funiculus, and also from its being closely applied, or even slightly adhering to the testa of the seed. But hence it is possible to reconcile the structure of these species with that of Fabago itself, in which the raphe seems to me to be external: and if this be really the case, Fabago differs from those Zygophylla of South Africa alluded to, merely in the more intimate union of the funiculus with the surface of the testa. Whether this observation might be extended to the other genera of the order, I have not yet attempted to ascertain.

Balanites Ægyptiaca, though not belonging to Zygophylleæ, may be here mentioned. The specimen is from Bornou, but, like all the other plants of that country, has no particular place of growth indicated, nor is there any

[1] In *Hooker's Flora Scotica*, p. 284.

observation respecting it. For a very full and interesting history of this plant, I may refer to M. Delile's 'Flore d'Egypte' (*p.* 77, *tab.* 28).

Of CISTINEÆ, three species were observed between Tripoli and Mourzuk.

The GERANIACEÆ of the collection consist of four species of *Erodium*, all of which were found on the same journey.

Of MALVACEÆ, considered as a class, there are twelve species in the herbarium. Only two of these are particularly deserving of notice. The first, *Adansonia digitata*, found in Soudan, where the tree is called Kouka, is described by Captain Clapperton; the second, *Melhania*
233] *Denhamii*, a new and remarkable species of the genus, differing from all the others in having its bracteæ regularly verticillated and, at the same time, longer and much broader than the divisions of the calyx.

A single species of VITIS is in the collection, from Bornou.

NEURADA PROSTRATA, generally referred to Rosaceæ, was found in Wady Ghrurbi.

TAMARISCINEÆ. A species of Tamarix, apparently not different from T. gallica, is the *Attil*, common in Fezzan, where, acccording to Dr. Oudney, it is the only shady tree.

LORANTHEÆ. A species of Loranthus, parasitical on the Acacia nilotica, was observed very commonly from Fezzan to Bornou.

LEGUMINOSÆ. Of this class the herbarium contains thirty-three species, among which there are hardly more than two undescribed, and these belonging to a well-established genus.

Of the order or tribe MIMOSEÆ only three species occur, namely, Acacia nilotica, Mimosa Habbas, and *Inga biglobosa*, or a species very nearly related to it. Of this last-named plant, I judge merely from ripe fruits adhering to the singular club-shaped receptacle, or axis of the spike. The specimens were collected in Soudan, and belonging to a tree of considerable importance to the inhabitants of that country, by whom it is called *Doura*. According to Captain Clapperton, "The seeds are roasted as we roast coffee, then bruised, and allowed to ferment in water; when they begin to become putrid, they are well washed and pounded; the powder made into cakes, somewhat in the fashion of our chocolate; they form an excellent sauce for all kinds of food. The farinaceous matter surrounding the seed is made into a pleasant drink, and they also make it into a sweetmeat." The Doura of Captain Clapperton is probably not specifically different from the Nitta mentioned by Park in his 'First Journey'; nor from Inga biglobosa of the 'Flore d'Oware' of M. De Beauvois, according to whom it is the Nety of Senegal; and he also well remarks that Inga biglobosa, described by Jacquin as a native of Martinico, has probably been introduced into that island by the Negroes, as he himself found it to have been in St. Domingo.

Inga Senegalensis of M. De Candolle (*Prodr.* 2, *p.* 442) may also belong to the same species.

It is possible, however, that some of the plants here mentioned, though very nearly related to each other, and having all the same remarkable club-shaped spike, may be specifically distinct; for it appears from specimens collected at Sierra Leone by Professor Afzelius that two [234
plants having this form of spike are known in that colony, and two species, with similar inflorescence, probably distinct from those of Africa, are described in the manuscript 'Flora Indica' of Dr. Roxburgh. All these plants possess characters fully sufficient to distinguish them from Inga, to which they have hitherto been referred. The new genus which they form, one of the most striking and beautiful in equinoctial Africa, I have named PARKIA,[1] as a tribute of

[1] PARKIA.

ORD. NAT. *Leguminosæ-Mimoseæ*: Cæsalpineis proximum genus.

CHAR. GEN. *Calyx* tubulosus ore bilabiato (⅔); æstivatione imbricata! *Petala* 5, subæqualia, supremo (paulo) latiore; æstivatione conniventi-imbricata. *Stamina* decem, hypogyna, monadelpha. *Legumen* polyspermum: *epicarpio* bivalvi; *endocarpio* in loculos monospermos sarcocarpio farinaceo tectos solubili.

Arbores (*Africanæ et Indiæ orientalis*) *inermes.* Folia *bipinnata, pinnis folio-*

respect to the memory of the celebrated traveller, by whom the fruit of this genus was observed in his first journey, and who, among other services rendered to botany, ascertained that the plant producing Gum Kino is a species of Pterocarpus.[1] I have formerly endeavoured to distinguish Mimoseæ from Cæsalpineæ, by the valvular æstivation of both its floral envelopes, and by the hypogynous insertion of its stamina. Instances of perigynous insertion of stamina have since been noticed by MM. Kunth and Auguste de St. Hilaire; but no exception has been yet pointed out to the
235] valvular æstivation of their calyx and corolla. Parkia, however, differs from other Mimoseæ, not only in its æstivation, which is imbricate, but in the very manifest irregularity of its calyx, and in the inequality of its petals, which, though less obvious, is still observable.

Erythrophleum, another genus indigenous to equinoctial Africa, which I have elsewhere[2] had occasion to notice, and then referred to Cæsalpineæ, more probably belongs to Mimoseæ, although its stamina are perigynous. In this genus both calyx and corolla are perfectly regular, and their æstivation, if not strictly valvular, is at least not manifestly imbricate, though the flower-buds are neither acute nor angular. In Erythrophleum and Parkia, there-

lisque multijugis; stipulis *minutis.* Spicæ *axillares, pedunculatæ, clavatæ, floribus inferioribus (dimidii cylindracei racheos) sæpe masculis.*

PARKIA *Africana,* pinnis sub-20-jugis, pinnulis sub-30-jugis obtusis intervalla æquantibus cicatricibus distinctis parallelis, glandula ad basin petioli, rachi communi eglandulosa, partialium jugis (2-3) summis glandula umbilicata.

Inga biglobosa, *Palis. de Beauv. Flore d'Oware,* 2, *p.* 53, *tab.* 90. *Sabine in Hortic. Soc. Transact.* 5, *p.* 444. *De Cand. Prodr.* 2, *p.* 442.

Inga Senegalensis. *De Cand. Prodr.* 2, *p.* 442.

Mimosa taxifolia. *Pers. Syn.* 2, *p.* 266, *n.* 110.

Nitta. *Park's First Journey*, *p.* 336—337.

[1] *Park's Second Journey*, *p.* cxxiv, where it is stated to be an undescribed species of that genus. Soon after that Narrative appeared, on comparing Mr. Park's specimen, which is in fruit only, with the figure published by Lamarck in his Illustrations (*tab.* 602, *f.* 4), and with M. Poiret's description (*Encyc. Meth. Botan.* 5, *p.* 728), I referred it to that author's *P. erinacea,* a name which is, I believe, adopted in the last edition of the Pharmacopœia of the London College. Dr. Hooker has since published a drawing of the same plant by the late Mr. Kummer, and, considering it a new species, has called it Pterocarpus Senegalensis. (*Gray's Travels in Western Africa,* *p.* 395, *tab.* D.)

[2] *Tuckey's Congo*, *p.* 430. (*Antè*, *p.* 111.)

fore, exceptions to all the assumed characters of Mimoseæ are found, and there is some approach in both genera to the habit of Cæsalpineæ. It is still possible, however, to distinguish, and it will certainly be expedient to preserve, these two tribes or orders. Abandoning divisions strictly natural, and so extensive as the tribes in question, merely because we may not be able to define them with precision, while it would imply, what is far from being the case, that our analysis of their structure is complete, would, at the same time, be fatal to many natural families of plants at present admitted, and among others to the universally received class to which these tribes belong. No clear character, at least, is pointed out in the late elaborate work of M. De Candolle,[1] by which Leguminosæ may be distinguished from Terebintaceæ and Rosaceæ, the orders supposed to be most nearly related to it. It is possible, however, that such characters, though hitherto overlooked, may really exist; and I shall endeavour to show that Leguminosæ, independent of the important but minute differences in the original structure and development of its ovulum, may still be distinguished at least from Rosaceæ.

In the character of Polygaleæ, which I published in 1814,[2] I marked the relation of the parts of the floral envelopes to the axis of the spike, or to the subtending bractea. I introduced this circumstance chiefly to contrast Polygaleæ with Leguminosæ, and to prove, as I conceived, that Securidaca, which had generally been referred to the latter family, really belonged to the former.

M. De Jussieu, who soon after published a character of Polygaleæ, entirely omitted this consideration, and continued to refer Securidaca to Leguminosæ. M. De Candolle, however, in the first volume of his 'Prodromus,' has adopted both the character and limits of Polygaleæ, which [236
I had proposed, though apparently not altogether satisfied with the description he himself has given of the divisions of the calyx and corolla.

The disposition of the parts of the floral envelopes, with

[1] *Mémoires sur la Famille des Legumineuses.*
[2] *Flinders's Voy. to Terra Austr.* 2, *p.* 542. (*Antè, pp.* 13, 14.)

reference to the axis of the spike, in Polygaleæ, namely, the fifth segment of the calyx being posterior or superior and the fifth petal anterior or inferior, is the usual relation in families the division of whose flower is quinary. This relation is in some cases inverted; one example of which I have formerly pointed out in Lobeliaceæ,[1] as I proposed to limit it, and a similar inversion exists in Leguminosæ. But this class also deviates from the more general arrangement of the parts of the flower with regard to each other. That arrangement consists, as I have long since remarked,[2] in the regular alternation of the divisions of the proximate organs of the complete flower. To this arrangement, indeed, many exceptions are well known; and M. De Candolle has given a table of all the possible deviations, but without stating how many of these have actually been observed.[3]

In Leguminosæ the deviation from the assumed regular arrangement consists in the single pistillum being placed opposite to the lower or anterior segment of the calyx.

In these two characters, namely, the relation of the calyx and corolla both to the simple pistillum and to the axis of the spike or to the bractea, Leguminosæ differ from Rosaceæ in which the more usual arrangements are found.

But in those Rosaceæ in which the pistillum is solitary and placed within the anterior petal, its relation to the axis of the spike is the same as that of Leguminosæ, in which it is within the anterior division of the calyx. And in all families, whether dicotyledonous or monocotyledonous, this, I believe, is uniformly the position of the simple solitary pistillum with regard to the spike or bractea.

The frequent reduction of Pistilla, in plants having the other parts of the flower complete in number, must have been generally remarked. But the order in which these abstractions of pistilla take place, or the relations of the reduced series to the other parts of the flower, have, as far as I know, never yet been particularly attended to. It will probably appear singular that the observation of these

[1] *Flinders's Austr.* 2, *p.* 560. (*Antè, p.* 32.)
[2] *Prodr. Flor. Nov. Holl.* 1, *p.* 558. [3] *Theor. elem. ed.* 2, *p.* 183.

relations in the reduced series of pistilla should have suggested the opinion, that in a complete flower, whose [237
parts are definite, the number of stamina and also of pistilla is equal to that of the divisions of the calyx and corolla united in Dicotyledones, and of both series of the perianthium in Monocotyledones.

This assumed complete number of stamina is actually the prevailing number in Monocotyledones; and though in Dicotyledones less frequent than what may be termed the symmetrical number, or that in which all the series are equal, is still found in decandrous and octandrous genera, and in the greater part of Leguminosæ. The tendency to the production of the complete number, where the symmetrical really exists, is manifested in genera belonging or related to those pentandrous families in which the stamina are opposite to the divisions of the corolla, as by Samolus related to Primulaceæ, and by Bæobotrys, having an analogous relation to Myrsineæ; for in both these genera, five additional imperfect stamina are found alternating with the fertile, and consequently occupying the place of the only stamina existing in most pentandrous families. Indications of this number may also be said to exist in the divisions of the hypogynous disk of many pentandrous orders.

With respect to the Pistilla, the complete number is equally rare in both the primary divisions of phænogamous plants. In Monocotyledones the symmetrical number is very general, while it is much less frequent in Dicotyledones, in which there is commonly a still further reduction.

Where the number of Pistilla in Dicotyledones is reduced to two, in a flower in which both calyx and corolla are present and their division quinary, one of these pistilla is placed within a division of the calyx, the other opposite to a petal or segment of the corolla. In other words, the addition to the solitary pistillum, (which is constantly anterior or exterior), is posterior or interior. This is the general position of the component parts of a bilocular ovarium, or an ovarium having two parietal placentæ; and in flowers whose division is quinary, I can recollect no other exceptions to it than in some genera of Dilleniaceæ.

It is particularly deserving of notice, that the common position of the cells of the bilocular pericarpium with relation to the axis of the spike was well known to Cæsalpinus, who expressly distinguished *Cruciferæ* from all other bilocular families by their peculiarity in this respect, the loculi in that family being placed right and left, instead of being anterior and posterior.[1]

238] On the subject of the position of the Pistilla in the other degrees of reduction from the symmetrical number, I shall not at present enter. But in reference to Leguminosæ I may remark that it would be of importance to ascertain the position of the Pistilla in the pentagynous Mimosea, stated to have been found in Brazil by M. Auguste De St. Hilaire.[2] Are these Pistilla placed opposite to the divisions of the calyx, as might probably be inferred from the position of the solitary Legumen in this class? Or are we to expect to find them opposite to the petals, which is the more usual relation, and their actual place in Cnestis, though the single ovarium of Connarus, a genus belonging to the same family, is seated within the anterior division of the calyx?

In the very few Leguminosæ in which the division of the flower is quaternary, namely, in certain species of Mimosa, the ovarium is still placed within one of the divisions of the calyx.

As to *Moringa*, which was originally referred to this class from a mistaken notion of its absolutely belonging to Guilandina, it is surely sufficiently different from all Leguminosæ, not only in its compound unilocular ovarium with three parietal placentæ, but also in its simple unilocular antheræ; and it appears to me to be an insulated genus, or family (*Moringeæ*), whose place in the natural series has not yet been determined.

CÆSALPINEÆ. Of this tribe, four species only occur in the collection. One of these is *Bauhinia rufescens* of Lamarck (*Illustr.* 329, *f.* 2); another is *Cassia* (*Senna*)

[1] *Cæsalp. de Plantis, p.* 327, *cap.* xv, *et p.* 351, *cap.* liii.
[2] *De Cand. Legum. p.* 52.

obovata, which, according to Dr. Oudney, grows wild in small quantities in Wady Ghrurbi.

PAPILIONACEÆ. Twenty-six species of this tribe are contained in the herbarium, none of which form new genera, and the only two species that appear to be unpublished belong to Indigofera.

Alhagi Maurorum, or *Agoul*, is abundant in Fezzan, where it forms excellent food for camels.

COMPOSITÆ. Of this class, thirty-six species exist in the collection. The far greater part of these were found in the vicinity of Tripoli and in the Desert. All of them appear to belong to established genera, and very few species are undescribed.

RUBIACEÆ. The herbarium contains only six species of this family, five of which, belonging to Spermacoce and [239
Hedyotis, were found in Bornou and Soudan; the sixth, a species of Galium, near Tripoli.

Of ASCLEPIADEÆ only three plants occur. One of these is a new species of Oxystelma, exactly resembling in its flowers O. esculentum of India, from which it differs in the form of its leaves, and in that of its fruit.[1] A species of Dœmia was found in the Desert; but the specimens are too imperfect to be ascertained.

Of *Apocineæ*, strictly so-called, there is no plant whatever in the collection; and of Gentianeæ, a single species only of Erythræa.

SESAMEÆ. An imperfect specimen of *Sesamum pterospermum*, of the catalogue of Mr. Salt's Abyssinian plants,[2] is in the collection from Bornou.

SAPOTEÆ. The only plant of this family in the herbarium is the *Micadania*, or Butter Tree of Soudan, particularly noticed by Captain Clapperton. The specimen, however, is very imperfect, consisting of detached leaves, an incomplete fruit, and a single ripe seed. On comparing these leaves with the specimen of Park's Shea Tree,[3] in the

[1] OXYSTELMA *Bornouense*, floribus racemosis, corollæ laciniis semiovatis, folliculis inflatis, foliis lanceolatis basi cordatis.

Obs. Inflorescentia et corolla omnino *O. esculenti*, a quo differt folliculis inflatis, et foliis omnibus basi cordatis.

[2] *Salt's Voy. to Abyss. append. p.* lxiii. (*Antè, p.* 94.)

[3] *Park's First Journey, pp.* 202 *and* 352.

Banksian herbarium, I have little doubt that they both belong to one and the same species. Whether this plant is really a Bassia, is not equally certain; and the seed at least agrees better with Vitellaria paradoxa of the younger Gærtner (*Carpol. tab.* 205) than with that of Bassia figured by his father (*de Fruct. et Sem. Pl. tab.* 104).

That the woody shell in the nuts of all Sapoteæ is really formed of the testa or outer membrane of the seed, as I have elsewhere stated[1] and not of a portion of the substance of the pericarpium, according to the late M. Richard and the younger Gærtner, is proved, not only by the aperture or micropyle being still visible on its surface, as M. Turpin has already shown in one case (*Ann. du Mus. d'Hist. Nat.* 7, *tab.* 11, *f.* 3); but also by the course and termination of the raphe, as exhibited in the younger Gærtner's figures of Calvaria and Sideroxylum (*Carpol. tabb.* 200, 201, *et* 202), and by the origin and ramification of the internal vessels.

240] SCROPHULARINÆ. Only six species of this family occur, none of which are unpublished.

OROBANCHE COMPACTA of Viviani was observed between Fezzan and Bornou.

Of CONVOLVULACEÆ there are five species, four of which belong to Bornou; the fifth is an aquatic Ipomœa, found creeping on the borders of a small lake near Tintuma. Possibly this plant may be Ipomœa aquatica of Forskal, and consequently Convolvulus repens of Vahl (*symb.* 1, *p.* 17). It is not, however, the plant so called by Linnæus, which proves, as I have elsewhere stated (*Prodr. Fl. Nov. Holl.* 1, *p.* 483) to be Calystegia sepium; nor does it belong to either of his synonymes. Our plant differs also from Vahl's description of his Convolvulus repens, in having constantly single-flowered peduncles, and leaves whose posterior lobes are rather acute than obtuse, and are quite entire. It is probably, therefore, distinct; and I have named it Ipomœa Clappertoni.[2]

[1] *Prod. Flor. Nov. Holl.* 1, *p.* 528.

[2] IPOMŒA *Clappertoni*, glaberrima repens, foliis sagittatis: lobis posticis acutiusculis integerrimis, pedunculis unifloris.

Among the few *Labiatæ*, there is a species of Lavandula, possibly distinct from but very nearly related to L. multifida. It was found on the mountains of Tarhona.

Of BORAGINEÆ, the herbarium includes eleven species, the greater part of which were collected near Tripoli, and all of them belong to well-established genera.

PRIMULACEÆ. Of this family two species of Anagallis occur in the collection, and of these A. cærulea was observed both near Tripoli and in Bornou.

SAMOLUS VALERANDI was also found near Tripoli, in Wady Sardalis in Fezzan, and in Bornou.

Of Dicotyledonous, or even of all phænogamous plants, *S. Valerandi* is perhaps the most widely diffused. It is a very general plant in Europe, has been found in several parts of North Africa, in Dr. Oudney's herbarium it is from Bornou, I have myself observed it at the Cape of Good Hope and in New South Wales, and it is also indigenous to North America.

The geographical distribution of the genus Samolus is equally remarkable. At present eight species are known, of which S. Valerandi is the only one indigenous to Europe [241
or which, indeed, has been found in the northern hemisphere, except the nearly related *S. ebracteatus* of Cuba. All the other species belong to the southern hemisphere, where *S. Valerandi* has also a very extensive range.

Of PLUMBAGINEÆ, there are three species of *Statice Taxanthema;* for the latter name may be preserved as belonging to a section, though hardly as that of a genus, so far at least as depends on inflorescence, which in both subdivisions of Statice is essentially similar, that of *Statice Armeria* being only more condensed. Of the three species in the herbarium, one appears to be unpublished.

Among the plants of the *Apetalous orders* in the collection, there are very few remarkable, and hardly any new species.

Gymnocarpus decandrum was observed by Dr. Oudney very commonly in gravelly deserts, on the route from Tripoli to Fezzan; and *Cornulaca monacantha* of M. Delile is said to be widely extended from Tripoli to Bornou, and to be excellent food for camels.

MONOCOTYLEDONES. The number of species belonging to this primary division contained in the herbarium is altogether seventy. But Gramineæ and Cyperaceæ being excluded, thirteen only remain, namely, three species of Juncus, a single Commelina, three Melanthaceæ, three Asphodeleæ, one species of Iris, and two Aroideæ, of which Pistia Stratiotes is one.

Of these thirteen plants, two appear to be unpublished, both of them belonging to Melanthaceæ. The first, a congener of Melanthium punctatum, which is also in the collection, was found in Fezzan.

The second is a species of *Colchicum*, very different from any hitherto described; and which yet, by Mr. Ritchie, who first observed it, is said to be common in the desert near Tripoli, where it was also found by Dr. Oudney.

This species, which I have named *Colchicum Ritchii*, is easily distinguished from all its congeners by having two cristæ or membranous processes, which are generally fimbriated, at the base of each segment of the perianthium, parallel to each other and to the intermediate filament. But this character, though excellent as a specific difference, is neither of generic importance, nor sufficient to authorise the formation of a separate section.[1]

242] Bulbocodium and Merendera, however, which, following Mr. Ker,[2] I consider as belonging to Colchicum, appear to me decidedly to form subgenera or sections, and in this opinion I am confirmed by having found a fourth section of the same genus. This fourth subgenus is established on HYPOXIS FASCICULARIS, a plant which has been seen by very few botanists, and which Linnæus introduced

[1] *Colchicum* (*Hermodactylus*) *Ritchii*, limbi laciniis basi intus bicristatis! fasciculo 2—multifloro, foliis linearibus.

Obs. Spathæ 2-8-floræ; limbi laciniæ vel lanceolatæ acutiusculæ vel oblongæ obtusæ; cristæ laciniarum omnium sæpe fimbriato-incisæ, exteriorum nunc integerrimæ. Ovula in singulis ovarii loculis biseriata, placentarum marginibus approximata; nec ut in C. autumnali quadriseriata.

[2] *Botan. Magaz.* 1028.

into his 'Species Plantarum,' and referred to Hypoxis, solely on the authority of the figure published in Dr. Russell's 'History of Aleppo.' In the Banksian Herbarium I have examined part of the original specimen of this species, found by Dr. Alexander Russell, and figured by Ehret in the work referred to, as well as more perfect specimens collected by Dr. Patrick Russell; and am satisfied that its ovarium is not in any degree adherent to the tube of the perianthium. I find also that Hypoxis fascicularis differs from Colchicum merely in having a simple unilocular ovarium, with a single parietal placenta and an undivided style, instead of the compound trilocular ovarium, with distinct or partially united styles, common to all the other sections of that genus.

A reduction, as in this case, to the solitary simple pistillum,[1] though existing in all Gramineæ and in certain genera of several other families of Monocotyledones, is yet comparatively rare in that primary division of phænogamous plants, and in the great class Liliaceæ, the present species of Colchicum offers, I believe, the only known example. [243
Yet this remarkable character is here so little influential, if I may so speak, that Hypoxis fascicularis very closely resembles some states of Colchicum Ritchii, and in the Banksian herbarium has actually been confounded with another species of the first or trigynous section of the genus.

To the first section, which includes *Colchicum Ritchii,*

[1] The late celebrated M. Richard, in his excellent 'Analyse du Fruit,' in pointing out the distinctions between a simple and compound pericarpium, produces that of Melanthaceæ as an example of the compound, in opposition to that of Commelineæ or of Junceæ, which, though equally multilocular, he considers as simple. A knowledge of the structure of Colchicum Monocaryum would, no doubt, have confirmed him in his opinion respecting Melanthaceæ.

It has always appeared to me surprising that a carpologist so profound as M. Richard, and whose notions of the composition of true dissepiments, and even of the analogy in placentation between multilocular and unilocular pericarpia, were, in a great degree, equally correct and original, should never have arrived at the knowledge of the common type of the organ or simple pistillum, to which all fruits, whether unilocular or multilocular, were reducible; and that he should, in the instance now cited, have attempted to distinguish into simple and compound two modifications of the latter so manifestly analogous, and which differ from each other only in the degree of coalescence of their component parts.

the subgeneric name *Hermodactylum* may, perhaps, be applied, while that established on Hypoxis fascicularis may be called *Monocaryum*.

The position of the pistillum in *Colchicum* (*Monoçaryum*) *fasciculare* is not easily determined. I believe it to be placed within the anterior segment of the outer series of the perianthium; but, from the great length of the tube, it is difficult to ascertain such a point in dried specimens. This, however, is the position in which I should expect it, both in reference to the usual relation of the solitary simple pistillum to the axis of the spike, or to the subtending bractea in all phænogamous plants, and also with regard to the constant relation of the parts of the compound pistillum to the divisions of the perianthium in Monocotyledones; for it is worthy of remark, that a difference in this relation may be said to exist in the two primary divisions of phænogamous plants—the pistilla when distinct, or their component parts when united, being in Dicotyledones usually placed opposite to the petals, when these are of equal number; while in Monocotyledones the cells of the trilocular ovarium are, I believe, uniformly opposite to the divisions of the outer series of the perianthium.

Cyperaceæ. Of twelve species of this family existing in the herbarium, six are referable to Cyperus, three to Fimbristylis, and three to Scirpus. Among these there is no remarkable, nor, I believe, any undescribed species. Of C. Papyrus, which, according to Captain Clapperton, grows in the Shary, there is no specimen in the collection.

Gramineæ. Of this extensive family, with which Dr. Oudney was more conversant than with any other, and to which, therefore, during the expedition, he probably paid greater attention, the herbarium contains forty-five species; and in dividing the order into two great tribes, as I have for-
244] merly proposed,[1] thirty of these species belong to *Poaceæ* and fifteen to *Paniceæ*. This relative proportion of these two tribes is considerably different from what might have

[1] *Flinder's Voy. to Terra Austr.* 2, *p.* 582. (*Antè*, *pp.* 57-8.)

been expected, in the climates in which the collection was formed; it seems, however, to be connected with the nature of the surface; for in the Great Desert the reduction of Paniceæ is still more remarkable; this tribe being to Poaceæ, in that region, in the proportion of only five to eighteen.

Dr. Oudney remarks, with respect to the grasses of the desert, that he observed no species with creeping roots; for a species of Arundo related to Phragmites, which he notices as the only exception, is not properly a desert plant.

Among the very few Gramineæ deserving particular notice, the first is Avena Forskalii of Vahl. The specimens in the herbarium which were collected in the Desert of Tintuma in some respects differ from all the others that I have seen of this variable species. In the Banksian herbarium there is an authentic specimen from Forskal; I have received from M. Delile specimens both of his *A. Forskalii* and *arundinacea*, described and figured in his 'Flore d'Egypte'; and am also in possession of others in somewhat different states, collected in Egypt by M. Nectoux and Dr. Sieber. From a comparison of all these specimens I am led to believe that A. Forskalii and arundinacea are not specifically distinct; and it is at least evident that *arundinacea* more nearly approaches to the plant of Forskal than that to which M. Delile has applied the name *Forskalii*.

This grass, which does not belong to Avena, is referable to Danthonia, from the structure of the outer valve of its perianthium. But Danthonia requires subdivision into several sections, of which, perhaps, our plant may be considered as forming one.

The character of the section established on *Danthonia Forskalii* would chiefly consist in the very remarkable obliquity of the joints of the locusta, which is, indeed, so great, that after their separation each flower seems to have at the base an almost vertically descendent spur; and as the inferior extremity of the upper joint is produced beyond the lower, a short calcar actually exists before separation, and

this calcar is equally manifest in the terminal rudiment of the locusta. The present, therefore, is a case of more remarkably oblique articulation in grasses than even that
245] existing in *Holcus acicularis* (Andropogon acicularis, *Retz*), which led to the formation of *Centrophorum*, a genus still admitted by Professor Sprengel,[1] and respecting the structure of which a very singular explanation has been lately offered by M. Raspail.[2] In one respect, the two cases differ. In *Danthonia* (*Centropodia*) *Forskalii*, the articulations being in the axis of the locusta or spicula, each flower appears to have this spur-like process; while in *Holcus* (*Rhaphis*) *acicularis*, the joint being in the peduncle or branch of the racemus, the spur is common to three locustæ.

Dr. Fischer, in whose herbarium the specimen was observed which led to the formation of Centrophorum, will probably recollect the communication made to him on the subject of that plant, of which Dr. Trinius himself has since corrected the characters. He retains it, however, as a distinct genus, for which he has adopted the name Rhaphis, given to it by Loureiro, by whom it was originally proposed on other, but not more satisfactory grounds.

TRIRAPHIS PUMILIO is the second plant of this family to be noticed. It is undescribed, and belongs to a genus of which the only two published species were found in the intratropical part of New Holland.[3] In several points of structure the African plant is very different from *T. pungens*, the first of these species; in some respects it approaches to *mollis*, the second species, especially in the inequality of its setæ or aristæ; but it differs from both in habit, and in having only one perfect flower in each locusta.[4]

Of PENNISETUM DICHOTOMUM (*Delile, Flore d'Egypte, p.* 15, *tab.* 8, *f.* 1), which, in several different states, is in the collection, it is remarked by Dr. Oudney that "it

[1] *Syst. Veg.* 1, *p.* 132. [2] *Annal. des Scien. Nat.* 4, *p.* 425.

[3] *Prodr. Flor. Nov. Holl.* 1, *p.* 185.

[4] *Triraphis Pumilio*, panicula coarctata abbreviata, locusta glumam vix superante 3-4-flora: flosculo infimo hermaphrodito; reliquis neutris univalvibus.

is a great annoyance to man and beast from the prickly calyx (involucrum);" and by Major Denham that from Aghedem to Woodie "it covered the surface of the country, and annoyed the travellers to misery;" he observes also that the seed is called *Kasheia,* and is eaten.

PANICUM TURGIDUM (*Forsk. Arab., p.* 18; *Delile,* [246
Flore d'Egypte, p. 19, *tab.* 19, *f.* 2) is also one of the most common grasses from Tripoli to Bornou.

Of ACOTYLEDONES, the only plant in the collection is *Acrostichum velleum,* found on the Tarhona mountains. Mr. Ritchie's herbarium contains, also, a single plant of the same family, namely *Grammitis Ceterach.*

The foregoing observations have extended much beyond the limits which the number and importance of the plants they relate to may seem to require. I still regret, however, that I cannot add a few remarks on such species as, although not in the herbarium, were observed, either indigenous or cultivated, in the countries visited by the mission, and for information respecting which I am indebted to Major Denham and Captain Clapperton. But it being determined no longer to delay the publication of the very interesting Narrative to which the observations already made will form an Appendix, I am unable at present to enter on this part of my subject.

GENERAL VIEW

OF THE

BOTANY OF SWAN RIVER.

BY

ROBERT BROWN, ESQ., F.R.S.

[*Extracted from the 'Journal of the Royal Geographical Society of London.' Vol. I, pp.* 17—21.]

LONDON:

1832.

GENERAL VIEW OF THE BOTANY

OF THE

VICINITY OF SWAN RIVER.

By R. BROWN, Esq., F.R.S.

Read November 22nd, 1830.

The vegetation of the banks of Swan River, and of the adjoining country to the southward, is at present known chiefly from the report of Mr. Charles Fraser, the botanical collector, who accompanied Captain Stirling in his examination of that district in 1827, and from collections of specimens which were then formed. [17

I have inspected, and in part examined, two of these collections; one of which I received from Mr. Fraser himself, through my friend Alexander Macleay, Esq., the Secretary of the Colony of New South Wales; for the second I am indebted to Captain Mangles.

The number of species in both collections does not exceed 140; and some dicotyledonous herbaceous tribes, as well as grasses, Cyperaceæ, and Orchideæ, are entirely wanting.

From materials so limited in extent, but few general observations can be hazarded on the vegetation of this portion of the south-west coast of New Holland.

The principal families of plants contained in the collections are *Proteaceæ; Myrtaceæ; Leguminosæ*, such especially as belong to *Decandrous Papilionaceæ*, and to the

Leafless Acaciæ; Epacrideæ; Goodenoviæ; and *Compositæ*. And the more conspicuous plants, not belonging to any of these families, and which greatly contribute to give a character to the landscape, are, *Kingia australis*, a species of *Xanthorrhæa;* a *Zamia*, nearly allied to, and perhaps not distinct from, *Z. spiralis* of the east coast, although it is said frequently to attain the height of thirty feet; a species of *Callitris;* one or two of *Casuarina;* an *Exocarpus*, probably not different from *E. cupressiformis;* and *Nuytsia floribunda*,[1] a plant hitherto referred to *Loranthus*, but sufficiently distinct in the texture and the form of its fruit, and now named in memory of the discoverer of that part of the coast to which this very singular tree is nearly limited.

If an opinion were to be formed of the nature of the country merely from the inspection of these collections, it certainly would be extremely unfavorable as to the quality of the soil; for not only do the prevailing families already enumerated, but the whole of the genera of those families, and even many of the species, agree with those found on the shores of King George's Sound, which, with the exception of a few patches of very small extent, seem absolutely incapable of cultivation.

The opinion so formed, however, would be necessarily modified in noticing the entire want in the collections of
18] tribes, all of which must be supposed to exist, and some even in considerable proportion, in the tract examined; in allowing for the unfavorable season when the herbarium was collected; in admitting the statements in Mr. Fraser's report, respecting the abundance and luxuriance of *Anthistiria australis*—the Kangaroo-grass of New South Wales; from the account given in the same report of the extraordinary size of some arborescent species of *Banksia*, which, in the neighbourhood of King George's Sound, generally form small trees only; and lastly, in adverting to the important fact stated by Captain Stirling in his despatch to Government—namely, that the stock had not only been

[1] Loranthus floribundus. *Labill. Nov. Holl.* i, *p.* 87, *t.* 113.

supported through nearly the whole of the dry season, but that most descriptions of it had even fattened on the natural herbage of the country.

From these more general observations I proceed to make a very few remarks, chiefly relating to the geographical distribution of some of the families or more interesting species, either contained in the herbarium, or distinctly noticed in Mr. Fraser's report.

The striking resemblance in general character, and the identity of many of the species with those of King George's Sound, have been already mentioned. But this portion of the shores of New Holland, extending from Swan River on the west coast to Middle Island, in 123° 10′ east long. on the south coast, may be said to contain the greatest proportion of those genera which form the chief peculiarities of New Holland vegetation.

In comparing the Flora of the district of Swan River with more distant regions of the same continent, it may be remarked, that probably not more than four or five species are common to this part of the west coast, and to the same parallel of the east coast of New Holland; and that even the existence of some of these species at Swan River is not altogether certain.

In the collections which I have examined there is no specimen of *Anthistiria australis*, or Kangaroo-grass of New South Wales; but as this valuable grass must have been well known to the botanical collector, and as it is perhaps the most general plant in New Holland, I have no hesitation in admitting its existence on the authority of Mr. Fraser's report.

Mesembryanthemum æquilaterale is neither contained in the herbarium, nor mentioned by the collector. I find, however, in one of the letters from Swan River, published by Mr. Cross, a plant noticed as a pot-herb, that, from the account of the writer, is probably this plant, which, next to *Anthistiria australis*, is perhaps the most widely diffused species in the Flora of New Holland.

The third species is *Pteris esculenta*, the only fern found by Mr. Fraser, and which is both general and abundant beyond the tropic in New Holland and in Van Diemen's Land.

19] The *Zamia*, already noticed, if not specifically different from *spiralis*, would furnish another example of a plant peculiar to New Holland, and very generally found in the extra-tropical parts of that continent. I had, however, myself observed on the south coast a *Zamia* of at least ten feet in height, which I suspected might be distinct from *Z. spiralis* of the neighbourhood of Port Jackson, and which is probably the same with that of Swan River.

The *Exocarpus* of the Swan River may possibly differ from *cupressiformis*, though there is nothing in the specimens to make it probable that it is specifically distinct. But *Exocarpus cupressiformis* is found very generally, not only in the southern parts of New Holland and Van Diemen's Land, but also within the tropic.

The last plant in the collection whose range is very extensive remaining to be noticed, I have not been able to distinguish from *Arenaria marina* of the shores of Europe.

Of the families existing in the vicinity of Swan River, the most striking, as well as the most extensive, is *Proteaceæ*, a tribe which, from its general dispersion, and the remarkable forms of its numerous genera and species, includes many of the chief peculiarities of the vegetation of New Holland.

In Mr. Fraser's collection, the principal genera of this order are—*Petrophila*, *Isopogon*, *Hakea*, and *Banksia*; and these are also the most abundant in the districts of King George's Sound and of Lucky Bay. The number of species of the two first-mentioned genera confirms the remark made in the Botanical Appendix to Captain Flinders's Voyage[1]—namely, that in New Holland, at the western extremity of the parallel of latitude in which the great mass of this order of plants is found, a closer resemblance is observable to the South African portion of the order than on the east coast, where those allied to the American part chiefly occur.

This is not the place to enter into a particular account of the new species of this family existing in the collections

[1] (*Antè*, p. 41.)

from Swan River. I may observe, however, that the number is considerable, and that their specific characters have been recently published.[1]

The *Myrtaceæ* of Swan River belong chiefly to *Melaleuca*, *Beaufortia*, *Calothamnus*, *Calythrix*, *Billottia*,[2] and *Eucalyptus*.

Of *Eucalyptus* the only species in the collection had been first found in Captain Flinders's voyage at King George's Sound, on the shores of which it was the only useful [20
timber tree, though there of very moderate size. I have named it *Eucalyptus calophylla*.

Mr. Fraser describes it as forming, on the banks of the Swan, a large forest tree, and erroneously refers it to *Angophora*, a genus which is limited to the east coast of New Holland. Other species of *Eucalyptus*, forming the timber of the country, are mentioned in the report, and considered to be some of the common gum-trees of Port Jackson, from which, however, I have no doubt they will prove to be distinct; for I am acquainted with no species of this genus common even to the east and south coasts of New Holland.

I shall conclude with a remark relating equally to the genus *Eucalyptus* and to the *Leafless Acaciæ*, several species of which are found in the collection. This observation I have formerly made in the Appendix to Captain Flinders's Voyage in the following terms[3]:—"These two genera are not only the most widely diffused, but by far the most extensive in Terra Australis, about 100 of each having already been observed; and if taken together, and considered with respect to the mass of vegetable matter they contain, calculated from the size as well as the number of individuals, are perhaps nearly equal to all the other plants of that country. They agree very generally also, though belonging to very different families, in a part of their economy, which contributes somewhat to the peculiar

[1] *Supp.* I, *Prodr. Flor. Nov. Holl.*

[2] A genus distinct from *Leptospermum*, to which the few species hitherto published, namely, *B. marginata*, *flexuosa*, and *linearifolia*, have been referred.

[3] (*Antè*, *p.* 62.)

character of the Australian forests, namely, in their leaves, or the parts performing the functions of leaves, being vertical, or presenting their margin, and not either surface, towards the stem: both surfaces having consequently the same relation to light.

"This economy, which uniformly takes place in the *Acaciæ*, is in them the consequence of the vertical dilatation of the foliaceous footstalk; while in *Eucalyptus*, where, though very general, it is by no means universal, it proceeds from the twisting of the footstalk of the leaf."

To this quotation it may be added that these two genera still more uniformly agree in the similarity of the opposite surfaces of their leaves. But this similarity is the indication of a more important fact—namely, the existence equally on both surfaces of the leaf, of those organs, for which, as I believe them to be in general imperforated, I have adopted the name of *cutaneous glands*, but which by most authors are denominated pores, or *stomata* of the *epidermis*.

In leaves, especially of trees and shrubs, these glands are generally found on the under surface only; while among arborescent plants in a very few instances, as in several *Coniferæ*, they are confined to the upper surface.

21] In addition to the two extensive New Holland tribes here mentioned, there are many other cases in which these organs occupy both paginæ; and I am inclined to think such cases more frequently occur on that continent than in any other part of the world. It is at least certain that on this microscopic character, of the equal existence of cutaneous glands on both surfaces of the leaf, depends that want of lustre which is so remarkable in the forests of New Holland.

BOTANICAL APPENDIX

TO

CAPTAIN STURT'S EXPEDITION

INTO

CENTRAL AUSTRALIA.

BY

ROBERT BROWN, ESQ., D.C.L., F.R.S., F.L.S., &c.

[*Extracted from the 'Narrative of an Expedition into Central Australia, during the years* 1844, 1845, *and* 1846,' *by Captain Charles Sturt, F.L.S., F.R.G.S. Vol. II, Appendix, pp.* 66—92.]

LONDON:

1849.

PLANTS OF CENTRAL AUSTRALIA. [66

MY friend, Captain Sturt, having placed at my disposal the Collection of Plants formed in his recent Expedition into the Southern Interior of Australia, I am desirous of giving some account of the principal novelties it contains.

The collection consists of about one hundred species, to which might be added, if they could be accurately determined, many other plants, chiefly trees, slightly mentioned in the interesting narrative, which is about to appear, and to which the present account will form an appendix. I may also observe, in reference to the limited number of species, that Captain Sturt and his companion, Mr. Brown, seem to have collected chiefly those plants that appeared to them new or striking, and of such the collection contains a considerable proportion.

In regard too to such forms as appear to constitute genera hitherto undescribed, it greatly exceeds the much more extensive herbarium, collected by Sir Thomas Mitchell in his last expedition, in which the only two plants proposed as in this respect new belong to genera already well established, namely, Delabechia to Brachychiton, and Linschotenia to Dampiera.

In Captain Sturt's collection, I have been obliged, from the incomplete state of the specimens, to omit several species, probably new, from the following account, in which the plants noticed, chiefly new genera and species, are arranged according to the order of families in the Pro- [67
dromus of De Candolle.

BLENNODIA.

Cruciferarum genus, prope Matthiolam.

Char. Gen.—*Calyx* clausus, foliolis lateralibus basi saccatis. *Petala* æqualia, laminis obovatis. *Stamina*: filamentis edentulis. *Ovarium* lineare. *Stylus* brevissimus. *Stigma* bilobum dilatatum. *Siliqua* linearis valvis convexiusculis, stigmate coronata, polysperma. *Semina* aptera pube fibroso-mucosa tecta! *Cotyledones* incumbentes.

Herba (v. Suffrutex) *erecta ramosa canescens, pube ramosa*; foliis *lato-linearibus remotè dentatis*; racemis *terminalibus*.

1. Blennodia *canescens*.

Loc. In arenosis depressis.

Desc. Suffruticosa, sesquipedalis, caule ramisque teretibus. Folia vix pollicaria paucidentata. Racemi multiflori, erecti, ebracteati. Flores albicantes. Calyx incano-pubescens. Petalorum ungues calyce paulo longiores. Stamina 6, tetradynama, filamentis linearibus membranaceis apice sensim angustato.

Obs. This plant has entirely the habit, and in many important points the structure of Matthiola, near which in a strictly natural method it must be placed; differing, however, in having incumbent cotyledons, and in the mucous covering of its seeds. The mucus proceeds from short tubes covering the whole surface of the testa, each containing a spiral fibre which seems to be distinct from the membrane of the tube. A structure essentially similar is known to occur generally in several families; to what extent or in what genera of Cruciferæ it may exist, I have not ascertained; it is not found, however, in those species of Matthiola which I have examined.

STURTIA. [68

Malvacearum genus, proximum Gossypio, affine etiam Senræ.

Char. Gen.—*Involucrum* triphyllum integerrimum. *Calyx* 5-dentatus, sinubus rotundatis. *Petala* cuneato-obovata, basi inæquilatera. *Columna* staminum polyandra. *Ovaria* 5, polysperma. *Styli* cohærentes. *Stigmata* distincta linearia. *Pericarpia*. . . *Semina*. . .

Suffrutex *orgyalis glaber; foliis petiolatis obovatis integerrimis; floribus pedunculatis solitariis.*

2. Sturtia *Gossypioides.*

Loc. " In the beds of the creeks on the Barrier Range." *D. Sturt.*

Desc. Suffrutex orgyalis glaber. Folia ramorum alterna, diametro unciali, trinervia; petiolo folium subæquanti, basi in stipulam subscariosam adnatam dilatato. Pedunculi vel potius rami floriferi suboppositifolii nec verè axillares uniflori, juxta apicem folio nano petiolato stipulis 2 distinctis stipato instructi. Involucrum foliaceum venosum, foliolis distinctis, cordatis, punctis nigricantibus glandulosis conspersis. Calyx dentibus acutis, sinubus rotundatis. Petala sesquipollicaria, uti calycis tubus glanduloso-punctata glandulis nigricantibus semi-immersis, purpurea basibus atropurpureis margine barbatis. Columna staminum e basi nuda super ad apicem usque antherifera: antheris reniformibus, loculis apice confluentibus. Pollen hispidum.

Obs. Sturtia is no doubt very nearly related to Gossypium, from which it differs in the entire and distinct leaves of its foliaceous involucrum, in the sharp teeth and broad rounded sinuses of the calyx, and possibly also in its fruit and seeds, which are, however, at present unknown. They agree in the texture and remarkable glands of the calyx, and in the structure of the columna staminum. Senra, which, like Sturtia, has the foliola of its three-leaved in- [69 volucrum distinct and entire, differs from it in having its calyx 5-fid with sharp sinuses, in the absence of glands, in the reduced number of stamina, and in its dispermous ovaria.

3. Tribulus (*Hystrix*) lanatus, foliis 8-10-jugis, fructibus undique tectis spinis subulatis longitudine inæqualibus: majoribus sparsis longitudinem cocci superantibus.

Loc. "In collinis arenosis. Lat. 26°." D. Sturt.

Desc. Herba diffusa, sericea, incana. Folium majus cuiusque paris 8-10-jugum,. foliolis ovatis. Flores magni. Calyx æstivatione leviter imbricatâ. Petala calyce duplo longiora. Stamina decem, antheris linearibus.

Obs. I. A species nearly related to T. Hystrix, found on the west coast of Australia, or on some of its islands, in the voyage of the Beagle, may be distinguished by the following character. *Tribulus* (*occidentalis*) sericeo-lanatus, foliis suboctojugis, coccis undique densè armatis: spinis omnibus conico-subulatis longitudine invicem æqualibus. These two species differ from all others in the uniform shape of the spines, which equally cover the whole external surface of the fruit.

Obs. II. The American species of the Linnean genus Tribulus are distinguishable from the rest of the published species, by having ten monospermous cocci, by their persistent calyx, and the absence of glands subtending the 5 filaments opposite to the sepals.

This tribe was originally separated as a genus by Scopoli, under the name of Kallstrœmia, which has been recently adopted by Endlicher.

Another tribe exists in the intratropical part of the Australian continent, to which, nearly 40 years ago, in the Bank-
70] sian Herbarium, I gave the generic name of Tribulopis and which may readily be distinguished by the following characters.

TRIBULOPIS.

Calyx 5-partitus deciduus. *Petala* 5. *Stamina* decem (nunc 5). *Filamenta* quinque, sepalis opposita, basi glandula stipata. *Ovaria* 5, monosperma. *Cocci*, præter tubercula 2 v. 4 baseos, læves.

Herbæ *annuæ prostratæ;* foliis *omnibus alternis!*

Tribulopis (*Solandri*) foliis bi-trijugis, foliolis subovatis inæquilateris, coccis basi quadrituberculatis.

Loc. In ora orientali intratropica Novæ Hollandiæ prope Endeavour River, anno 1770. DD. Banks et Solander.

Tribulopis (*angustifolia*), foliis 3-4 jugis (raro bijugis), foliolis linearibus, tuberculis baseos coccorum abbreviatis.

Loc. Ad fundum sinus Carpentariæ annis 1802 et 3. R. Brown.

Tribulopis (*pentandra*), foliis bijugis, foliolis oblongo-lanceolatis pari superiore duplo majore, floribus pentandris, petalis lanceolatis.

Loc. In insulis juxta fundum sinus Carpentariæ anno 1803. R. Brown.

4. Crotalaria (*Sturtii*) tomentosa, foliis simplicibus ovalibus utrinque sericeo-tomentosis, petiolis apice geniculatis, racemis terminalibus multifloris.

Loc. "On the top of the ridges in pure sand, from S. Lat. 28° to 26°." D. Sturt.

Desc. Frutex 2-3-pedalis (D. Sturt). Folia alterna, ovata passim ovalia, obtusa, sesquipollicem longa, utrinque velutina; petiolus teres basi vix crassiore apice curvato. Racemus terminalis; pedicellis approximatis calycem vix æquantibus apice bibracteatis. Flores sesquipollicares. Calyx 5-fidus; laciniis lanceato-linearibus acutis subæqualibus tubum paulo superantibus. Corolla sordidè flava, calyce plus duplo major. Vexillum magnum, basi simplici nec auriculata, late ovatum, acutum. Alæ vexillo fere dimidio

breviores, basi semicordata. Carina longitudine vexilli,
71] acuminata, basi gibbosa, ibique aperta marginibus tomentosis. Stamina 10 diadelpha, simplex et novemfidum. Antheræ quinque majores lineares, juxta basin affixæ; quinque reliquæ ovatæ, linearibus triplo breviores, incumbentes. Ovarium lineare, multi-ovulatum. Stylus extra medium et præsertim latere interiore barbatum. Stigma obtusum. Legumen desideratur.

Obs. A species very nearly related to C. Sturtii, having flowers of nearly equal size, and of the same colour and proportion of parts, found in 1818, by Mr. Cunningham, on the north-west coast of Australia, and since in Captains Wickham and Stokes' Voyage of the Beagle; may be distinguished by the following character:—*Crotalaria* (*Cunninghamii*) tomentosa, foliis simplicibus ovali-obovatis utrinque sericeo-tomentosis, petiolis apice curvatis, pedunculis axillaribus unifloris.

5. Clianthus (*Dampieri*) herbaceus prostratus sericeo-villosissimus, foliolis oppositis (rarissime alternis) oblongis passim lineari-oblongis obovatisve, pedunculis erectis scapiformibus, floribus subumbellatis, calycibus 5-fidis sinubus acutis, ovariis (leguminibusque immaturis) sericeis.

Clianthus Oxleyi *A. Cunningham in Hort. Soc. Transac. II series, vol.* 1, *p.* 522.

Donia speciosa *Don, Gen. Syst. vol.* 2, *p.* 468.

Clianthus Dampieri *Cunningham, loc. cit.*

Colutea Novæ Hollandiæ, &c., *Woodward in Dampier's Voy. vol.* 3, *p.* 111, *tab.* 4, *f.* 2.

Loc. "In ascending the Barrier Range near the Darling, about 500 feet above the river." D. Sturt.

Obs. In July, 1817, Mr. Allan Cunningham, who accompanied Mr. Oxley in his first expedition into the Western Interior of New South Wales, found his Clianthus Oxleyi on the eastern shore of Regent's Lake, on the River Lachlan. The same plant was observed on the GawlerRange,
72] not far from the head of Spencer's Gulf, by Mr. Eyre in 1839, and more recently by Captain Sturt, on his Barrier Range near the Darling. I have examined specimens from

all these localities, and am satisfied that they belong to one and the same species.

In March (not May), 1818, Mr. Cunningham, who accompanied Captain King in his voyages of survey of the coasts of New Holland, found on one of the islands of Dampier's Archipelago, a plant which he then regarded as identical with that of Regent's Lake. This appears from the following passage of his MS. Journal:

"I was not a little surprised to find Kennedya speciosa (his original name for Clianthus Oxleyi), a plant discovered in July, 1817, on sterile, bleak, open flats, near Regent's Lake, on the River Lachlan, in lat. 33° 13′ S. and long. 146° 40′ E. It is not common; I could see only three plants, of which one was in flower." "This island is the Isle Malus of the French." Mr. Cunningham was not then aware of the figure and description in Dampier above referred to, which, however, in his communication to the Horticultural Society in 1834, he quotes for the plant of the Isle Malus, then regarded by him as a distinct species from his Clianthus Oxleyi of the River Lachlan. To this opinion he was probably in part led by the article Donia or Clianthus, in Don's System of Gardening and Botany, vol. 2, p. 468, in which a third species of the genus is introduced, founded on a specimen in Mr. Lambert's Herbarium, said to have been discovered at Curlew River, by Captain King. This species, named Clianthus Dampieri by Cunningham, he characterises as having leaves of a slightly different form, but its principal distinction is in its having racemes instead of umbels; at the same time he confidently refers to Dampier's figure and description, both of which prove the flowers to be umbellate, as he describes those of his Clianthus Oxleyi to be. But as the flowers in this last plant [73
are never strictly umbellate, and as I have met with specimens in which they are rather corymbose, I have no hesitation in referring Dampier's specimen, which many years ago I examined at Oxford, as well as Cunningham's, to Clianthus Dampieri. This specimen, however, cannot now be found in his Herbarium, as Mr. Heward, to whom he bequeathed his

collections, informs me; nor can I trace Mr. Lambert's plant, his Herbarium having been dispersed.

Since the preceding observations were written, I have seen in Sir William Hooker's Herbarium two specimens of a Clianthus, found by Mr. Bynoe, on the north-west coast of Australia, in the voyage of the Beagle. These specimens, I have no doubt, are identical with Dampier's plant, and they agree both in the form of leaves and in their subumbellate inflorescence with the plant of the Lachlan, Darling, and the Gawler Range. From the form of the half-ripe pods of one of these specimens, I am inclined to believe that this plant, at present referred to Clianthus, will, when its ripe pods are known, prove to be sufficiently different from the original New Zealand species to form a distinct genus, to which, if such should be the case, the generic name Eremocharis may be given, as it is one of the greatest ornaments of the desert regions of the interior of Australia, as well as of the sterile islands of the North-west coast.

CLIDANTHERA.

Char. Gen.—*Calyx* 5-fidus. *Petala* longitudine subæqualia. *Stamina* diadelpha: *antheræ* uniformes; loculis apice confluentibus, valvula contraria ab apice ad basin separanti dehiscentes! *Ovarium* monospermum. *Stylus* subulatus. *Stigma* obtusum. *Legumen* ovatum, lenticulari-compressum, echinatum.

74] Herba, v. Suffrutex, *glabra, glandulosa; ramulis angulatis.* Folia *cum impari pinnata*; *foliolis oppositis, subtus glandulosis.* Stipulæ *parvæ, basi petioli adnatæ.* Flores *spicati, parvi, albicantes.*

Obs. Subgenus forsan Psoraleæ, cui habitu simile, foliis calycibusque pariter glandulosis; diversum dehiscentia insolita antherarum!

6. Clidanthera *psoralioides.*

Loc. Suffrutex bipedalis in paludosis. D. Sturt.

Desc. Herba, vel suffrutex, erecta, bipedalis, glabriuscula. Ramuli angulati. Folia cum impari pinnata, 4-5-juga; foliola opposita, lanceolata, subtus glandulis crebris parvis manifestis, marginibus scabris. Spicæ densæ, multifloræ. Calyx 5-fidus, parum inæqualis, acutus, extus glandulis dense conspersus. Corolla: *Vexillum* lamina oblonga subconduplicata nec explanata, basi simplici absque auriculis; ungue abbreviato. *Alæ* vexillo paulo breviores, carinam æquantes, laminis oblongis, auriculo baseos brevi. *Carinæ petala* alis conformes. Stamina diadelpha, simplex et novemfidum; antheræ subrotundæ v. reniformes, valvula ventrali anthera dimidio minore subrotunda. Ovarium hispidum ovulo reniformi. Legumen basi calyce subemarcido cinctum, echinatum. Semen reniforme, absque strophiola; integumento duplici. Embryo viridis; cotyledones obovatæ, accumbentes.

Obs. This plant, which in some respects resembles certain species of Glycyrrhiza, appears to be not unfrequent in the southern interior. It was found in one of the early expeditions of Sir Thomas Mitchell, and Mrs. (Capt.) Grey observed it on the flats of the Murray.

7. Swainsona (*grandiflora*) suffruticosa pubescens, foliis 8-10-jugis inexpansis incano-tomentosis; foliolis oblongis obtusis retusisve: adultis semiglabratis: rachi subincana, racemo multifloro folium superante, bracteolis lanceato-linearibus acutis æquantibus tubum calycis albo-lanati [75
quinquefidi: laciniis acutissimis longitudine ferè tubi, vexillo bicalloso.

Loc. "Common on the rich alluvial flats of the Murray and Darling." D. Sturt.

Obs. This plant is, perhaps, not specifically distinct from S. Greyana, Lindl. Bot. Regist. 1846, tab. 66, of which the figure is a good representation of S. grandiflora in every respect, except in the form and proportions of the teeth of the calyx and lateral bracteæ. In these points it exactly agrees with complete specimens, for which I am indebted to Mrs. Grey, from the banks of the Murray, and Mr. Eyre's station (Moorundi), about 98 miles from Adelaide, where it was first found in November, 1841. The following characters, if constant, will sufficiently distinguish it from S. grandiflora.

Swainsona (*Greyana*) suffruticosa pubescens, foliis 5-9-jugis inexpansis incano-tomentosis; foliolis oblongis obtusis retusisve: adultis semiglabratis: rachi subincana, racemis multifloris folio longioribus, bracteis lateralibus lanceato-linearibus brevioribus tubo calycis albo-lanati quinquedentati: dentibus obtusiusculis tubo dimidio brevioribus, vexillo bicalloso.

In the second edition of Hortus Kewensis (vol. 4, p. 326), I excluded from the generic character of Swainsona the calli of the vexillum, having observed two Australian species where they were wanting, but which in every other respect appeared to me referable to this genus; for the same reason I continue to introduce the calli, where they exist, into the specific characters, as was done in Hortus Kewensis, l. c. In the generic character of Swainsona, given in De Candolle's Prodromus (vol. 2, p. 271), the calli of vexillum are transferred to the calyx; this can only be regarded as an oversight, which perhaps has been cor-
76] rected by the author himself, and which, so far as I know, has never been adopted in any more recent work in which the generic character of Swainsona is given.

8. Swainsona? (*laxa*) glabra, caule ramoso, foliis 6-7-jugis; foliolis oblongo-ovalibus obtusis, racemis elongatis laxis, pedicellis calyce glabro quinquedentato brevioribus, bracteolis subulatis, vexillo ecalloso.

Loc. Statio nulla indicata, in Herb. D. Sturt.

Obs. There is something in the aspect of this plant not entirely agreeing with the other species of the genus; and as the fruit is unknown, and the flowers yellow, I refer it with a doubt to Swainsona.

PENTADYNAMIS.

Char. Gen.—*Calyx* 5-fidus subæqualis. *Vexillum* explanatum, callo baseos laminæ in unguem decurrenti. *Carina* obtusa, basin versus gibba, longitudine alarum. *Stamina* diadelpha; *antheris* 5 majoribus linearibus, reliquis ovatis. *Ovarium* polyspermum. *Stylus* e basi arcuata porrectus, postice barbatus. *Legumen* compressum.

Herba (Suffrutex sec. D. Sturt), *bipedalis sericeo-incana; caule angulato erecto.* Folia *ternata; foliolis sessilibus, linearibus, obtusis.* Flores *racemosi, flavi.*

9. Pentadynamis *incana.*

Loc. "On sand-hills with Crotalaria Sturtii." D. Sturt.

Desc. Herba erecta, ramosa, sericeo-incana. Folia alterna, ternata; petiolo elongato, teretiusculo, foliolo terminali longiore vix unciali. Racemi multiflori, erecti; pedicelli subæquantes calycem. Bracteolæ subulatæ, infra apicem pedicelli, basin calycis attingentes. Calyx 5-fidus; laciniis acutis tubum æquantibus. Corolla flava, calyce plus duplo longior. Vexillum explanatum, basi absque auriculis sed callo in unguem decurrenti ibique barbato auctum. Carina infra medium gibba pro receptione baseos [77 styli. Staminum antheræ majores lineares, basi vel juxta basin affixæ; 5 minores ovatæ, incumbentes. Ovarium lineare, pubescens. Stigma terminale, obtusum. Legumen immaturum incanum, stylo e basi arcuata porrecto terminatum, calyce subemarcido subtensum.

Obs. In the collection of the plants of his last expedition, presented to the British Museum by Sir Thomas Mitchell, there is a plant which seems to belong to the genus Pentadynamis, which is probably, therefore, one of the species of Vigna, described by Mr. Bentham.

10. Cassia (*Sturtii*), tomentoso-incana, foliis 4-jugis foliolis lanceolato-linearibus planis: glandula depressa inter

par infimum, racemo corymboso paucifloro cum pedunculo suo folium paulo superante v. æquante, calyce tomentoso.

Loc. "In sandy brushes of the Western interior." D. Sturt.

Obs. Species proxima C. artemisiæfoliæ De Cand. Prodr. quæ Cassia glaucescens Cunningh. MSS. 1817, cui foliola teretiuscula, et racemus corymbosus cum pedunculo suo folio brevior.

11. Cassia (*canaliculata*), cinerascens pube tenuissima, foliis 2-jugis (raro 1-jugis) foliolis angustato-linearibus canaliculatis: glandula inter par inferius et dum unijuga inter terminale, calycibus glabriusculis, racemis corymbosis paucifloris folio brevioribus.

Loc. "In the bed of the creeks of the Barrier Range, about thirty-six miles from the Darling, in lat. 32° S." D. Sturt.

Obs. Proxima C. eremophilæ Cunningh. MSS. quæ sequentibus notis a Cassia phyllodinea et C. zygophylla, *Benth.* facile distinguenda.

Cassia (*eremophila*), glabra, foliis unijugis rarò passim
78] bijugis; foliolis linearibus canaliculatis latitudine racheos linearis aversæ, corymbis paucifloris folio brevioribus.

Loc. In desertis prope fluvium Lachlan, anno 1817, detexit D. Cunningham.

Cassia (*zygophylla*), glabra foliis unijugis; foliolis linearibus planis rachi duplo latioribus, corymbis paucifloris folio brevioribus.

Cassia zygophylla, *Benth. in Mitch. trop. Austr. p.* 288.

Another species nearly related to C. zygophylla is readily distinguished by the following character:

Cassia (*platypoda*), glabra, foliis unijugis; foliolis linearibus apiculo recurvo duplo angustioribus rachi aversa lanceolato-lineari.

Loc. Juxta fluvium Murray, anno 1841, detexit Domina Grey.

12. Cassia (*phyllodinea*), canescens pube arctissimè adpressa, phyllodiis aphyllis linearibus planis falcatis aversis, calycibus glabris, legumine plano-compresso.

Loc. In Herbario D. Sturt specimen exstat nulla stationis aut loci indicatione, sed eandem speciem ad fundum sinus Spencer's Gulf dicti in sterilibus apricis anno 1802 legi.

Desc. Frutex quadripedalis, ramosissimus. Phyllodia semper aphylla, aversa, linearia, acuta, basi attenuata, plus minusvè falcato-incurva, biuncialia, $\frac{1}{8}$ circiter unciæ lata exstipulata, paginis pube arctissime adpressa canescentibus, margine superiore glandula unica depressa obsoleta. Flores flavi, in umbella axillari 2-3 flora.

Obs. Cassia phyllodinea is one of the very few species of the genus, which, like the far greater part of New Holland Acaciæ, lose their compound leaves, and are reduced to the footstalk, or phyllodium, as it is then called, and
which generally becomes foliaceous by vertical compression [79
and dilatation. A manifest vertical compression takes place in this species of Cassia.

A second species, Cassia circinata of Benth. in Mitch. trop. Austr. p. 384, is equally reduced to its footstalk, but which is without manifest vertical compression. To this species may perhaps be referred Cassia linearis of Cunningham MSS., discovered by him in 1817, but which appears to differ in having a single prominent gland about the middle of its phyllodium; Bentham's plant being entirely eglandular.

These two, or possibly three species, belong to the desert tracts of the South Australian interior. In the same regions we have another tribe of Cassiæ closely allied to the aphyllous species; they have only one pair of foliola which are caducous, and whose persistent footstalk is more or less vertically compressed. Along with these, and nearly related to them, are found several species of Cassia, having from two to four or five pairs of foliola which are narrow, but their footstalks are without vertical compression, and their foliola are caducous, chiefly in those, however, which have only two pairs.

PETALOSTYLIS.

Cæsalpinearum genus, Labicheæ proximum.

Char. Gen.—*Calyx* 5-phyllus, æqualis. *Petala* 5 subæqualia, patentia. *Stamina: Filamenta* quinque sepalis opposita, quorum *tria antherifera*, antheris basifixis linearibus, *duo reliqua* castrata. *Ovarium* oligospermum. *Stylus* maximus, petaloideus, trilobus, lobo medio longiore axi incrassata desinente in *stigma* obtusum simplex!

Frutex *glaber, erectus*. Folia *alterna, pinnata cum impari, foliolis alternis*. Racemi *axillares, pauciflori*. Flores *flavi*.

80] 13. Petalostylis *Labicheoides*.

Loc. "In the bed of a creek along with Sturtia." D. Sturt.

Obs. Eadem omnino species exstat inter plantas in Insulis Archipelagi Dampieri juxta oram septentrio-occidentalem Novæ Hollandiæ in itinere navis Beagle dictæ lectas.

Desc. Frutex facie fere Cassiæ et Labicheæ. Folia alterna, cum impari pinnata, foliolis alternis brevissimè petiolatis oblongo-lanceolatis cum mucronulo terminali paulo majore. Stipulæ parvæ caducæ. Racemi pauciflori, axillares, folio breviores. Alabastrum ovali-oblongum acutiusculum. Calyx viridis, sepalis subæqualibus oblongis acutis, æstivatione imbricatis. Petala quinque subæqualia, oblonga, flava, astivatione imbricata, sepalis sesquilongiora. Stamina 3 antherifera æqualia, filamentis abbreviatis, antheris acutis bilocularibus, loculis sulco longitudinali insculptis; 2 reliqua rudimenta parva subfiliformia. Ovarium sessile, lineare, 3-4-spermum. Stylus lobo medio triplo longiore, oblongo-lanceolato, lobis lateralibus auriculiformibus semiovatis obtusis. Stigma imberbe.

Obs. The structure of the style, which forms the only important character of this genus, so far as the specimens enable me to judge, is so remarkable and peculiar, as to render it necessary to state, that I have found it quite uniform in all the flowers I have examined; namely, in four immediately before, and in three after expansion.

PODOCOMA.

Char. Gen.—*Involucrum* imbricatum, foliolis angustis acutis. *Ligulæ* pluriseriales, angustissimæ, femineæ. *Flosculi* pauciores hermaphrodito-masculi. Ligularum pappo capillari, stipitato, denticulato. *Receptaculum* epaleatum.

Herba *humilis, setosa;* caule *densè foliato;* folia *petiolata, cuneata, incisa, setis albis conspersa.*

14. Podocoma *cuneifolia.* [81

Loc. In Herbario D. Sturt absque ulla indicatione loci vel stationis.

Obs. This plant appears to be generically distinct from Erigeron, particularly in its stipitate pappus. The specimens, however, are so incomplete, that I am unable to determine whether what I have considered stem, may not be a branch only.

LEICHARDTIA.

Char. Gen.—*Calyx* 5-partitus. *Corolla* urceolata; tubo intus imberbi; fauce annulo integerrimo incrassata. *Corona staminea* 5-phylla, foliolis antheris oppositis, iisque brevioribus, indivisis. *Antheræ* membrana (brevi) terminatæ. *Massæ Pollinis* erectæ basi affixæ. *Stigma* vix divisum.

Suffrutex *volubilis;* foliis *linearibus, fascicularibus, extra-alaribus;* folliculis *ventricosis ovato-oblongis.*

15. Leichardtia *australis.*

Doubah *Mitchell, trop. Austr. p.* 85.

Loc. "Common on the Murray, and in the interior." D. Sturt.

Desc. Suffrutex pubescens, subcinereus; ramis striatis nec omnino teretibus. Folia sesquipollicaria, linearia, acuta. Fasciculi multiflori. Calycis foliola obtusa, pube tenui cinerascentia. Corolla glabra; tubo absque squamulis denticulisve, ventricoso; limbo vix longitudine tubi, laciniis conniventibus sinistrorsum imbricatis. Coronæ foliola e basi dilatata adnata linearia, indivisa. Massæ Pollinis (Pollinia) lineares.

Obs. Doubah was originally found by Sir T. Mitchell, but with fruit only, in one of his journeys, and also in his last expedition; and, according to him, the natives eat the seed-vessel entire, preferring it roasted. Captain Sturt, on the other hand, observes that the natives of the districts where he found it eat only the pulpy seed-vessel, rejecting the seeds.

82] 16. Jasminum *lineare.* Br. prodr. 1, p. 521.

Jasminum Mitchellii. *Lindl. in Mitch. trop. Austr. p.* 365.

Obs. In Captain Sturt's collection there are perfect specimens of this plant, on which a few remarks may be here introduced, chiefly referring to its very general existence in the sterile regions of the interior of Southern Australia, and even extending to the north-west coast.

The species was established on specimens which I collected in 1802, in the sterile exposed tract at the head of Spencer's Gulf. With these I have compared and found identical Mr. A. Cunningham's specimens gathered in the vicinity of the Lachlan, in 1817; Captain Sturt's, in his earlier expeditions, from the Darling; those of Sir Thomas Mitchell, in his different journeys; and specimens collected in one of the islands of Dampier's Archipelago. In this great extent of range, it exactly agrees with a still more remarkable plant, and one much less likely to belong to a desert country, namely, Clianthus Dampieri.

I have considered Jasminum Mitchellii as hardly a variety of J. lineare, the character of this supposed species depending on its smooth leaves, and its axillary nearly sessile corymbi or fasciculi, which are much shorter than their subtending leaves; but even in the specimen contained in the collection presented to the British Museum by Sir Thomas Mitchell, the young branches, as well as the pedunculus and pedicelli, are covered with similar pubescence, and in the same degree as that of J. lineare; the specimens from Dampier's Archipelago have leaves equally smooth, but have the inflorescence of J. lineare; and I have specimens of J. lineare in which, with the usual pubescence of that species, the inflorescence is that of Mitchellii. Among Sir Thos. Mitchell's collection at the Museum, there is a Jasminum not noticed by Professor [83 Lindley, which, though very nearly related to J. lineare, and possibly a variety only, may be distinguished by the following character.

Jasminum (*micranthum*) cinereo-pubescens, foliis ternatis; foliolis lanceato-linearibus, pedunculis axillaribus 1-3 floris, corollæ laciniis obtusis dimidio tubi brevioribus.

17. Goodenia (*cycloptera*) ramosissima pubescens, foliis radicalibus serrato-incisis; caulinis lanceolato-ellipticis obsoletè serratis in petiolum attenuatis, pedunculis axillaribus unifloris folia subæquantibus, seminibus orbiculatis membrana angusta cinctis.

Loc. Indicatio nulla stationis in Herb. D. Sturt.

18. Scævola (*depauperata*), erecta ramosissima, ramis alternis; ultimis oppositis divaricatis, foliis minimis sublinearibus: ramorum alternis ramulorum oppositis, pedunculis e dichotomiis ramulorum solitariis unifloris.

Loc. "In salt ground, in lat. 26° S." D. Sturt.

Desc. Herbacea, vix suffruticosa, adulta glabriuscula, erecta, ramosissima. Rami ramulique angulati; ultimi oppositi, indivisi, divaricati, apice diphylli, foliis minimis et rudimento minuto floris abortivi. Folia sessilia, linearia, acuta, brevissima, ramos subtendentia alterna, ramulos ultimos brachiatos opposita. Pedunculi e dichotomiis ramulorum ultimorum penultimorumque solitarii, uniflori, ebrac-

teati. Calyx: limbo supero quinquepartito; laciniis lineari-lanceatis, æqualibus, pubescentibus. Corolla: tubo hinc ad basin usque fisso; limbo unilabiato, 5-partito; laciniis lanceolatis, æqualibus, marginibus angustis induplicatis, extus uti tubus pubescentibus, intus glabris trinerviis, nervo medio venoso. Stamina: filamenta distincta, anguste linearia, glabra, axi incrassata; antheræ liberæ, lineares, imberbes, basi affixæ, loculis longitudinaliter dehiscentibus. Ovarium biloculare? loculis monospermis, ovulis erectis. Stylus cylindraceus, glaber. Stigmatis indusium margine
84] ciliatum et extus pilis copiosis longis strictis acutis albis tectum v. cinctum.

19. Eremophila (*Cunninghamii*) arborescens, foliis alternis linearibus mucronulo recurvo, sepalis fructûs unguiculatis eglandulosis, corolla extus glabra.

Eremophila? arborescens, Cunningh. MSS. 1817.

Eremodendron Cunninghami, De Cand. prodr. xi, p. 713. Delessert ic. select. vol. v, p. 43, tab. 100 (ubi error in num. ovulorum).

Loc. "In the sandy bushes of the low western interior, not beyond lat. 29° S." D. Sturt.

Obs. The genus Eremophila was founded on very unsatisfactory materials, namely, on two species, E. oppositifolia and alternifolia, which I found growing in the same sandy desert at the head of Spencer's Gulf in 1802, the only combining character being the scariose calyx, which I inferred must have been enlarged after flowering. This, however, proves not to be the case in E. alternifolia, which Mrs. Grey has found in flower towards the head of St. Vincent's Gulf; and from analogy with other species since discovered, it probably takes place only in a slight degree in E. oppositifolia, whose expanded flowers have not yet been seen.

In 1817 Mr. Cunningham, in Oxley's first expedition, discovered a third and very remarkable species in flower and unripe fruit, which he referred, with a doubt, to Eremophila, and which M. Alphonse De Candolle has recently separated, but as it seems to me on very insufficient grounds,

with the generic name of Eremodendron, established entirely on Mr. Cunningham's specimens. A fourth species has lately been described by Mr. Bentham in Sir Thos. Mitchell's narrative of his Journey into Tropical Australia; and [85
some account of a fifth is given in the following article.

These five species may be arranged in four sections, distinguished by the following characters:

α. Folia opposita; sepala unguiculata.

Eremophila oppositifolia. *Br. prodr.* 1, *p.* 518.

β. Folia alterna; sepala unguiculata, eglandulosa; antheræ exsertæ.

E. Cunninghamii.

γ. Folia alterna; sepala brevè unguiculata, eglandulosa; stamina inclusa.

Eremophila Mitchelli. *Benth. in Mitch. trop. Austr. p.* 31.

Eremophila Sturtii.

δ. Folia alterna glanduloso-tuberculata, sepala cuneato-obovata, sessilia, glandulosa.

E. alternifolia. *Br. prodr.* 1, *p.* 518.

This last species might be separated from Eremophila; it is not, however, referable to Stenochilus, with some of whose species it nearly agrees in corolla, but from all of which it differs in its glandular scariose calyx.

20. Eremophila (*Sturtii*), pubescens, foliis angustè linearibus apiculo recurvo, corollis extus pubescentibus limbo intus barbato, staminibus inclusis.

Loc. "On the Darling; flowers purplish, sweet-scented." D. Sturt.

Desc. Frutex orgyalis (D. Sturt.). Calyx 5-partitus, æqualis; sepalis obovato-oblongis, basi angustioribus sed in unguem vix attenuatis, membranaceis, uninerviis, venosis. Corolla bilabiata, tubo amplo recto, labiis obtusis, extus pubescens, intus hinc (inferius) barbata. Labium superius tripartitum; lobo medio bifido (e duobus conflato); laciniis omnibus obtusis; inferius obcordatum bilobum lobis rotundatis, densius barbatum. Stamina quatuor didynama, omnino inclusa. Filamenta glabra. Antheræ reniformes, loculis apice confluentibus. Ovarium densè lanatum. Stylus glaber. Stigma indivisum, apice styli vix crassius.

86] Obs. Species proxima E. Mitchelli *Benth. in Mitch. trop. Austr.* p. 31.

21. Stenochilus *longifolius.* Br. prodr. 1, p. 517.

Stenochilus pubiflorus. *Benth. in Mitch. trop. Aust. p.* 273.

Stenochilus salicinus. *Benth. in Mitch. trop. Austr. p.* 251.

Loc. Nulla stationis indicatio.

22. Stenochilus *maculatus, Ker in Bot. Regist. tab.* 647. *Cunningh. MSS.* 1847.

β. Stenochilus curvipes. *Benth. in Mitch. trop. Austr. p.* 221. Varietas S. maculati, sepalorum acumine paulo breviore.

Obs. M. Alphonse De Candolle, in Prodr. xi, p. 715, refers S. ochroleucus of Cunningh. MSS. 1817, as a variety to S. maculatus; it is, however, very distinct, having a short erect peduncle like that of S. glaber, to which it is much more nearly related, differing chiefly in its being slightly pubescent.

23. Grevillea (Eugrevillea) *Sturtii,* foliis indivisis (nonnullis raro bifidis) augustè linearibus elongatis uninerviis: marginibus arctè revolutis, racemis oblongis cylindraceisve: rachi pedicellis perianthiisque inexpansis glutinoso-pubescentibus, ovario sessili, stylo glabro.

Loc. "On sand-hills in lat. 27° S." D. Sturt.

Desc. Arbor 15-pedalis (Sturt). Rami teretes, pube arctè adpressa persistenti incani. Folia 6-10-pollices longa, vix tres lineas lata, subter pubescentia incana, super tandem glabrata. Thyrsus terminalis, 2-4 uncialis, rachi pedicellisque pube erecta nec adpressa secretione glutinosa intermista. Flores aurantiaci.

Obs. In the collection presented to the British Museum by Sir Thomas Mitchell, of the plants of his last expedition, there is a very perfect specimen, in flower, of Grevillea Sturtii.

The following observations respecting the Grevilleæ of the same collection may not be without interest.

Grevillea Mitchellii, *Hooker, in Mitch. Trop. Austr. p.* [87
265, proves to be Gr. Chrysodendrum, *prodr. fl. Nov. Holl. p.* 379, the specific name of which was not derived from the colour of the under surface of the leaves, which is, indeed, nearly white, but from the numerous orange-coloured racemes, rendering this tree conspicuous at a great distance.

Grevillea longistyla and G. juncea of the same narrative both belong to that section of the genus which I have named Plagiopoda.

A single specimen, in most respects resembling Gr. longistyla, of which possibly it may be a variety, but which at least deserves notice, has all its leaves pinnatifid, instead of being undivided. It may be distinguished by the following character:—*Grevillea* (*Plagiopoda*) *neglecta*, foliis pinnatifidis subtus niveis; laciniis linearibus, stylis glabris.

A single specimen also exists of Grevillea (or Hakea) lorea, *prodr. flor. Nov. Holl. p.* 380, but without fructification.

24. Grevillea (Cycloptera?) *lineata*, foliis indivisis lineari-ensiformibus enerviis subter striis decem paucioribus elevatis uniformibus interstitia bis-terve latitudine superantibus, cicatrice insertionis latiore quam longa utrinque obtusa, racemis terminalibus alternis, pistillis semuncia brevioribus stigmate conico.

Loc. "It takes the place of the gum-tree (Eucalyptus) in the creeks about lat. 29° 30′ S." D. Sturt.

Obs. It is difficult to distinguish this species, which, according to Captain Sturt, forms a tree about 20 feet in height, from Grevillea striata. I have endeavoured to do so in the above specific difference, contrasted with which the leaves of G. striata have always more than 10 striæ, which are hardly twice the breadth of the pubescent in- [88
terstices, and the cicatrices of whose leaves are longer than broad, and more or less acute, both above and below. This is a source of character which in the supplement to the Prodr. Floræ Novæ Hollandiæ, I have employed in a

few cases both in Grevillea and Hakea, but which I believe to be important, as it not only expresses a difference of form, but also in general of vascular arrangement.

25. Ptilotus (*latifolius*) capitulis globosis, bracteis propriis calycem superantibus, foliis ovatis petiolatis.

Loc. "In lat. 26° S." D. Sturt.

Desc. Herba diffusa, ramosa, incana. Folia alterna, petiolata, latè ovata, integerrima. Capitula ramos terminantia, solitaria vel duo approximata. Bracteæ laterales scariosæ, sessiles, latè ovatæ, enerviæ. Perianthium; foliolis subæqualibus, lana implexa alba basi tectis, ante expansionem ungue nervoso tunc brevissimo, post anthesin laminam scariosam enervem fere æquante. Stamina 5 antherifera; filamenta basi in cyathulum edentulum connata. Antheræ biloculares, loculis utrinque distinctis medio solum conjunctis. Ovarium monospermum, glabrum. Stylus filiformis, glaber. Stigma capitatum, parvum. Utriculus evalvis, ruptilis.

Obs. I was at first inclined to consider this plant as a genus distinct from Ptilotus, more, however, from the remarkable difference in habit than from any important distinction in the flower, for its character would have chiefly consisted in the great size of its lateral bracteæ, and in the form of its antheræ.

In a small collection formed during the voyage of Captains Wickham and Stokes, there is a plant very nearly related to, and perhaps not specifically distinct from, Ptilotus latifolius, but having narrower leaves. It was found on one of the islands of Dampier's Archipelago.

89] 26. Neurachne (*paradoxa*) glaberrima, culmo dichotomo, foliis rameis abbreviatis, fasciculis paucifloris, glumis perianthiisque imberbibus valvula exteriore cujusve floris septemnervia.

Loc. Nulla indicatio loci v. stationis, in Herbario D. Sturt.

Desc. Gramen junceum, facie potius Cyperaceæ cujusdam. Folia radicalia in specimine unico viso defuere; ramos subtendentia abbreviata, vagina aperta ipsum folium superante; floralia subspathiformia sed foliacea nec membranacea. Fasciculi pauciflori: spiculæ cum pedunculo brevissimo articulatæ et solubiles, et subtensæ bractea nervosa carinata ejusdem circiter longitudinis. Gluma bivalvis biflora, nervosa, acuta, mutica; valvulæ subæquales septemnerviæ; exterioris nervis tribus axin occupantibus sed distinctis reliquis per paria a marginibus et axilibus subæquidistantibus; interioris nervis æquidistantibus, externis margini approximatis. Perianthium inferius (exterius), bivalve neutrum; valvula exterior septemnervis, exteriori glumæ similis textura forma et longitudine; valvula interior (superior) angustior pauloque brevior, dinervis, nervis alatis marginibus veris latis induplicatis. Perianthium superius hermaphroditum, paulo brevius, pergamineo-membranaceum, nervis dilutè viridibus; valvula exterior quinquenervis, acuta, concava; interior ejusdem fere longitudinis, dinervis. Stamina 3, filamentis linearibus. Ovarium oblongum, imberbe. Styli duo. Stigmata plumosa, pallida?

Obs. Neurachne paradoxa, founded on a single specimen, imperfect in its leaves and stem, but sufficiently complete in its parts of fructification, differs materially in habit from the original species, N. alopecuroidea, as well as from N. Mitchelliana of Nees, while these two species differ widely from each other in several important points of structure.

In undertaking to give some account of the more re- [90
markable plants of Captain Sturt's collection, it was my intention to have entered in some detail into the general character of the vegetation of the interior of Australia, south of the Tropic.

I am now obliged to relinquish my original intention, so far as relates to detail, but shall still offer a few general remarks on the subject.

These remarks will probably be better understood if I refer, in the first place, to some observations published in 1814, in the Botanical Appendix to Captain Flinders's Voyage.[1]

[1] *Antè, p.* 61.

From the knowledge I then had of New Holland, or Australian vegetation, I stated that its chief peculiarities existed in the greatest degree in a parallel, included between 33° and 35° S. lat. which I therefore called the principal parallel, but that these peculiarities or characteristic tribes were found chiefly at its western and eastern extremities, being remarkably diminished in that intermediate portion, included between 133° and 138°, E. long. These observations related entirely to the shores of Australia, its interior being at that period altogether unknown; and the species of Australian plants, with which I was then acquainted, did not exceed 4200. Since that time great additions have been made to the number, chiefly by Mr. Allan Cunningham, in his various journeys from Port Jackson, and on the shores of the North and North-west coasts during the voyages of Captain King whom he accompanied; by Messrs. William Baxter, James Drummond, and M. Preiss, at the western extremity of the principal parallel, and by Mr. Ronald Gunn in Van Diemen's Land. It is probable that I may be considered as underrating these additions, when I venture to state them as only be-
91] tween two and three thousand; and that the whole number of Australian plants at present known, does not exceed, but rather falls short of 7000 species.

These additions, whatever their amount may be, confirm my original statement respecting the distribution of the characteristic tribes of the New Holland Flora; some additional breadth might perhaps be given to the principal parallel, and the extent of the peculiar families may now be stated as much greater at or near its western, than at its eastern extremity.

With the vegetation of the extra-tropical interior of Australia we are now in some degree acquainted, chiefly from the collections formed by the late Mr. Allan Cunningham, and Charles Fraser, in Oxley's two expeditions from Port Jackson into the western interior, in 1817 and 1818; from Captain Sturt's early expeditions, in which the rivers Darling, Murrumbidgee, and Murray, were discovered; from those of Sir Thomas Mitchell, who never failed to form extensive collections of plants of the regions he visited; and lastly, from Captain Sturt's present collection.

The whole number of plants collected in these various expeditions may be estimated at about 700 or 750 species; and the general character of the vegetation, especially of the extensive sterile regions, very nearly resembles that of the heads of the two great inlets of the south coast, particularly that of Spencer's Gulf; the same or a still greater diminution of the characteristic tribes of the general Australian Flora being observable. Of these characteristic tribes, hardly any considerable proportion is found, except of Eucalyptus, and even that genus seems to be much reduced in the number of species; of the leafless Acaciæ, which appear to exist in nearly their usual proportion; and of Callitris and Casuarina. The extensive families of Epacrideæ, Stylideæ, Restiaceæ, and the tribe of Decandrous [92
Papilionaceæ, hardly exist, and the still more characteristic and extensive family of Proteaceæ is reduced to a few species of Grevillea, Hakea, and Persoonia.

Nor are there any extensive families peculiar to these regions; the only characteristic tribes being that small section of aphyllous, or nearly aphyllous Cassiæ, which I have particularly adverted to in my account of some of the species belonging to Captain Sturt's collection; and several genera of Myoporinæ, particularly Eremophila and Stenochilus. Both these tribes appear to be confined to the interior, or to the two great gulfs of the South coast, which may be termed the outlets or direct continuation of the southern interior; several of the species observed at the head of Spencer's Gulf also existing in nearly the same meridian, several degrees to the northward. It is not a little remarkable that nearly the same general character of vegetation appears to exist in the sterile islands of Dampier's Archipelago, on the North-west coast, where even some of the species which probably exist through the whole of the southern interior are found; of these the most striking instances are, Clianthus Dampieri, and Jasminum lineare, and to establish this extensive range of these two species was my object in entering so minutely into their history in the preceding account.

A still greater reduction of the peculiarities of New Holland vegetation takes place in the islands of the South coast.

PART II.

STRUCTURAL AND PHYSIOLOGICAL MEMOIRS.

SOME OBSERVATIONS

ON THE

PARTS OF FRUCTIFICATION IN MOSSES;

WITH

CHARACTERS AND DESCRIPTIONS

OF

TWO NEW GENERA OF THAT ORDER.

BY

MR. ROBERT BROWN, LIBR. LINN. SOC.

READ JUNE 20TH, 1809.

[*Extracted from 'The Transactions of the Linnean Society of London.' Vol. X, pp.* 312—324.]

LONDON:

1811.

SOME OBSERVATIONS, &c. [312

THE account which the celebrated Hedwig has given of the sexes of Mosses seems to be founded on so ample an induction, and is now so generally received, that it must be [un]necessary to notice the arguments which mere theoretical botanists have from time to time produced against it. There is, however, one author, Mons. Palisot Beauvois, who has not only objected to the account of Hedwig, but has proposed a theory of his own, and who, consequently, appealing to actual observations, and appearing to have particularly studied, specifically at least, this tribe of plants, merits some attention. The earliest account of Mons. Beauvois' theory is to be found in the observations added to the order Musci, in the "Genera Plantarum" of Jussieu; and it was soon after more fully given by the author himself in a Memoir on the Sexual Organs of Mosses, published in the third volume of the American Philosophical Transactions: since that time he has, in his different works, occasionally treated of the same subject, and has lately repeated the substance of his original essay, in the introduction to his "*Prodrome des Cinquième et Sixième Familles de l'Æthiogamie*," published at Paris in 1805, a translation of which is given by my friend Mr. Konig, in the second volume of the Annals of Botany. To this work, as it must be in the hand of [313
every scientific botanist, I refer for a full account of M. Beauvois' hypothesis, and confine myself to observing, that what is generally called the capsule of mosses, is by him considered as the containing organ of both sexes; that the granules which Hedwig supposes to be seeds, he regards as pollen; the real seeds according to him being imbedded in

the substance of that body which occupies the centre of the capsule, and to which botanists have given the name of *columnula* or *columella*. The supposed seeds of this author, however, having entirely escaped the two most acute and experienced observers in this department of botany, Schmidel and Hedwig, in all the species of which they have given dissections, it might fairly be concluded that they are not of universal existence, and this alone would be sufficient perhaps to overturn the hypothesis. But it would be more satisfactory, if, while the accuracy of these excellent observers was confirmed in other instances, the cause of that appearance, which I apprehend has misled M. Beauvois, could at the same time be pointed out. The species more particularly described and figured by him in the American Transactions, is *Hypnum velutinum;* which therefore, had it been in a proper state, I should have preferred as the subject of my examination; but as he asserts that his observations were repeated, and with similar results, on all the species of mosses found in the neighbourhood of Paris and Lisle, I have chosen *Funaria hygrometrica,* perhaps the most general plant in existence; which therefore must have been examined by him, and is within the reach of every one.

As, according to M. Beauvois, the action of the pollen on the seeds does not take place till the separation of the operculum, he probably did not conceive it necessary to observe the capsule until it had acquired its full size, and was in fact nearly ripe, or, as he terms it, in blossom. At
314] this period he examined under the microscope a transverse section of the capsule, in which, as appears both from his description and figure, he found a dense stratum of granular matter, which he considered to be pollen, situated immediately within the inner membrane; while in the substance occupying the centre, which he describes as reticulated, he observed scattered granules, in size and appearance like those of the pollen already mentioned: these he regards as the genuine seeds, and the containing organ he calls the capsule.

It is remarkable that he nowhere expressly states the

manner in which this capsule bursts : but it may be inferred, from the use he assigns to the peristomium, that he supposes it to eject its contents by the upper extremity : for, if the bursting were lateral, the seeds would at once come into contact with the pollen : but though impregnation would in this way more certainly be accomplished, the motions of the ciliæ could no longer be considered as in any degree assisting it.

Desirous to examine an object as nearly similar as possible to that on which the hypothesis appears to be founded, I in the first place made a transverse section of the full-grown but green capsule of *Funaria hygrometrica ;* and, I confess, was both surprised and disappointed to find it, under the microscope, exactly resembling M. Beauvois' figure [18]. But little reflection, however, was necessary to show that these scattered granules might either have been forced into the pulpy central substance, by the pressure necessarily applied to the stratum of pollen in making the section, or, what is more probable, been carried over its surface by the cutting instrument, which had previously passed through this stratum. Accordingly, by repeated immersion in water, and more readily still by the careful application of a small hair-pencil, the greater part of the granules was removed. [315
A transverse section at an earlier stage of the capsule, before the falling of the calyptra, exhibited, as I expected, fewer granules on the substance of the columella, and which were removable in like manner. Lastly, by a longitudinal section, in which, if well performed, the scalpel could not be supposed to carry any part of the pollen over the surface of the columella, I obtained a distinct view of this part, perfectly free from these supposed seeds, and evidently consisting of large cells filled with an uniform pulpy substance ; a continuation of which occupied the cavity of the operculum.

From these observations, even added to those of Schmidel and Hedwig, though they seem conclusive against the hypothesis of M. Beauvois, I by no means pretend to reason strictly respecting the whole order : on the contrary, from the conversations I have had with my ingenious and accurate

friend, Mr. Francis Bauer, as well as from some observations of my own, I am disposed to believe that considerable diversities may exist in the placentation of mosses: that in some cases the seeds may be formed in a much greater portion of the columnula than in others: and it is even not improbable that in certain cases its whole substance may be converted into seeds; or, to speak more accurately, that it may produce seeds even to the centre, and that the cells in which they were probably formed may be re-absorbed. This I am inclined to think is the case in *Phascum alternifolium* of Dickson, in the ripe capsule of which there is hardly the vestige of a columnula; and I have observed the same structure in two new species of *Anodontium* of Bridel; which, if it equally exists in the only species of this genus hitherto described, would perhaps considerably strengthen its character. In these cases the inner membrane is also
316] evanescent; and such a structure, it may be remarked, equally militates against M. Beauvois' theory, whether we suppose the columella to have existed at an earlier stage, in the usual form, or not.

As to this organ being tubular, and discharging its contents by the top, it is neither consistent with what has been already observed, nor with the appearance of its remains in the ripe capsule: but, admitting for a moment its tubular nature, there are certain mosses in which no discharge could possibly take place in the way described; the column being elongated even to the apex of the operculum, to which it often continues to adhere, as in *Buxbaumia*, and in the first of the two new genera which I now proceed to describe.

DAWSONIA.

Peristomium penicillatum, ciliis numerosissimis capillaribus rectis æqualibus e capsulæ parietibus columellâque (!) ortis.

Capsula hinc plana, indè convexa.

Calyptra exterior e villis implexis, *interior* apice scabra.

Muscus *hinc arctè affinis* Polytricho, *quocum foliis, floribus masculis, et calyptrâ penitùs convenit; indè aliquo modo* Buxbaumiæ *accedens, præsertim figurâ capsulæ, et structurâ columellæ.* Peristomio *autem ab omnibus diversissimus.*

Dawsonia polytrichoides.

Tab. 11 [XXIII].[1] *Fig.* 1.

Patria. Novæ Hollandiæ ora orientalis, extra tropicum.

Statio. Ripæ subumbrosæ rivulorum, ad radices montium, in vicinitate Portûs Jackson.

Desc. *Cæspites* laxi, amorphi. *Radiculæ* tenuissimæ, tomenti instar, caudicem descendentem brevem inves- [317 tientes. *Caulis* simplicissimus, erectus, strictus, 2—3-uncialis, basi reliquiis foliorum squamatus, suprà densè foliatus. *Folia*, e basi dilatatâ semiamplexicauli membranaceâ fuscâ, lineari-subulata, opaca, viridia, marginibus longitudinaliter dorsoque apicis denticulatis, spinulis sursum crebrioribus majoribusque, concaviuscula, patula, siccatione appressa, canaliculata, superiora vix semuncialia, inferiora sensim breviora.

Masculi Flores terminales, discoidei. *Folia perigonialia* cuneato-orbiculata, mucronata, integerrima, semimembranacea, exteriora sensim majora. *Fila succulenta* numerosa, articulata, basi attenuata. *Antheræ* flosculi singuli 6-8, cylindraceæ, brevissimè pedicellatæ.

Femineus Flos in distincto individuo. *Seta* terminalis, solitaria, erecta, lævis, nitens, rufo-fusca, caule ter brevior, foliis terminalibus duplò longior. *Vaginula* cylindracea, stricta, glabra, tegmine pilorum calyptræ exterioris instar instructa.

Calyptra duplex: *exterior* constans pilis intertextis dimidio inferiore tenui flexuoso pallido ramuloso edentulo, superiore ferrugineo stricto denticulato: *interior* membranacea straminea, capsulæ maturæ subulata, suprà longitudinaliter fissa, apice solum denticulata.

Capsula nutans, angulum ferè rectum cum setâ efformans,

[1] The figures within brackets refer to the numbering of this and subsequent plates in the 'Linnean Transactions.'—Ed.

ovata, per lentem reticulata, areolis subrotundis, sordidè fusca, lævis, nonnitens, suprà plana marginibus acutis, subtùs modicè convexa ore coarctato, marginato. Apophysis nulla.

Operculum conico-cylindraceum, capsulâ brevius, apice lateris superioris in mucronem levissimè incurvum producto, basi incrassatâ, cum calyptrâ sæpissimè deciduum.

Peristomium penicillum densum album referens, longitudine circiter dimidii capsulæ, formatum *Ciliis* indeterminatim numerosissimis (200 et ultrà) capillaribus inarti-
818] culatis æqualibus rectis albis opacis, pluribus e capsulæ parietibus ortum ducentibus, centralibus (circiter 50) columellam terminantibus!

Membrana interior capsulæ maturæ exteriori approximata, vasculisque numerosis connexa.

Columella longitudine capsulæ maturæ, in quâ latiuscula, corrugata, colli brevis margine incrassatâ, intra cilias desinens in processum filiformem solidum indivisum apicem operculi attingentem eique arctiùs adhærentem.

Semina minutissima, lævia, in cumulo viridia, seorsùm hyalina.

Obs. I. I have named this remarkable genus in honour of my esteemed friend Dawson Turner, Esq., a gentleman eminently distinguished in every part of cryptogamic botany, and from whom, after he has finished the incomparable work on *Fuci*, in which he is now engaged, we may expect a general history of mosses.

Obs. II. The strict relationship between *Dawsonia* and *Polytrichum* in most respects, and the striking dissimilarity of their peristomiums, may tend, perhaps, in some degree to lessen our confidence in the characters derived from that part; for there seems in this case but little analogy between the two structures. The better to understand that of *Polytrichum*, I was induced along with Mr. Turner to examine it in the unripe capsule: in this state the cavity of the operculum was found completely filled with a cellular pulp, similar to that composing the columella, of which it appeared evidently to be a continuation; to the surface of this pulp the teeth of the peristomium were closely pressed, but did not adhere: by degrees the pulp dries up, and in the ripe capsule leaves only the membrane or tympanum of an inorganic appearance, and firmly cohering with the teeth by the inner side of their apices. It does not therefore [319 properly belong to the operculum, though in some cases it may adhere to it, as does the analogous process of the columella in *Dawsonia* and in several other mosses.

The affinity of *Dawsonia* to *Buxbaumia* is certainly less strict than to *Polytrichum*, and rests chiefly on the similarity of the figure of the capsule, and in the central process of the columella, which is still more evident in *Buxbaumia*, where it forms part of the Linnean generic character, though unaccountably overlooked by Schmidel in his masterly dissertation; but, if I mistake not, actually represented by him [in fig. 14, b[1]], and confounded with the peristomium, which in this case, I suppose, had adhered to the operculum, as I have repeatedly found it to do, and thus escaped his notice. Hedwig considers the plaited membrane which constitutes the peristomium of *Buxbaumia*, as derived from the inner membrane of the capsule, and quotes the figure just mentioned of Schmidel in proof of this origin. In both species, however, I find it arising from the exterior membrane, though considerably within its margin, which in *Buxbaumia aphylla* is said by Hedwig to be divided into teeth,—an appearance I could not observe in the few ripe capsules I have dissected. In other respects, the two species seem essentially to agree, and therefore ought not to be separated, as Ehrhart and some late writers have done. The generic character comprehending both, I would propose to alter in the following manner.

BUXBAUMIA.

Capsula obliqua, hinc convexior, vel gibba.

Peristomium intra marginem, quandoque dentatum, membranæ exterioris ortum, tubulosum, plicatum, apice apertum.

[1] Schmidel, *Dissertationes Botanici Argumenti.*

320]
LEPTOSTOMUM.

Capsula oblonga, exsulca; *Operculo* hemisphærico, mutico.

Peristomium simplex, membranaceum, annulare, planum, indivisum, e membrana interiori ortum.

Musci *dense cæspitosi*. Caules *erecti, annotino-ramosi*. Folia *undique modicè patentia, latiuscula, nervo valido, marginibus integris, revolutis, pilo (quandoque ramoso?) terminata*. Seta *terminalis*. Capsula *erecta v. inclinans, basi in apophysin obconicam attenuata, ore coarctato*. Calyptra *glabra, lævis, caduca*.

1. *L. inclinans*, foliis ovato-oblongis obtusis; pilo simplici, capsulis inclinatis obovato-oblongis.

Tab. 11 [XXIII]. Fig. 2.

Patria. Insula Van-Diemen.

Statio. Rupes et saxa ad latus orientale prope summitatem Montis Tabularis lat. aust. 43°, elevatione supra mare 3000 ad 3500 ped.

Desc. Muscus lætè virens 2-3-uncialis. Caules parùm divisi, infrà tomento denso ferrugineo vestiti, suprà confertim foliati. Folia concaviuscula, per lentem minutissime punctato-areolata, pilo tortili ipso folio quater breviore. Seta fusca, lævis. Vaginula infrà stipata adductoribus pluribus filisque succulentis capillaribus articulatis.

2. *L. erectum*, foliis oblongo-parabolicis obtusis; pilo simplici, capsulis erectis oblongis.

Patria. Novæ Hollandiæ ora orientalis, extra tropicum.

Statio. Rupes prope fluviorum ripas, in regione montanâ; ad fluvios Hawkesbury et Grose.

Desc. Muscus 2-3-uncialis. Caules simplices et subra-
321] mosi, infrà tomento ferrugineo vestiti, suprà densè foliati. Folia siccatione parùm curvata, et simul adpressa.

Seta elongata, fusca, lævis. Capsula æquilatera. Operculum delapsum fuit.

3. *L. gracile,* foliis ovato-oblongis acutiusculis; pilo simplici folium dimidium æquante, capsulis oblongis æquilateris inclinatis.

PATRIA. Nova Zelandia.

STATIO. Umbrosa humida (?) ad Dusky Bay, *Dom. Arch. Menzies.*

DESC. Caules subramosi. Folia siccatione adpressa, areolato-punctata. Seta elongata, lævis. Vaginula cylindracea, filis succosis adductoribusque numerosis cincta.

4. *L. Menziesii,* foliis oblongo-lanceolatis acutis; pilo simplici folio quater breviore, capsulis oblongis inclinatis arcuato-recurvis.

PATRIA. Americæ Australis Staten-land, ubi anno 1787 detexit *Dom. Arch. Menzies,* cujus amicitiæ hanc et præcedentem speciem debeo.

STATIO. - - - -

DESC. Muscus lætè virens, sesquiuncialis. Caules subsimplices, basi ferrugineo-tomentosi, suprà confertim foliati. Folia erecto-patentia, siccatione adpressa, minutissimè areolata v. punctata. Seta caulem sæpiùs superans, erecta, fusca, lævis. Capsula subfalcata ad angulum acutum rariusve ferè rectum inclinans.

OBS. The plants which I have referred to this genus are all natives of the southern hemisphere, and in their habit, in which there is something peculiar, strictly agree with each other, and with *Bryum macrocarpum* of Hedwig. [322
In three of the four species here described, I have had the opportunity of removing the operculum without having been able in any case to observe an external peristomium, which, from the appearance of these plants, might be expected to exist, and which Hedwig has figured in his *Bryum macrocarpum.* Of this plant I have only seen specimens that had lost the operculum: the mouth of the capsule, however, seemed to be very perfect, and was fur-

nished with a membrane, exactly as in the species here described, but I could not perceive any remains of external teeth. In opposition to such authority, however, I do not venture to add it to this genus, to which in every other respect it seems to belong.

The character of *Leptostomum*, derived from the undivided annular process of the inner membrane of the capsule, may to many appear too minute, and perhaps unimportant; and had it been observed in one species alone, I should not have ventured on that account to distinguish it as a genus: but finding it in four species, accompanied too with a habit widely different from that of *Gymnostomum*, to which these plants must otherwise be referred, I have not hesitated to employ it. As, however, Hedwig has actually figured and described an external peristomium in his *Bryum macrocarpum*, whose striking resemblance to *Leptostomum* has been already noticed, there may be still some reason to doubt the sufficiency of the generic character, and it may seem somewhat improbable that Mosses of such a habit should be really destitute of an outer peristomium. But, without questioning the accuracy of Hedwig in this instance, I may be permitted to observe, that the outer peristomium which he has figured in *Bryum macrocarpum* is extremely unlike that of any other genus where the fringe
323] is double: and it may perhaps in some degree tend to strengthen the character of *Leptostomum*, to advert to what appears to be really the case in certain species of *Pterogonium*, in one of which[1] Mr. Hooker has already described the fringe as derived solely from the inner membrane; and I have collected, on the mountains of Van Diemen's Island, a moss with a peristomium decidedly of like origin; a circumstance that appeared to me so remarkable, that I had actually described it as a distinct genus, before I was aware of the similar structure of the Nepal plant described by Mr. Hooker; or of the probability, from Hedwig's own figures, that some at least of his *Pterogonia* were of the same structure; a point that I have not at present

[1] Pterogonium declinatum. *Trans. Linn. Soc.* ix, *p.* 309.

the means of determining, but which I beg leave to recommend to the attention of those botanists who are provided with perfect specimens of the published *Pterogonia.*

Explicatio Tabulæ 11 (XXIII).

Fig. 1. Dawsonia polytrichoides. *a.* Mascula planta magnitudine naturali. *b.* Discus masc. auctus. *c.* Ejusdem flos unicus. *d.* Idem absque folio perigoniali, magisque auctus. *e.* Anthera et filum succulentum maximè aucta. *f.* Femineæ plantæ magn. nat. *g.* Vaginula cum foliis perichætialibus auctis. *h.* Capsula cum calyptrâ exteriori. *i.* Pili calyptræ exterioris magis aucti. *j.* Capsula cum operculo et calyptrâ interiori. *k. l.* Capsula deoperculata cum peristomio. *m.* Capsulæ sectio ejusdem figuram insertionemque ciliarum ostendens. *o.* Calyptra interior. *p.* Operculum cum columellæ processu [324
filiformi. *q.* Columella ciliis suis terminata. *r.* Semina. *s.* Ciliæ peristomii auctæ.

Fig. 2. Leptostomum inclinans, magnitudine naturali. *α.* Ejusdem capsula aucta cum membranâ annulari. *β.* Operculum. *γ.* Idem a basi visum cum annulo cohærenti.

ON

SOME REMARKABLE DEVIATIONS FROM THE USUAL STRUCTURE OF SEEDS AND FRUITS.

BY

ROBERT BROWN, Esq., F.R.S., Libr. L.S.

Read March 5th, 1816.

[*Extracted from the 'Transactions of the Linnean Society of London.' Vol. XII, pp.* 143—151.]

LONDON:

1818.

ON [143

SOME REMARKABLE DEVIATIONS, &c.

The principal part of the following paper was read to the Society in March, 1813. It was then withdrawn with a view of rendering it more perfect by additional facts, which I hoped I might be able to collect. Since that time I have not had it in my power to pay much attention to the subject. As, however, the facts formerly stated appear to me of some importance, and are as yet unpublished, I take the liberty of again submitting them to the Society, along with a few additional instances of anomalies in the structure of seeds and fruits, hardly less remarkable than those contained in the original essay.

It is, I believe, generally admitted by physiological botanists, that the seeds of plants are never produced absolutely naked :—in other words, that the integument through some point or process of which impregnation takes place, cannot properly be considered as part of the seed itself.

That such a covering, distinct from the seed, really exists, may in most, perhaps in all, cases be satisfactorily shown by a careful examination of the unimpregnated ovarium, to a part only of whose cavity the ovulum will be found to be attached.

There are, however, many cases where soon after fecunpation, and more remarkably still in the ripe fruit, this integument acquires so complete and intimate a cohesion [144
with the proper coat of the seed as to be no longer either separable or distinguishable from it.

But systematic botanists have generally agreed to term a

naked seed not only this kind of fruit, but every monospermous pericarpium bearing a general resemblance to a seed, and whose outer covering, though distinct from the nucleus, is only ruptured after germination commences.

For the purposes of an artificial arrangement this language may perhaps be sufficiently accurate; but in determining the affinities of plants, it is necessary to express by appropriate terms those differences which are no less important than real.

Of the fruits improperly called naked seeds, there are two principal kinds: the first, in which the pericarpium is distinct from the seed, is termed *Akena* by Richard in his excellent *Analyse du Fruit;* the second, in which the pericarpium coheres with the seed, is the *Caryopsis* of the same author.

An *Akena* (or *Achenium*), even in a separate state, may in general be readily determined. But it is not always equally easy to distinguish a *Caryopsis* from a seed. It may indeed be done in certain cases, as in Grasses, by attending to its surface, in which two distinct and distant cicatrices are observable; the one indicating the point of attachment to the parent plant, the other that by which it was fecundated. In certain other tribes, however, this criterion cannot be had recourse to, the surface of the *Caryopsis* exhibiting but one areola or cicatrix, which includes the closely approximated points of attachment and impregnation: in such cases, the true nature of the fruit can only be determined by its examination in an earlier stage.

But although it must be admitted that an ovulum is never produced without a covering, through some part of
145] which it is impregnated; it is still possible to conceive a case in which a ripe seed may be considered as truly naked while retaining its attachment to the parent plant; and this not subsequent to germination, but even preceding the formation of the embryo. For if we suppose, as the immediate effect of impregnation, a swelling of the ovulum without a corresponding enlargement of the ovarium, the consequence will obviously be a premature rupture of the ovarium, and the production of a seed provided with its proper integuments only.

I am not aware that such an economy has hitherto been described; I have observed it, however, in several plants belonging to very different families, and of essentially different structures.

The first of these is *Leontice thalictroides* of Linnæus, *Caulophyllum thalictroides* of Michaux, who has founded his new genus on a difference of fruit, the nature of which he has entirely misunderstood. It is remarkable that its real structure should have escaped so accurate an observer as M. Richard, through whose hands it is generally understood Michaux's work passed previous to its publication; but the fact may at least serve to show how entirely unexpected such an economy must have been even to that excellent carpologist.

My observations were made in the summer of 1812, on a plant of *Leontice thalictroides*, which flowered and ripened fruit in the royal gardens at Kew. An examination of the unimpregnated ovarium proved it to be in every respect of the same structure with that of the other species of *Leontice*; and essentially the same with the whole order of *Berberides*, to which this genus belongs. A careful inspection of the fruit, in different states, proved also that the "Drupa stipitata" of Michaux is in reality a naked seed, that in a very early stage had burst its pericarpium, the withered re- [146
mains of which were in most cases visible at the base of the ripe seed. The first error of Michaux naturally led to a series of mistakes; and the naked seed being considered by him as a drupa, the albumen, which is of a horny texture, is described as a "nux cornea crassissima," and the embryo itself as the seed.

But although this account of the fruit of *Leontice thalictroides* be in no respect similar to that given by Michaux, it may perhaps be considered by some as still differing sufficiently from *Leontice* to authorise the establishment of a distinct genus; and that, therefore, the name *Caulophyllum* may be retained, and its character derived from the remarkable circumstance described, namely, the early rupture of its pericarpium. I believe, however, it will be found more expedient to reduce it again to *Leontice*.

For, in the first place, its habit is entirely that of the original species of the genus. And secondly, though the pericarpium of *Leontice Leontopetalum,* which is the type of the genus, remains shut until the ripening of the seeds, and attains a size more than sufficient for the mere purpose of containing them ; yet in *Leontice altaica,* a species in other respects more nearly approaching to *L. Leontopetalum* than to *L. thalictroides,* the pericarpium, though it enlarges considerably after impregnation, is ruptured by the seeds long before they have arrived at maturity.

The accompanying drawing, for which I am indebted to my friend Mr. Ferdinand Bauer, will materially assist in explaining the singular economy now described; and may also perhaps render more intelligible the account I proceed to give of the second instance in which I have observed an analogous structure, but to illustrate which I have at present no drawing prepared.

147] This second instance occurs in *Peliosanthes Teta* of Andrews's Repository and the Botanical Magazine.

In this monocotyledonous plant, which in 1812 nearly ripened seed in Mr. Lambert's collection at Boyton, the ovarium coheres with the tube of the perianthium or corolla, and has originally three cells, each containing two ovula. Soon after impregnation has taken place, from one to three of these ovula rapidly increase in size, by their pressure prevent the development of the others, and rupture the ovarium, which remains, but little enlarged at the base of the fruit, consisting of from one to three naked berry-like seeds.

In the Botanical Magazine Mr. Ker, in describing a second species of *Peliosanthes,*[1] takes the opportunity of altering in some respects the character of the genus he had previously given, and of adding a description of its supposed pericarpium, from an inspection, as it seems, of the unripe fruit of *Peliosanthes Teta.* It is evident, however, that he is not aware of its real structure; and consequently does not succeed in reconciling its appearance with the unquestionable fact of its having "germen inferum."

[1] Botan. Magaz. 1532.

There are some cases in which this early opening of the ovarium, instead of being, as in the preceding instances, an irregular bursting, apparently caused by the pressure of the enlarged ovula, is a regular dehiscence in the direction of the suture. Of this *Sterculia platanifolia* and *S. colorata* are remarkable examples; their folliculi after opening, which takes place long before the maturity of the seeds, acquiring the form and texture of leaves, to whose thickened margins the ovula continue firmly attached until they ripen. Another example of this early and regular dehiscence occurs in an undescribed genus of the same family, which differs from *Sterculia platanifolia* in its pericarpium having a terminal wing and a single seed.

In the specimens of a plant lately sent from Brazil by [148
Mr. Sellow, I observe a similar economy. In this case the ovarium, which is originally unilocular with five parietal placentæ, soon after fecundation opens regularly into five equal foliaceous valves, to the inner surface of each of which an indefinite number of ovula are attached.

The genus *Reseda*, whose capsule opens at top at a very early period, may be considered as affording another instance, though much less remarkable, of the same anomaly. And it is possible that this may be the real structure in certain cases of which a very different view has been taken.

In the instances of naked seeds now given, the bursting of the pericarpium precedes the distinct formation of the embryo, while the proper coats of the seed remain entire till after its separation from the parent plant, and germination has commenced.

It may not be uninteresting to contrast this economy with that of the Mangroves and other plants of tropical countries, which grow on the shores, and within the influence of the tide. In many of these the embryo, long before the seed loses its original attachment, acquires a very considerable size; and the first effect of this unusual development is the rupture, in most cases succeeded by the complete absorption or disappearance, of the proper integument of the seed. In some instances the develop-

ment proceeds still further, and the pericarpium itself is perforated by the embryo, which, while preserving its connection with the parent plant, often attains the length of from eighteen inches to two feet. This happens in *Rhizophora* and *Bruguiera*, or the Mangroves properly so called. In some of the spurious Mangroves, as *Avicennia* and *Ægiceras*, a lesser degree of development takes place, and in general their pericarpia remain entire till they have dropped from the tree. In both cases the final cause of
149] the economy is sufficiently evident; a greater than ordinary evolution of the embryo being necessary to ensure its vegetation in the unfavorable circumstances in which it is unavoidably placed.

But an analogous structure exists in other plants, where the final cause is less apparent, as in certain species of *Eugenia*, in which the integument of the seed is completely absorbed before its separation from the parent plant, and while the pericarpium remains entire.

An economy no less remarkable than that of the Mangroves, but of a nature diametrically opposite, takes place in the bulb-like seeds of certain liliaceous plants, especially of *Pancratium*, *Crinum* and *Amaryllis;* in some of whose species the seed separates from the plant, and even from the pericarpium, before the embryo becomes visible. This observation respecting some of these seeds was, I believe, first made by Mr. Salisbury; and in such as I have myself examined, I have found the fact connected with one no less interesting, namely, an unusual vascularity in the fleshy substance.

I have in another place,[1] in speaking of this substance, which constitutes the mass of the seed, and in a central cavity of which the future embryo is formed, stated it to be destitute of vessels, and entirely composed of cellular texture. But on a more careful inspection, of those seeds at least in which the separation precedes the visible formation of the embryo, I now find very distinct spiral vessels: —these enter at the umbilicus, ramify in a regular manner in the substance of the fleshy mass, and appear to have a

[1] Prodr. Flor. Nov. Holland. p. 297.

certain relation to the central cavity where the embryo is afterwards formed, and which, filled with a glairy fluid, is distinctly visible before the separation of the seed. It is a curious consequence of this tardy evolution of the embryo, which in some cases does not become visible unless the [150
seed be placed in a situation favorable to germination, that very different directions may be given to its radicular extremity, according to circumstances which we have it in our power to regulate.

There is a fourth kind of anomaly in the structure of certain seeds, which, as I have formerly described it,[1] I shall here notice in a few words. It is that which takes place in certain *Aroideæ*, especially in some species of *Caladium*. In these, the nucleus of the seed is not properly a monocotyledonous embryo, but has an appearance and économy more nearly resembling those of the tuber of a root; for, instead of being distinguishable into a cotyledon, a plumula and radicula, and of germinating in a determinate manner and from a single point, it is composed of a mass whose internal structure is uniform, and on the surface of which frequently more than one germinating point is observable.

None of these anomalies appear to me materially to lessen the importance of the characters derived from the seeds of plants; but they evidently render a minute attention to every circumstance absolutely necessary in all attempts either to deduce affinities or establish genera from this source; and they especially demonstrate the necessity of carefully ascertaining the state of the unimpregnated ovarium; for, while its structure remains unknown, that of the ripe fruit can never be thoroughly understood.

[1] Prodr. Flor. Nov. Holl. p. 335.

EXPLANATION OF PLATE 12 (VII). [151

A.—A branch of the panicle of LEONTICE THALICTROIDES *Linn.* (Caulophyllum thalictroides *Michaux*), of the natural size.

B.—The same magnified, to show at 1, the early rupture of the ovarium, the ovula as yet but little enlarged and only in part protruded: at 2, the same parts in a more advanced state; one seed being nearly ripe, supported by its elongated and thickened umbilical cord; a second ovulum considerably increased in size, but abortive; and the remains of the ruptured ovarium—somewhat enlarged.

C and *D.*—Two longitudinal sections of the nearly ripe seed; exhibiting the vascular cord continued from the axis of the funiculus umbilicalis to the apex of the seed; the remarkable process of the inner integument at the umbilicus (of which another view is given separately at *E*); and the unripe embryo nearly in contact with this process, and as yet undivided.

AN ACCOUNT

OF

A NEW GENUS OF PLANTS,

NAMED

RAFFLESIA.

BY

ROBERT BROWN, F.R.S.,

CORRESPONDING MEMBER OF THE ROYAL INSTITUTE OF FRANCE, AND OF THE ROYAL ACADEMIES OF SCIENCES OF BERLIN AND MUNICH; MEMBER OF THE IMPERIAL ACADEMY NATURÆ CURIOSORUM; OF THE LITERARY AND PHILOSOPHICAL SOCIETY OF NEW YORK; OF THE CAMBRIDGE PHILOSOPHICAL SOCIETY; OF THE WERNERIAN SOCIETY OF EDINBURGH, AND OF THE WETTERAVIAN NATURAL HISTORY SOCIETY. LIBRARIAN TO THE LINNEAN SOCIETY.

[*Reprinted from the 'Transactions of the Linnean Society.' Vol. XIII, pp.* 201—234.]

LONDON.

APRIL, 1821.

ACCOUNT OF A NEW GENUS OF PLANTS, [201

NAMED

RAFFLESIA.

Read to the Linnean Society, June 30th, 1820.

It is now nearly eighteen months since some account of a flower of extraordinary size was received by my lamented friend and patron the late revered President of the Royal Society, from Sir Stamford Raffles, Governor of the East India Company's establishments in Sumatra.

This gigantic flower, which forms the subject of the present communication, was discovered in 1818 on Sir Stamford's first journey from Bencoolen into the interior. In that journey he was accompanied by a naturalist of great zeal and acquirements, the late Dr. Joseph Arnold, a member of this Society, from whose researches, aided by the friendship and influence of the Governor, in an island so favorably situated and so imperfectly explored as Sumatra, the greatest expectations had been formed. But these expectations were never to be realised; for the same letter which gave the account of the gigantic flower, brought also the intelligence of Dr. Arnold's death.

As in this letter many important particulars are stated respecting the plant which I am about to describe, and a just tribute is paid to the merits of the naturalist by whom it was discovered, I shall introduce my account by the following extract.

"Bencoolen; 13*th August*, 1818.

"You will lament to hear that we have lost Dr. Arnold: he fell a sacrifice to his exertions on my first tour into the interior, and died of fever about a fortnight ago.

202] "It is impossible I can do justice to his memory by any feeble encomiums I may pass on his character; he was in everything what he should have been, devoted to science and the acquisition of knowledge, and aiming only at usefulness.

"I had hoped, instead of the melancholy event I have now to communicate, that we should have been able to send you an account of our many interesting discoveries from the hand of Dr. Arnold. At the period of his death he had not done much; all was arrangement for extensive acquirement in every branch of natural history. I shall go on with the collections as well as I can, and hereafter communicate with you respecting them, and in the mean time content myself with giving you the best account I can of the largest and most magnificent flower which, as far as we know, has yet been described. Fortunately I have found part of a letter from poor Arnold to some unknown friend, written while he was on board ship, and a short time before his death, from which the following is an extract.

"After giving an account of our journey to Passummah, he thus proceeds:

"'But here (at Pulo Lebbar on the Manna River, two days' journey inland of Manna) I rejoice to tell you I happened to meet with what I consider as the greatest prodigy of the vegetable world. I had ventured some way from the party, when one of the Malay servants came running to me with wonder in his eyes, and said, "Come with me, sir, come! a flower, very large, beautiful, wonderful!" I immediately went with the man about a hundred yards in the jungle, and he pointed to a flower growing close to the ground under the bushes, which was truly astonishing. My first impulse was to cut it up and carry it to the hut. I therefore seized the Malay's parang (a sort of instrument

like a woodman's chopping-hook), and finding that it sprang from a small root which ran horizontally (about as large as two fingers, or a little more), I soon detached it and re- [203
moved it to our hut. To tell you the truth, had I been alone, and had there been no witnesses, I should I think have been fearful of mentioning the dimensions of this flower, so much does it exceed every flower I have ever seen or heard of; but I had Sir Stamford and Lady Raffles with me, and a Mr. Palsgrave, a respectable man resident at Manna, who, though equally astonished with myself, yet are able to testify as to the truth.

"'The whole flower was of a very thick substance, the petals and nectary being in but few places less than a quarter of an inch thick, and in some places three quarters of an inch; the substance of it was very succulent. When I first saw it a swarm of flies were hovering over the mouth of the nectary, and apparently laying their eggs in the substance of it. It had precisely the smell of tainted beef. The calyx consisted of several roundish, dark-brown, concave leaves, which seemed to be indefinite in number, and were unequal in size. There were five petals attached to the nectary, which were thick, and covered with protuberances of a yellowish-white, varying in size, the interstices being of a brick-red colour. The nectarium was cyathiform, becoming narrower towards the top. The centre of the nectarium gave rise to a large pistil, which I can hardly describe, at the top of which were about twenty processes, somewhat curved and sharp at the end, resembling a cow's horns; there were as many smaller very short processes. A little more than half way down, a brown cord about the size of common whipcord, but quite smooth, surrounded what perhaps is the germen, and a little below it was another cord somewhat moniliform.

"'Now for the dimensions, which are the most astonishing part of the flower. It measured a full yard across; the petals, which were subrotund, being twelve inches from the base to the apex, and it being about a foot from the insertion of the one petal to the opposite one; Sir Stam- [204
ford, Lady Raffles and myself taking immediate measures

to be accurate in this respect, by pinning four large sheets of paper together, and cutting them to the precise size of the flower. The nectarium, in the opinion of all of us, would hold twelve pints, and the weight of this prodigy we calculated to be fifteen pounds.

"'I have said nothing about the stamina; in fact, I am not certain of the part I ought to call stamina. If the moniliform cord surrounding the base of the pistil were sessile anthers, it must be a polyandrous plant; but I am uncertain what the large germen contained; perhaps there might be concealed anthers within it.

"'It was not examined on the spot, as it was intended to preserve it in spirits and examine it at more leisure; but from the neglect of the persons to whom it was entrusted the petals were destroyed by insects, the only part that retained its form being the pistil, which was put in spirits along with two large buds of the same flower, which I found attached to the same root; each of these is about as large as two fists.

"'There were no leaves or branches to this plant; so that it is probable that the stems bearing leaves issue forth at a different period of the year. The sóil where this plant grew was very rich, and covered with the excrement of elephants.

"'A guide from the interior of the country said that such flowers were rare, but that he had seen several, and that the natives called them *Krûbût.*

"'I have now nearly finished a coloured drawing of it on as large drawing-paper as I could procure, but it is still considerably under the natural size; and I propose also to make another drawing of the pistil removed from the nectarium.

"'I have now, I believe, given you as detailed an account of this prodigious plant as the subject admits of;
205] indeed it is all I know of it. I would draw your attention, however, to the very great porosity of the root, to which the buds are attached.

"'I have seen nothing resembling this plant in any of my books; but yesterday, in looking over Dr. Horsfield's immense collections of the plants of Java, I find something which perhaps may approach to it; at any rate the buds of the flower he has represented grow from the root precisely in the same manner: his drawing, however, has a branch of leaves, and I do not observe any satisfactory dissections. He considers it as a new genus; but the difference of the two plants appears from this, that his full-blown flower is about three inches across, whereas mine is three feet.'"

Sir Stamford proceeds:

"Dr. Arnold did not live to return to Bencoolen, nor to fulfil the intentions expressed in the above extract; but we have finished the drawing of the whole flower, and it is now forwarded under charge of Dr. Horsfield, to whom I have also entrusted the pistil and buds.

"I shall make exertions for procuring another specimen, with which I hope we shall be more fortunate.

(Signed) "T. S. RAFFLES.

"To the Right Honorable
Sir JOSEPH BANKS, Bart., G.C.B., &c. &c."

The drawing of the expanded flower, and the specimens mentioned in the preceding extract, were brought to England by Dr. Horsfield; and, having been put into my hands, I proceeded without delay to examine the smaller flower-bud. In this examination the antheræ, although not at first obvious, were soon discovered, but no part was found which could be considered either as a perfect pistillum or as indicating the probable nature or even the exact place of the ovarium. The remains of the expanded flower [206
exhibited the same structure; and the larger bud, which was examined by Mr. Bauer, whose beautiful drawings of it form the most valuable part of the present communication, proved also to be male.

These materials, it must be admitted, are insufficient even for the satisfactory establishment of the proposed new genus, and in my opinion do not enable us absolutely to determine its place in the natural system.

The curiosity of botanists, however, has been so much

excited by the discovery of a flower of such extraordinary dimensions, the male flower is in many respects so singular, and its structure is so admirably illustrated by Mr. Bauer's drawings, that, accompanied by them, even the present incomplete account will probably be thought worthy of a place in the Society's Transactions.

Its publication is the less objectionable, as it may still be a considerable time before the plant is met with in all its states; and however unsatisfactory our present materials may be, either for determining its affinities, or the equally important question, whether it be parasitic on the root to which it is attached, there can be no doubt that it forms a genus abundantly distinct from any that has hitherto been described.

It is proposed, in honour of Sir Stamford Raffles, to call this genus Rafflesia, the name I am persuaded that Dr. Arnold himself would have chosen had he lived to publish an account of it; and it may in the mean time be distinguished by the following characters.

307] RAFFLESIA.

Perianthium monophyllum, coloratum; *tubo* ventricoso; *corona faucis* annulari, indivisa; *limbo* quinquepartito, æquali.

Mas. *Columna* (inclusa): *limbo apicis* reclinato, subtus simplici serie polyandro; *disco* processibus (concentricis) tecto.

Antheræ sessiles, subglobosæ, cellulosæ, poro apicis dehiscentes.

Fem. - - - - - -

Rafflesia Arnoldi.

Tab. 13—20 (XV—XXII.)

Descriptio.

E *Radice* lignea horizontali tereti, lævi, crassitie fere et structura interiore omnino radicis Vitis viniferæ (*tab.* 20 (22), *f.* 8) ortum ducit *Flos* unicus, ante expansionem, dum bracteis imbricatis adhuc inclusus, brassicæ minori figura et magnitudine similis (*tab.* 14 (16)); cum radice parum dilatata connexus *Basi* (*tab.* 15 (17)) modicè convexa, abbreviata, insignita lineolis numerosis, elevatis, nigricantibus, plerisque reticulatim confluentibus, nonnullis brevioribus distinctis, omnibus sulco longitudinali tenui per axin exaratis, apothecia Opegraphæ æmulantibus, superioribus desinentibus in *annulum* modicè elevatum exsulcum, ejusdem fere substantiæ, definientem basin reticulatam.

Bracteæ (*tab.* 14 (16)) supra annulum baseos reticulatæ, numerosæ, densè imbricatæ, subrotundæ, coriaceæ, glaberrimæ, integerrimæ, venis vix vel parum emersis, ramosis, distinctis, nec anastomosantibus, infra apicem evanescentibus, lata basi insertæ ibique crassæ, versus apicem sensim tenuiores, subfoliaceæ; intimæ e latiore basi, $\frac{1}{5}$ usque ad $\frac{1}{4}$ circuli æquante.

Perianthium (*tab.* 13 (15)) intra bracteas sessile, monophyllum, coloratum, ante expansionem depresso-sphæroideum (*tab.* 16 et 17 (18 et 19)). *Tubus* ventricosus, [208
abbreviato-urceolatus, extus lævis, intus ramentis filiformibus simplicibus passimque parum divisis densè tectus. *Faux: corona* annulari integerrima, intus ornata areolis numerosis, convexiusculis, subrotundis transversim paulo latioribus, superioribus omnino lævibus, reliquis margine inferiore aucto ramentis filiformibus brevibus. *Limbus* quinque-partitus (diametro tripedali), *laciniis* æqualibus, (patentibus reflexisve) rotundatis, integerrimis, *extus* lævibus, præter venas parum elevatas, numerosas, dichotomas, passim anastomosantes, ad apicem usque attingentes; *intus* verrucis numerosis, subrotundis, sparsis, inæqualibus, interstitiis lævibus: *æstivatione* arctè imbricatis, exterioribus interiores utroque margine equitantibus (*tab.* 17 (19)).

Columna centralis (*tab.* 18 (20) et 19 (21), *fig.* 1) staminifera, cavitatem tubi perianthii ferè omnino replens, inclusa, solida, carnosa, intus cum substantia ipsius baseos reticulatæ extus cum tubi superficie ramentacea continua; prope basin aucta *annulis* duobus modicè elevatis, rotundatis, ante

expansionem approximatis (*tab.* 19 (21), *f.* 1, 2), in expanso flore remotioribus (*tab.* 20 (22), *f.* 2), *inferiore* paulo crassiore, striis leviter depressis numerosis rugoso, *superiore* exsulco, punctis minutis elevatis inæquali: supra annulum superiorem lævis et sensim angustata in *collum* brevissimum, insculptum *excavationibus* (*tab.* 19 (21), *f.* 2) numero antherarum iisque oppositis, basi angustatis, longitudinaliter elevato-striatis, interstitiis subcarinatis, carinis marginibusque ciliatis: *apex* dilatatus, cujus *discus* planiusculus, tectus *processibus* numerosis carnosis leviter incurvis subcorniformibus, simplicibus apiceve parum divisis, in seriebus pluribus concentricis, interioribus plus minus irregulariter dispositis, nonnullis minoribus sæpe minimis sparsim intermixtis, majorum singulis fasciculo vasculari centrali tenui instructis, omnibus lævibus, præter apices lobulorum qui
209] sæpe hispiduli vel minutè penicillati; *limbus* solutus reclinatus, e basi recurvata, subtus punctis parvis elevatis quandoque piliferis inæquali, adscendens, margine erecto-conniventi, indiviso tenuiter crenulato, substantia et superficie processibus disci similis, intus fasciculis vascularibus simplici serie dispositis et ad basin antheræ singulæ flexura notabili instructis (*tab.* 19 (21), *f.* 2, 3, 7, 8, et *t.* 20 (22), *f.* 6).

Antheræ (*tab.* 19 (21), *f.* 4—8, et *t.* 20 (22), *f.* 4—6) simplici serie dispositæ, æquidistantes, 35 circiter, vix 40, sessiles, excavationibus dimidiæ recurvatæ limbi, cum iis colli continuis, lata basi insertæ, semiimmersæ, apicibus deorsum spectantibus, in respondentibus cavitatibus colli receptis, ovato-globosæ, pisi magnitudine, apice depressione unica centrali demum aperiente umbilicatæ, cellulosæ, cellulis indefinitè numerosis, subconcentricis, longitudinalibus, exterioribus versus apicem conniventibus, passim confluentibus et quandoque transversim interruptis, plenis *Polline* (*tab.* 19 (21), *f.* 9) minuto, sphærico, simplici, lævi.

Pistilli rudimenta nulla certa; processus enim corniculati apicis columnæ staminiferæ, in circulis pluribus concentricis dispositi atque singuli fasciculo vasculari centrali donati, dubiæ naturæ sunt.

To the foregoing description of *Rafflesia* it is necessary to add some observations explanatory of structure; and I shall also offer a few conjectures on certain points of the economy of the plant, and on its affinities.

The great apparent simplicity in the internal structure of every part, especially in a flower of such enormous size, is in the first place deserving of notice.

This observation particularly applies to the *Column*, which is found to consist of a uniform cellular texture, with a very small proportion of vessels. The *cells* or utriculi are [210
nearly sphærical, slightly angular from mutual pressure, and, in the specimens examined at least, easily separable from each other without laceration. I have not been able to detect perforations on any part of their surface; but extremely minute granules, originally contained in great abundance in the cells, and frequently found adhering to their parietes, may readily be mistaken for pores.

The structure of *vessels* either in the column, perianthium or bracteæ, in all of which they are apparently similar, has not been satisfactorily ascertained. They may be supposed to approach most nearly to the ligneous, though certainly unaccompanied by spiral vessels, which do not appear to exist in any part of the plant.

The same internal structure is continued below the origin of the bracteæ, down to the line at which the vessels of the root appear to terminate, and where an evident change takes place (*plate* 18 (20) and 20 (22), *f.* 1).

The *Perianthium* and *Bracteæ* in their cellular texture very nearly agree with the column, except that in their more foliaceous parts the cells are considerably elongated.

I have not found in any part of their surface, or in that of the column, those areolæ universally considered as cuticular pores, and which, though of very general occurrence, do not perhaps exist in the imperfectly developed leaves of plants parasitic on roots.

In the external composition of the column, the part most deserving of attention is the *Anthera;* for in apparent origin, as well as in form and structure, it presents the most singular modification of stamen that has yet been observed.

It appears to me of importance to inquire into the real relation which so remarkable a structure bears to the more ordinary states of Anthera.

211] A satisfactory determination of this point, while it would certainly assist in explaining the nature of the other parts of the column, might also in some degree lead to correct notions of the affinities of the genus; and the question is perhaps sufficiently interesting, even independent of these results.

In this inquiry, it is necessary in the first place to take a general view of the principal forms of Antheræ in phænogamous plants; all of which, however different they may appear, I consider as modifications of one common structure.

In this assumed regular structure or type of Anthera, I suppose it to consist of two parallel folliculi or *thecæ*, fixed by their whole length to the margins of a compressed filament: each *theca* being originally filled with a pulpy substance, on the surface or in the cells of which the pollen is produced; and having its cavity divided longitudinally into two equal cells, the subdivision being indicated externally by a depression or furrow, which is also the line of dehiscence.[1]

[1] A certain degree of resemblance between this supposed regular state of Anthera, and that which in a former essay (on Compositæ, *Linn. Soc. Transact.* xii, *p.* 89) I have considered as the type of Pistillum in phænogamous plants, will probably be admitted; and both structures have, as it appears to me, an evident relation to the *Leaf*, from whose modifications all the parts of the flower seem to be formed.

This hypothesis of the formation of the flower may be considered as having originated with Linnæus in his *Prolepsis Plantarum*, though he has not very clearly stated it, and has also connected it with other speculations, which have since been generally abandoned. It is, however, more distinctly proposed by Professor Link (in *Philos. Bot. Prodr. p.* 141), and very recently has been again brought forward, with some modifications, by M. Aubert du Petit Thouars.

In adopting the hypothesis as stated by Professor Link, I shall, without entering at present into its explanation or defence, offer two observations in illustration of it, founded on considerations that have not been before adverted to.

My first observation is, that the principal point in which the antheræ and

The structure now described actually exists in many [212 families of plants; and the principal deviations from it

ovaria agree, consists in their essential parts, namely, the pollen and ovula, being produced on the margins of the modified leaf.

In the *Antheræ*, which are seldom compound, and whose thecæ are usually distinct, the marginal production of pollen is generally obvious.

In the *Ovaria*, however, where, with very few exceptions, the same arrangement of ovula really exists, it is never apparent, but is always more or less concealed either by the approximation and union of the opposite margins of the simple pistillum, and of the compound when multilocular; or in the unilocular pistillum with several parietal placentæ by the union of the corresponding margins of its component parts.

The few cases of apparent exception, where the ovula are inserted over the whole or greater part of the internal surface of the ovarium, occur either in the compound pistillum, as in *Nymphæa* and *Nuphar;* or in the simple pistillum, as in *Butomeæ* of Richard; and in *Lardizabaleæ*, an order of plants sufficiently distinct in this remarkable character alone, and differing also in the structure of embryo and in habit, from *Menispermeæ*, to which the genera composing it (*Lardizabala* and *Stauntonia*) have hitherto been referred.

The marginal production of ovula, though always concealed in the ordinary or complete state of the Ovarium, not unfrequently becomes apparent where its formation is in some degree imperfect, and is most evident in those deviations from regular structure, where stamina are changed, more or less completely, into pistilla. Thus, in the case of the nearly distinct or simple pistillum, it is shown by this kind of montrosity in *Sempervivum tectorum;* in the compound multilocular pistillum, by that of *Tropæolum majus;* and in the compound pistillum with parietal placentæ, by similar changes in *Cheiranthus Cheiri*, *Cochlearia armoracia, Papaver nudicaule,* and *Salix oleifolia.*

In all the cases now quoted, and in several others with which I am acquainted, it is ascertained that a single stamen is converted into a simple pistillum, or into one of the constituent parts of the compound organ: a fact which, in my opinion, establishes the proposed type of Ovarium.

I have entered thus slightly at present into the proof of this type, derived from these deviations from regular structure, partly on account of an observation which I find in the second edition of the excellent *Théorie Elémentaire de la Botanique* of Professor De Candolle, to whom, in 1816, I had shown drawings of most of the instances of monstrosity now mentioned. To these drawings, and to my deductions from them with regard to the structure of pistillum, I suppose the ingenious author alludes, in the passage in question. His views, however, on this subject differ considerably from mine, which he does not seem to have been aware were already published (*Linn. Soc. Trans. l. c.*)

My second observation relates to the more important differences between the antheræ and ovaria, independent of their essential parts.

In the Anthera the vascularity, with relation to that of the Leaf, may be said to be diminished without being otherwise sensibly modified; the pollen is formed in a cellular substance apparently destitute of vessels; and is always produced internally, or under the proper membrane of the secreting organ.

In the Ovarium, on the other hand, the vascularity, compared with that of the Leaf, is in general rather modified than diminished; the principal vessels occupying the margins or lines of production, and giving off branches towards the axis, whose vascularity is frequently reduced. The ovula constantly arise from vascular cords, and, with reference to the supposed original state of the ovarium, are uniformly produced externally; though by the union of its parts,

may be stated to depend either on a reduced or increased development of the parts enumerated, on differences in the manner of bursting, or on the confluence of two or more antheræ.

Reduced development may consist merely in the approximation of the thecæ, consequent on the narrowing or entire absence of the connecting portion of the filament, which is one of the most common states of anthera; in their partial confluence, generally at the upper extremity; their paral-
213] lelism either continuing, which is also not unfrequent; or accompanied by various degrees of divergence, as in many genera of *Labiatæ;* in their complete confluence while they remain parallel, as in *Epacrideæ, Polygaleæ,* and in some genera of *Acanthaceæ;* and lastly, in the imperfect production or entire suppression of one of the thecæ, as in *Westringia, Anisomeles,* and *Maranteæ.*

Increased development may in like manner be confined to the dilatation, elongation, or division of the connecting portion of the filament, of which examples occur in many *Scitamineæ, Orchideæ* and *Acanthaceæ;* it may consist in
214] the elongation of the thecæ either above or below the connecting filament; in an increased number of divisions of each theca by longitudinal, transverse, or oblique processes of the receptacle of the pollen, as in several genera of *Orchideæ* and *Laurinæ;* or in the persistence of part of the cells in which the pollen is formed, as in *Ægiceras.*

Reduced and increased development of different parts may co-exist in the same organ, as in the bifid or incumbent anthera with contiguous thecæ; in the extraordinary dilatation of the connecting portion of the filament, while one of the thecæ is abortive or imperfect, as in the greater number of *Salviæ;* or in the thecæ being confluent, while

whether in the simple or compound state, they become always inclosed, and, before fecundation at least, are completely protected from the direct action of light and of the atmosphere.

In *Coniferæ* and *Cycadeæ,* however, according to the view I am disposed to take of them (*Tuckey's Congo, append. p.* 454, *antè, p.* 138), this is not entirely the case. But these two families will perhaps be found to differ from all other phænogamous plants in the more simple structure both of their ovaria and antheræ.

the polliniferous cells are at the same time persistent, as in certain species of *Viscum*.

The deviations from the regular mode of bureting are also numerous; in some cases consisting either in the aperture being confined to a definite portion, generally the upper extremity, of the longitudinal furrow, as in *Dillenia* and *Solanum;* in the apex of each theca being produced beyond the receptacle of the pollen into a tube opening at top, as in several *Ericinæ;* or in the two thecæ being confluent at the apex, and bursting by a common foramen or tube, as in *Tetratheca.* In other cases a separation of determinate portions of the membrane takes place, either the whole length of the theca, as in *Hamamelideæ* and *Berberideæ;* or corresponding with its subdivisions, as in several *Laurinæ;* or lastly, having no obvious relation to internal structure, as in certain species of *Rhizophora.*

The regular structure may also be altered or disguised by the union of two or more stamina; the thecæ of each anthera either remaining distinct and parallel, as in *Myristica, Canella,* and in several *Aroideæ;* being divaricate and united, as in *Cissampelos;* or absolutely separate, by division of the filament, as in *Conospermum* and *Synaphea.*

It is unnecessary for my present purpose to enter into [215
a more minute account of the various structures of stamina, most of which appear to me easily reducible to the type here assumed.

The precise relation of the anthera of *Rafflesia*, however, to this type is so far from being obvious, that at least three different opinions may be formed respecting it.

According to one of these, each actual anthera would be considered as composed of several united stamina. But in adopting this opinion, which is suggested solely by the existence and disposition of the cells of the anthera, it seems also necessary to consider the apparently simple flower of *Rafflesia* as in reality compound, and analogous to the spike of an *Aroidea;* the pistilla, if present, being consequently to be looked for not in the centre but in the circumference. On attending, however, to the whole external structure of the flower, as well as to the disposition of vessels, this supposition will, I conclude, appear still more improbable than that in support of which it is adduced.

A second opinion, diametrically opposite to the former, would regard the anthera of *Rafflesia* as only half a regular anthera, whose two thecæ are separated by portions of the united filaments, which, being produced beyond the antheræ, together form the crenated limb of the column.

This view, though less paradoxical than the first, will hardly be considered as affording so probable an explanation of structure as the third opinion; according to which each anthera would be regarded as complete, made up of two united thecæ, opening by a common foramen, and internally subdivided into numerous vertical cells by persistent portions of the confluent receptacles of the pollen; a structure not perhaps essentially different from that of certain antheræ more obviously reducible to the supposed type.

Even in adopting this opinion, a question would still
216] remain respecting the limb of the column under which the antheræ are inserted; namely, whether it is to be viewed as an imperfectly developed stigma, or as made up of processes of the united filaments. In support of the former supposition the nearly similar relation of the sexual organs in certain *Asarinæ* may be adduced; and in favour of the latter, not only their disposition and form in other plants of the same natural family, but also the vascular structure of the column itself; the limb deriving its vessels from branches of the same fasciculi that supply the antheræ (*plate* 18 (20), *f.* 1). If this latter view, however, of the origin of the limb were admitted, it might be considered not altogether improbable, that even the corniculate processes of the disk of the column, each of which has a central vascular cord, are of the same nature. For if, on the other hand, these processes are to be regarded as imperfect styles or stigmata, their number and disposition would indicate a structure of ovarium to be found only in families to which it is not probable at least that *Rafflesia* can be nearly related, as *Annonaceæ* and the singular genus

Eupomatia,[1] which I have placed near that natural order.

Another point to be inquired into connected with the same subject is, in what manner the impregnation of the female flower is likely to be effected by antheræ so completely concealed as those of *Rafflesia* seem to be in all states of the flower; for it does not appear either that they can ever become exposed by a change in the direction of the limb under which they are inserted, or even that this part of the column in any stage projects beyond the tube of the perianthium.

It is probable, therefore, that the assistance of insects is absolutely necessary; and it is not unlikely, both as connected with that mode of impregnation and from the structure of the anthera itself, that in *Rafflesia* the same economy obtains as in the stamina of certain *Aroideæ*, in which it has been observed that a continued secretion and [217
discharge of pollen takes place from the same cell; the whole quantity produced greatly exceeding the size of the secreting organ.

The passage of the pollen to the bottom of the flower, where it is more easily accessible to insects, seems likewise to be provided for, not only by the direction of the antheræ, but also by the form of the corresponding cavities in the neck of the column, in the upper part of which they are immersed.

That insects are really necessary to the impregnation of *Rafflesia*, is confirmed by Dr. Arnold's statement respecting the odour of the plant, by which they may be supposed to be attracted, and also by the fact of the swarms actually seen hovering about and settling in the expanded flower.

The structure of *Rafflesia* is at present too imperfectly known to enable us to determine its place in the natural system. I shall, however, offer some observations on this question, which can hardly be dismissed without examination.

As to which of the two primary divisions of phænoga-

[1] Flinders's *Voyage*, ii, *p.* 597 (*Antè*, *p.* 73), *tab.* 2.

mous plants the genus belongs, it may, I think, without hesitation be referred to *Dicotyledones;* yet if the plant is parasitic, and consequently no argument on this subject to be derived from the structure of the root, which is exactly that of the Vine,[1] its exclusion from *Monocotyledones* would rest on no other grounds, that I am able to state, than the quinary division of the perianthium, which in other respects also bears a considerable resemblance to that of certain dicotyledonous orders, the number of stamina, and the ramification of vessels in the bracteæ.

Assuming, however, that *Rafflesia* belongs to *Dicotyledones,*
218] and considering the foliaceous scales which cover the unexpanded flower, both from their indefinite number and imbricate insertion as bracteæ, and consequently the floral envelope as simple, its comparison with the families of this primary division would be limited to such as are apetalous; either absolutely as *Asarinæ;* those of a nature intermediate between the apetalous and polypetalous, in which the segments of the perianthium are generally, though not always, disposed in a double series, as *Passifloreæ, Cucurbitaceæ,* and *Homalinæ*; or those which have a simple coloured floral envelope, but are decidedly related to polypetalous families, as *Sterculiaceæ.*

With *Asarinæ,* the only truly apetalous order to which it seems necessary to compare it, *Rafflesia* has several points of resemblance, especially in the structure of the central column. In *Aristolochia* the antheræ, though only six in number, are in like manner sessile, and inserted near the apex of a column formed by the union of stamina and pistillum. The mere difference in the number of stamina seems to be of no importance in the present question, there being twelve in *Asarum;* and in *Thottea,* a genus certainly belonging to this family, though referred by Rottböll to *Contortæ,*[2] the stamina are not only still more numerous, but are disposed in a double circular series one above the other; an arrangement which may perhaps be considered

[1] Compare the magnified section of the Root, *tab.* 20 (22), *f.* 8, with that of the Vine in Grew's Anat. *tab.* 17.

[2] Thottea grandiflora. *Rottböll in Nov. Act. Soc. Reg. Hafn.* ii, *p.* 529, *tab.* 2.

analogous to the concentric series of processes in the apex of the column of *Rafflesia*.

In all these genera of *Asarinæ* and in *Bragantia* of Loureiro, which is also referable to the same order, the flowers are hermaphrodite; but in *Cytinus*, which, if not absolutely belonging to this order, is at least very nearly related to it, they are diclinous.

The affinity is also in some degree confirmed by the appearance of the inner surface of the tube of the perianthium of some *Asarinæ*, especially *Aristolochia grandiflora*, and by the thickening or annular projection of the faux in the [219
same plant, as well as in a new species of *Bragantia* discovered in Java by Dr. Horsfield.

It may also be noticed in support of it, that some of the largest flowers which were known before the discovery of *Rafflesia* belong to *Asarinæ*, as those of *Aristolochia grandiflora*, and particularly *Aristolochia cordiflora* of Mutis, which, according to M. Bonpland, are sixteen inches in diameter, or nearly half that of our plant.[1]

The first objection that occurs to this approximation is the ternary division of the perianthium in the regular flowered genera of *Asarinæ*, opposed to its quinary division in *Rafflesia*; but in *Cytinus* it is divided into four segments, a number more generally connected in natural families with five than with three.

A second objection would exist, if it be considered more probable that the ovarium of *Rafflesia* is superior, or free, than inferior, or cohering with the tube of the perianthium.

There is indeed nothing in the structure of the column itself indicating the particular position of the ovarium. But if it be admitted, that a base of a form equally calculated for support should exist in the female flower, as is found in the male, it might perhaps be considered somewhat more probable that such a base should be connected with a superior than with an inferior ovarium.

Even admitting this objection, however, it would be considerably weakened, on the one hand, by allowing that

[1] *Humboldt Bonpl. et Kunth Nov. Gen. et Sp.* ii, *p.* 118.

Nepenthes, which has a superior ovarium, is related to *Asarinæ*, as I am inclined to believe; and on the other, by considering *Homalinæ*, whose ovarium is inferior, as allied to *Passifloreæ*, the order with which I shall now proceed to compare *Rafflesia*.

The comparison is suggested by the obvious resemblance between the perianthium of our genus, and that of certain
220] species of *Passiflora* itself; or of other genera of the order, as *Deidamia*, in which the inner series of segments is wanting. Thus, they agree essentially, and even remarkably, in æstivation of perianthium: the corona of *Rafflesia* may be compared with that of *Murucuia*, and the two annular elevations at the base of the column with the processes of like origin and nearly similar form in some species of *Passiflora*. The affinity is also supported by the position of the stamina on a central column.

The peculiar structure of antheræ in *Rafflesia* can hardly be regarded as an objection of much weight to the proposed association; and it will at least almost equally apply to any other family with which this genus may be compared.

If the concentric processes on the disk of the column in our plant are to be regarded as indications of the number and disposition of pistilla, or of the internal structure of ovarium in the female flower, they present a formidable objection to its affinity with *Passifloreæ*, in all of which the ovarium is unilocular with parietal placentæ. If, however, these processes were considered as inner series of imperfect stamina, the objection derived from their number and arrangement merely, would be comparatively slight; for in some genera of *Passifloreæ*, particularly in *Smeathmannia*,[1] the stamina are also numerous and perhaps even indefinite.

[1] As *Smeathmannia* forms a very remarkable addition to the order in which I have proposed to place it, and is still unpublished; I shall here give its characters, and add a few remarks in support of this arrangement.

SMEATHMANNIA. *Soland. Mss. in Biblioth. Banks.*

Ord. Nat. Passifloreæ. *Br. in Tuckey's Congo, p.* 439. (*Antè p.* 121.)

Syst. Linn. Polyandria Pentagynia.

CHAR. GEN. *Perianthium* duplex, utrumque 5-partitum; *exterius* semicalycinum persistens; *interius* petaloideum marcescens. *Urceolus* simplex, membranaceus, ex ipsa basi perianthii. *Stamina* numerosa, distincta, apici columnæ

It has been already remarked, that there is nothing [221
in the structure of the column in *Rafflesia* to enable us to
determine the position of the ovarium in the female flower; [222
but that from another consideration there seems a somewhat greater probability of its being superior. If, however, it were even inferior, the objection to the affinity in question would not be insuperable, the relatiouship of *Homalinæ* to *Passifloreæ* being admitted.

If *Napoleona* or *Belvisia* be really allied to *Passifloreæ*,

brevissimæ genitalium inserta. *Styli* 5. *Stigmata* peltata. *Capsula* inflata, quinquevalvis. *Semina* axibus valvularum inserta.

Frutices (*forsan decumbentes*). Folia *alterna simplicia subdentata*, stipulis *lateralibus* (*utrinque solitariis geminisve*) *distinctis, callosis.* Flores *axillares subsolitarii, pedunculis, quandoque brevissimis, basi bracteolatis.* Urceolus *abbreviatus, ore denticulato.* Filamenta *simplici serie, viginti circiter.* Antheræ *incumbentes, lineares.* Capsula *chartacea.* Semina *axibus filiformibus valvularum subsimplici serie inserta, pedicellata, punctata, omnino* Passifloræ.

Patria. Africa æquinoctialis.

1. S. *pubescens,* ramis tomentosis, foliis oblongo-ovatis basi obtusis: adultis pube rara conspersis, urceolo barbato.

Smeathmannia pubescens. *Solander l. c.*

Loc. Nat. Guinea, prope Sierra Leone, *Smeathman, Afzelius.*

2. S. *lævigata,* ramis glabris, foliis oblongis ovatisve basi acutis: adultis glaberrimis utrinque nitidis, urceolo imberbi inciso.

Smeathmannia lævigata. *Soland. l. c.*

Loc. Nat. Guinea, prope Sierra Leone, *Smeathman, Afzelius, Purdie.*

3. S. *media,* ramis glabris, foliis obovato-oblongis basi obtusis: adultis utrinque glabris subopacis.

Loc. Nat. Guinea, prope Sierra Leone, *Smeathman.*

Forsan varietas *S. lævigatæ.*

The affinity of *Smeathmannia* to *Paropsia* of M. du Petit Thouars will probably be admitted without hesitation; and its exact agreement in fruit in every important point, both with this genus and with *Modecca,* seems to leave no doubt of its belonging to *Passifloreæ,* with which it agrees in habit even better than *Paropsia,* and certainly much more nearly than *Malesherbia,* considered by M. de Jussieu (in *Flor. Peruv.* iii, *p.* xix) as belonging to the same family.

Smeathmannia differs then from the other genera of *Passifloreæ* solely in its greater number of stamina, which, however, may not be really indefinite; and an approach to this structure is already known to exist in an unpublished genus (*Thompsonia*) discovered in Madagascar by Mr. Thompson, of which the habit is entirely that of *Deidamia,* and whose stamina are equal in number to the divisions of both series of the perianthium.

But from *Smeathmannia* the transition is easy to *Ryania,* which differs chiefly in its still greater number of stamina, in the want of petals or inner series of perianthium, in the single style being only slightly divided, and in the form of its placentæ.

And *Ryania,* although it has a superior ovarium, may even be supposed to be related to *Asteranthos* and *Belvisia,* if the fruit of these two genera should prove to be unilocular with several parietal placentæ.

which is very doubtful, however, and can only be determined by an examination of the fruit, it may also be compared with *Rafflesia*. At first sight this singular genus seems to resemble our plant in several respects, particularly in the manner of insertion of its sessile flower into the branch, in the bracteæ surrounding the ovarium, the confluence and dilatation of its filaments, and in the existence of a double corona. But some of these points are obviously unimportant; and the comparison between the corona of the great flower and the double corolla of *Belvisia* will probably be considered paradoxical.[1]

It seems unnecessary to compare *Rafflesia* with *Cucurbitaceæ*, to which it could only be considered as approaching, if its affinity to *Aphyteia* should appear probable, and the relationship of that genus to *Cucurbitaceæ*, suggested chiefly by the structure of antheræ, were at the same time admitted.

223] The points of agreement between *Rafflesia* and *Sterculiaceæ* are the division and form of the coloured perianthium, the sessile antheræ terminating a column, and the separation of sexes.

[1] M. de Beauvois, in his account of *Napoleona* (*Flore d'Oware* ii, *p.* 32), has mentioned a genus allied to it, which has been since published by M. Desfontaines under the name of *Asteranthos*. These two genera are without doubt nearly related; and, even independent of the structure of fruit, which in both remains to be ascertained, possess sufficient characters to separate them from every known family, as M. de Jussieu is disposed to think; and certainly from *Symploceæ*, where M. Desfontaines has placed them.

In adopting the generic name proposed by M. Desvaux for *Napoleona*, this order may be called

BELVISEÆ.

Calyx monophyllus, limbo diviso, persistens. *Corolla?* monopetala, plicata, (multiloba vel indivisa; simplex v. duplex) decidua. *Stamina* vel definita v. indefinita; basi corollæ inserta. *Ovarium* inferum. *Stylus* 1. *Stigma* lobatum v. angulatum. *Pericarpium* baccatum, polyspermum.

Frutices (Africæ æquinoctialis; an etiam Brasiliæ?) *foliis alternis integerrimis exstipulatis, floribus axillaribus lateralibusve solitariis.*

BELVISIA, *Desvaux in Journal de Botanique appliq.* iv, *p.* 130.

Napoleona, *Palisot de Beauvois Flore d'Oware* ii, *p.* 29.

Calyx 5-*fidus*. *Corolla?* duplex; *exterior* indivisa; *interior* (e staminibus sterilibus connatis formata?) multifida. *Stamina: Filamenta* 5 dilatata biantherifera.

ASTERANTHOS, *Desfont. in Mem. du Mus.* vi, *p.* 9, *tab.* 3.

Calyx multidentatus. *Corolla?* simplex multiloba. *Stamina* indefinite numerosa distincta.

On these resemblances, however, I am not disposed to insist; and I am even persuaded that there is here no real affinity; though I confess I have no other objections to state to it than the valvular æstivation of the perianthium, and the absence both of the corona and of the annular elevations at the base of the column in *Sterculiaceæ*.

To conclude this part of my subject, I am inclined to think that *Rafflesia*, when its structure is completely known, will be found to approach either to *Asarinæ* or *Passifloreæ*; and that, from our present imperfect materials, notwithstanding the very slight affinity generally supposed to exist between these two orders, it cannot be absolutely determined to which of them it is most nearly allied.

The only question that remains to be examined respecting *Rafflesia* is, whether the flower with its enveloping bracteæ and reticulate base do not together form a complete plant parasitic on the root from which it springs?

That such was probably the case, occurred to me on [224
first inspecting the flower bud; the opinion being suggested not only by the direct origin of the flower from the root, but more particularly by the disposition, texture and colour of the bracteæ; in which it so nearly resembles certain plants known to be parasites, as *Cytinus*, *Cynomorium*, *Caldasia* of Mutis,[1] *Balanophora*, and *Sarcophyte*.

In this opinion I was confirmed on seeing the figure of the plant mentioned in Dr. Arnold's letter, as probably related to the Great Flower, though not more than three inches in diameter.

The plant in question, which had been found in Java by Dr. Horsfield several years before the discovery of *Rafflesia*

[1] In the Journal of Science, vol. iii. p. 127, from El Semanario del Nuevo Reyno de Granada, for 1810. To this genus belong *Cynomorium jamaicense*, and perhaps *cayanense* of Swartz, an unpublished species from Brazil, and some other plants of equinoctial America. Before the appearance of *Caldasia* in the Journal of Science, I was aware that these plants formed a genus very distinct from *Cynomorium* (Journal of Science, iii, p. 129), but I had not given it a name, which is still wanting, that of *Caldasia* having long been applied to a very different and well known genus.

The new name, however, may be left to M. Richard, who is about to publish, and who will no doubt illustrate with his usual accuracy, the plants formerly referred to *Cynomorium*, of one of the species of which (*C. cayanense*) he is himself the discoverer.

Arnoldi, only, however, in the unexpanded state, is represented in the figure referred to as springing from a horizontal root in the same manner as the Great Flower; like which also it is enveloped in numerous imbricate bracteæ, as having a perianthium of the same general appearance, with indications of a similar entire annular process or corona at the mouth of the tube, a pustular inner surface, and a central column terminated by numerous acute processes.
225] It is therefore unquestionably a second species of the same genus:[1] but the branch with leaves, which, though separately represented in the drawing, is considered as proceeding from the same root, appears to me, on an examination of the specimen figured, to belong to a species of *Vitis*: and on mentioning my supposition respecting the Great Flower to Dr. Horsfield, he informed me he had observed this second species of the genus also connected with leaves of a different kind, and which seemed likewise to be those of a *Vitis*.[2]

Even with all the evidence now produced, I confess I was inclined, on a more minute examination of the buds of *Rafflesia Arnoldi*, to give up the opinion of its being a parasite; on considering, first, the great regularity of the reticulate base, which, externally at least, seemed to be merely a dilatation of the bark of the root: secondly, the nearly imperceptible change of structure from the cortical part of the base to the bracteæ in contact with its upper elevated margin: thirdly, the remarkable change of direction and increased ramification of the vessels of the root at the point of dilatation; a modification of structure which must probably have taken place at a very early stage of

[1] This second species may be named *Rafflesia Horsfieldii*, from the very meritorious naturalist by whom it was discovered. At present, however, the two species are to be distinguished only by the great difference in the size of their flowers; those of the one being nearly three feet, of the other hardly three inches in diameter.

[2] Isert (in *Reise nach Guinea, p.* 283) mentions a plant observed by him in equinoctial Africa, parasitic on the roots of trees, consisting, according to the very slight notice he has given of it, almost entirely of a single flower of a red colour, which he refers to the Linnean class Icosandria, and compares in appearance, I suppose in the young state, to the half of a Pine-cone. It is not unlikely that this plant also may be really allied to *Rafflesia*, the smaller species of which it probably resembles in appearance.

its growth: and lastly, on finding these vessels in some cases penetrating the base of the column itself (*plate* 20 (22), *f.* 1).

But to judge of the validity of these objections, it became necessary to examine the nature of this connection in plants known to be parasitic on roots; in those especially, which [226 in several other respects resemble *Rafflesia*, as *Cytinus*, *Aphyteia*, *Cynomorium*, and *Balanophora*. On this subject I cannot find that a single observation has hitherto been made, at least with respect to the genera now mentioned. Sufficient materials, indeed, for such an investigation are hardly to be expected in collections, in which the parasite is most frequently separated from the root; and even when found in connection with it, is generally in a state too far advanced to afford the desired information. I consider myself fortunate, therefore, in having obtained specimens of several species where the union is preserved; and the result of the examination of these, though not completely satisfactory, has been to lead me back to my first opinion, namely, that the Great Flower is really a parasite, and that the root on which it is found probably belongs to a species of *Vitis*.

An account of some of the more remarkable of this class of parasitic plants, to which a few years ago I had paid particular attention, may hereafter form the subject of a separate communication. At present I shall confine myself to such general observations on the class as relate to the question respecting *Rafflesia*.

In the first place, plants parasitic on roots are chiefly distinguishable by the imperfect development of their leaves and the entire absence of green colour; an observation which, as applying to the whole tribe, was I believe originally made by Linnæus.[1] In both these points they agree with *Rafflesia*.

A second observation which may be made respecting them is, that their seeds are small, and their embryo not only minute, but apparently imperfectly developed; in some cases being absolutely undivided, and probably acoty-

[1] *Fungus Melitensis*, p. 3. *Amœn. Acad.* iv, p. 353.

ledonous, even in plants which, from their other characters, are referable to dicotyledonous, or at least to monocotyledonous families.

227] In these points the structure of *Rafflesia* remains to be ascertained. In the mean time, however, if it be considered as a parasite, and as likely to agree with the other plants of the tribe in the state of its embryo, it may be remarked, with reference to the question of its affinities, that such a structure would approximate it rather to *Asarinæ* than to *Passifloreæ*.

My principal and concluding observation relates to the modes of union between the stock and the parasite. These vary in the different genera and species of the tribe, which may be divided into such as are entirely dependent on the stock during the whole of their existence, and such as in their more advanced state produce roots of their own.

Among those that are in all stages absolutely parasitic, to which division *Rafflesia* would probably belong, very great differences also exist in the mode of connection. In some of those that I have examined, especially two species of *Balanophora*,[1] the nature of this connection is such, as can only be explained on the supposition that the germinating seed of the parasite excites a specific action in the stock the result of which is the formation of a structure, either wholly or in part, derived from the root, and adapted to the support and protection of the undeveloped parasite; analogous therefore to the production of galls by the puncture of insects.

On this supposition, the connection between the flower of *Rafflesia* and the root from which it springs, though considerably different from any that I have yet met with, may also be explained. But until either precisely the same kind of union shall have been observed in plants known to
228] be parasitic, or, which would be still more satisfactory, until the leaves and fructification belonging to the root to which *Rafflesia* is attached shall have been found, its being

[1] *Balanophora fungosa* of Forster, and *B. dioica*, an unpublished species, lately sent by Dr. Wallich from Nepaul, where it was discovered by Dr. Hamilton, and also found in Java by Dr. Horsfield.

a parasite, though highly probable, cannot be considered as absolutely ascertained.[1]

ADDITIONAL OBSERVATIONS.

Read November 21st, 1820.

SINCE my paper on *Rafflesia*, or the Great Flower of Sumatra, was read to the Society, further information respecting it has been received from Sir Stamford Raffles and Mr. Jack, which will form an important addition to my former account.

Sir Stamford, in a letter to Mr. Marsden, states the following particulars:

"I find the *Krûbût* or Great Flower to be much more general and more extensively known than I expected. In some districts it is simply called *Ambun Ambun*. It seems to spring from the horizontal roots of those immense Climbers, which are attached like cables to the largest trees in the forest. We have not yet met with the leaves. The fruit also is still a desideratum. It is said to be a many-seeded berry, the seeds being found in connection with the processes on the summit of the pistillum. I have had buds brought in from Manna, Sillibar, the interior of Bencoolen and Laye: and in two or three months we expect the full-blown flower. It takes three months from the first appearance of the bud to the full expansion of the flower; and the flower appears but once a year, at the conclusion of the rainy season."

The first communication from my friend Mr. Jack con- [229
sisted of a description of recent flower-buds, at that time regarded by him as hermaphrodite, but which he has since ascertained to be male. It is unnecessary to introduce this

[1] Annals of Philosophy for September 1820, p. 225.

description here, as it essentially agrees with that already given, and may also be considered as superseded by the important information contained in the following letter, which I have more recently received from the same accurate botanist.

BENCOOLEN, *June* 2, 1820.

"MY DEAR SIR,—Since I wrote you last I have ascertained several particulars respecting the Gigantic Flower of Sumatra, additional to those contained in the account forwarded by Sir Stamford Raffles to Mr. Marsden, and by him communicated to you, which it may be interesting to you to know.

"Numerous specimens, in every stage of growth, have been sent from various parts of the country, which have enabled me to ascertain and confirm every essential point. The first and most unexpected discovery is, that it has no stem of its own, but is parasitic on the roots and stems of a ligneous species of *Cissus* with ternate and quinate leaves: I have not ascertained the species.[1] It appears to take its origin in some crack or hollow of the stem, and soon shows itself in the form of a round knob, which, when cut through exhibits the infant flower enveloped in numerous bracteal sheaths, which successively open and wither away as the flower enlarges, until, at the time of full expansion, there are but a very few remaining, which have somewhat the appearance of a broken calyx. The flowers I find to be unisexual, which I did not before suspect, and consequently diœcious. The male I have already described. The female
230] differs very little in appearance from it, but totally wants the globular anthers, which are disposed in a circle round the lower side of the rim or margin of the central column of the male.

"In the centre of this column or pistillum in the female are perceived a number of fissures traversing its substance without order or regularity, and their surfaces are covered

[1] Mr. Jack has since determined it to be *Cissus angustifolia* of Roxburgh. *Fl. Ind.* i, *p.* 427.

with innumerable minute seeds. The flower rots away not long after expansion, and the seeds are mixed with the pulpy mass.

"The male and female flowers can be distinguished by a section not only when mature, but at every stage of their progress. I have made drawings of every essential part, which I hope soon to be able to send home, together with a further account than I have yet had leisure to make.

"I remain, &c.,

"WILLIAM JACK."

The two principal desiderata respecting *Rafflesia*, namely the satisfactory proof of its being a parasite, and the discovery of the female flower, are now therefore supplied.

Additional information, however, on several points is still wanting to complete the history of this extraordinary plant.

Thus, it would be interesting, by a careful examination of the buds in every stage, to trace the changes produced in the root by the action of the parasite, and especially to ascertain the early state of the reticulate base, which I have ventured to consider as a production of an intermediate nature, partly derived from the root itself, and which I suppose will be found to exist before the bracteæ become visible.

Further details are also wanting respecting the circumstance of its being found both on the roots and stems of the *Cissus* or *Vitis*,[1] no instance being, I believe, at present, [231
known of parasites on roots, which likewise originate from other parts of the plant.

Many important particulars remain to be ascertained respecting the *Pistillum*.

From Mr. Jack's account it appears that the seeds are found in the substance of the column; in other words, that the ovarium is superior. But as I have formerly remarked, that in the male flower the same internal structure seems to be continued below the apparent base of the column, it is

[1] As these two genera differ from each other merely in number of parts, I have formerly proposed to unite them under the name of *Vitis*. (*Tuckey's Congo*, *p.* 465. *Antè*, *p.* 151.)

possible that in the female the production of seeds may extend to an equal depth; the ovarium would then become essentially inferior, as far at least as regards the question of the affinity of the plant. This point would be determined by a description of the unimpregnated ovarium, a knowledge of whose structure is also wanting to enable us to understand the nature of the ripe fruit, and especially the origin and direction of the fissures, on the surfaces of which the seeds are produced.

It is desirable likewise to have a more particular description of the *Stigma*, to which Mr. Jack seems to refer both the corniculate processes of the disk, and the undivided limb of the column. These parts in the male flower have no evident papulose or secreting surface; for the hispid tips of the processes can hardly be regarded as such. But it is not likely that in the female flower they are equally destitute of this, which is the ordinary surface of a stigma; and it appears to me more probable that such a surface should be confined to a definite portion, probably the tips, of the corniculate processes, than that it should extend over every part of the apex of the column.

Whatever may be the fact, my conjecture respecting these processes being possibly imperfect stamina is completely set aside; though it is still difficult to connect their number and arrangement with the supposed structure of ovarium.

232] Until these points are ascertained, and the seeds have been examined, the question of the affinities of the genus will probably remain undetermined. In the mean time it may be remarked, that as far as the structure of the fruit of *Rafflesia* is yet understood, it may be considered as in some degree confirming the proposed association of the genus with *Asarinæ*; especially with *Cytinus*, in which the ovarium is unilocular, with numerous parietal placentæ extending nearly to the centre of the cavity, and having their surfaces covered with minute ovula.

From the appearance of the ripe fruit of *Aphyteia*, a similar structure may be supposed to exist also in that genus, of which, however, the unimpregnated ovarium has not been examined. But these two genera are parasitic on roots, and have also their stigmata remarkably developed; and although *Rafflesia* probably differs from both of them in having a superior ovarium, I have endeavoured to show that this difference alone would not form an insuperable objection to their affinity.

EXPLANATION OF THE PLATES RELATING TO RAFFLESIA ARNOLDI.

Plate 13 (XV.)

The expanded Flower reduced to somewhat less than $\frac{1}{3}$ of its natural size; the scale given on the plate being too long by nearly $\frac{1}{7}$.

Plate 14 (XVI).

A Flower-bud covered with its bracteæ, of the natural size.

Plate 15 (XVII). [233

The underside of the same Bud; to show the root, the reticulate base with the circular elevation in which it terminates, and the origin of the outer bracteæ. Natural size.

Plate 16 (XVIII).

Flower-bud, of which the bracteæ, whose insertions are shown, are removed. Natural size.

Plate 17 (XIX).

A different view of the Bud in the same state, to show the æstivation and veins of the segments of the perianthium. Natural size.

Plate 18 (XX).

Fig. 1. A vertical section of the Bud deprived of its bracteæ: exhibiting the principal vessels of the column and perianthium, and the structure of the root, especially the change in the direction, increased ramification and termination of its vessels at the base of the parasite. Natural size.

Fig. 2. One half of the vertically-divided perianthium of the same Bud, in which the internal surface of the tube, corona and segments is shown. Natural size.

Plate 19 (XXI).

Fig. 1. A Flower-bud, its bracteæ and perianthium being removed, to show the column with the two annular processes at its base. Natural size.

Fig. 2. A portion (about $\frac{1}{3}$) of the column, of which part of the limb is removed, to show the cavities of the neck, into which the antheræ are received. Natural size.

Fig. 3. The portion of the Limb removed from fig. 2, with its antheræ immersed in their proper cavities. Natural size.

234] Fig. 4. An Anthera, magnified three diameters, as are figures 5, 6, 7, and 8.

Fig. 5. A transverse section of the same above the middle.

Fig. 6. A transverse section of the same below the middle.

Figs. 7, 8. Vertical sections of the same.

Fig. 9. Pollen, magnified 200 diameters.

Plate 20 (XXII).

Fig. 1. A vertical section of part of the base of the smaller Flower-bud, showing the vessels of the root, some of which appear to penetrate the substance of the parasite. Natural size.

Figs. 2, 3. Portions of the column of the expanded Flower, nearly corresponding with those of the Bud, in *Pl.* 19 (21), *f.* 2 and 3. Natural size.

Fig. 4. Anthera of the expanded Flower, magnified 3 diameters, as are figures 5 and 6.

Fig. 5. Transverse section of the same below the middle.

Fig. 6. Vertical section of the same.

Fig. 7. Pollen of the expanded Flower, magnified 200 diameters.

Fig. 8. A transverse section of the Root, magnified 3 diameters.

ON THE

FEMALE FLOWER AND FRUIT

OF

RAFFLESIA ARNOLDI

AND ON

HYDNORA AFRICANA.

BY

ROBERT BROWN, ESQ.,

D.C.L. OXF.; LL.D. EDIN.; F.R.S.; HON. MEMB. R.S. EDIN. AND R.I. ACAD.; V.P.L.S.
ONE OF THE EIGHT FOREIGN ASSOCIATES OF THE ACADEMIES OF SCIENCES IN
THE ROYAL INSTITUTES OF FRANCE AND OF HOLLAND; FOREIGN MEMBER
OF THE IMPERIAL AND ROYAL ACADEMIES OF SCIENCES OF RUSSIA,
PRUSSIA, SWEDEN, NAPLES, BAVARIA, DENMARK, BELGIUM,
BOLOGNA, UPSALA, AND OF THE IMPERIAL ACADEMY
OF NATURALISTS, ETC.

[*Reprinted from the 'Transactions of the Linnean Society.'*
Vol. XIX, pp. 221—247.]

LONDON:

1844.

ON THE

FEMALE FLOWER AND FRUIT

OF

RAFFLESIA ARNOLDI, &c.

READ JUNE 17TH, 1834.

THE principal object of the present communication is to complete, as far as my materials enable me, the history of *Rafflesia Arnoldi,* the male flower of which is described and figured in the thirteenth volume of the Society's Transactions.

The specimens from which this additional information has been obtained, as well as those formerly described, were received from the late Sir Stamford Raffles; and for the drawings so beautifully representing their structure, I am indebted to the same distinguished botanical painter and naturalist, who obligingly supplied those already published.

In my former essay some observations were made on the affinities of *Rafflesia,* a subject on which I could not then speak with much confidence. From such knowledge as I possessed, however, I ventured to state that this genus appeared to be most nearly allied to *Asarinæ,* and especially to *Cytinus,* on the one hand, and on the other to *Aphyteia* or *Hydnora,* an equally remarkable parasite of South Africa, but the structure of which was at that time very imperfectly understood.

An examination of complete specimens of *Hydnora africana* has confirmed this view; and as there are points in its structure which seem to throw some light on one of the most difficult questions respecting *Rafflesia*, I have included an account of this genus in the present paper.

The accompanying drawings of *Hydnora africana*, which so admirably display its structure, were kindly made from these specimens by my lamented friend and fellow-traveller Mr. Ferdinand Bauer, when he revisited England in 1824; they were probably the last drawings he ever made of an
222] equally interesting and difficult botanical subject, and I consider them his best.[1]

Since the publication of my former memoir, much light has been thrown on the structure and economy of *Rafflesia*, chiefly by Dr. Blume, who in his 'Flora Javæ' has given a very full history of a nearly related species, his *Rafflesia Patma*, as well as of *Brugmansia*, a parasite of similar economy, very distinct as a genus, but evidently belonging to the same natural family. Before, however, noticing more particularly what has been done by others, I shall resume the subject where I left it at the conclusion of my former memoir, in adverting to those points which I then regarded as the principal desiderata in the botanical history of this extraordinary plant.

The first of these related to the reticulate base, which I ventured to consider a production of an intermediate kind, or rather as one derived from the stock or root of the Vine,

[1] Since this paper was read, the Linnean Society have had to lament the loss of Francis Bauer, who died in 1841 at the advanced age of eighty-three. Like his brother Ferdinand, he continued, till within a short time of his death, to take the same interest in those scientific investigations which formed the constant occupation and the chief pleasure of a long life.

The figures of *Rafflesia* and *Hydnora*, which so admirably illustrate, and form the more valuable part of this communication, are among the best specimens of the unrivalled talent of the two brothers Francis and Ferdinand Bauer, who, as botanical painters, equally united the minute accuracy of the naturalist with the skill of the artist.

To this brief note I may be permitted to add how fortunate I consider myself in having so long enjoyed the friendship and so often been indebted for the important assistance of these two distinguished men, whose merits in the branch of art which they cultivated have never been equalled, and to both of whom the illustrations of the present paper, so happily connected, may form an appropriate monument, the work of their own hands.

but excited and determined in its form and nature by the specific stimulus of the parasite. I expected, therefore, to find it existing in the form of a covering to the bracteæ in the early state, as in *Cytinus*. This point has been fully confirmed, and is well shown in Mr. Bauer's drawings of the very young buds.[1] From the same figures it appears that the parasite is occasionally found on the stems of the Vine, as Dr. Jack had stated, but which seemed to me to require confirmation.

Of the structure of the female flower of *Rafflesia* I [223
judged entirely from Dr. Jack's account in his letter published in my former essay; and respecting this structure several important points, which even his subsequent description in the 'Malayan Miscellany' did not supply, were regarded as undetermined.

Whether the ovarium is wholly distinct from the calyx or cohering with it at the base, was the first of these points which required further examination. The specimens now prove it to be chiefly superior or free in the flowering state, and wholly so in the ripe fruit.

The internal structure of the ovarium, especially the origin and arrangement of the numerous ovuliferous surfaces or placentæ, I considered one of the principal desiderata. Dr. Jack's account of these placentæ, which, as far as it extends, is essentially correct, is confirmed by Dr. Blume's description and figures of *Rafflesia Patma*, as well as by the more complete drawings which accompany the present paper. The important question, however, namely the analogy of this apparently singular arrangement with ordinary structure, may be regarded as still in some degree obscure.

The transverse section of the ovarium presenting an indefinite number of cavities irregular in form, having no apparent order, and over the whole of whose surfaces the ovula are inserted, is hardly reconcileable to the generally received notions of the type of the female organ; and as

[1] That the whole of this covering belongs to the stock, is proved by its containing those raphides or acicular crystals which are so abundant in the root of the *Vitis* or *Cissus*, and which are altogether wanting in the parasite.

these cavities exist to the same extent and with similar irregularity from centre to circumference, they may with equal probability be considered as originating from the axis or from the parietes of the ovarium. The vertical section too, if viewed without reference to the external development of the top of the column, exhibits a structure equally anomalous. If, however, the corniculate processes terminating the disc of the column be regarded as styles, which is surely the most obvious and not an improbable view, their arrangement would lead to the supposition that the ovarium is composed of several concentric circular series of simple pistilla, each having its proper placenta, bearing ovula over its whole surface. But the structure is so much obscured by the complete confluence of the supposed component parts, that this view might not at once present itself. It is readily suggested, however, by the seemingly analogous structure of *Hydnora*, in which the cylindrical placentæ, whose number is considerable and apparently indefinite, are all pendulous from the top of the cavity, neither cohering
224] with its sides or base, wholly distinct from each other, and uniformly and densely covered with ovula.

But although this is the most obvious view suggested by *Hydnora*, a more careful examination, especially as to the relation of stigmata to placentæ, leads to a very different notion of the composition of the ovarium in that genus: for as the placentæ correspond with, and may be said to be continuations of the subdivisions of the stigmata, and as these stigmata appear to be three in number, each with numerous subdivisions diverging from the circumference towards the centre of the ovarium, and each of these subdivisions bearing one or more placentæ pendulous from its internal surface, the ovarium of *Hydnora* may be regarded as composed of three confluent pistilla, having placentæ really parietal, but only produced at the top of the cavity; the sides of which, however, exhibit no indication whatever of composition.

Between this most remarkable structure of *Hydnora* and that of *Cytinus* there is some, though not perhaps a very obvious analogy, each of the strictly parietal placentæ in the latter being subdivided into distinct lobes, as in many *Orchideæ*, a family which *Cytinus* also resembles in the structure of the seed, and probably in the mode of impregnation, though so widely different in almost every other respect.

It would certainly be difficult to reduce *Rafflesia* to the view here taken of the formation of the compound ovarium in these two genera; and it may therefore, perhaps, be said, that although the structure of *Hydnora*, in one important particular, suggests or confirms the more probable notion of the composition of ovarium in *Rafflesia*, as already stated,[1] it is in other respects very distinct.

Another point, which in my former paper I considered [225
doubtful, namely the seat or limit of the stigmata, is not even now satisfactorily established; for the slender processes forming the hispid tips of the supposed styles, which have so much the appearance of the ultimate divisions of stigma, are merely hairs of a very simple structure, and exactly resembling those found in other parts of the column;

[1] My confidence in this hypothesis respecting *Rafflesia* is greatly lessened on considering the structure of the female flower of a lately discovered species of the genus, namely, *Rafflesia Cumingii* or *Manillana*, in which the style-like processes terminating the column are much fewer in number, and so arranged as to form a single circular series of about ten, not very distant from the limb, with only from one to three processes within it, which are placed near the centre, while the irregular cavities in the ovarium are evidently much more numerous, and in arrangement have no apparent relation to that of the supposed styles, there being as great complexity in the centre as towards the circumference. These relations between styles and ovarial cavities seem, according to the figures of *Rafflesia Patma*, to be reversed in that species, its styles being apparently more numerous than the cavities of the ovarium; and as even in *Rafflesia Arnoldi* their correspondence is far from obvious, it would seem that the number and arrangement of these processes afford no satisfactory evidence of the composition of the ovarium in any known species of the genus. But if the placentation of *Rafflesia Arnoldi* and *Cumingii*, notwithstanding the objections stated in the text (p. 404), be considered parietal, as Blume has described it in *R. Patma*, and as from his figures it seems actually to be in *Brugmansia*, there would still be no means of determining the exact degree of composition of ovarium in *Rafflesia;* for in no species of the genus is there the slightest indication afforded by the arrangement of cavities or ramification of the assumed placentæ, to mark any definite number of component parts. Similar objections apply with equal force to the adoption of that opinion which regards placentation as in all cases central or derived from the axis.

In conclusion, therefore, it may perhaps be said that *Rafflesia*, in the structure both of ovarium and antheræ, is not obviously reconcileable to any hypothesis hitherto proposed to account either for the origin or for a common type of the sexual organs of Phænogamous plants.

though in several of the specimens examined they were greatly altered in appearance, from a coating of mucous matter taken up and again deposited by the spirit in which the specimens were preserved.[1] A slight difference, indeed, seems to exist between the tissue of the apices of the styles and the other parts of their surface; hardly sufficient, however, to prove it to be stigma, though this is no doubt the probable seat of that organ.

The next point of importance in the female flower of *Rafflesia* is the structure and gradual development of the ovula. These, in the earliest state observed, consist of merely conical or subcylindrical papillæ, having a perfectly smooth surface as well as uniform internal substance.

The first perceptible change taking place in the papilla is a slight contraction at its summit, the upper minute contracted apex being the rudiment of the nucleus. Immediately below this contracted portion a dilatation is soon observable, which, gradually enlarging and becoming slightly hollowed, forms a cup in which the nucleus, also proportionally increased in size, is partly immersed. This cup, the rudiment of the future integument, continues gradually to enlarge, until it completely covers and extends considerably beyond the nucleus, but without cohering with it. If a transverse section is made near the slightly depressed
226] apex of this integument, an extremely minute perforation or capillary channel, extending to the free apex of the included nucleus, may be observed.

This account of the gradual development of the ovulum of *Rafflesia,* I believe, is in every essential point applicable to Phænogamous plants generally, except that here one coat only is developed. It is, however, in some important points different from the description given by M. Mirbel, who considers the nucleus in its earliest state as included in the integuments, which in the next stage open and dilate so as to leave it entirely exposed; they then, as he supposes, remain quiescent until the nucleus has considerably enlarged,

[1] See Mr. Bauer's representation of the hairs in this state, TAB. 22 (XXIII), figs. 3, 4, 5, 6, and 7.

when they again become active and increase in size until they once more completely cover it.

While the development, as I have here described it, of the nucleus and its integument in *Rafflesia* is going on, another change is at the same time gradually taking place, namely, at first a slight bending, which at last ends in a complete inversion, in the direction of the nucleus and its integument in regard to the placenta, with which, in this advanced stage, the perforated apex of the latter is nearly or absolutely in contact.

In this change of direction, the ovulum of *Rafflesia* resembles that of the far greater part of Phænogamous plants: the change, however, is effected in a way which is much less common, the curvature in *Rafflesia* taking place solely in the upper part of the funiculus, the direction of the inverted ovulum being parallel with, but distinct from, the portion below the curvature; whereas in Phænogamous plants generally, the curvature is produced in that part of the funiculus which is connate with the testa or outer integument. For this difference, a reason, perhaps, may be assigned; the integument which generally forms the testa or outer coat being in *Rafflesia* entirely wanting, or only indicated by the remarkable dilatation of the apex of the funiculus.[1]

In the more essential points of structure, the ovula of *Hydnora* and *Cytinus* agree with that of *Rafflesia*. They differ, however, in both these genera in retaining their original direction.

In *Hydnora* I have ascertained the perforation of the single integument and the position of the included nucleus, [227
but the very earliest stages I have not yet distinctly seen; while in *Cytinus*, in addition to the coat analogous to that of *Rafflesia* and *Hydnora*, a two-lobed or bipartite membrane is observable.

Of these three genera, I have hitherto observed the pollen

[1] The earlier production of the inner of the two coats generally present in the ovula of Phænogamous plants, and the absence of the outer in this and several other cases, will probably be considered a valid objection to the terminology of M. Mirbel.

or mucous tubes only in *Cytinus*, in which they pass along the surfaces of a definite number of cylindrical cords existing in the style until they reach the cavity of the ovarium, when they follow the direction of the placentæ and become mixed with the ovula, to which I have not yet, however, found them actually attached.[1]

The structure of the pericarpium and the ripe seed of *Rafflesia* have been satisfactorily ascertained from the examination of a single fruit found among the numerous flower-buds in various states which were received from Sumatra by Sir Stamford Raffles long after his return to England. In this fruit, which is very accurately represented of the natural size in Mr. Bauer's figure, the column, deprived entirely of its style-like processes, had become a compact fleshy mass, having deep fissures on its surface dividing it into nearly square lobes, somewhat resembling the surface of the dilated base of *Testudinaria*, and within, like the ovarium, exhibiting irregular cavities, whose surfaces were thickly covered with minute seeds.

These seeds, which are also beautifully shown in Mr. Bauer's figures, differ but little in form from the ovula of the expanded but unimpregnated flower; they are considerably larger, however, and the apex of the funiculus is still more dilated. From their great hardness, as well as from their internal structure, they appear to be quite ripe; and it is worthy of remark, that of the many thousands contained in the fruit, the very considerable portion seen were of uniform size and appearance.

The testa or outer integument, which is evidently that existing in the unimpregnated ovarium, is of such hardness and thickness that it may be termed a nut; it is of a chestnut colour, its surface regularly reticulate and deeply pitted, a depression occupying the centre of each areola.

The inner integument is a thin light-coloured membrane,
228] very slightly areolated, and of uniform surface. Within

[1] In a few cases where the supposed pollen-tubes were present I found them applied to the apices of the enlarged ovula. In some instances I have met with only a very loose tissue, consisting of elongated cells mixed with mucus, forming cords descending from the stigmata, and reaching to, but not extending beyond, the origin of the placentæ.

this the nucleus, of similar form and dimensions, seems to be more firmly attached at its upper extremity to the coat by a short and very slender funiculus.

The nucleus separated from its coat has an areolated surface, and at first appears to be entirely composed of a loose and uniform cellular tissue. But on a more careful examination this substance is found to contain another cellular body, of nearly cylindrical form, adhering with some firmness to the upper extremity of the including cellular mass, whose vertical axis it occupies for nearly three fourths of its length.

This inner body, which I regard as the *embryo*, consists of large cells, disposed nearly, but not with absolute regularity, in two longitudinal series, and so transparent, that it may be safely affirmed that there is no included body nor any perceptible difference in the contents of any of the component cells.

This account of the embryo differs in some respects from Mr. Bauer's representation of it, especially as to the point of attachment, and in the distinct appearance and transparency of cells.[1]

The seed of *Hydnora* in many essential points resembles that of *Rafflesia*. Its nucleus consists of a dense albumen, the cells of which are so disposed as to exhibit, when slightly magnified, a kind of radiation in whatever direction it is cut. This albumen is much denser than that of *Rafflesia*, the greater density arising, perhaps, from the unusual thickness of the walls of each cell, its cavity bearing so small a proportion to the supposed external dimensions of the cell as to give it the appearance of a nucleus or more opaque central body.[2]

Enclosed in the albumen a perfectly spherical embryo is

[1] I have therefore added to TAB. 23 (XXV), a circumscribed figure, marked R. Br., in which I have endeavoured to represent (but not very successfully) the structure as I have seen it.

[2] But these supposed cells with thickened walls, admitting them to have been originally distinct, are in the ripe seed nearly or entirely obliterated, so that the substance of the cartilaginous albumen consists of a uniform, semi-transparent mass, in which the more opaque nuclei or cells, containing minute granular matter, are, as it were, immersed.

found, consisting entirely of a more minute and much less dense cellular tissue. On the surface of this embryo I have observed no point marking original attachment, nor any
229] indication of a channel connecting it with the surface of the albumen, in the centre of which it is seated.

In *Cytinus*, in which I believe I have at length found ripe fruits, the seeds are extremely minute, and generally retain at their base the bipartite membrane more distinctly observable in the unimpregnated ovulum. To this membrane the name of arillus may be given; but it may also, and, perhaps, with greater probability, be considered the imperfect production of the testa or outer membrane.

The seed itself is elliptical, with a slight inequality at top indicating the depression or perforation observable in the ovulum. The single integument of the seed is easily separable from the nucleus, and by moderate pressure splits longitudinally and with great regularity into two equal portions; in texture it is a crustaceous membrane, indistinctly reticulate, the areolæ, when very highly magnified, appearing to be minutely dotted with a semi-opaque centre.

The nucleus, corresponding exactly in size and form with the integument, has its surface also reticulate, but the areolæ are not dotted; and it appears, as far as I can ascertain in so minute a body, to consist of a uniform cellular tissue, very exactly resembling the nucleus of an Orchideous plant.

The result of the comparison now made, and which might be extended to other points of structure of *Rafflesia*, *Brugmansia*, *Hydnora* and *Cytinus*, seems to be, that these four genera, notwithstanding several important differences, form a natural family to which the name of RAFFLESIACEÆ may be given; and that this family is again divisible into three tribes or sections:

The first *Rafflesieæ*, consisting of *Rafflesia* and *Brugmansia*, is distinguishable by the ovarium being either in part or wholly superior to the origin of the calyx, in its composition or internal structure, in the placentation and direction of the ovula, in the structure of the seed and in that of the antheræ.

The second section, *Hydnoreæ*, formed of *Hydnora* alone, is characterised by its completely adherent ovarium, singularly divided stigmata, the peculiar origin and structure of its pendulous placentæ, its embryo enclosed and seated in the centre of a dense albumen, and by the arrangement and structure of its antheræ.

In the third section, or *Cytineæ*, the placentæ are parietal, the ovarium is connate with the calyx, and the cellular [230
undivided embryo forms the whole mass of the seed, or is apparently destitute of albumen.[1]

That this third section is nearly related to *Asarinæ* seems to me unquestionable; if, therefore, its affinity to *Hydnora* and *Rafflesia* be admitted, the place of this singular family would be nearly established.

That *Rafflesia*, *Hydnora* and *Cytinus* do not essentially differ from many of the more perfectly developed Phænogamous plants in their vascular structure, I have satisfactorily ascertained, and there is no sufficient reason to doubt that the same observation may be extended to *Brugmansia*.

In my former paper, in treating of the composition of the

[1] To the third section of *Rufflesiaceæ*, *Apodanthes* and *Pilostyles* may perhaps be referred. These genera indeed agree with *Cytinus* in their unilocular ovarium with parietal placentation, in their cellular undivided embryo forming the whole mass of the seed, and in their adherent or semi-adherent ovarium, whose cavity in *Pilostyles* extends even below the insertion of the bracteæ. The existence of petals, however, in both, and especially in *Apodanthes*, will probably be considered as an objection of some weight to their absolute union with *Cytineæ;* and there is even an important difference in their placentation, the ovula being produced equally over the whole surface of the ovarial cavity, while in *Cytinus* the placentæ are distinct, definite in number, and subdivided into numerous lobes, nearly as in *Orchideæ*.

Whether *Apodanthes* and *Pilostyles* are to be included in the same genus, as Professor Endlicher (in Gen. p. 76) first conjectured, and as Mr. Gardner has more recently (in Hooker Ic., new ser. vol. iii, tab. 644) endeavoured to prove though not improbable, must, I think, remain somewhat doubtful so long as we are unacquainted with the male flower of *Apodanthes*. In the mean time this genus may be distinguished from *Pilostyles* by the singular insertion of its petals, which also differ remarkably in texture from the quadrifid persistent calyx, and by the two bracteæ of the flower being seated below the origin of an angular ovarial cavity, and which, after the falling off of the parasite, remain attached to the stock.

vascular bundles existing in various parts of *Rafflesia,* I too hastily assumed the absence of spiral vessels, the expression used evidently implying that I had satisfied myself of their non-existence in the fasciculi or bundles examined; instead of which I should only have stated that I had not been able to find them.

The absence of spiral vessels has since been affirmed by Dr. Blume with respect to his *Rhizantheæ,* consisting of *Rafflesia* and *Brugmansia;* and still more recently by Messrs. Endlicher and Lindley, who, overlooking probably the very positive statement of Dr. von Martius respecting *Langsdorfia,* have equally denied the existence of spiral vessels in *Balanophoreæ*; and partly, perhaps chiefly, determined by this supposed conformity and peculiarity of structure, have referred *Rafflesiaceæ* and *Balanophoreæ* to the same natural class.

231] I have in the first place to correct my own error respecting *Rafflesia,* in various parts of the female flower of which I have found spiral vessels of the ordinary structure, consisting of a single, easily unrolled fibre; and on re-examining the same specimen of the male flower respecting which my former assertion was made, I found these vessels equally distinct. Professor Meyer has already stated their existence in the procumbent stems or rhizomata of *Hydnora triceps;* in which I have also found them in *Hydnora africana,* as well as in other parts of the same species; and in *Cytinus* they are still more obvious.

I may also add, that wherever I had specimens of *Balanophoreæ* in a fit state for minute examination, I have never failed to find spiral vessels in various parts of their tissue, particularly in *Cynomorium coccineum* and *Helosis guianensis.*[1]

[1] Although in *Rafflesiaceæ* and in the genera at present referred to *Balanophoreæ* spiral vessels undoubtedly exist, in the greater number, indeed, sparingly, but in some cases in hardly reduced proportion, it may still perhaps be alleged, by those botanists who have proposed to unite both families into one natural class, that the vascular system of all these parasites is uniform and more simple than that of the far greater part of Phænogamous plants; that the spiral or slight modifications of it is the only form of vessel hitherto observed in any of them; and that the large tubes or vessels, with frequent contractions, corresponding imperfect diaphragms, and variously marked surface, which have

I may hereafter have an opportunity of entering fully [232 into the question whether *Rafflesiaceæ* and *Balanophoreæ*

received several names, as vasa porosa, punctata, vasiform cellular tissue, dotted ducts, &c., and which are so conspicuous in the majority of arborescent Phænogamous plants, have never been observed in any part strictly belonging to these parasites. But even admitting the non-existence of the large vessels here referred to, their absence will hardly be regarded as a sufficient reason for the union into one class of the two families in question, especially when it is considered—

First. That conformity in vascular structure, even when accompanied by peculiarity of tissue, does not always indicate, much less determine, botanical affinity. This is strikingly exemplified in *Coniferæ* and *Winteraneæ,* two families which, though so nearly agreeing in the uniformity and peculiarity of their vessels, and in both of which the large tubes referred to are wanting, yet differ so widely from each other in their organs of reproduction and in their leaves, that they may be regarded as placed at opposite extremities of the scale of *Dicotyledones.*

Secondly. That uniformity of vascular structure is not always found in strictly natural families. Thus many *tropical woody climbers* exhibit remarkable peculiarities of vascular arrangement not existing in the greater part of the families to which they respectively belong, but which peculiarities appear to have no influence whatever in modifying their reproductive organs.

Thus also in *Myzodendron*[1] the whole woody tissue consists of vasa scalariformia, a peculiar structure, and very different from that of all the other genera belonging to *Loranthaceæ,* to which this genus has been referred, and to which, though it does not absolutely belong, it is nearly related. Even this peculiar structure of the stems of *Myzodendron* admits of considerable modifications in the different species of the genus, which is strikingly exemplified in comparing the loose vascular tissue with large and singularly constructed medullary rays of *M. brachystachyum* and *quadriflorum* with the more minute vessels and extremely narrow rays of *M. punctulatum.*

I may also notice that in *Tillandsia usneoides,* as well as in the nearly related species of that genus, the capillary stems are destitute even of spiral vessels, though in *Bromeliaceæ* generally the ordinary vascular system is found.

Whatever may be the state of vessels in the fully developed parasites belonging to *Rafflesiaceæ,* it appears to me that at least *Rafflesia* in its very early

---

1 *Myzodendron* of Banks and Solander, from μυζέω or μύζω *sugo,* and δένδρον, has been changed to *Misodendron* by De Candolle and all following systematic writers; no doubt merely from a mistake as to the intended derivation. *Myzodendron,* hitherto referred to *Loranthaceæ,* to which it is certainly closely allied, especially through *Antidaphne* of Pœppig, appears to me to have characters sufficient to distinguish it as, at least, a suborder or tribe (*Myzodendreæ*), namely, the structure of its ovarium, in which it approaches to *Santalaceæ,* having three ovula suspended from the apex of a central placenta, only one of which ripens; the entire absence of floral envelope in the male; the singular feathery appendages of the female flower and fruit compensating in the dispersion and subsequent adhesion of its seeds, which are destitute of that viscidity existing in those of the parasitic *Loranthaceæ;* and lastly, the embryo being undivided, with its dilated and exserted radicle enclosed in a semi-transparent covering, a continuation of the membrane lining the cavity of the albumen in which the embryo is lodged.

233] form merely different orders of the same natural class, in giving an account of a new and remarkable genus of the latter family.[1]

At present I shall only remark, that the sole remaining character employed to unite these two families and supposed to distinguish them from all others, namely, the simple or

stages is entirely cellular, and that this continues to be the case not only until that mutual adaptation of parasite and stock which enables the former to complete its development has taken place, but until the first indications of its future structure have become perceptible. It may also be remarked, that even after the formation of vessels in the parasite is obvious, the direct union between *Rafflesia* and the *Vitis* continues to be chiefly if not entirely cellular, the connection consisting in a slight mutual penetration or indentation of the two substances, whose cells are easily distinguishable.

I may here advert to one of the most difficult points in the economy of *Rafflesiaceæ*, namely, by what means their minute embryos, which are at the same time of an extremely loose texture, are enabled to penetrate through the bark of the plants on which they vegetate, so as to account for such appearances as those exhibited in the nascent *Rafflesia Arnoldi* represented in Tab. 25 (XXVI), A, in which I have been unable to trace any perceptible communication with the surface, and where the parasite seems rather to grow out of than into the stock.

Connected with this point a question may also arise, whether the earliest effort of the seed after its deposition in the proper nidus, by whatever means this is effected, may not consist in the formation of a cellular tissue extending laterally under the bark of the stock and capable of producing the fully developed parasite.

This question might not occur in regard to *Rafflesia* and *Brugmansia*, in both of which the individual plants are in general sufficiently distant on the root of the *Vitis* to make it probable that each developed parasite is produced from a distinct seed. But in *Pilostyles*, and even *Cytinus*, where they are closely approximated, their possible origin from one common basis or thallus is more readily suggested, especially on considering that in the former genus, which is diœcious, each group of parasites is generally, perhaps always, exclusively of one sex; and that these groups, often of great density, not unfrequently surround completely the branch of the stock. But although this view did occur to me as not very improbable, and as tending to remove some of the apparent difficulties, I have never been able to trace any substance decidedly distinct from the proper tissue of the stock; there are, however, some appearances favouring the hypothesis in both genera, especially in *Pilostyles*, but which require careful examination in the living plants.

[1] This genus, which was first found by Francis Masson, is the *Mystropetalon* of Mr. Harvey (in South Afr. Gen. p. 418), who has described two species, from both of which Masson's plant is perhaps distinct.

I may here advert to a note at p. 225 of my former memoir (in Linn. Soc. Trans. vol. xiii), [*Antè, p.* 390] in which I thought it not improbable that a parasite briefly noticed by Isert (in Reise nach Guinea, p. 283) might be related to *Rafflesia*. I have now, however, reason to believe that Isert's plant is the *Thonningia sanguinea* of Vahl (in Act. Soc. Hist. Nat. Hafn. t. vi, p. 124, t. 6, and Schumacher, Guineische Plant. p. 431), a genus nearly related to, if really distinct from *Balanophora*.

acotyledonous embryo, exists equally in *Orchideæ*. And if it be employed along with those characters connected with their peculiar economy, namely, the imperfect development of leaves, the want of stomata and absence of green colour, the class cannot be limited to *Rafflesiaceæ* and *Balanophoreæ*, for an embryo of exactly the same kind exists in *Orobanche*, and other, perhaps all other, genera parasitic on roots, a remark which I made, though not with sufficient precision, in my former essay.[1] But such a classification, though founded on seemingly very important technical characters, would hardly be received in a strictly natural arrangement, and it seems to me quite as paradoxical to approximate two such genera as *Rafflesia* and *Balanophora*.

RAFFLESIA ARNOLDI.

Rafflesia Arnoldi, *R. B. in Linn. Soc. Transact.* vol. xiii, p. 201, tabs. 15—22 (*Antè, p.* 374, *tabs.* 13—20) Mas.

Rafflesia Titan, *Jack in Malayan Miscell., Append. to* vol. i.

Desc. Planta feminea masculæ omninò similis in- [234
sertione, bracteis et perianthio.

Columna quæ figura, stylis disci et limbo elevato indiviso apicis, necnon annulo duplici bascos cum mascula per singula puncta convenit; ab eadem differt *externè* rudimentis solum minutis papillæformibus polline destitutis antherarum, et loco cavitatum antheris maris respondentium sulci tantum lineares angusti nec profundi: *internè* ovario processibus indefinitè numerosis irregulariter confluentibus in cavitatibus labyrinthi speciem formantibus diviso.

Ovula numerosissima parietibus cavitatum ovarii sine ordine sparsa, primò nucleo papilliformi sessili nudo, mox basi attenuato in funiculum rectum, apice incrassatum primordium annulare integumenti simplicis quod sensim auctum demum nucleum omninò includit apice perforato, funi-

[1] *Antè, p.* 391-2.

culoque simul elongato extra medium arctè recurvato et incrassato.

Pericarpium, bracteis, calyce, apiceque patelliformi styligero columnæ delapsis, denudatum, omninò superum vel liberum, subovatum carnosum crassum altè rimosum indehiscens, cavitatibus sicut in ovario indefinitè numerosis inordinatis amorphis polyspermis.

Semina pedicellata, funiculi dimidio *inferiore* cylindraceo cellulari molli pallido: *superiore* maximè incrassato arctè recurvato subovato castaneo lacunoso solido duro. *Semen ipsum* ovatum vix diametro apicis dilatati funiculi, castaneum altè lacunosum.

Integumentum exterius crasso-crustaceum subnucamentaceum; *interius* membranaceum pallidum lacunis exterioris leviter impressum.

Albumen magnitudine integumenti interioris laxè cellulosum aqueo-pallidum.

Embryo e cellulis subduplici serie ordinatis iis albuminis majoribus constans, ex apice albuminis ortus, ejusque dimidio longior.

HYDNORA AFRICANA.

Hydnora africana, *Thunb. in Act. Holm.* (1775), vol. xxxvi, p. 69, tab. 2. *E. Meyer in Nov. Act. Acad. Cæsar. Nat. Curios.* vol. xvi, par. 2, p. 775, tab. 58.

Planta Aphyteja, *Resp. Achar. cum. tab.* (1776). *Amœn. Acad.* vol. viii, p. 310.

Aphyteia Hydnora, *Harv. Gen. South Afr.* p. 299.

Loc. Nat. Africa Australis parasitica in radicibus Euphorbiæ succulentæ cujusdam secundum Thunberg et Drege; et quandoque Cotyledonis orbiculatæ auct. D. Mundt *in Harvey, South Afr. Gen.* p. 299. Nuperrimè etiam in Africa boreali detecta, fid. sp. asserv. in Museo Parisiensi.

Desc. Primordia sunt Caules e dilatata radice plantæ sustinentis orti, humifusi v. sæpius semisepulti, angulati (4-5-6-goni) digitum crassi simplices v. sæpius ramosi, solidi angulis tuberculatis, tuberculis approximatis obtusis, apice sæpe rimoso, quasi dehiscenti sed nunquam fibras exse- [235 renti; intra corticem strato paulo laxiore magisque colorato, centro densiore e cellulis præsertim conflato et fasciculis tenuibus parcis vasorum instructo.

E tuberculo plurimum aucto exsertus est *Flos* erectus basi in pedunculum abbreviatum intùs vasculosum sensim paulo angustatus, penitus ebracteatus.

Perianthium monophyllum, tubulosum, subinfundibuliforme, carnosum, extùs (uti pedunculus) rimis plurimis superficialibus in areolis subrotundis plus minus angulatis squamas primo intuitu quodammodo referentibus divisum et quasi leprosum. *Tubus* intus glaberrimus sed sæpe transversim subrugosus. *Limbus* tubum subæquans tripartitus (rarissimè 4-partitus) æqualis, æstivatione induplicato-valvata; laciniis primum latere hiantibus, apicibus diutius cohærentibus demum distinctis, modicè patentibus ovalibus oblongisve obtusiusculis, marginibus veris latè et obliquè induplicatis majorem partem disci apicemque omninò occultantibus extùs ramentis numerosis subulatis conspersis marginalibusque elongatis ciliatis; singulis disco lævi e majore parte tecto *pulvinulo* adnato oblongo carnoso, sæpè longitudinaliter striato, apice marginibus laciniarum ibi coalitis occultato acutiusculo, basi obtusa subcordata.

Columna staminea infra medium tubi orta, brevissima, annulum efformans altè trilobum, lobis laciniis limbi oppositis rotundatis obtusis. *Antheris* indefinitè numerosis, connectivo communi crasso carnoso penitus adnatis, parallelo-approximatis, elongato-linearibus, bicruribus, crure altero plurimarum postico altero antico, nonnullis quasi pressione reliquarum et præcipuè iis ad ortum loborum columnæ sitis abbreviatis sæpiùs in dorso, rariùs in ventre lobi obviis; omnibus primum bilocularibus sed sulco unico longitudinali dehiscentibus.

Pollen simplex sphæroideum læve.

Ovarium totum adherens, parietibus cavitatis lævibus.

Stigma discum apicis ovarii occupans, sessile depressum trilobum; lobis iis annuli staminei et laciniis limbi perianthii oppositis; singulis striis linearibus numerosis, e peripheria cordata lobi centrum versus plus minus divergentibus, respondentibus totidem lamellis planis arctè approximatis sed ad cavitatem ovarii usque distinctis, ibique manifestiùs separatis et placentiferis.

Placentæ indefinitè numerosæ, una pluresve e superficie interiore lamellæ singulæ stigmatis ortæ, ideoque omnes ex apice ovarii pendulæ, cylindraceæ, dimidium cavitatis, cujus parietes læves omninò steriles, superantes, undique ovulis densè tectæ.

Ovula primum sessilia papillæformia uniformia, dein subcylindracea, brevè pedicellata, apice obtuso depresso, v. perforato v. membrana semitransparente tecto, nucleo incluso manifesto.

Pericarpium perianthio toto supero et annulo stamineo delapsis denudatum, stigmate persistenti apice clausum, sphæroideum magnitudine pomi minoris, areolis squamas
236] æmulantibus inæquale quasi leprosum, carnosum crassum indehiscens, cavitate placentis undique seminiferis densè repleta.

Semina subglobosa, pedicello brevi quandoque subnullo insidentia.

Integumentum exterius crasso-membranaceum subpulposum areolatum cellulis minutè granulatis: *interius* albumini arctè adherens.

Albumen densum, subcartilagineum, aqueo-pallidum, per lentem modicè augentem structura quasi radiata, sed magis auctum constare videtur substantia densa semitransparenti alba nec in cellulas manifestè divisa, sed undique farcta corpusculis celluliformibus figura variis, in serie extima majoribus oblongo-obovatis, reliquis minoribus vix symmetricè positis, omnibus semiopacis e membrana materia minutè granulosa repleta formatis.

Substantia densa Albuminis uniformis forsan e cellulis parietibus incrassatis et obliteratis, singulis nucleo (corpusculo) semiopaco fœtis.

Embryo in centro albuminis parvus subglobosus aqueopallidus e cellulis numerosis parvis mollibus, materia minutè granulosa repletis, ab albumine facilè separabilis, et absque ulla manifesta communicatione cum ejusdem peripheria vel ope suspensorii, v. canalis intermedio.

EXPLANATION OF THE PLATES.

RAFFLESIA ARNOLDI.

TAB. 21 (XXII).

Fig. 1. A female flower-bud, with the roots of the *Vitis* (or *Cissus*) vertically divided, which shows the numerous irregular cavities of the ovarium chiefly, if not entirely, above the insertion of bracteæ and calyx, and the vascular lines continued from the walls of the cavities through the upper solid part of the column into the axes of the style-like processes :—natural size.

Fig. 2. A female flower-bud in the same stage of development, the bracteæ and calyx entirely removed, to show its outward resemblance to the male flower-bud (figured in Linn. Trans. vol. xiii, TAB. XXI) (*Antè, t.* 19) :—natural size.

TAB. 22 (XXIII).

Fig. 1. A small segment of the column, of which part of the elevated undivided limb is removed, to show the narrow furrows of the sides of the column corresponding in number with the rudiments of antheræ, seen in

Fig. 2, which is the portion of the limb removed from fig. 1 :—natural size. [237

Fig. 3. The upper half of one of the styles of the disc, with its terminating hairs :—magnified 10 diameters.

Fig. 4. A portion of fig. 3, somewhat more highly magnified (20 diameters), vertically divided.

Figs. 5, 6, and 7. Some of the hairs still more highly magnified, which, according to Mr. Bauer, have a secreting surface seen in fig. 7, and which in figs. 5 and 6 is covered with the secretion, consisting of spherical particles enveloped in mucus at fig. 8 :—magnified 100 diameters (but see observations respecting them in page 225). (*Antè, p.* 405.)

Figs. 9 and 10. Longitudinal and transverse sections of a style :—magnified 50 diameters.

Fig. 11. A transverse section of half the ovarium, to show the numerous irregularly ramified cavities, and the arrangement of vascular cords belonging to the bracteæ and calyx :—natural size.

Fig. 12. A small portion of the ovarium, with the ovula covering the surface of the cavities, and the vascular lines passing through the axes of the parietes : —magnified 20 diameters.

Figs. 13—18. Ovula in various stages (the earliest observed are not represented) :—magnified 100 diameters.

Tab. 23 (XXIV).

Fig. 1. A ripe pericarpium, of the natural size, the calyx, bracteæ and apex of the column being deciduous.

Fig. 2. The same divided vertically, and showing the thickness of the densely-fleshy and deeply-furrowed covering, and also that the whole of the ovarial cavity is above the insertion of bracteæ and calyx.

Tab. 24 (XXV).

Fig. 1. A small portion of the wall of two adjoining cavities, the surfaces covered with numerous seeds, all of equal size:—magnified 20 diameters.

Fig. 2. A seed with its funiculus, of which the lower erect portion is filiform, the recurved upper half being of the same texture, colour and surface with the seed, which it somewhat exceeds in thickness:—magnified 100 diameters.

Fig. 3. The same divided longitudinally, to show the structure of the seed (according to Mr. Bauer), and that the enlarged apex of the funiculus is solid:—magnified 100 diameters.

Fig. 4. The nucleus of the seed taken out of its thick nut-like outer covering:—magnified 100 diameters.

Fig. 5. The same nucleus, whose membranous coat is separated by pressure,
238] to show the albumen:—magnified 100 diameters.

Fig. 6. The denuded loosely-cellular albumen.

Fig. 7. A portion of the albumen, exhibiting the embryo, its surface and lateral origin, according to Mr. Bauer:—magnified 100 diameters.

Fig. R. Br. is a longitudinal section of the albumen, exhibiting R. Brown's view of the origin, form, and surface of the embryo.

Tab. 25 (XXVI).

Fig. 1. A branch of the *Vitis*, on which are four very young buds of *Rafflesia Arnoldi*:—natural size. Of these,

a. (not separately figured) is merely a very slight swelling, caused by the nascent parasite, but before its parts are distinguishable.

A. (also separately figured, vertically divided and moderately magnified), the youngest parasite whose parts are distinguishable, deeply seated, entirely enclosed, and before its cortical covering corresponds with it in form.

B. (in like manner separately figured, divided and magnified), in which the parasite is entirely enclosed in its reticular covering.

C. In which the reticular covering has burst, vertically divided and magnified.

HYDNORA AFRICANA.

Tab. 26 (XXVII).

Fig. 1. A flower of *Hydnora africana*, with its very short base.

ig. 2. The same longitudinally divided:—both of the natural size.

Tab. 27 (XXVIII).

Fig. 1. Transverse section of a part of the tube of the perianthium, to show the three-lobed columna staminea:—moderately magnified.

Fig. 2. The inner surface of one of the three lobes of the column or antheral annulus.

Fig. 3. Outer surface of the same:—both magnified in the same degree with fig. 1.

Fig. 4. Vertical section of a portion of one of the lobes of the columna staminea, to show the thickness and texture of the common connective.

Fig. 5. Transverse section of the same, which shows the original bilocularity of each anthera:—both more highly magnified.

Fig. 6. Grains of pollen, still more highly magnified.

Fig. 7. Transverse section of the flower, to show the form and surface of [239
the stigma (of which the three primary divisions are opposite to the lobes of the columna staminea):—magnified in the same degree with fig. 1.

Fig. 8. A portion of the stigma, which shows its composition.

Fig. 9. A transverse section about the middle of the same:—both magnified somewhat more than figs. 2 and 3.

Fig. 10. A vertical section of the stigma, showing that the divisions of its surface extend quite through to the cavity of the ovarium, separating it into an equal number of lamellæ, from the inner terminations of which the placentæ are pendulous:—more highly magnified than the preceding figures.

Fig. 11. A small portion of the same, still more highly magnified.

Fig. 12. A transverse section, more highly magnified than fig. 11, with its densely crowded ovula arising from every part of its surface.

Fig. 13. Three ovula more highly magnified than fig. 12, to show the pedicellus or attenuated base and depressed or perforated apex.

Tab. 28 (XXIX).

A ripe fruit (fig. 1), with the stock (the root of the supposed *Euphorbia*, fig. 3) and the decumbent angular branched stems of the parasite, from the thickened trunk of which the ripe fruit originates at fig. 2, and from a branch of which a very young flower-bud proceeds:—natural size.

Tab. 29 (XXX).

Fig. 1. The same ripe fruit vertically divided, with the prostrate thickened stem of the parasite and the root of the supposed *Euphorbia*, whose woody fibres and vessels appear to penetrate deeply into the substance of the thickened stem:—natural size.

Fig. 2. A portion of the fruit transversely divided.

Fig. 3. A transverse section of one of the placentæ, with the ripe seeds:—slightly magnified.

Fig. 4. Two seeds, more highly magnified than fig. 3.

Fig. 5. A seed, magnified in the same degree as fig. 4, and vertically divided, which exhibits the albumen more distinctly radiating than I have ever found it.

Fig. 6. A seed deprived of its outer coat.

Fig. 7. The same transversely divided, which, as well as fig. 5, shows the central globular embryo.

240]
SUPPLEMENT.

To render the account of *Rafflesia Arnoldi* more complete, I shall add the distinguishing characters of the order, tribes, genera and species of *Rafflesiaceæ* with which I am acquainted. These characters, which form the chief part of the present supplement, as well as the notes to the original communication, have been written since November last.

The paper itself is printed as it was read in June 1834, a very few slight alterations, and those chiefly verbal, excepted.[1]

[1] The following brief abstract was published in the Philosophical Magazine for July, 1834:—

"LINNEAN SOCIETY.

"June 17.—A paper was read 'On the Female Flower and Fruit of *Rafflesia*, with Observations on its Affinities, and on the Structure of *Hydnora*.' By Robert Brown, Esq., V.P.L.S.

"The author's principal object in this paper is to complete his account of *Rafflesia Arnoldi*, the male flower of which he described in a former communication, published in the 13th volume of the Society's Transactions; and, in connection with the question of its place in a natural arrangement, he introduces a more detailed description and figures of *Hydnora africana*, than have hitherto been given. The drawings of *Rafflesia* which accompany the paper are by Francis Bauer, Esq., and those of *Hydnora* by the late Mr. Ferdinand Bauer.

"From a comparison of *Rafflesia* with *Hydnora* and *Cytinus*, he is confirmed in the opinion expressed in his former paper, but founded on less satisfactory evidence, that these three genera (to which *Brugmansia* of Blume is now to be added), notwithstanding several remarkable peculiarities in each, may all be referred to the same natural family; and this family, named by him *Rafflesiaceæ*, he continues to regard as being most nearly allied to *Asarinæ*.

"He does not, however, admit an arrangement lately proposed by M. Endlicher, and adopted by Mr. Lindley, by whom these genera are included in the same natural class with *Balanophoreæ* of Richard; an approximation founded on their agreement in the structure of embryo, and on the assumed absence of spiral vessels. On this subject he remarks, that in having a homogeneous or acotyledonous embryo, they essentially accord, not only with many other plants, parasitical on roots, which it has never been proposed to unite with them, as *Orobanche*, &c., but also with *Orchideæ*, their association with which would be still more paradoxical. And with respect to the supposed peculiarity in their vascular structure, he states that he has found spiral vessels not only in *Rafflesia* (in which he had formerly denied their existence), and in *Hydnora* and *Cytinus*, but likewise in all the *Balanophoreæ* examined by him, particularly *Cynomorium* and *Helosis*, as Dr. von Martius had long since done in *Langsdorfia*, and Professor Meyer very recently in *Hydnora*.

"In his observations on the ovulum of *Rafflesia*, he gives a view of its early

I have also to state, that an extensive and highly important essay, entitled, "An Attempt to analyse *Rhizantheæ*," by Mr. William Griffith, has been read during the present season before the Linnean Society, of which an abstract is given in the Proceedings. From this essay I have here introduced the character of *Sapria*, a new genus belonging to *Rafflesiaceæ*; and have ventured to propose an alteration of the trivial name from *Himalayana* to *Griffithii*, in honour of the discoverer of this interesting addition to the tribe *Rafflesieæ*, whose species, with one exception, have names similarly derived. [241

RAFFLESIACEÆ.

CHAR. DIFF. ORD. *Perianthium* monophyllum regulare.

Corolla nulla.

Stamina: Antheræ numerosæ, simplici serie.

Ovarium: placentis pluribus polyspermis, ovulis orthotropis (sed in quibusdam recurvatione apicis, penitus vel partim, liberi funiculi quasi anatropis).

Pericarpium indehiscens polyspermum.

Embryo indivisus (cum v. absque albumine).

Parasiticæ *radicibus rariusve in ramis plantarum dicotyledonearum.*

stages of development, and which he extends to Phænogamous plants generally, in some respects different from that taken by M. Mirbel, who considers the nucleus of the ovulum, in its earliest state, as inclosed in its coats, which gradually open until they have attained their maximum of expansion, when they again contract around the nucleus, and, at the same time, by elongating, completely inclose it. Mr. Brown, on the other hand, regards the earliest stage of the nucleus as merely a contraction taking place in the apex of a pre-existing papilla, whose surface, as well as substance, is originally uniform, and that its coats are of subsequent formation, each coat consisting, at first, merely of an annular thickening at the base of the nucleus, which, by gradual elongation, it entirely covers before impregnation takes place.

"But this mode of development of the ovulum, he remarks, though very general, is not without exception; for in many, perhaps in all, *Asclepiadeæ* and *Apocineæ*, the ovulum continues a uniform cellular tissue, exhibiting no distinction of parts until after the application of the pollen tube to a definite part of its surface, when an internal separation or included nucleus first becomes visible."—See a translation of this abstract in *Annal. des Sc. Nat* ser. 2de, tom. i, p. 369.

Obs. Huic ordini appendendæ *Apodanthes* et *Pilostyles*, quæ a Rafflesiaceis Corolla tetrapetala et Antheris 2-3-seriatis diversæ; necnon quod in caule aut ramis solùm nec unquam in radicibus parasiticæ: attamen pluribus notis *Cytineis* conveniunt.

242] RAFFLESIEÆ.

Char. Diff. Trib. *Perianthium* 5-10-fidum.

Antheræ sub apice dilatato columnæ simplici serie adnatæ, distinctæ, poro unico v. duplici dehiscentes.

Ovarium placentis confluentibus v. distinctis undique ovuliferis.

Pericarpium (semiadherens v. liberum) carnosum.

Semina recurvata funiculo apice dilatato.

Embryo albumine inclusus axilis, albumine brevior.

Parasiticæ *in radicibus specierum* Vitis *v.* Cissi. Flos *subsessilis, bracteis imbricatis venosis.*

Rafflesia, *R. Br.*

Char. Diff. Gen. *Perianthium* 5-fidum, æstivatione imbricata, corona faucis annulari indivisa.

Columna genitalium apice dilatato patelliformi: disco processibus (stylis?) numerosis styliformibus!; limbo elevato indiviso.

Antheræ multicellulosæ, poro unico dehiscentes.

Rafflesia (Patma) hermaphrodita, antheris viginti pluribus, stylis indefinitè numerosis confertis, perianthii tubo intus lævi (diametro floris sesquipedali-bipedali).

Rafflesia Patma, *Blume*, *Flor. Jav.* p. 8, tabb. 1-3.

Loc. Nat. Crescit in umbrosis Insulæ parvæ *Nusa Kambangan*, Javæ ab austro vicinæ. *Blume.*

R. (Arnoldi) dioica, antheris viginti pluribus, stylis indefinitè numerosis confertis, annulo baseos columnæ duplici, perianthii tubo intus ramentaceo (diametro floris bi-tripedali).

Rafflesia Arnoldi, *R. Br. in Linn. Soc. Trans.* vol. xiii, p. 207. (*Antè, p.* 374).

Loc. Nat. In sylvis umbrosis Sumatræ, anno 1818 detexit b. J. Arnold, M. D.

R. (Horsfieldii) dioica? stylis indefinitè numerosis: centralibus confertis (diametro floris semipedali).

Loc. Nat. Java, D. Horsfield, qui Alabastra solum detexit et depingi curavit.

Obs. Species dubia a sequente diversa numero et ordinatione stylorum (fid. ic. ined. D. Horsfield).

R. (Cumingii) dioica, antheris 10-12, stylis antheras [243
numero vix superantibus abbreviatis: exterioribus (sæpius 10) simplici serie; interioribus (1-3) invicem subæquidistantibus, annulo baseos columnæ unico, perianthii tubo intus ramentaceo (diametro floris semipedali).

Rafflesia Manillana, *Teschem. in Boston Journ. Nat. Hist.* vol. iv, p. 63, tab. 6, *mas.*

Loc. Nat. In Samar, Insula Philippinarum; ubi primum legit *D. Cuming*, v. s. et in sp. vin. asserv.

Desc. Planta dioica *R. Arnoldi* multoties minor, diametro floris expansi sexpollicari, cæterum ante expansionem *externè* similis ut ovum ovo, indusio e cortice *Vitis* aut *Cissi* formato rugoso sed vix reticulato; *internè* convenit corona faucis indivisa tubo intus ramentaceo: differt annulo baseos columnæ unico (exteriore *R. Arnoldi* deficiente), antheris maris paucioribus (10-12), stylis utriusque sexus vix antheras numero superantibus, haud confertis sed subsimplici serie circulari propiùs limbo quam centro dispositis, cum nonnullis (1-3) centralibus invicem distinctis et ferè æquidistantibus, omnibus abbreviatis crassitie dimidium longitudinis subæquante, apice pilis brevibus acutis rigidulis barbato: femina absque antherarum rudimentis: ovarii cavitatibus stylos manifestè superantibus et tam numerosis in centro ac versus peripheriam ut in *R. Arnoldi*.

Obs. I. The trivial name *Manillana*, given to this species by Mr. Teschemacher, who has described and figured the male flower, can hardly be retained for a plant not known to grow in Luzon, of which Manilla is the capital, but in the island of Samar, where it was first found by Mr. Hugh Cuming. I have named it, therefore, in honour of the discoverer,—a change which is not likely to be objected to, as Mr. Teschemacher (*loc. cit.*) expresses his readiness to adopt any name Mr. Cuming may wish it to retain.

Obs. II. In the general tissue of this species each cell has an extremely small, round, opake nucleus. In a transverse section of the column both of the male and female flower, the central part appears to be somewhat more solid; and each of the cells, of which it seems to be entirely formed, contains a large nucleus, easily separable, of a somewhat oval shape, and apparently consisting of a membrane including minute granular matter, which renders it opake. In the surrounding somewhat looser substance of the column, there seems to be an oval cell within each outer or mother cell, occupying the greater part of its cavity, with less granular matter, and having frequently a minute round nucleus. The parietes of the placentæ have in each simple
244] cell a small nucleus like that of the general tissue and of the outer portion of the column.

Sapria, *Griffith in Proceed. Linn. Soc.* p. 217.

Char. Diff. Gen. *Flores* dioici. *Perianthium* 10-fidum! duplici serie imbricatum, corona faucis indivisa.

Columna apice dilatato concavo e centro conum indivisum exserenti.

Antheræ sub apice dilatato columnæ, simplici serie adnatæ, 2-3-cellulosæ, poro unico dehiscentes.

Ovarium inferum, placentis indefinitè numerosis (parietalibus, ovulis anatropis, *Griffith*).

Sapria Griffithii.

Sapria Himalayana, *Griffith, loc. cit.*

Loc. Nat. In radicibus *Vitis* v. *Cissi* cujusdam in sylvis umbrosis Montium Mishmee jugi Himalayani, anno 1836 detexit *D. Griffith*.

BRUGMANSIA, *Blume, Flor. Jav.*

CHAR. DIFF. GEN. *Flores* hermaphroditi.

Perianthium 5-fidum, laciniis bi-trifidis, æstivatione valvata apicibus arctè inflexis.

Columna supernè dilatata apice excavato indiviso.

Antheræ sub apice dilatato simplici serie adnatæ, poro duplici dehiscentes!

Ovarium adherens, placentis indefinitè numerosis (parietalibus, *Blume*).

BRUGMANSIA ZIPPELII, *Blume, Flor. Javæ*, p. 15, tabb. 3-6.

Loc. Nat. In provinciâ Buitenzorg Javæ occidentalis, primus reperit Hortulanus Zippel. *Blume, loc. cit.*

HYDNOREÆ.

CHAR. TRIB. *Perianthium* trifidum, æstivatione valvata.

Stamina tubo perianthii inserta.

Antheræ numerosæ, longitudinaliter dehiscentes, connatæ in *annulum* trilobum cujus lobi perianthii laciniis oppositi.

Ovarium inferum: *Stigma* sessile depressum trilobum, lobis singulis formatis e lamellis pluribus appositis ad ovarii cavitatem usque distinctis ibique placentiferis! *Placentæ* ab apice ovarii (stigmatis lamellis) pendulæ, subcylindraceæ, undique ovulis numerosissimis orthotropis tectæ.

Pericarpium calvum, carnosum, cavitate placentis undique seminiferis pleno.

Semina: Embryo globosus in centro! albuminis cartilaginei.

HYDNORA, *Thunb.* [245

CHAR. GEN. idem ac tribus.

HYDNORA (AFRICANA) hermaphrodita, perianthii laciniis

latè induplicatis margine (angulo induplicationis) ciliatis apicibus demùm liberis, antheris bicruribus aversis (crure altero postico altero antico).

Hydnora africana, *Thunb. in Act. Holm.* vol. xxxvi, p. 69, tab. 2.

Loc. Nat. In Africâ australi primum detexit *Thunberg*.

H. (TRICEPS) hermaphrodita, perianthii laciniis supernè dilatatis connatisque infernè hiantibus margine nudis, antheris omninò posticis.

Hydnora triceps, *Meyer in Nov. Act. Acad. Nat. Curios.* vol. xvi, par. 2, p. 779.

Loc. Nat. In Africâ australi. *D. Drege.*

H. (AMERICANA) dioica, perianthii laciniis liberis nudis: marginibus induplicatis angustissimis, antheris posticis.

Loc. Nat. Exemplar unicum in Herb. D. Hooker in Americâ australi lectum vidi.

CYTINEÆ.

CHAR. DIFF. TRIB. *Flores* diclines. *Perianthium* 4-8-fidum, æstivatione imbricata.

MAS. *Antheræ* in apice columnæ simplici serie, definitæ, biloculares loculis parallelo-appositis longitudinaliter dehiscentibus.

FEM. *Ovarium* totum adherens uniloculare, placentis parietalibus definitis (8-16) distinctis, per paria approximatis, lobatis. *Stylus* 1. *Stigma* radiato-lobatum. *Embryo* exalbuminosus, indivisus, homogeneus.

CYTINUS, *Linn.*

CHAR. GEN. id. ac tribus.

CYTINUS (HYPOCISTIS) spica androgyna, perianthio quadrifido: laciniis extus tenuissimè pubescentibus.

Cytinus Hypocistis, *Linn.*

Thyrsine, *Gledit. verm. Abhand.* i, p. 199, tab. 2.
Loc. Nat. Europa australis et Africa borealis.

C. (DIOICUS) spicis dioicis paucifloris, floribus bibracteatis pedunculatis, perianthio sexfido: laciniis extus hispidulis. [246
Cytinus dioicus, *Juss. in Annal. du Mus.* xii, p. 443. *Hook. Ic.* vol. iv, tab. 336.
Phelypæa sanguinea, *Thunb. Nov. Gen.* pars 5ta, p. 93.
Aphyteia multiceps, *Burch. Trav.* vol. i, p. 213, fid. exempl. in herb. auctoris visi.
Loc. Nat. Africa australis.

C. (AMERICANUS) spicis dioicis multifloris, floribus sessilibus absque bracteis lateralibus, perianthio octofido patentissimo.
Loc. Nat. America æquinoctialis. *D. Barclay,* v. exemplaria mas. pl. in sp. vin. asserv.
OBS. Mascula planta solum visa. *Spicæ* densæ. *Perianthia* sessilia sæpius octo-quandoque novem-fida, laciniis patentissimis. *Columna* staminea teres. *Antheræ* 8-9, biloculares posticæ, loculis appositis longitudinaliter dehiscentibus, connectivis basi connatis extra medium distinctis singulisque in cuspidem subulatam productis. *Pollen* simplex. Nulla vestigia ovarii.

APODANTHEÆ.

CHAR. TRIB. *Flores* dioici. *Perianthium* 4-fidum v. 4-partitum, æstivatione imbricatum persistens.
Corolla 4-petala! decidua.
MAS. *Antheræ* infra apicem dilatatum indivisum columnæ bi-triseriatæ! sessiles uniloculares.
FEM. *Ovarium* adherens, uniloculare, ovulis orthotropis, per totam superficiem cavitatis sparsis. *Stigma* capitatum.
Pericarpium baccatum, inferum v. semisuperum.
Embryo exalbuminosus, indivisus, homogeneus.
Parasiticæ *in caulibus et ramis* (*nunquam in radicibus*) *plantarum dicotyledonearum.*

Apodanthes, *Poiteau.*

Char. Gen. *Perianthium* monophyllum 4-fidum, bibracteatum.

Petala ipso ovario (altius quam perianthium quod textura diversum) inserta.

Mas. . . . . . .

Fem. *Ovarium* semisuperum.

Pericarpium carnosum, cavitate tetragona.

Semina: testa nucamentacea lacunosa (funiculo nucleum æquante v. superante).

247] Apodanthes Caseariæ, *Poiteau in Annal. des Sc. Nat.* iii, p. 422, tab. 26, fig. 1.

Loc. Nat. Guiana gallica, in caulibus Caseariæ spec. macrophyllæ, *Vahl*, proximæ. *Poiteau*, v. in sp. vin. asserv. in Mus. Paris.

Pilostyles, *Guillemin.*

Frostia, *Bert. ined. et Endl. gen.* n. 725.

Char. Gen. *Perianthium* 4-partitum, superum.

Petala textura ferè perianthii et bractearum.

Ovarium cavitate infra insertionem bractearum superiorum producta.

Pericarpium cavitate absque angulis.

Pilostyles (Berterii) bracteis sepalisque ovato-oblongis margine nudis, stigmate papuloso apice depresso-umbilicato!

Pilostyles Berterii, *Guillem. in Annal. Sc. Nat.* ser. 2, vol. ii, p. 21, tab. 1.

Apodanthes Berterii, *Gardn. in Hook. Ic.* new ser. vol. iii, tab. 655, A.

Loc. Nat. Chili, *Bertero* et *Bridges*, in Adesmiis parasitica, v. s.

P. (Blanchetii) bracteis sepalisque subrotundis margin ciliatis! stigmate apice convexo.

Apodanthes Blanchetii, *Gardn. loc. cit.* B.

Loc. Nat. Brasilia, *Blanchet*, in Bauhiniæ sp. parasitica, v. s.

Obs. Mas hujusce v. maxime affinis speciei vidi in Museo Vindobonensi a. b. *Pohl* in Brasiliâ lecta, in ramis Bauhiniæ cujusdam parasitica.

P. (Calliandræ) bracteis sepalisque ovatis margine nudis, stigmate ovato-conico apice convexo.

Apodanthes Calliandræ, *Gardn. in Hook. Ic.* new ser. vol. iii, tab. 644.

Loc. Nat. In provinciâ Goyaz Brasiliæ, in caulibus Calliandræ sp. *D. Gardner*, v. s.

CHARACTER AND DESCRIPTION

OF

KINGIA;

A NEW GENUS OF PLANTS

FOUND ON THE

SOUTH-WEST COAST OF NEW HOLLAND.

WITH OBSERVATIONS

ON THE

STRUCTURE OF ITS UNIMPREGNATED OVULUM, AND ON THE FEMALE FLOWER OF CYCADEÆ AND CONIFERÆ.

BY

R. BROWN, Esq., F.R.S.L. & E., F.L.S.

[*Reprinted from the 'Narrative of a Survey of the Intertropical and Western Coasts of Australia performed between the years* 1818 *and* 1822. *By Captain Philip P. King, R.N., F.R.S., F.L.S.' Vol. II, pp.* 534—565.]

LONDON.

1827.

28

CHARACTER AND DESCRIPTION [534

OF

KINGIA, &c.

(Read before the Linnean Society of London, Nov. 1 and 15, 1825.)

IN the Botanical Appendix to the *Voyage to Terra Australis,* I have mentioned a plant of very remarkable appearance, observed in the year 1801, near the shores of King George the Third's Sound, in Mr. Westall's view of which, published in Captain Flinders' Narrative, it is introduced.

The plant in question was then found with only the imperfect remains of fructification: I judged of its affinities, therefore, merely from its habit, and as in this respect it entirely agrees with Xanthorrhœa, included the short notice given of it in my remarks on Asphodeleæ, to which that genus was referred.[1] Mr. Cunningham, the botanist attached to Captain King's voyages, who examined the plant in the same place of growth, in February, 1818, and in December, 1821, was not more fortunate than myself. Captain King, however, in his last visit to King George's Sound, in November, 1822, observed it with ripe seeds: and at length Mr. William Baxter, whose attention I had particularly directed to this plant, found it on the shores of the same port in 1823, both in flower and fruit. To this zealous collector, and to his liberal employer, Mr. Henchman, I am indebted for complete specimens of its fructifi- [535

[1] *Flinders' Voy.* vol. ii, *p.* 576. (*Antè, p.* 51.)

cation, which enable me to establish it as a genus distinct from any yet described.

To this new genus I have given the name of my friend Captain King, who, during his important surveys of the Coasts of New Holland, formed valuable collections in several departments of Natural History, and on all occasions gave every assistance in his power to Mr. Cunningham, the indefatigable botanist who accompanied him. The name is also intended as a mark of respect to the memory of the late Captain Philip Gidley King, who, as Governor of New South Wales, materially forwarded the objects of Captain Flinders' voyage; and to whose friendship Mr. Ferdinand Bauer and myself were indebted for important assistance in our pursuits while we remained in that colony.

KINGIA.

Ord. Nat. *Junceæ* prope Dasypogon, Calectasiam et Xerotem.

Char. Gen. *Perianthium* sexpartitum, regulare, glumaceum, persistens. *Stamina* sex, fere hypogyna: *Antheris* basi affixis. *Ovarium* triloculare, loculis monospermis; *ovulis* adscendentibus. *Stylus* 1. *Stigma* tridentatum. *Pericarpium* exsuccum, indehiscens, monospermum, perianthio scarioso cinctum.

Planta *facie* Xanthorrhœæ elatioris. Caudex *arborescens cicatricibus basibusve foliorum exasperatus?* Folia *caudicem terminantia confertissima longissima, figura et dispositione* Xanthorrhœæ. Pedunculi *numerosi foliis breviores, bracteis vaginantibus imbricatis tecti, floriferi terminales erecti, mox, caudice parum elongato foliisque novellis productis, laterales, et divaricati vel deflexi, terminati capitulo denso globoso floribus tribracteatis.*

Kingia *Australis*. Tab. C.[1]

Desc. *Caudex* arborescens erectus simplicissimus cylin-

[1] See note at p. 187.

draceus, 6—18-pedes altus, crassitie femoris. *Folia* caudicem terminantia numerosissima patula, apicibus ar- [536 cuato-recurvis, lorea, solida, ancipitia apice teretiusculo, novella undique tecta pilis adpressis strictis acutis lævibus, angulis lateralibus et ventrali retrorsum scabris. *Pedunculi* numerosi teretes 8—12-pollicares crassitie digiti, vaginis integris brevibus imbricatis hinc in foliolum subulatum productis tecti. *Capitulum* globosum, floridum magnitudine pruni minoris, fructiferum pomum parvum æquans. *Flores* undique densè imbricati, tribracteati, sessiles. *Bractea exterior* lanceolata brevè acuminata planiuscula erecta, extus villosa intus glabra, post lapsum fructus persistens: *duæ laterales* angusto-naviculares, acutissimæ, carina lateribusque villosis, longitudine fere exterioris, simul cum perianthio fructifero, separatim tamen, dilabentibus. *Perianthium* sexpartitum regulare subæquale glumaceum: *foliola* lanceolata acutissima disco nervoso nervis immersis simplicissimis, antica et postica plana, lateralia complicata lateribus inæqualibus, omnia basi subangustata, extus longitudinaliter sed extra medium præcipue villosa, intus glaberrima, æstivatione imbricata. *Stamina* sex subæqualia, æstivatione stricta filamentis sensim elongantibus: *Filamenta* fere hypogyna ipsis basibus foliolorum perianthii quibus opposita leviter adhærentia, filiformia glabra teretia: *Antheræ* stantes, ante dehiscentiam lineares obtusæ filamento paulo latiores, defloratæ subulatæ vix crassitie filamenti, loculis parallelo-contiguis connectivo dorsali angusto adnatis, axi ventrali longitudinaliter dehiscentibus, lobulis baseos brevibus acutis subadnatis: *Pollen* simplex brevè ovale læve. *Pistillum: Ovarium* sessile disco nullo squamulisve cinctum, lanceolatum trigonoanceps villosum, triloculare, loculis monospermis. *Ovula* erecta fundo anguli interioris loculi paulo supra basin suam inserta, obovata lenticulari-compressa, aptera: *Testa* in ipsa basi acutiusculâ foramine minuto perforata: *Membrana interna* respectu testæ inversa, hujusce nempe apici lata basi inserta, ovata apice angustato aperto foramen testæ obturante: *Nucleus* cavitati membranæ conformis, ejusdem basi insertus, cæterum liber, pulposus solidus, apice acu-

tiusculo lævi aperturam membranæ internæ attingente. *Stylus* trigonus strictus, infra villosus, dimidio superiore glabro, altitudine staminum, iisdem paulo præcocior, exsertus nempe dum illa adhuc inclusa. *Stigmata* tria brevis-
537] sima acuta denticuliformia. *Pericarpium* exsuccum, indehiscens, villosum, basi styli aristatum, perianthio scarioso et filamentis emarcidis cinctum, abortione monospermum. *Semen* turgidum obovatum retusum, integumento (testâ) simplici membranaceo aqueo-pallido, hinc (intus) fere a basi acutiuscula, *raphe* fusca verticem retusum attingente ibique in *chalazam* parvam concolorem ampliata. *Albumen* semini conforme densè carnosum album. *Embryo* monocotyledoneus, aqueo-pallidus subglobosus, extremitate inferiore (radiculari) acuta, in ipsa basi seminis situs, semi-immersus, nec albumine omnino inclusus.

Tab. C. fig. 1. KINGIÆ AUSTRALIS pedunculus capitulo florido terminatus; fig. 2, capitulum fructiferum; 3, sectio transversalis pedunculi; 4, folium: hæ magnitudine naturali, sequentes omnes plus minus auctæ sunt; 5, flos; 6, stamen: 7, anthera antice et; 8, eadem postice visa; 9, pistillum; 10, ovarii sectio transversalis; 11, ejusdem portio longitudinaliter secta exhibens ovulum adscendens cavitatem loculi replens; 12, ovulum ita longitudinaliter sectum ut membrana interna solummodo ejusque insertio in apice cavitatis testæ visa sit; 13, ovuli sectio longitudinalis profundius ducta exhibens membranam internam et nucleum ex ejusdem basi ortum; 14, bracteæ capituli fructiferi; 15, pericarpium perianthio filamentisque persistentibus cinctum; 16, pericarpium perianthio avulso filamentorum basibus relictis; 17, semen.

OBS. I.—It remains to be ascertained, whether in this genus a resin is secreted by the bases of the lower leaves, as in Xanthorrhœa; and whether, which is probable, it agrees also, in the internal structure of its stem with that genus. In Xanthorrhœa the direction of fibres or vessels of the caudex seems at first sight to resemble in some degree the dicotyledonous arrangement, but in reality much more nearly approaches to that of Dracæna Draco, allowance being made for the greater number, and extreme narrowness of leaves, to which all the radiating [538 vessels belong.[1]

Obs. II.—I have placed Kingia in the natural order Junceæ along with Dasypogon, Calectasia and Xerotes, genera peculiar to New Holland, and of which the two former have hitherto been observed only, along with it, on the shores of King George's Sound.

The striking resemblance of Kingia, in caudex and leaves, to Xanthorrhœa, cannot fail to suggest its affinity to that genus also. Although this affinity is not confirmed by a minute comparison of the parts of fructification, a sufficient agreement is still manifest to strengthen the doubts formerly expressed of the importance of those characters by which I attempted to define certain families, of the great class Liliaceæ.

In addition, however, to the difference in texture of the outer coat of the seed, and in those other points, on which I then chiefly depended in distinguishing Junceæ from Asphodeleæ, a more important character in Junceæ exists in the position of the embryo, whose radicle points always to the base of the seed, the external umbilicus being placed in the axis of the inner or ventral surface, either immediately above the base as in Kingia, or towards the middle, as in Xerotes.

Obs. III.—*On the structure of the* Unimpregnated [539 Ovulum *in Phænogamous Plants.*

The description which I have given of the Ovulum of Kingia, though essentially different from the accounts hitherto published of that organ before fecundation, in

[1] My knowledge of this remarkable structure of Xanthorrhœa is chiefly derived from specimens of the caudex of one of the larger species of the genus, brought from Port Jackson, and deposited in the collection at the Jardin du Roi of Paris by M. Gaudichaud, the very intelligent botanist who was attached to Captain De Freycinet's voyage.

reality agrees with its ordinary structure in Phænogamous plants.

I shall endeavour to establish these two points; namely, the agreement of this description with the usual structure of the Ovulum, and its essential difference from the accounts of other observers, as briefly as possible at present; intending hereafter to treat the subject at greater length, and also with other views.

I have formerly more than once[1] adverted to the structure of the Ovulum, chiefly as to the indications it affords, even before fecundation, of the place and direction of the future Embryo. These remarks, however, which were certainly very brief, seem entirely to have escaped the notice of those authors who have since written on the same subject.

In the Botanical Appendix to the Account of Captain Flinders' Voyage, published in 1814, the following decription of the Ovulum of *Cephalotus follicularis* is given "Ovulum erectum, intra testam membranaceam continens sacculum pendulum, magnitudine cavitatis testæ," and in reference to this description, I have in the same place remarked that, from "the structure of the Ovulum, even in the unimpregnated state, I entertain no doubt that the radicle of the Embryo points to the umbilicus."[2]

My attention had been first directed to this subject in 1809, in consequence of the opinion I had then formed
540] of the function of the Chalaza in seeds;[3] and some time before the publication of the observation now quoted, I had ascertained that in Phænogamous plants the unimpregnated Ovulum very generally consisted of two concentric membranes or coats, enclosing a Nucleus of a pulpy cellular texture. I had observed also that the inner coat had no connection either with the outer or with the nucleus, except at its origin; and that with relation to the outer coat it was generally inverted, while it always agreed in direction with the nucleus. And,

[1] *Flinders' Voy.* ii, p. 601 (*Antè, p.* 77), and *Linn. Soc. Transac.* xii. p. 136.
[2] *Flinders' Voy.* loc. cit.
[3] *Linn. Soc. Transac.* x, p. 35.

lastly, that at the apex of the nucleus the radicle of the future Embryo would constantly be found.

On these grounds my opinion respecting the Embryo of Cephalotus was formed. In describing the ovulum in this genus, I employed, indeed, the less correct term, "sacculus," which, however, sufficiently expressed the appearance of the included body in the specimens examined, and served to denote my uncertainty in this case as to the presence of the inner membrane.

I was at that time also aware of the existence, in several plants, of a foramen in the coats of the Ovulum, always distinct from, and in some cases diametrically opposite to, the external umbilicus, and which I had in no instance found cohering either directly with the parietes of the Ovarium, or with any process derived from them. But, as I was then unable to detect this foramen in many of the plants which I had examined, I did not attach sufficient importance to it; and in judging of the direction of the Embryo, entirely depended on ascertaining the apex of the nucleus, either directly by dissection, or indirectly from the vascular cord of the outer membrane; the termination of this cord affording a sure indication of the origin of the inner membrane, and consequently of the base of the nucleus, the position of whose apex is therefore readily de- [541
termined.

In this state of my knowledge the subject was taken up in 1818, by my lamented friend the late Mr. Thomas Smith, who, eminently qualified for an investigation where minute accuracy and great experience in microscopical observation were necessary, succeeded in ascertaining the very general existence of the foramen in the membranes of the Ovulum. But as the foramina in these membranes invariably correspond both with each other and with the apex of the nucleus, a test of the direction of the future Embryo was consequently found nearly as universal, and more obvious than that which I had previously employed.

To determine in what degree this account of the vegetable Ovulum differs from those hitherto given, and in some measure, that its corectness may be judged of, I shall proceed to state the various observations that have been actually made, and the opinions that have been formed on the subject as briefly as I am able, taking them in chronological order.

In 1672, Grew[1] describes in the outer coat of the seeds of many Leguminous plants a small foramen, placed opposite to the radicle of the Embryo, which, he adds, is "not a hole casually made, or by the breaking off of the stalk," but formed for purposes afterwards stated to be the aeration of the Embryo, and facilitating the passage of its radicle in germination. It appears that he did not consider this foramen in the testa as always present, the functions which he ascribes to it being performed in cases where it is not found, either, according to him, by the hilum itself, or in hard fruits, by an aperture in the stone or shell.

542] In another part of his work[2] he describes and figures, in the early state of the Ovulum, two coats, of which the outer is the testa; the other, his "middle membrane," is evidently what I have termed nucleus, whose origin in the Ovulum of the Apricot he has distinctly represented and described.

Malpighi, in 1675,[3] gives the same account of the early state of the Ovulum; his "secundinæ externæ" being the testa, and his chorion the nucleus. He has not, however, distinguished, though he appears to have seen, the foramen of Grew, from the fenestra and fenestella, and these, to which he assigns the same functions, are merely his terms for the hilum.

In 1694, Camerarius, in his admirable essay on the sexes of plants,[4] proposes, as queries merely, various modes in which either the entire grains of pollen, or their particles after bursting may be supposed to reach and act upon the unimpregnated Ovula, which he had himself carefully observed. With his usual candour, however, he acknowledges his obligation on this subject to Malpighi, to whose more detailed account of them he refers.

[1] *Anatomy of Veget. begun* p. 3. *Anat. of Plants*, p. 2.
[2] *Anat. of Plants*, p. 210, tab. 80. [3] *Anatome Plant.* p. 75, et 80.
[4] *Rudolphi Jacobi Camerarii de sexu plantarum epistola*, p. 8. 46, et seq.

Mr. Samuel Morland, in 1703,[1] in extending Leeuwenhoek's hypothesis of generation to plants, assumes the existence of an aperture in the Ovulum, through which it is impregnated. It appears, indeed, that he had not actually observed this aperture before fecundation, but inferred its existence generally and at that period, from having, as he says, "discovered in the seeds of beans, peas, and Phaseoli, just under one end of what we call the eye, a manifest perforation, which leads directly to the seminal [543
plant," and by which he supposes the embryo to have entered. This perforation is evidently the foramen discovered in the seeds of Leguminous plants by Grew, of whose observations respecting it he takes no notice, though he quotes him in another part of his subject.

In 1704, Etienne François Geoffroy,[2] and in 1711, his brother Claude Joseph Geoffroy,[3] in support of the same hypothesis, state the general existence of an aperture in the unimpregnated vegetable Ovulum. It is not, however, probable that these authors had really seen this aperture in the early state of the Ovulum in any case, but rather that they had merely advanced from the observation of Grew, and the conjecture founded on it by Morland whose hypothesis they adopt without acknowledgment, to the unqualified assertion of its existence, in all cases. For it is to be remarked, that they take no notice of what had previously been observed or asserted on the more important parts of their subject, while several passages are evidently copied, and the whole account of the original state and development of the Ovulum is literally translated from Camerarius's Essay. Nor does the younger Geoffroy mention the earlier publication of his brother, from which his own memoir is in great part manifestly derived.

In 1718, Vaillant,[4] who rejects the vermicular hypothesis of generation, supposes the influence of the Pollen to consist in an aura, conveyed by the tracheæ of the style to the

[1] *Philosoph. Transact.* vol. xxiii, n. 287, p. 1474.

[2] *Quæstio Medica an Hominis primordia Vermis?* in auctoris *Tractatu de Materia Medica*, tom. i. p. 123.

[3] *Mem. de l'Acad. des Sc. de Paris*, 1711, p. 210.

[4] *Discours sur la Structure des Fleurs*, p. 20.

ovula, which it enters, if I rightly understand him, by the
544] funiculus umbilicalis: at the same time he seems to admit the existence of the aperture in the coat.

In 1745, Needham,[1] and in 1770, Gleichen,[2] adopt the hypothesis of Morland, somewhat modified, however, as they consider the particles in the grains of Pollen, not the grains themselves, to be the embryos, and that they enter the ovula by the umbilical cord.

Adanson, in 1763,[3] states the Embryo to exist before fecundation, and that it receives its first excitement from a vapour or aura proceeding from the Pollen, conveyed to it through the tracheæ of the style, and entering the Ovulum by the umbilical cord.

Spallanzani,[1] who appears to have carefully examined the unimpregnated Ovula of a considerable variety of plants, found it in general to be a homogeneous, spongy, or gelatinous body; but in two Cucurbitaceæ to consist of a nucleus surrounded by three coats. Of these coats he rightly supposes the outermost to be merely the epidermis of the middle membrane or testa. Of the relative direction of the testa and inner coat in the two plants in question he takes no notice, nor does he in any case mention an aperture in the Ovulum.

Gærtner, who, in the preface to his celebrated work, displays great erudition in every branch of his subject, can hardly, however, be considered an original observer in this part. He describes the unimpregnated Ovulum as a pulpy homogeneous globule, whose epidermis, then scarcely distinguishable, separates in a more advanced stage, and becomes the testa of the seed, the inner membrane of which is
545] entirely the product of fecundation.[5] He asserts also that the Embryo constantly appears at that point of the ovulum where the ultimate branches of the umbilical vessels perforate the inner membrane; and therefore mistakes the apex for the base of the nucleus.

[1] *New Microscopical Discoveries*, p. 60.

[2] *Observ. Microscop.* p. 45, *et* 61, § cxviii.

[3] *Fam. des Plant.* tom. i, p. 121.

[4] *Fisica Anim. e Veget.* tom. iii, p. 309—332.

[5] *Gært. de Fruct. et Sem.* i, p. 57, 59, *et* 61.

In 1806 Mons. Turpin[1] published a memoir on the organ by which the fecundating fluid is introduced into the vegetable ovulum. The substance of this memoir is, that in all Phænogamous plants fecundation takes place through a cord or fasciculus of vessels entering the outer coat of the ovulum, at a point distinct from, but, at the period of impregnation, closely approximated to the umbilicus; and to the cicatrix of this cord, which itself is soon obliterated, he gives the name of Micropyle: that the ovulum has two coats each having its proper umbilicus, or, as he terms it, omphalode; that these coats in general correspond in direction; that more rarely the inner membrane is, with relation to the outer, inverted; and that towards the origin of the inner membrane the radicle of the embryo uniformly points.

It is singular that a botanist, so ingenious and experienced as M. Turpin, should, on this subject, instead of appealing in every case to the unimpregnated ovulum, have apparently contented himself with an examination of the ripe seed. Hence, however, he has formed an erroneous opinion of the nature and origin, and in some plants of the situation, of the micropyle itself, and hence also he has in all cases mistaken the apex for the base of the nucleus.

A minute examination of the early state of the ovulum does not seem to have entered into the plan of the late celebrated M. Richard, when in 1808 he published his valuable and original *Analyse du Fruit*. The ovulum has, [546
according to him, but one covering, which in the ripe seed he calls episperm. He considers the centre of the hilum as the base, and the chalaza, where it exists, as the natural apex of the seed.

M. Mirbel, in 1815, though admitting the existence of the foramen or micropyle of the testa,[2] describes the ovulum as receiving by the hilum both nourishing and fecundating vessels,[3] and as consisting of a uniform parenchyma, in which the embryo appears at first a minute point, gradually converting more or less of the surrounding tissue into its

[1] *Annal. du Mus. d'Hist. Nat.* vii, p. 199.
[2] *Elém. de Physiol. Vég. et de Bot.* tom. i, p. 49.
[3] *Id.* tom. i, p. 314.

own substance; the coats and albumen of the seed being formed of that portion which remains.[1]

In the same year, M. Auguste de Saint Hiliare[2] shows that the micropyle is not always approximated to the umbilicus; that in some plants it is situated at the opposite extremity of the ovulum, and that in all cases it corresponds with the radicle of the embryo. This excellent botanist, at the same time, adopts M. Turpin's opinion, that the micropyle is the cicatrix of a vascular cord, and even gives instances of its connection with the parietes of the ovarium; mistaking, as I believe, contact, which in some plants unquestionably takes place, and in one family, namely, Plumbagineæ, in a very remarkable manner, but only after a certain peried, for original cohesion, or organic connection, which I have not met with in any case.

In 1815 also appeared the masterly dissertation of Professor Ludolf Christian Treviranus, on the development of 547] the vegetable Embryo,[3] in which he describes the ovulum before fecundation as having two coats; but of these, his inner coat is evidently the middle membrane of Grew, the chorion of Malpighi, or what I have termed nucleus.

In 1822, Mons. Dutrochet, unacquainted, as it would seem, with the dissertation of Professor Treviranus, published his observations on the same subject.[4] In what regards the structure of the ovulum, he essentially agrees with that author, and has equally overlooked the inner membrane.

It is remarkable that neither of these observers should have noticed the foramen in the testa. And as they do not even mention the well-known essays of MM. Turpin and Auguste de St. Hilaire on the micropyle, it may be presumed that they were not disposed to adopt the statements of these authors respecting it.

Professor Link, in his *Philosophia Botanica*, published in 1824, adopts the account given by Treviranus, of the

[1] *Id.* loc. cit.
[2] *Mém. du Mus. d'Hist. Nat.* ii, p. 270, *et seq.*
[3] *Entwick. des Embryo im Pflanzen-Ey.*
[4] *Mém. du Mus. d'Hist. Nat.* tom. viii, p. 241, *et seq.*

coats of the ovulum before impregnation;[1] and of M. Turpin, as to the situation of the micropyle, and its being the cicatrix of a vascular cord. Yet he seems not to admit the function ascribed to it, and asserts that it is in many cases wanting.[2]

The account which I have given of the structure of the vegetable ovulum differs essentially from all those now quoted, and I am not acquainted with any other observations of importance respecting it.

Of the authors referred to, it may be remarked, that those who have most particularly attended to the ovulum externally, have not always examined it at a sufficiently [548 early period, and have confined themselves to its surface: that those who have most minutely examined its internal structure, have trusted too much to sections merely, and have neglected its appearance externally: and that those who have not at all examined it in the early stage have given the most correct account of its surface. This account was founded on a very limited observation of ripe seeds, generalized and extended to the unimpregnated ovulum, in connexion with an hypothesis then very commonly received: but this hypothesis being soon after abandoned, their statement respecting the ovulum was rejected along with it.

In the ovulum of Kingia, the inner membrane, with relation to the external umbilicus, is inverted; and this, as I have already observed, though in direct opposition to M. Turpin's account, is the usual structure of the organ. There are, however, several families in each of the two primary divisions of phænogamous plants, in which the inner membrane, and consequently the nucleus, agrees in direction with the testa. In such cases the external umbilicus alone affords a certain indication of the position of the future embryo.

It is an obvious consequence of what has been already stated, that the radicle of the embryo can never point directly to the external umbilicus or hilum, though this is

[1] *Elém. Philos. Bot.* p. 338. [2] *Id.* p. 340.

said to be generally the case by the most celebrated carpologists.

Another observation may be made, less obviously a consequence of the structure described, but equally at variance with many of the published accounts and figures of seeds, namely, that the radicle is never absolutely enclosed in the albumen; but, in the recent state, is either immediately in contact with the inner membrane of the seed, or this contact is established by means of a process generally very
549] short, but sometimes of great length, and which indeed in all cases may be regarded as an elongation of its own substance. From this rule I have found one apparent deviation, but in a case altogether so peculiar, that it can hardly be considered as setting it aside.

It is necessary to observe, that I am acquainted with exceptions to the structure of the ovulum as I have here described it. In Compositæ its coats seem to be imperforated, and hardly separable, either from each other or from the nucleus. In this family, therefore, the direction of the embryo can only be judged of from the vessels of the testa.[1] And in Lemna I have found an apparent inversion of the embryo with relation to the apex of the nucleus. In this genus, however, such other peculiarities of structure and economy exist, that, paradoxical as the assertion may seem, I consider the exception rather as confirming than lessening the importance of the character.

It may perhaps be unnecessary to remark, that the raphe, or vascular cord of the outer coat, almost universally belongs to that side of the ovulum which is next the placenta. But it is at least deserving of notice, that the very few apparent exceptions to this rule evidently tend to confirm it. The most remarkable of these exceptions occur in those species of Euonymus, which, contrary to the usual structure of the genus and family they belong to, have pendulous ovula; and, as I have long since noticed, in the perfect ovula only of Abelia.[2] In these, and in the other cases in which the raphe is on the outer side, or that most remote

[1] *Linn. Soc. Transact.* xii, p. 136. [2] Abel's *China*, p. 377.

from the placenta, the ovula are in reality resupinate; an economy apparently essential to their development.

The distinct origins and different directions of the
nourishing vessels and channel through which fecunda- [550
tion took place in the ovulum, may still be seen in many of
those ripe seeds that are winged, and either present their
margins to the placenta, as in Proteaceæ, or have the plane
of the wing at right angles to it, as in several Liliaceæ.
These organs are visible also in some of those seeds that
have their testa produced at both ends beyond the inner
membrane, as Nepenthes; a structure which proves the
outer coat of scobiform seeds, as they are called, to be really
testa, and not arillus, as it has often been termed.

The importance of distinguishing between the membranes of the unimpregnated ovulum and those of the ripe seed, must be sufficiently evident from what has been already stated. But this distinction has been necessarily neglected by two classes of observers. The first consisting of those, among whom are several of the most eminent carpologists, who have regarded the coats of the seed as products of fecundation. The second, of those authors who, professing to give an account of the ovulum itself, have made their observations chiefly, or entirely, on the ripe seed, the coats of which they must consequently have supposed to be formed before impregnation.

The consideration of the *arillus*, which is of rare occurrence, is never complete, and whose development takes place chiefly after fecundation, might here, perhaps, be entirely omitted. It is, however, worthy of remark, that in the early stage of the ovulum, this envelope is in general hardly visible even in those cases where, as in Hibbertia volubilis, it attains the greatest size in the ripe seed; nor does it in any case, with which I am acquainted, cover the foramen of the testa until after fecundation.

The *testa*, or outer coat of the seed, is very generally
formed by the outer membrane of the ovulum; and in most
cases where the nucleus is inverted, which is the more [551
usual structure, its origin may be satisfactorily determined;
either by the hilum being more or less lateral, while the

foramen is terminal; or more obviously, and with greater certainty, where the *raphe* is visible, this vascular cord uniformly belonging to the outer membrane of the ovulum. The *chalaza*, properly so called, though merely the termination of the raphe, affords a less certain character, for in many plants it is hardly visible on the inner surface of the testa, but is intimately united with the areola of insertion of the inner membrane or of the nucleus, to one or other of which it then seems entirely to belong. In those cases where the testa agrees in direction with the nucleus, I am not acquainted with any character by which it can be absolutely distinguished from the inner membrane in the ripe seed; but as a few plants are already known, in which the outer membrane is originally incomplete, its entire absence, even before fecundation, is conceivable; and some possible cases of such a structure will be mentioned hereafter.

There are several cases known, some of which I have formerly noticed,[1] of the complete obliteration of the testa in the ripe seed; and on the other hand it appears to constitute the greater part of the substance of the bulb-like seeds of many Liliaceæ, where it no doubt performs also the function of albumen, from which, however, it is readily distinguished by its vascularity.[2] But the most remarkable deviation from the usual structure and economy of the outer membrane of the ovulum, both in its earliest stage and in the ripe fruit, that I have yet met with, occurs in Banksia and Dryandra. In these two genera I have ascertained that the inner membrane of the ovulum, before fecundation,
552] is entirely exposed, the outer membrane being even then open its whole length; and that the outer membranes of the two collateral ovula, which are originally distinct, cohere in a more advanced stage by their corresponding surfaces, and together constitute the anomalous dissepiment of the capsule; the inner membrane of the ovulum consequently forming the outer coat of the seed.

The *inner membrane* of the ovulum, however, in general

[1] *Linn. Soc. Transact.* xii, p. 149. (*Antè*, *p.* 364.) [2] Ibid.

appears to be of greater importance as connected with fecundation, than as affording protection to the nucleus at a more advanced period. For in many cases, before impregnation, its perforated apex projects beyond the aperture of the testa, and in some plants puts on the appearance of an obtuse, or even dilated stigma; while in the ripe seed it is often either entirely obliterated, or exists only as a thin film, which might readily be mistaken for the epidermis of a third membrane then frequently observable.

This *third coat* is formed by the proper membrane or cuticle of the Nucleus, from whose substance in the unimpregnated ovulum it is never, I believe, separable, and at that period is very rarely visible. In the ripe seed it is distinguishable from the inner membrane only by its apex, which is never perforated, is generally acute and more deeply coloured, or even sphacelated.

The membrane of the nucleus usually constitutes the innermost coat of the seed. But in a few plants an additional coat, apparently originating in the inner membrane of Grew, the vesicula colliquamenti or amnios of Malpighi, also exists.

In general the Amnios, after fecundation, gradually enlarges, till at length it displaces or absorbs the whole substance of the nucleus, containing in the ripe seed both the embryo and albumen, where the latter continues to exist. In such cases, however, its proper membrane is commonly [553
obliterated, and its place supplied either by that of the nucleus, by the inner membrane of the ovulum, or, where both these are evanescent, by the testa itself.

In other cases the albumen is formed by a deposition of granular matter in the cells of the nucleus. In some of these cases the membrane of the amnios seems to be persistent, forming even in the ripe seed a proper coat for the embryo, the original attachment of whose radicle to the apex of this coat may also continue. This, at least, seems to me the most probable explanation of the structure of true Nymphæaceæ, namely, Nuphar, Nymphæa, Euryale, Hydropeltis, and Cabomba, notwithstanding their very remarkable germination, as observed and figured in Nymphæa and Nuphar by Tittmann.[1]

In support of this explanation, which differs from all those yet given, I may here advert to an observation published many years ago, though it seems to have escaped every author who has since written on the subject, namely, that before the maturity of the seed in Nymphæaceæ, the sacculus contains along with the embryo a (pulpy or semi-fluid) substance, which I then called Vitellus, applying at that time this name to every body interposed between the albumen and embryo.[2] The opinion receives some confirmation also from the existence of an extremely fine filament, hitherto overlooked, which, originating from the centre of the lower surface of the sacculus, and passing through the hollow axis of the Albumen, probably connects this coat of the Embryo in an early stage with the base of the nucleus.

554] The same explanation of structure applies to the seeds of Piperaceæ and Saururus; and other instances occur of the persistence either of the membrane or of the substance of the amnios in the ripe seed.

It may be concluded from the whole account which I have given of the structure of the ovulum, that the more important changes consequent to real, or even to spurious fecundation, must take place within the nucleus; and that the albumen, properly so called, may be formed either by a deposition or secretion of granular matter in the utriculi of the amnios, or in those of the nucleus itself, or lastly, that two substances having these distinct origins, and very different textures, may coexist in the ripe seed, as is pr̂obably the case in Scitamineæ.

On the subject of the ovulum, as contained in an ovarium, I shall at present make but one other remark, which forms a necessary introduction to the observations that follow.

[1] *Keimung der Pflanzen*, p. 19 et 27, tab. 3 et 4.
[2] *Prodr. Flor. Nov. Holl.* i, p. 306.

On the Structure of the Female Flower in CYCADEÆ *and* CONIFERÆ.

That the apex of the nucleus is the point of the ovulum where impregnation takes place, is at least highly probable, both from the constancy in the appearance of the embryo at that point, and from the very general inversion of the nucleus; for by this inversion its apex is brought nearly, or absolutely, into contact with that part of the parietes of the ovarium, by which the influence of the pollen may be supposed to be communicated. In several of those families of plants, however, in which the nucleus is not inverted, and the placentæ are polyspermous, as Cistineæ,[1] it is difficult to comprehend in what manner this influence can [555
reach its apex externally, except on the supposition, not hastily to be admitted, of an impregnating aura filling the cavity of the ovarium; or by the complete separation of the fecundating tubes from the placentæ, which, however, in such cases I have never been able to detect.

It would entirely remove the doubts that may exist respecting the point of impregnation, if cases could be produced where the ovarium was either altogether wanting, or so imperfectly formed, that the ovulum itself became directly exposed to the action of the pollen, or its fovilla; its apex, as well as the orifice of its immediate covering, being modified and developed to adapt them to this economy.

But such, I believe, is the real explanation of the structure of Cycadeæ, of Coniferæ, of Ephedra, and even of Gnetum, of which Thoa of Aublet is a species.

To this view the most formidable objection would be removed, were it admitted, in conformity with the preceding observations, that the apex of the nucleus, or supposed

[1] This structure of ovulum, indicated by that of the seed, as characterising and defining the limits of Cistineæ (namely, Cistus, Helianthemum, Hudsonia, and Lechea), I communicated to Dr. Hooker, by whom it is noticed in his *Flora Scotica* (p. 284), published in 1821; where, however, an observation is added respecting Gærtner's description of Cistus and Helianthemum, for which I am not accountable.

point of impregnation, has no organic connection with the parietes of the ovarium, In support of it, also, as far as regards the direct action of the pollen on the ovulum, numerous instances of analogous economy in the animal kingdom may be adduced.

The similarity of the female flower in Cycadeæ and Coniferæ to the ovulum of other phænogamous plants, as I have described it, is indeed sufficiently obvious to render the opinion here advanced not altogether improbable. But the
556] proof of its correctness must chiefly rest on a resemblance, in every essential point, being established, between the inner body in the supposed female flower in these tribes, and the nucleus of the ovulum in ordinary structures; not only in the early stage, but also in the whole series of changes consequent to fecundation. Now, as far as I have yet examined, there is nearly a complete agreement in all these respects. I am not entirely satisfied, however, with the observations I have hitherto been able to make on a subject naturally difficult, and to which I have not till lately attended with my present view.

The facts most likely to be produced as arguments against this view of the structure of Coniferæ, are the unequal and apparently secreting surface of the apex of the supposed nucleus in most cases; its occasional projection beyond the orifice of the outer coat; its cohesion with that coat by a considerable portion of its surface, and the not unfrequent division of the orifice of the coat. Yet most of these peculiarities of structure might perhaps be adduced in support of the opinion advanced, being apparent adaptations to the supposed economy.

There is one fact that will hardly be brought forward as an objection, and which yet seems to me to present a difficulty, to this opinion; namely, the greater simplicity in Cycadeæ, and in the principal part of Coniferæ, of the supposed ovulum which consists of a nucleus and one coat only, compared with the organ as generally existing when enclosed in an ovarium. The want of uniformity in this respect may even be stated as another difficulty, for in some genera of Coniferæ the ovulum appears to be complete.

In Ephedra, indeed, where the nucleus is provided with two envelopes, the outer may, perhaps, be supposed rather analogous to the calyx, or involucrum of the male flower, than as belonging to the ovulum; but in Gnetum, [557
where three envelopes exist, two of these may, with great probability, be regarded as coats of the nucleus; while in Podocarpus and Dacrydium, the outer cupula, as I formerly termed it,[1] may also, perhaps, be viewed as the testa of the ovulum. To this view, as far as relates to Dacrydium, the longitudinal fissure of the outer coat in the early stage, and its state in the ripe fruit, in which it forms only a partial covering, may be objected.[2] But these objections are, in a great measure, removed by the analogous structure already described in Banksia and Dryandra.

The plurality of embryos sometimes occurring in Coniferæ, and which, in Cycadeæ, seems even to be the natural structure, may also, perhaps, be supposed to form an objection to the present opinion, though to me it appears rather an argument in its favour.

Upon the whole, the objections to which the view here taken of the structure of these two families is still liable, seem to me, as far as I am aware of them, much less important than those that may be brought against the other opinions that have been advanced, and still divide botanists on this subject.

According to the earliest of these opinions, the female flower of Cycadeæ and Coniferæ is a monospermous pistillum, having no proper floral envelope.

To this structure, however, Pinus itself was long considered by many botanists as presenting an exception.

Linnæus has expressed himself so obscurely in the natural character which he has given of this genus, that I find it difficult to determine what his opinion of its structure really was. I am inclined, however, to believe it to have been [558
much nearer the truth than is generally supposed; judging of it from a comparison of his essential with his artificial

[1] *Flinders' Voy.* vol. ii, p. 573 (*antè, p.* 47). [2] *Id.* loc. cit.

generic character, and from an observation recorded in his *Prælectiones*, published by Giseke.[1]

But the first clear account that I have met with, of the real structure of Pinus, as far as regards the direction, or base and apex of the female flowers, is given, in 1767, by Trew, who describes them in the following manner:—" Singula semina vel potius germina stigmati tanquam organo feminino gaudent,"[2] and his figure of the female flower of the Larch, in which the stigmata project beyond the base of the scale, removes all doubt respecting his meaning.

In 1789, M. de Jussieu, in the character of his genus Abies,[3] gives a similar account of structure, though somewhat less clearly as well as less decidedly expressed. In the observations that follow, he suggests, as not improbable, a very different view, founded on the supposed analogy with Araucaria, whose structure was then misunderstood; namely, that the inner scale of the female amentum is a bilocular ovarium, of which the outer scale is the style. But this, according to Sir James Smith,[4] was also Linnæus's opinion; and it is the view adopted in Mr. Lambert's splendid monograph of the genus published in 1803.

In the same year in which Mr. Lambert's work appeared, Schkuhr[5] describes, and very distinctly figures, the female flower of Pinus, exactly as it was understood by Trew, whose opinion was probably unknown to him.

559] In 1807 a memoir on this subject, by Mr. Salisbury, was published,[6] in which an account of structure is given, in no important particular different from that of Trew and Schkuhr, with whose observations he appears to have been unacquainted.

M. Mirbel, in 1809,[7] held the same opinion, both with respect to Pinus and to the whole natural family. But in 1812, in conjunction with M. Schoubert,[8] he proposed a

1 *Prælect. in Ord. Nat.* p. 589.
2 *Nov. Act. Acad. Nat. Curios.* iii, p. 453, tab. 13, fig. 23.
3 *Gen. Pl.* p. 414.
4 Rees's *Cyclop. art. Pinus.*
5 *Botan. Handb.* iii, p. 276, tab. 308.
6 *Linn. Soc. Transact.* viii, p. 308.
7 *Ann. du Mus. d'Hist. Nat.* tom. xv, p. 473.
8 *Nouv. Bulletin des Sc.* tom. iii, pp. 73, 85, *et* 121.

very different view of the structure of Cycadeæ and Coniferæ, stating, that in their female flowers there is not only a minute cohering perianthium present, but an external additional envelope, to which he has given the name of cupula.

In 1814 I adopted this view, as far at least as regards the manner of impregnation, and stated some facts in support of it.[1] But, on reconsidering the subject in connection with what I had ascertained respecting the vegetable ovulum, I soon after altogether abandoned this opinion, without, however, venturing explicitly to state that now advanced, and which had then suggested itself.[2]

It is well known that the late M. Richard had prepared a very valuable memoir on these two families of plants; and he appears, from some observations lately published by his son, M. Achille Richard,[3] to have formed an opinion respecting their structure somewhat different from that of M. Mirbel, whose cupula is, according to him, the perianthium, more or less cohering with the included pistillum. He was probably led to this view on ascertaining, which I had also done, that the common account of the structure of [560 Ephedra was incorrect,[4] its supposed style being in reality the elongated tubular apex of a membranous envelope, and the included body being evidently analogous to that in other genera of Coniferæ.

To the earliest of the opinions here quoted, that which considers the female flower of Coniferæ and Cycadeæ as a naked pistillum, there are two principal objections. The first of these arises from the perforation of the pistillum, and the exposure of that point of the ovulum where the embryo is formed to the direct action of the pollen; the second from the too great simplicity of structure of the supposed ovulum, which I have shown accords better with that of the nucleus as existing in ordinary cases.

To the opinions of MM. Richard and Mirbel, the first

[1] *Flinders' Voy.* ii, 572 (*antè, p.* 46).
[2] *Tuckey's Congo*, p. 454 (*antè, p.* 138), *et Linn. Soc. Transact.* vol. xiii, p. 213 (*antè, p.* 380, *note*).
[3] *Dict. Class. d'Hist. Nat.* tom. iv, p. 395, *et* tom. v, p. 216.
[4] *Id.* tom. vi, p. 208.

objection does not apply, but the second acquires such additional weight, as to render those opinions much less probable, it seems to me, than that which I have endeavoured to support.

In supposing the correctness of this opinion to be admitted, a question connected with it, and of some importance, would still remain, namely, whether in Cycadeæ and Coniferæ the ovula are produced on an ovarium of reduced functions and altered appearance, or on a rachis or receptacle. In other words, in employing the language of an hypothesis, which, with some alterations, I have elsewhere attempted to explain and defend, respecting the formation of the sexual organs in Phænogamous plants,[1] whether the ovula in these two families originate in a modified leaf, or proceed directly from the stem.

561] Were I to adopt the former supposition, or that best agreeing with the hypothesis in question, I should certainly apply it, in the first place, to Cycas, in which the female spadix bears so striking a resemblance to a partially altered frond or leaf, producing marginal ovula in one part, and in another being divided into segments, in some cases nearly resembling those of the ordinary frond.

But the analogy of the female spadix of Cycas to that of Zamia is sufficiently obvious; and from the spadix of Zamia to the fruit-bearing squama of Coniferæ, strictly so called, namely, of Agathis or Dammara, Cunninghamia, Pinus, and even Araucaria, the transition is not difficult. This view is applicable, though less manifestly, also to Cupressinæ; and might even be extended to Podocarpus and Dacrydium. But the structure of these two genera admits likewise of another explanation, to which I have already adverted.

If, however, the ovula in Cycadeæ and Coniferæ be really produced on the surface of an ovarium, it might, perhaps, though not necessarily, be expected that their male flowers should differ from those of all other phænogamous plants, and in this difference exhibit some analogy to the

[1] *Linn. Soc Transact.* vol. xiii, p. 211 (*antè, p.* 378).

structure of the female flower. But in Cycadeæ, at least, and especially in Zamia, the resemblance between the male and female spadices is so great, that if the female be analogous to an ovarium, the partial male spadix must be considered as a single anthera, producing on its surface either naked grains of pollen, or pollen subdivided into masses, each furnished with its proper membrane.

Both these views may at present, perhaps, appear equally paradoxical; yet the former was entertained by Linnæus, who expresses himself on the subject in the following terms, "Pulvis floridus in Cycade minime pro Antheris agnoscendus est sed pro nudo polline, quod unusquisque qui un- [562
quam pollen antherarum in plantis examinavit fatebitur."[1] That this opinion, so confidently held by Linnæus, was never adopted by any other botanist, seems in part to have arisen from his having extended it to dorsiferous Ferns. Limited to Cycadeæ, however, it does not appear to me so very improbable as to deserve to be rejected without examination. It receives, at least, some support from the separation, in several cases, especially in the American Zamiæ, of the grains into two distinct, and sometimes nearly marginal, masses, representing, as it may be supposed, the lobes of an anthera; and also from their approximation in definite numbers, generally in fours, analogous to the quaternary union of the grains of pollen, not unfrequent in the antheræ of several other families of plants. The great size of the supposed grains of pollen, with the thickening and regular bursting of their membrane, may be said to be circumstances obviously connected with their production and persistence on the surface of an anthera, distant from the female flower; and with this economy, a corresponding enlargement of the contained particles or fovilla might also be expected. On examining these particles, however, I find them not only equal in size to the grains of pollen of many antheræ, but being elliptical and marked on one side with a longitudinal furrow, they have that form which is one of the most common in the simple pollen of phænogamous plants. To suppose, therefore, merely on the grounds already stated,

[1] *Mém. de l'Acad. des Scien. de Paris*, 1775, p. 518.

that these particles are analogous to the fovilla, and the
containing organs to the grains of pollen in antheræ of the
usual structure, would be entirely gratuitous. It is, at
the same time, deserving of remark, that were this view
563] adopted on more satisfactory grounds, a corresponding
development might then be said to exist in the essential
parts of the male and female organs. The increased development in the ovulum would not consist so much in the
unusual form and thickening of the coat, a part of secondary importance, and whose nature is disputed, as in the
state of the nucleus of the seed, respecting which there is
no difference of opinion; and where the plurality of embryos,
or at least the existence and regular arrangement of the
cells in which they are formed, is the uniform structure in
the family.

The second view suggested, in which the anthera in Cycadeæ is considered as producing on its surface an indefinite number of pollen masses, each enclosed in its proper membrane, would derive its only support from a few remote analogies; as from those antheræ, whose loculi are subdivided into a definite, or more rarely an indefinite, number of cells, and especially from the structure of the stamina of Viscum album.

I may remark, that the opinion of M. Richard,[1] who considers these grains, or masses, as unilocular antheræ, each of which constitutes a male flower, seems to be attended with nearly equal difficulties.

The analogy between the male and female organs in
Coniferæ, the existence of an open ovarium being assumed,
is at first sight more apparent than in Cycadeæ. In Coniferæ, however, the pollen is certainly not naked, but is
enclosed in a membrane similar to the lobe of an ordinary
anthera. And in those genera in which each squama of
the amentum produces two marginal lobes only, as Pinus,
Podocarpus, Dacrydium, Salisburia, and Phyllocladus, it
nearly resembles the more general form of the antheræ
564] in other Phænogamous plants. But the difficulty occurs
in those genera which have an increased number of lobes

[1] *Dict. Class. d'Hist. Nat.* tom. v. p. 216.

on each squama, as Agathis and Araucaria, where their number is considerable and apparently indefinite, and more particularly still in Cunninghamia, or Belis,[1] in which the lobes, though only three in number, agree in this respect, as well as in insertion and direction, with the ovula. The supposition, that in such cases all the lobes of each squama are cells of one and the same anthera, receives but little support either from the origin and arrangement of the lobes themselves, or from the structure of other phænogamous plants: the only cases of apparent, though doubtful, analogy that I can at present recollect occurring in Aphyteia, and perhaps in some Cucurbitaceæ.

That part of my subject, therefore, which relates to the analogy between the male and female flowers in Cycadeæ and Coniferæ, I consider the least satisfactory, both in regard to the immediate question of the existence of an anomalous ovarium in these families, and to the hypothesis repeatedly referred to, of the origin of the sexual organs of all phænogamous plants.

In concluding this digression, I have to express my regret that it should have so far exceeded the limits [565 proper for its introduction into the present work. In giving an account, however, of the genus of plants to which it is annexed, I had to describe a structure, of whose nature and importance it was necessary I should show myself aware; and circumstances have occurred while I was engaged in preparing this account, which determined me to enter much more fully into the subject than I had originally intended.

[1] In communicating specimens of this plant to the late M. Richard, for his intended monograph of Coniferæ, I added some remarks on its structure, agreeing with those here made. I at the same time requested that, if he objected to Mr. Salisbury's Belis as liable to be confounded with Bellis, the genus might be named Cunninghamia, to commemorate the merits of *Mr. James Cunningham*, an excellent observer in his time, by whom this plant was discovered; and in honour of *Mr. Allan Cunningham*, the very deserving botanist who accompanied Mr. Oxley in his first expedition into the interior of New South Wales, and Captain King in all his voyages of survey of the Coasts of New Holland.

on each side, as in Agathis and Araucaria, where their number is [illegible] indefinite, and more [illegible] with in [illegible] in which th[illegible] only three to [illegible] in this respect, as well as in insertion and [illegible] with the ovule. The supposition [illegible] the lobes of each [illegible] of [illegible] and [illegible] themselves, [illegible] from the [illegible] of the lobes themselves, [illegible] from the structure of other phænogamous plants. In [illegible] cases of [illegible], though doubtful, [illegible] present [illegible] occurring in Ailanthus, and [illegible] Cucurbitaceæ.

[illegible] that [illegible] subject [illegible] relates to the [illegible] between the male and female flowers of [illegible] and [illegible] the [illegible] satisfactory, such [illegible] of the [illegible] of the structure of [illegible] and [illegible] of the [illegible] of [illegible] plant.

In [illegible] this [illegible], I [illegible] regrets [illegible] about [illegible] for [illegible] for its [illegible] the [illegible] regard, however, of the [illegible] to which [illegible] had to [illegible] and [illegible] of [illegible] greatly [illegible] this [illegible] me to [illegible] were fully [illegible] originally [illegible]

[illegible]

A BRIEF ACCOUNT OF MICROSCOPICAL OBSERVATIONS

Made in the Months of June, July, and August, 1827,

ON THE PARTICLES CONTAINED IN THE POLLEN OF PLANTS;

AND

ON THE GENERAL EXISTENCE OF ACTIVE MOLECULES

IN ORGANIC AND INORGANIC BODIES.

BY

ROBERT BROWN,

F.R.S., HON. M.R.S.E. AND R.I. ACAD., V.P.L.S.,

MEMBER OF THE ROYAL ACADEMY OF SCIENCES OF SWEDEN, OF THE ROYAL SOCIETY OF DENMARK, AND OF THE IMPERIAL ACADEMY NATURÆ CURIOSORUM; CORRESPONDING MEMBER OF THE ROYAL INSTITUTES OF FRANCE AND OF THE NETHERLANDS, OF THE IMPERIAL ACADEMY OF SCIENCES AT ST. PETERSBURG, AND OF THE ROYAL ACADEMIES OF PRUSSIA AND BAVARIA, ETC.

[*Not Published.*]

MICROSCOPICAL OBSERVATIONS. [3

THE observations, of which it is my intention to give a summary in the following pages, have all been made with a simple microscope, and indeed with one and the same lens, the focal length of which is about $\frac{1}{32}$nd of an inch.[1]

The examination of the unimpregnated vegetable Ovulum, an account of which was published early in 1826,[2] led me to attend more minutely than I had before done to the structure of the Pollen, and to inquire into its mode of action on the Pistillum in Phænogamous plants.

In the Essay referred to, it was shown that the apex of the nucleus of the Ovulum, the point which is universally the seat of the future Embryo, was very generally brought into contact with the terminations of the probable channels of fecundation; these being either the surface of the placenta, the extremity of the descending processes of the style,

[1] This double convex lens, which has been several years in my possession, I obtained from Mr. Bancks, optician, in the Strand. After I had made considerable progress in the inquiry, I explained the nature of my subject to Mr. Dollond, who obligingly made for me a simple pocket microscope, having very delicate adjustment, and furnished with excellent lenses, two of which are of much higher power than that above mentioned. To these I have often had recourse, and with great advantage, in investigating several minute points. But to give greater consistency to my statements, and to bring the subject as much as possible within the reach of general observation, I continued to employ throughout the whole of the inquiry the same lens with which it was commenced.

[2] In the Botanical Appendix to Captain King's Voyages to Australia, vol. ii, p. 534, *et seq.* (*antè p.* 435).

or more rarely, a part of the surface of the umbilical cord. It also appeared, however, from some of the facts noticed in the same Essay, that there were cases in which the Particles contained in the grains of pollen could hardly be conveyed
4] to that point of the ovulum through the vessels or cellular tissue of the ovarium; and the knowledge of these cases, as well as of the structure and economy of the antheræ in Asclepiadeæ, had led me to doubt the correctness of observations made by Stiles and Gleichen upwards of sixty years ago, as well as of some very recent statements, respecting the mode of action of the pollen in the process of impregnation.

It was not until late in the autumn of 1826 that I could attend to this subject; and the season was too far advanced to enable me to pursue the investigation. Finding, however, in one of the few plants then examined, the figure of the particles contained in the grains of pollen clearly discernible, and that figure not spherical but oblong, I expected, with some confidence, to meet with plants in other respects more favorable to the inquiry, in which these particles, from peculiarity of form, might be traced through their whole course: and thus, perhaps, the question determined whether they in any case reach the apex of the ovulum, or whether their direct action is limited to other parts of the female organ.

My inquiry on this point was commenced in June 1827, and the first plant examined proved in some respects remarkably well adapted to the object in view.

This plant was *Clarckia pulchella,* of which the grains of pollen, taken from antheræ full grown, but before bursting, were filled with particles or granules of unusually large size, varying from nearly $\frac{1}{4000}$th to about $\frac{1}{5000}$th of an inch in length, and of a figure between cylindrical and oblong, perhaps slightly flattened, and having rounded and equal extremities. While examining the form of these particles immersed in water, I observed many of them very evidently in motion; their motion consisting not only of a change of place in the fluid, manifested by alterations in their relative positions, but also not unfrequently of a change of form in

the particle itself; a contraction or curvature taking place repeatedly about the middle of one side, accompanied by a corresponding swelling or convexity on the opposite side of the particle. In a few instances the particle was seen to turn on its longer axis. These motions were such as to satisfy me, after frequently repeated observation, that they arose neither from currents in the fluid, nor from its [5
gradual evaporation, but belonged to the particle itself.

Grains of pollen of the same plant taken from antheræ immediately after bursting, contained similar subcylindrical particles, in reduced numbers, however, and mixed with other particles, at least as numerous, of much smaller size, apparently spherical, and in rapid oscillatory motion.

These smaller particles, or Molecules as I shall term them, when first seen, I considered to be some of the cylindrical particles swimming vertically in the fluid. But frequent and careful examination lessened my confidence in this supposition; and on continuing to observe them until the water had entirely evaporated, both the cylindrical particles and spherical molecules were found on the stage of the microscope.

In extending my observations to many other plants of the same natural family, namely *Onagrariæ*, the same general form and similar motions of particles were ascertained to exist, especially in the various species of Œnothera, which I examined. I found also in their grains of pollen taken from the antheræ immediately after bursting, a manifest reduction in the proportion of the cylindrical or oblong particles, and a corresponding increase in that of the molecules, in a less remarkable degree, however, than in Clarckia.

This appearance, or rather the great increase in the number of the molecules, and the reduction in that of the cylindrical particles, before the grain of pollen could possibly have come in contact with the stigma,—were perplexing circumstances in this stage of the inquiry, and certainly not favorable to the supposition of the cylindrical particles acting directly on the ovulum; an opinion which I was inclined to adopt when I first saw them in motion. These circumstances, however, induced me to multiply my observations, and I accordingly examined numerous species of many of the more important and remarkable families of the two great primary divisions of Phænogamous plants.

In all these plants particles were found, which in the different families or genera, varied in form from oblong to spherical, having manifest motions similar to those already described: except that the change of form in the oval and
6] oblong particles was generally less obvious than in Onagrariæ, and in the spherical particle was in no degree observable.[1] In a great proportion of these plants I also remarked the same reduction of the larger particles, and a corresponding increase in the molecules after the bursting of the antheræ: the molecule, of apparently uniform size and form, being then always present; and in some cases, indeed, no other particles were observed, either in this or in any earlier stage of the secreting organ.

In many plants belonging to several different families, but especially to Gramineæ, the membrane of the grain of pollen is so transparent that the motion of the larger particles within the entire grain was distinctly visible; and it was manifest also at the more transparent angles, and in some cases even in the body of the grain in Onagrariæ.

In *Asclepiadeæ*, strictly so called, the mass of pollen filling each cell of the anthera is in no stage separable into distinct grains; but within, its tesselated or cellular membrane is filled with spherical particles, commonly of two sizes. Both these kinds of particles when immersed in water are generally seen in vivid motion; but the apparent motions of the larger particle might in these cases perhaps be caused by the rapid oscillation of the more numerous molecules. The mass of pollen in this tribe of plants never bursts, but merely connects itself by a determinate point, which is not unfrequently semitransparent, to a process of nearly similar consistence, derived from the gland of the corresponding angle of the stigma.

[1] In *Lolium perenne*, however, which I have more recently examined, though the particle was oval and of smaller size than in Onagrariæ, this change of form was at least as remarkable, consisting in an equal contraction in the middle of each side, so as to divide it into two nearly orbicular portions.

In *Periploceæ*, and in a few *Apocineæ*, the pollen, which in these plants is separable into compound grains filled with spherical moving particles, is applied to processes of the stigma, analogous to those of Asclepiadeæ. A similar economy exists in *Orchideæ*, in which the pollen masses are always, at least in the early stage, granular; the grains, whether simple or compound, containing minute, nearly spherical particles, but the whole mass being, with [7
very few exceptions, connected by a determinate point of its surface with the stigma, or a glandular process of that organ.

Having found motion in the particles of the pollen of all the living plants which I had examined, I was led next to inquire whether this property continued after the death of the plant, and for what length of time it was retained.

In plants, either dried or immersed in spirit for a few days only, the particles of pollen of both kinds were found in motion equally evident with that observed in the living plant; specimens of several plants, some of which had been dried and preserved in an herbarium for upwards of twenty years, and others not less than a century, still exhibited the molecules or smaller spherical particles in considerable numbers, and in evident motion, along with a few of the larger particles, whose motions were much less manifest, and in some cases not observable.[1]

In this stage of the investigation having found, as I believed, a peculiar character in the motions of the particles of pollen in water, it occurred to me to appeal to this peculiarity as a test in certain families of Cryptogamous plants, namely, Mosses, and the genus Equisetum,

[1] While this sheet was passing through the press I have examined the pollen of several flowers which have been immersed in weak spirit about eleven months, particularly of *Viola tricolor*, *Zizania aquatica*, and *Zea Mays*; and in all these plants the peculiar particles of the pollen, which are oval or short oblong, though somewhat reduced in number, retain their form perfectly, and exhibit evident motion, though I think not so vivid as in those belonging to the living plant. In *Viola tricolor*, in which, as well as in other species of the same natural section of the genus, the pollen has a very remarkable form, the grain on immersion in nitric acid still discharged its contents by its four angles, though with less force than in the recent plant.

in which the existence of sexual organs had not been universally admitted.

In the supposed stamina of both these families, namely, in the cylindrical antheræ or pollen of Mosses, and on the surface of the four spathulate bodies surrounding the naked ovulum, as it may be considered, of Equisetum, I found minute spherical particles, apparently of the same size with the molecule described in Onagrariæ, and having equally
8] vivid motion on immersion in water; and this motion was still observable in specimens both of Mosses and of Equiseta, which had been dried upwards of one hundred years.

The very unexpected fact of seeming vitality retained by these minute particles so long after the death of the plant would not perhaps have materially lessened my confidence in the supposed peculiarity. But I at the same time observed, that on bruising the ovula or seeds of Equisetum, which at first happened accidentally, I so greatly increased the number of moving particles, that the source of the added quantity could not be doubted. I found also that on bruising first the floral leaves of Mosses, and then all other parts of those plants, that I readily obtained similar particles, not in equal quantity indeed, but equally in motion. My supposed test of the male organ was therefore necessarily abandoned.

Reflecting on all the facts with which I had now become acquainted, I was disposed to believe that the minute spherical particles or Molecules of apparently uniform size, first seen in the advanced state of the pollen of Onagrariæ, and most other Phænogamous plants,—then in the antheræ of Mosses and on the surface of the bodies regarded as the stamina of Equisetum,—and lastly in bruised portions of other parts of the same plants, were in reality the supposed constituent or elementary Molecules of organic bodies, first so considered by Buffon and Needham, then by Wrisberg with greater precision, soon after and still more particularly by Müller, and, very recently, by Dr. Milne Edwards, who has revived the doctrine and supported it with much interesting detail. I now therefore expected to find these molecules in all organic bodies: and accordingly on examining the various animal and vegetable tissues, whether living or dead, they were always found to exist; and merely by bruising these substances in water, I never failed to disengage the molecules in sufficient numbers to ascertain their apparent identity in size, form, and motion, with the smaller particles of the grains of pollen.

I examined also various products of organic bodies, particularly the gum resins, and substances of vegetable origin, extending my inquiry even to pit-coal; and in all these [9
bodies Molecules were found in abundance. I remark here also, partly as a caution to those who may hereafter engage in the same inquiry, that the dust or soot deposited on all bodies in such quantity, especially in London, is entirely composed of these molecules.

One of the substances examined, was a specimen of fossil wood, found in Wiltshire oolite, in a state to burn with flame; and as I found these molecules abundantly, and in motion in this specimen, I supposed that their existence, though in smaller quantity, might be ascertained in mineralized vegetable remains. With this view a minute portion of silicified wood, which exhibited the structure of Coniferæ, was bruised, and spherical particles, or molecules in all respects like those so frequently mentioned, were readily obtained from it; in such quantity, however, that the whole substance of the petrifaction seemed to be formed of them. But hence I inferred that these molecules were not limited to organic bodies, nor even to their products.

To establish the correctness of the inference, and to ascertain to what extent the molecules existed in mineral bodies, became the next object of inquiry. The first substance examined was a minute fragment of window-glass, from which, when merely bruised on the stage of the microscope, I readily and copiously obtained molecules agreeing in size, form, and motion with those which I had already seen.

I then proceeded to examine, and with similar results, such minerals as I either had at hand or could readily obtain, including several of the simple earths and metals, with many of their combinations.

Rocks of all ages, including those in which organic remains have never been found, yielded the molecules in abundance. Their existence was ascertained in each of the constituent minerals of granite, a fragment of the Sphinx being one of the specimens examined.

To mention all the mineral substances in which I have found these molecules, would be tedious; and I shall confine myself in this summary to an enumeration of a few of the most remarkable. These were both of aqueous and igneous origin, as travertine, stalactites, lava, obsidian,
10] pumice, volcanic ashes, and meteorites from various localities.[1] Of metals I may mention manganese, nickel, plumbago, bismuth, antimony, and arsenic. In a word, in every mineral which I could reduce to a powder, sufficiently fine to be temporarily suspended in water, I found these molecules more or less copiously; and in some cases, more particularly in siliceous crystals, the whole body submitted to examination appeared to be composed of them.

In many of the substances examined, especially those of a fibrous structure, as asbestus, actinolite, tremolite, zeolite, and even steatite, along with the spherical molecules, other corpuscles were found, like short fibres somewhat moniliform, whose transverse diameter appeared not to exceed that of the molecule, of which they seemed to be primary combinations. These fibrils, when of such length as to be probably composed of not more than four or five molecules, and still more evidently when formed of two or three only, were generally in motion, as least as vivid as that of the simple molecule itself; and which from the fibril often changing its position in the fluid, and from its occasional bending, might be said to be somewhat vermicular.

In other bodies which did not exhibit these fibrils, oval particles of a size about equal to two molecules, and which were also conjectured to be primary combinations of these, were not unfrequently met with, and in motion generally more vivid than that of the simple molecule; their motion consisting in turning usually on their longer axis, and then

[1] I have since found the molecules in the sand-tubes, formed by lightning, from Drig in Cumberland.

often appearing to be flattened. Such oval particles were found to be numerous and extremely active in white arsenic.

As mineral bodies which had been fused contained the moving molecules as abundantly as those of alluvial deposits, I was desirous of ascertaining whether the mobility of the particles existing in organic bodies was in any degree affected by the application of intense heat to the containing substance. With this view small portions of wood, both living and dead, linen, paper, cotton, wool, silk, hair, and muscular fibres, were exposed to the flame of a candle or burned in platina forceps, heated by the blowpipe; and in all these bodies so heated, quenched in water, and immediately submitted to examination, the molecules were found, and in as evident motion as those obtained from the same substances before burning.

In some of the vegetable bodies burned in this manner, in addition to the simple molecules, primary combinations of these were observed, consisting of fibrils having transverse contractions, corresponding in number, as I conjectured, with that of the molecules composing them; and those fibrils, when not consisting of a greater number than four or five molecules, exhibited motion resembling in kind and vivacity that of the mineral fibrils already described, while longer fibrils of the same apparent diameter were at rest.

The substance found to yield these active fibrils in the largest proportion and in the most vivid motion was the mucous coat interposed between the skin and muscles of the haddock, especially after coagulation by heat.

The fine powder produced on the under surface of the fronds of several Ferns, particularly of *Acrostichum calomelanos*, and the species nearly related to it, was found to be entirely composed of simple molecules and their primary fibre-like compounds, both of them being evidently in motion.

There are three points of great importance which I was anxious to ascertain respecting these molecules, namely, their form, whether they are of uniform size, and their absolute magnitude. I am not, however, entirely satisfied with what I have been able to determine on any of these points.

As to form, I have stated the molecule to be spherical, and this I have done with some confidence; the apparent exceptions which occurred admitting, as it seems to me, of being explained by supposing such particles to be compounds. This supposition in some of the cases is indeed hardly reconcileable with their apparent size, and requires for its support the further admission that, in combination, the figure of the molecule may be altered. In the particles formerly considered as primary combinations of molecules, a certain change of form must also be allowed; and even the simple molecule itself has sometimes appeared to me when in motion to have been slightly modified in this respect.

12] My manner of estimating the absolute magnitude and uniformity in size of the molecules, found in the various bodies submitted to examination, was by placing them on a micrometer divided to five thousandths of an inch, the lines of which were very distinct; or more rarely on one divided to ten thousandths, with fainter lines, not readily visible without the application of plumbago, as employed by Dr. Wollaston, but which in my subject was inadmissible.

The results so obtained can only be regarded as approximations, on which, perhaps, for an obvious reason, much reliance will not be placed. From the number and degree of accordance of my observations, however, I am upon the whole disposed to believe the simple molecule to be of uniform size, though as existing in various substances and examined in circumstances more or less favorable, it is necessary to state that its diameter appeared to vary from $\frac{1}{15,000}$th to $\frac{1}{20,000}$th of an inch.[1]

I shall not at present enter into additional details, nor

[1] While this sheet was passing through the press, Mr. Dollond, at my request, obligingly examined the supposed pollen of *Equisetum virgatum* with his compound achromatic microscope, having in its focus a glass divided into 10,000ths of an inch, upon which the object was placed; and although the greater number of particles or molecules seen were about 1-20,000th, yet the smallest did not exceed 1-30,000th of an inch.

shall I hazard any conjectures whatever respecting these molecules, which appear to be of such general existence in inorganic as well as in organic bodies; and it is only further necessary to mention the principal substances from which I have not been able to obtain them. These are oil, resin, wax and sulphur, such of the metals as I could not reduce to that minute state of division necessary for their separation, and finally, bodies soluble in water.

In returning to the subject with which my investigation commenced, and which was indeed the only object I originally had in view, I had still to examine into the probable mode of action of the larger or peculiar particles of the pollen, which, though in many cases diminished in number before the grain could possibly have been applied to the stigma, and particularly in Clarckia, the plant first examined, were yet in many other plants found in less diminished proportion, [13 and might in nearly all cases be supposed to exist in sufficient quantity to form the essential agents in the process of fecundation.

I was now therefore to inquire, whether their action was confined to the external organ, or whether it were possible to follow them to the nucleus of the ovulum itself. My endeavours, however, to trace them through the tissue of the style in plants well suited for this investigation, both from the size and form of the particles, and the development of the female parts, particularly Onagrariæ, was not attended with success; and neither in this nor in any other tribe examined, have I ever been able to find them in any part of the female organ except the stigma. Even in those families in which I have supposed the ovulum to be naked, namely, Cycadeæ and Coniferæ, I am inclined to think that the direct action of these particles, or of the pollen containing them, is exerted rather on the orifice of the proper membrane than on the apex of the included nucleus; an opinion which is in part founded on the partial withering confined to one side of the orifice of that membrane in the larch,—an appearance which I have remarked for several years.

To observers not aware of the existence of the elementary

active molecules, so easily separated by pressure from all vegetable tissues, and which are disengaged and become more or less manifest in the incipient decay of semitransparent parts, it would not be difficult to trace granules through the whole length of the style : and as these granules are not always visible in the early and entire state of the organ, they would naturally be supposed to be derived from the pollen, in those cases at least in which its contained particles are not remarkably different in size and form from the molecule.

It is necessary also to observe that in many, perhaps I might say in most plants, in addition to the molecules separable from the stigma and style before the application of the pollen, other granules of greater size are obtained by pressure, which in some cases closely resemble the particles of the pollen in the same plants, and in a few cases even exceed them in size: these particles may be considered as
14] primary combinations of the molecules, analogous to those already noticed in mineral bodies and in various organic tissues.

From the account formerly given of Asclepiadeæ, Periploceæ, and Orchideæ, and particularly from what was observed of Asclepiadeæ, it is difficult to imagine, in this family at least, that there can be an actual transmission of particles from the mass of pollen, which does not burst, through the processes of the stigma; and even in these processes I have never been able to observe them, though they are in general sufficiently transparent to show the particles were they present. But if this be a correct statement of the structure of the sexual organs in Asclepiadeæ, the question respecting this family would no longer be, whether the particles in the pollen were transmitted through the stigma and style to the ovula, but rather whether even actual contact of these particles with the surface of the stigma were necessary to impregnation.

Finally, it may be remarked that those cases already adverted to, in which the apex of the nucleus of the ovulum, the supposed point of impregnation, is never brought into contact with the probable channels of fecundation, are more unfavorable to the opinion of the transmission of the particles of the pollen to the ovulum, than to that which considers the direct action of these particles as confined to the external parts of the female organ.

The observations, of which I have now given a brief account, were made in the months of June, July, and August, 1827. Those relating merely to the form and motion of the peculiar particles of the pollen were stated, and several of the objects shown, during these months, to many of my friends, particularly to Messrs. Bauer and Bicheno, Dr. Bostock, Dr. Fitton, Mr. E. Forster, Dr. Henderson, Sir Everard Home, Captain Home, Dr. Horsfield, Mr. Kœnig, M. Lagasca, Mr. Lindley, Dr. Maton, Mr. Menzies, Dr. Prout, Mr. Renouard, Dr. Roget, Mr. Stokes, and Dr. Wollaston; and the general existence of the active molecules in inorganic as well as organic bodies, their apparent indestructibility by heat, and several of the facts respecting the primary combinations of the molecules were communicated to Dr. Wollaston and Mr. Stokes in the last week of August.

None of these gentlemen are here appealed to for the [15
correctness of any of the statements made; my sole object in citing them being to prove from the period and general extent of the communication, that my observations were made within the dates given in the title of the present summary.

The facts ascertained respecting the motion of the particles of the pollen were never considered by me as wholly original; this motion having, as I knew, been obscurely seen by Needham, and distinctly by Gleichen, who not only observed the motion of the particles in water after the bursting of the pollen, but in several cases marked their change of place within the entire grain. He has not, however, given any satisfactory account either of the forms or of the motions of these particles, and in some cases appears to have confounded them with the elementary molecule, whose existence he was not aware of.

Before I engaged in the inquiry in 1827, I was acquainted only with the abstract given by M. Adolphe

Brongniart himself, of a very elaborate and valuable memoir, entitled "*Recherches sur la Génération et le Développement de l'Embryon dans les Végétaux Phanérogames*," which he had then read before the Academy of Sciences of Paris, and has since published in the *Annales des Sciences Naturelles*.

Neither in the abstract referred to, nor in the body of the memoir which M. Brongniart has with great candour given in its original state, are there any observations, appearing of importance even to the author himself, on the motion or form of the particles; and the attempt to trace these particles to the ovulum with so imperfect a knowledge of their distinguishing characters could hardly be expected to prove satisfactory. Late in the autumn of 1827, however, M. Brongniart having at his command a microscope constructed by Amici, the celebrated professor of Modena, he was enabled to ascertain many important facts on both these points, the result of which he has given in the notes annexed to his memoir. On the general accuracy of his observations on the motions, form, and size of the granules, as he terms the particles, I place great reliance. But in attempting to trace these particles through their whole course, he has overlooked two points of the greatest importance in the investigation.

16] For, in the first place, he was evidently unacquainted with the fact that the active spherical molecules generally exist in the grain of pollen along with its proper particles; nor does it appear from any part of his memoir that he was aware of the existence of molecules having spontaneous or inherent motion and distinct from the peculiar particles of the pollen, though he has doubtless seen them, and in some cases, as it seems to me, described them as those particles.

Secondly, he has been satisfied with the external appearance of the parts in coming to his conclusion, that no particles capable of motion exist in the style or stigma before impregnation.

That both simple molecules and larger particles of different form, and equally capable of motion, do exist in these parts, before the application of the pollen to the stigma can possibly take place, in many of the plants submitted by him to examination, may easily be ascertained; particularly in *Antirrhinum majus*, of which he has given a figure in a more advanced state, representing these molecules or particles, which he supposes to have been derived from the grains of pollen, adhering to the stigma.

There are some other points respecting the grains of pollen and their contained particles in which I also differ from M. Brongniart, namely, in his supposition that the particles are not formed in the grain itself, but in the cavity of the anthera; in his assertion respecting the presence of pores on the surface of the grain in its early state, through which the particles formed in the anthera pass into its cavity; and lastly, on the existence of a membrane forming the coat of his boyau or mass of cylindrical form ejected from the grain of pollen.

I reserve, however, my observations on these and several other topics connected with the subject of the present inquiry for the more detailed account, which it is my intention to give.

July 30th, 1828.

ADDITIONAL REMARKS ON ACTIVE MOLECULES.

By ROBERT BROWN, F.R.S.

About twelve months ago I printed an account of Microscopical Observations made in the summer of 1827, on the Particles contained in the Pollen of Plants; and on the general Existence of active Molecules in Organic and Inorganic Bodies.

In the present Supplement to that account my objects are, to explain and modify a few of its statements, to advert to some of the remarks that have been made, either on the correctness or originality of the observations, and to the causes that have been considered sufficient for the explanation of the phenomena.

In the first place, I have to notice an erroneous assertion of more than one writer, namely, that I have stated the active Molecules to be animated. This mistake has probably arisen from my having communicated the facts in the same order in which they occurred, accompanied by the views which presented themselves in the different stages of the investigation; and in one case, from my having adopted the language, in referring to the opinion, of another inquirer into the first branch of the subject.

2] Although I endeavoured strictly to confine myself to the statement of the facts observed, yet in speaking of the active Molecules, I have not been able, in all cases, to avoid the introduction of hypothesis; for such is the supposition that the equally active particles of greater size, and frequently of very different form, are primary compounds of these Molecules,—a supposition which, though professedly conjectural, I regret having so much insisted on, especially as it may seem connected with the opinion of the absolute identity of the Molecules, from whatever source derived.

On this latter subject, the only two points that I endeavoured to ascertain were their size and figure: and although I was, upon the whole, inclined to think that in these respects the Molecules were similar from whatever substances obtained, yet the evidence then adduced in support of the supposition was far from satisfactory; and I may add, that I am still less satisfied now that such is the fact. But even had the uniformity of the Molecules in those two points been absolutely established, it did not necessarily follow, nor have I anywhere stated, as has been imputed to me, that they also agreed in all their other properties and functions.

I have remarked that certain substances, namely, sulphur, resin, and wax, did not yield active particles, which, however, proceeded merely from defective manipulation; for I have since readily obtained them from all these bodies: at the same time I ought to notice that their existence in sulphur was previously mentioned to me by my friend Mr. Lister.

In prosecuting the inquiry subsequent to the publication of my Observations, I have chiefly employed the simple microscope mentioned in the Pamphlet as having been made for me by Mr. Dollond, and of which the three lenses that I have generally used, are of a 40th, 60th, and 70th of an inch focus.

Many of the observations have been repeated and confirmed with other simple microscopes having lenses of similar powers, and also with the best achromatic compound microscopes, either in my own possession or belonging to my friends.

The result of the inquiry at present essentially agrees with that which may be collected from my printed account, and may be here briefly stated in the following terms; namely,

That extremely minute particles of solid matter, whether obtained from organic or inorganic substances, when suspended in pure water, or in some other aqueous fluids, exhibit motions for which I am unable to account, and which from their irregularity and seeming independence resemble in a remarkable degree the less rapid motions of some of the simplest animalcules of infusions. That the smallest moving particles observed, and which I have termed Active Molecules, appear to be spherical, or nearly so, and to be between 1-20,000dth and 1-30,000dth of an inch in diameter; and that other particles of considerably greater and various size, and either of similar or of very different figure, also present analogous motions in like circumstances.

I have formerly stated my belief that these motions of the particles neither arose from currents in the fluid containing them, nor depended on that intestine motion which may be supposed to accompany its evaporation.

These causes of motion, however, either singly or combined

with others,—as, the attractions and repulsions among the particles themselves, their unstable equilibrium in the fluid in which they are suspended, their hygrometrical or capillary action, and in some cases the disengagement of volatile matter, or of minute air bubbles,—have been considered by several writers as sufficiently accounting for the appearances. Some of the alleged causes here stated, with others which I have considered it unnecessary to mention, are not likely to be overlooked or to deceive observers of any experience in microscopical researches; and the insufficiency of the most important of those enumerated may, I think, be satisfactorily shown by means of a very simple experiment.

This experiment consists in reducing the drop of water containing the particles to microscopic minuteness, and prolonging its existence by immersing it in a transparent fluid of inferior specific gravity, with which it is not miscible, and in which evaporation is extremely slow. If to almond-oil, which is a fluid having these properties, a considerably
4] smaller proportion of water, duly impregnated with particles, be added, and the two fluids shaken or triturated together, drops of water of various sizes, from 1-50th to 1-2000dth of an inch in diameter, will be immediately produced. Of these, the most minute necessarily contain but few particles, and some may be occasionally observed with one particle only. In this manner minute drops, which if exposed to the air would be dissipated in less than a minute, may be retained for more than an hour. But in all the drops thus formed and protected, the motion of the particles takes place with undiminished activity, while the principal causes assigned for that motion, namely, evaporation, and their mutual attraction and repulsion, are either materially reduced or absolutely null.

It may here be remarked, that those currents from centre to circumference, at first hardly perceptible, then more obvious, and at last very rapid, which constantly exist in drops exposed to the air, and disturb or entirely overcome the proper motion of the particles, are wholly prevented in drops of small size immersed in oil,—a fact which, however, is only apparent in those drops that are flattened, in consequence of being nearly or absolutely in contact with the stage of the microscope.

That the motion of the particles is not produced by any cause acting on the surface of the drop, may be proved by an inversion of the experiment; for by mixing a very small proportion of oil with the water containing the particles, microscopic drops of oil of extreme minuteness, some of them not exceeding in size the particles themselves, will be found on the surface of the drop of water, and nearly or altogether at rest; while the particles in the centre or towards the bottom of the drop continue to move with their usual degree of activity.

By means of the contrivance now described for reducing the size and prolonging the existence of the drops containing the particles, which, simple as it is, did not till very lately occur to me, a greater command of the subject is obtained, sufficient perhaps to enable us to ascertain the real cause of the motions in question.

Of the few experiments which I have made since this manner of observing was adopted, some appear to me so curious, that I do not venture to state them until they are [5
verified by frequent and careful repetition.

I shall conclude these supplementary remarks to my former Observations, by noticing the degree in which I consider those observations to have been anticipated.

That molecular was sometimes confounded with animalcular motion by several of the earlier microscopical observers, appears extremely probable from various passages in the writings of Leeuwenhoek, as well as from a very interesting Paper by Stephen Gray, published in the 19th volume of the Philosophical Transactions.

Needham also, and Buffon, with whom the hypothesis of organic particles originated, seem to have not unfrequently fallen into the same mistake. And I am inclined to believe that Spallanzani, notwithstanding one of his statements respecting them, has under the head of *Anima-*

letti d'ultimo ordine included the active Molecules as well as true Animalcules.

I may next mention that Gleichen, the discoverer of the motions of the Particles of the Pollen, also observed similar motions in the particles of the ovulum of Zea Mays.

Wrisberg and Müller, who adopted in part Buffon's hypothesis, state the globules, of which they suppose all organic bodies formed, to be capable of motion; and Müller distinguishes these moving organic globules from real Animalcules, with which, he adds, they have been confounded by some very respectable observers.

In 1814 Dr. James Drummond, of Belfast, published in the 7th volume of the Transactions of the Royal Society of Edinburgh, a very valuable Paper, entitled "On certain Appearances observed in the Dissection of the Eyes of Fishes."

In this Essay, which I regret I was entirely unacquainted with when I printed the account of my Observations, the author gives an account of the very remarkable motions of the spicula which form the silvery part of the choroid coat of the eyes of fishes.

These spicula were examined with a simple microscope, and
6] as opaque objects, a strong light being thrown upon the drop of water in which they were suspended. The appearances are minutely described, and very ingenious reasoning employed to show that, to account for the motions, the least improbable conjecture is to suppose the spicula animated.

As these bodies were seen by reflected and not by transmitted light, a very correct idea of their actual motions could hardly be obtained; and with the low magnifying powers necessarily employed with the instrument and in the manner described, the more minute nearly spherical particles or active Molecules which, when higher powers were used, I have always found in abundance along with the spicula, entirely escaped observation.

Dr. Drummond's researches were strictly limited to the spicula of the eyes and scales of fishes; and as he does not appear to have suspected that particles having analogous motions might exist in other organized bodies, and far less in inorganic matter, I consider myself anticipated by this acute observer only to the same extent as by Gleichen, and in a much less degree than by Müller, whose statements have been already alluded to.

All the observers now mentioned have confined themselves to the examination of the particles of organic bodies. In 1819, however, Mr. Bywater, of Liverpool, published an account of Microscopical Observations, in which it is stated that not only organic tissues, but also inorganic substances, consist of what he terms animated or irritable particles.

A second edition of this Essay appeared in 1828, probably altered in some points, but it may be supposed agreeing essentially in its statements with the edition of 1819, which I have never seen, and of the existence of which I was ignorant when I published my pamphlet.

From the edition of 1828, which I have but lately met with, it appears that Mr. Bywater employed a compound microscope of the construction called Culpepper's, that the object was examined in a bright sunshine, and the light from the mirror thrown so obliquely on the stage as to give a blue colour to the infusion.

The first experiment I here subjoin in his own words. [7

"A small portion of flour must be placed on a slip of glass, and mixed with a drop of water, then instantly applied to the microscope; and if stirred and viewed by a bright sun, as already described, it will appear evidently filled with innumerable small linear bodies, writhing and twisting about with extreme activity."

Similar bodies, and equally in motion, were obtained from animal and vegetable tissues, from vegetable mould, from sandstone after being made red hot, from coal, ashes, and other inorganic bodies.

I believe that in thus stating the manner in which Mr. Bywater's experiments were conducted, I have enabled microscopical observers to judge of the extent and kind of optical illusion to which he was liable, and of which he does not seem to have been aware. I have only to add, that it is not here a question of priority; for if his observations are to be depended on, mine must be entirely set aside.

July 28th, 1829.

OBSERVATIONS

ON THE

ORGANS AND MODE OF FECUNDATION

IN

ORCHIDEÆ AND ASCLEPIADEÆ.

BY

ROBERT BROWN, ESQ.,

D.C.L., F.R.S., HON. M.R.S. EDIN., AND R.I. ACAD. V.P.L.S.;

FOREIGN MEMBER OF THE ROYAL ACADEMY OF SCIENCES IN THE INSTITUTE OF FRANCE; OF THE IMPERIAL ACADEMY OF SCIENCES OF RUSSIA; THE ROYAL ACADEMIES OF SCIENCES OF SWEDEN AND BAVARIA; OF THE FIRST CLASS OF THE ROYAL INSTITUTE OF HOLLAND; THE ROYAL SOCIETY OF DENMARK; AND THE IMPERIAL ACADEMY OF NATURALISTS; CORRESPONDING MEMBER OF THE ROYAL ACADEMIES OF SCIENCES OF PRUSSIA AND BELGIUM, ETC., ETC.

[*Reprinted from the 'Transactions of the Linnean Society.' Vol. XVI, pp.* 685—745.]

LONDON:

1833.

ON THE

ORGANS AND MODE OF FECUNDATION

IN

ORCHIDEÆ AND ASCLEPIADEÆ.

READ NOVEMBER 1ST AND 15TH, 1831.*

IN the Essay now submitted to the Society, my principal object is to give an account of some observations, made chiefly in the course of the present year, on the structure and economy of the sexual organs in Orchideæ and Asclepiadeæ,—the two families of phænogamous plants which have hitherto presented the most important objections to the prevailing theories of vegetable fecundation.

But before entering on this account, it is necessary to notice the various opinions that have been held respecting the mode of impregnation in both families: and in concluding the subject of Orchideæ, I shall advert to a few other points of structure in that natural order.

[1] [This portion of the Memoir was originally printed for private distribution in October, 1831. The additions made to it when reprinted in the 'Linnean Transactions,' consist chiefly of the references to the authors quoted, of three notes at pp. 495, 496 and 497, and of the plates and their explanations. The alterations are merely verbal, with the exception of a passage at pp. 522-4, beneath which I have appended the corresponding passage of the first impression in a note.—EDIT.]

ORCHIDEÆ.

The authors whose opinions or conjectures on the mode of impregnation in Orchideæ I have to notice, may be divided into such as have considered the direct application of the pollen to the stigma as necessary: and those who,—from certain peculiarities in the structure and relative position of the sexual organs in this family,—have regarded the direct contact of these parts as in many cases difficult
686] or altogether improbable, and have consequently had recourse to other explanations of the function.

In 1760, Haller, the earliest writer of the first class, in describing his Epipactis, states that the antheræ or pollen masses, after leaving the cells in which they are originally inclosed, are retained by the process called by him sustentaculum, the rostellum of Richard, from which they readily fall upon the stigma.[1] He adds, that both in this genus and in Orchis the stigma communicates by a fovea or channel with the ovarium.

But as in 1742 he correctly describes the stigma of Orchis,[2] and in his account of Epipactis[3] notices also the gland derived, as he says, from the sustentaculum, and which is introduced between and connects the pollen masses, his opinion on the subject, though not expressed, is distinctly implied even at that period; or as indeed it may be said to have been so early as 1736,[4] when he first described the channel communicating with the ovarium, and considered it as being in the place of a style.

In 1763, Adanson[5] states that the pollen masses are projected on the stigma, of which his description is at least as satisfactory as that of some very recent writers on the subject. He also describes the flower of an Orchideous plant as being monandrous, with a bilocular anthera, containing pollen which coheres in masses (a view of structure

[1] *Orchid. class. constitut. in Act. Helvet.* iv, p. 100.
[2] *Hall. Enum.* p. 262. [3] *Id.* p. 274.
[4] *Meth. stud. bot.* p. 21. [5] *Fam. des Plant.* ii, p. 69.

first entertained, but not published, by Bernard de Jussieu);[1] and he correctly marks the relation both of the stamen and placentæ of the ovarium to the divisions of the perianthium.

In 1777, Curtis, in the Flora Londinensis in his figure and account of *Ophrys apifera*, correctly delineates and describes the pollen masses, called by him antheræ, the [687
glands at their base inclosed in distinct cuculli or bursiculæ, and the stigma, with the surface of which he represents the masses as coming in contact.

In his second volume, the two lateral adnate lobes of the stigma, and the auriculæ of the column of *Orchis mascula*, are distinctly shown; and these auriculæ, now generally denominated rudimentary stamina, are also delineated in some other species of Orchis afterwards figured in the same work.

In 1793, Christian Konrad Sprengel[2] asserts that the pollen masses are applied directly to the secreting or viscid surface on the front of the column, in other words to the stigma, and that insects are generally the agents in this operation.

In 1799, J. K. Wachter[3] supports the same opinion, as far as regards the necessity of direct contact of the pollen masses with the female organ; and this observer was the first who succeeded in artificially impregnating an Orchideous plant, by applying the pollen to the stigma of *Habenaria bifolia*.

In 1799 also, or beginning of 1800, Schkuhr[4] takes the same view of the subject, and states that the pollen masses, which resist the action of common moisture, are readily dissolved by the viscid fluid of the stigma.

In 1800 Swartz,[5] in adopting the same opinion, notices various ways in which the application of the pollen may be effected in the different tribes of this family, repeats the statement of Schkuhr on the solvent power of the stigma, and in *Bletia Tankervilliæ* describes ducts

1 *Juss. gen. pl.* p. 66.
2 *Entd. Geheim.* p. 401.
3 *Römer, Archiv.* ii, p. 209.
4 *Handbuch* iii, p. 192.
5 *Act. Holm.* 1800, p. 134.

which convey the absorbed fluid from that organ to the ovarium.

In 1804, Salisbury[1] asserts that he had succeeded in
688] impregnating many species belonging to different tribes of Orchideæ, by applying the pollen masses to the stigma, whose channel communicating with the cavity of the ovarium, and first noticed by Haller, he also describes..

In 1827, Professor L. C. Treviranus[2] published an account of several experiments made by him in 1824, which satisfactorily prove that impregnation in this family may be effected by the direct application of the pollen to the stigma.

About the end of 1830 a letter from Professor Amici[3] to M. Mirbel was published, in which that distinguished microscopical observer asserts that in many phænogamous plants the pollen tubes, or *boyaux*, penetrate through the style into the cavity of the ovarium, and are applied directly to the ovula.

In this important communication Orchideæ are not mentioned, but M. Adolphe Brongniart in a note states that he himself has seen the production of *boyaux* or pollen tubes even in this family; that here, however, as well as in all the other tribes in which he had examined these tubes, he found them to terminate in the tissue of the stigma.

Of the second class of authors the earliest is Linnæus,[4] who, in 1764, not satisfied either with his own or any other description then given of the stigma, inquires whether the influence of the pollen may not be communicated internally to the ovarium.

In 1770, Schmidel,[5] in an account which he gives of a species of Epipactis, describes and figures the upper lip of the stigma, the rostellum of Richard, with its gland both before and after the bursting of the anthera; and as he

[1] *Linn. Soc. Transact.* vii, p. 29. [2] *Zeitschrift f. Physiol.* ii, p. 225.
[3] *Annal. des Sc. Nat.* xxi, p. 329.
[4] *Prælect. in Ord. Nat. ed. Giseke*, p. 182.
[5] *Gesn. Op. Bot. hist. plant.* fasc. ii, p. 15, tab. 19.

denominates that part, before the pollen masses are [689
attached to it, "stigma virgineum," it may be considered as belonging to the same class.

Koelreuter, the next writer in point of time, and whose essay was published before Linnæus's query appeared, states, in 1775,[1] that the pollen masses, which he denominates naked antheræ, impart their fecundating matter to the surface of the cells of the true anthera, regarded by him consequently as stigma, and that through this surface it is absorbed and conveyed to the ovarium.

In 1787, Dr. Jonathan Stokes[2] conjectures that in Orchideæ, as well as in Asclepiadeæ, the male influence, or principle of arrangement, as it is termed by John Hunter, may be conveyed to the embryo without the intervention of air: a repetition certainly of Linnæus's conjecture, with which, however, as it was not published till 1791, he could not have been acquainted.

In 1791, Batsch[3] states that in Orchis and Ophrys,—and his observation may be extended at least to all Satyrinæ or Ophrydeæ,—the only way in which the mass of pollen can act on the ovarium, is by the retrogradation of the impregnating power through the pedunculus or caudicula of the pollen mass to the gland beneath it, which he is disposed to refer rather to the stigma than to the anthera.

The late Professor Richard, in 1802,[4] expressly says that fecundation is operated in Orchideæ and Asclepiadeæ without a change of place in the stamina; his opinion therefore must be considered identical with that of Batsch, and extended to the whole order.

It might perhaps be inferred from the description which I gave of Orchideæ in a work published in 1810,[5] that my
opinion respecting the mode of impregnation agreed with [690
that of Batsch and Richard, though it is not there actually expressed, nor indeed very clearly in another publication of nearly the same date,[6] in which I had adverted to this

1 *Act. Phys. Palat.* iii, p. 55.
2 *With. Bot. Arrang.* 2nd ed. ii, p. 964.
3 *Botanische Bemerk.* i, p. 3.
4 *Dict. de Botan. par Bulliard,* ed. 2, p. 56.
5 *Prodr. Flor. Nov. Holl.* i, p. 310.
6 *Linn. Soc. Transact.* x, p. 19.

family. But I have since on several occasions more explicitly stated that opinion, which, until lately, I always considered the most probable hypothesis on the subject. At the same time its probability in this family appeared to me somewhat less than in Asclepiadeæ. For in Orchideæ a secreting surface in the female organ, apparently destined to act on the pollen without the intervention of any other part, is manifest; and some direct evidence of the fact existed, though not then considered satisfactory. In Asclepiadeæ, however, I entertained hardly any doubt on the subject; the only apparently secreting surface of the stigma in that family being occupied by the supposed conductors of the male influence, and no evidence whatever, with which I was acquainted, existing of its action through any other channel.

In 1816 or 1818 I received from the late celebrated Aubert du Petit Thouars some printed sheets of an intended work on Orchideæ, which, with a few alterations, was completed and published in 1822.[1]

From the unfinished work, as well as that which was afterwards published, it appears that this ingenious botanist considered the glutinous substance connecting the grains or lobules of pollen as the "aura seminalis" or fecundating matter; that the elastic pedicel of the pollen mass, existing in part of the family, but according to him not formed before expansion, consists of this gluten; and that in the expanded flower the gluten which has escaped from the pollen is, in all cases, in communication with the stigma.

He describes the stigma as forming on the surface of
691] the column a glutinous disk, from which a central thread or cord of the same nature is continued through the style to the cavity of the ovarium, where it divides into three branches, and that each of these is again subdivided into two. The six branches thus formed, are closely applied to the parietes of the ovarium, run down on each side of the corresponding placenta to its base, each giving off nume-

[1] *Hist. des Orchid.* p. 14.

rous ramuli, which spread themselves among the ovula, and separate them into irregular groups.

Hence, according to this author, a communication is established between the anthera and the ovula, which he adds are impregnated through their surface, and not, as he supposes to be the case in other families, through their funiculus or point of attachment to the placenta.

The remarkable account of the stigma here quoted, though coming from so distinguished and original an observer, and one who had particularly studied this family of plants, seems either to have been entirely overlooked, or in some degree discredited by more recent writers, none of whom, as far as I can find, have even alluded to it. And I confess it entirely escaped me until after I had made the observations which will be stated in the present essay, and which confirm its accuracy as to the existence and course of the parietal cords, though not as to their nature and origin.

In 1824 Professor Link[1] expresses his opinion that the rostellum of Richard is without doubt the true stigma.

In 1829 Mr. Lindley,[2] who for several years has particularly studied and has lately published part of a valuable systematic work on Orchideous Plants, states that in this family impregnation takes effect by absorption from the pollen masses through their gland into the stigmatic channel.

In 1830, in his Introduction to the Natural System of Botany, the same statement is repeated; and in this [692
work it also appears that he regards the glands to which the pollen masses become attached in Ophrydeæ as derived from the stamen, and not belonging to the stigma,[3] as in 1810 I had described them. It would even appear, from a passage in his systematic work[4] published in the same

[1] *Philos. Bot.* p. 298. [2] *Synops. Brit. Flor.* p. 256.

[3] "The pollen is not less curious. Now we have it in separate grains, as in other plants, but cohering to a mesh-work of cellular tissue, which is collected into a sort of central elastic strap; now the granules cohere in small angular indefinite masses, and the central elastic strap becomes more apparent, has a glandular extremity, which is often reclined in a peculiar pouch especially destined for its protection."—*Introduct. to Nat. Syst. of Bot.* p. 263.

[4] *Gen. and Sp. of Orchid.* Part I, p. 3.

year, that he considers the analogous glands, existing in most other tribes of Orchideæ, as equally belonging to the stamen: in his "Introduction," however, he refers them to the stigma in all cases except in Ophrydeæ.

Towards the end of 1830 the first part of Mr. Francis Bauer's Illustrations of Orchideous Plants edited by Mr. Lindley, was published.

From this work, of the importance and beauty of which it is impossible to speak too highly, it may be collected that Mr. Bauer's opinion or theory of impregnation in Orchideæ does not materially differ from that of Batsch, Richard, and other more recent writers. From one of the figures it appears that this theory had occurred to him as early as 1792; and in another figure, bearing the same date, he has accurately represented the structure of the grains of pollen in a plant belonging to Ophrydeæ, a structure which I had not ascertained in that tribe till 1806. Although Mr. Bauer's theory is essentially the same as that of Batsch and Richard, yet there are some points in which it may be considered peculiar; and chiefly in his supposing impregnation to take effect long before the ex-
693] pansion of the flower, at a time when the sexual organs are so placed with relation to each other that the fecundating matter, believed by him to pass from the pollen mass through its caudicula, where that part exists, to the gland attached to it, may be readily communicated to the stigma, with which the gland is then either in absolute contact or closely approximated. The more important points of this account may be extended to nearly the whole order, but is strictly applicable only to Satyrinæ or Ophrydeæ, a tribe in which Mr. Bauer seems, with Mr. Lindley, to consider the glands as belonging to the stamen and not to the stigma.[1] In those genera of this tribe in which the glands

[1] In the second part of Mr. Bauer's Illustrations, which has appeared since this paper was read, the explanation of Tab. 3, fig. 6, is corrected in the following manner:

"*For* 6. A pollen mass with its caudicula and gland taken out of the anther;

"*Read* 6. A pollen mass with its caudicula and the internal socket of the stigmatic gland."

It is evident, indeed, in the second part of the Illustrations, from figs. 8, 9, 11, and 12, of Tab. 12, representing details of *Satyrium pustulatum*, and the

are included in a pouch or bursicula, he describes and figures perforations in the back of the pouch, through which the fecundating matter is communicated from the glands to the stigma; and one of the figures is intended to represent a gland in the act of parting with the fecundating matter.

It is impossible to jüdge correctly of Mr. Bauer's theory until all the proofs and arguments in its favour are adduced. I may observe, however, that those already published are by no means satisfactory to me.

For, in the first place, in the very early stage in which, [694
according to this theory, impregnation is supposed to be effected, it appears to me that the pollen is not in a state to impart its fecundating matter, nor the stigma to receive it; and it may be added, though this is of less weight, that the ovula have neither acquired the usual degree of development, nor that position which they afterwards take, and which gives the apex of the nucleus or point of impregnation the proper direction, with regard to the supposed impregnating surface.

Secondly, in the figure which may be said to exhibit a demonstration of the correctness of the theory,—in that, namely, representing the gland in the act of parting with the fecundating matter,—the magnifying power employed (which is only fifteen times) is surely insufficient for the establishment of a fact of this kind; while the disengagement of minute granules, which no doubt often takes place when the gland is immersed in water, may readily be accounted for in another way.[1]

drawings of which were made in 1800, that Mr. Bauer must, from that time at least, have correctly understood the origin of the glands in Ophrydeæ. There is nothing, however, in any of the figures in Tab. 3 of the first part at variance with their explanations, from which I judged of his opinion. It may therefore be concluded that Mr. Bauer had not examined these explanations before their publication.

This second observation ought not now to be taken into account, as in the second part of Mr. Bauer's Illustrations the following correction occurs respecting the figure alluded to (Tab. 3. fig. 8).

"This is in some measure an ideal figure to represent in what way the fecundating matter is supposed to leave the caudicula and stigmatic gland; for this reason there has been no attention paid to preserving a proportion between the pollen mass and the fecundating matter."

I may here, however, remark, that it was evidently not my intention, in the

Thirdly, I have never been able to find those perforations, represented by Mr. Bauer, in the bursiculæ of Orchis and Ophrys, and the existence of which in these genera is essential to his hypothesis.

And, lastly, the appearance of the stigma in *Bletia Tan-*
695] *kervilliæ*, after impregnation, as he believes, according to my view of the subject would rather prove that it was in a state capable of acting upon, but had not yet received the fecundating matter from, the anthera.

In thus venturing to differ from so accurate and experienced an observer as Mr. Bauer on a subject which he has for many years minutely studied and so beautifully illustrated, I am well aware how great a risk I incur of being myself found in error.

I am very desirous, however, that the perusal of this sketch of the various statements that have appeared on the question of impregnation, with the greater part of which he is at present probably unacquainted, should induce him to re-examine the facts and arguments by which his own opinion on this subject is supported. He will thus either succeed in establishing his theory on more satisfactory grounds, or, if the examination should prove unfavourable, he will, I am persuaded, from his well-known candour, as readily abandon it.

The notice here given of the opinions of botanists on impregnation in Orchideæ brings the subject down to the spring of the present year, when from circumstances, which I may hereafter have occasion to advert to, my attention was directed to this family of plants, the particular study of which I had for a long time discontinued.

In reviewing notes respecting them, made many years ago, I found some points merely hinted at, or imperfectly made out, which seemed deserving of further examination; and in the course of these inquiries, other observations of at least equal importance suggested themselves.

observation in question, to throw any doubt on the correctness of Mr. Bauer's figure, being aware that very minute granular matter, separating from the gland when immersed in water, is actually visible with a lens of about half an inch focus. I objected to it only as a satisfactory proof of the theory referred to.

I now proceed to state, in some cases briefly, in others at greater length, the results of this investigation.

The first question that occupied me was, the relation which the lateral and generally rudimentary stamina bear [696 to the other parts of the flower.

Into this subject I had in part entered in my Observations on Apostasia, published by Dr. Wallich in his 'Plantæ Asiaticæ Rariores,'[1] and had then considered it probable that in all cases these Stamina, in whatever state of development they were found, belonged to a different series from the middle and usually fertile stamen; in other words, were placed opposite to the two lateral divisions of the inner series of the perianthium. In 1810, however, when I first advanced my hypothesis of the true nature of these processes of the column, I supposed, though the opinion was not then expressed, that they formed the complement of the outer series of stamina; a view which has been since very generally adopted, especially by Dr. Von Martius, who has given it in a stenographic formula, and by Mr. Lindley, who has exhibited the relative position of parts in this family in a diagram.[2] A careful examination of the structure of the column in various tribes of the order, chiefly by means of transverse sections, has fully confirmed the opinion I entertained when treating of Apostasia; and more particularly established the fact in Cypripedium, in which these lateral stamina are perfectly developed.

On the hypothesis of rudimentary stamina I may remark, that it presented itself to me some time before the publication of the Prodromus Floræ Novæ Hollandiæ; and my belief is, that until the appearance of that work this view had not been taken by any other observer in England. Mr. Bauer at least, in a recent conversation on the subject, readily admitted, with his usual candour, that although acquainted with a case of accidental development, the general view had not occurred to him until stated by me.

In my mind it arose from contrasting the structure of [697 Cypripedium with those genera of New Holland Orchideæ —Diuris, Prasophyllum, and others—in which the lateral

[1] Vol. i, p. 74.

[2] *Introduct. to Nat. Syst.* p. 264.

processes or appendages of the column are so remarkably developed; and I afterwards, in searching for additional confirmations of the hypothesis, believed I had found such in the more minute lateral auriculæ of the column present in most Ophrydeæ.

These auriculæ, however, though they might serve to confirm, would hardly have suggested the hypothesis, at the period especially of which I speak. They had indeed until then been altogether overlooked, except by Malpighi,[1] by Curtis in his Flora Londinensis, perhaps in Walcott's Flora Britannica, and by Mr. Bauer, whom they were not likely to escape.

In my recent observations on Apostasia, referred to, I noticed a singular monstrosity of *Habenaria bifolia*, which, if such deviations from ordinary structure are always to be trusted, would throw great doubt on the hypothesis being applicable to these auriculæ of Ophrydeæ. For in this case, in which three antheræ are formed, auriculæ not only exist on the middle or ordinary stamen, but one is also found on the upper side of each of the lateral antheræ, which are here opposite to two divisions of the outer series of the perianthium. I have lately met with another instance of a similar monstrosity equally unfavourable; and I may add that this doubt is still further strengthened by my not being able to find vascular cords connected with these auriculæ in the only plants of Ophrydeæ in which I have carefully examined, with this object, the structure of the column, namely, *Orchis Morio*, *mascula*, and *latifolia*.

I do not indeed regard the absence of vessels as a complete proof of these auriculæ not being rudimentary stamina. But I may remark, that in the other tribes of Orchideæ, in
698] many of whose genera analogous processes are found, and in which tribes alone cases of their complete development have hitherto been observed, vessels not only generally exist in these processes, but may be traced to their expected origins, namely, into those cords which also supply the inner lateral divisions of the perianthium.

Although not necessarily connected with my subject, I

[1] *Op. Om.* tab. 25, fig. 142.

may here advert to the remarkable monstrosity in the flowers of an *Ophrys* described and figured by M. His[1] upwards of two years before the appearance of my Prodromus. This account I did not meet with till after that part of the volume relating to Orchideæ was printed; and I have here only to observe respecting it, that neither the monstrosity itself, consisting of the conversion into stamina of the three inner divisions of the perianthium, nor the author's speculation founded on it, has any connection with my opinion which relates to the processes of the column.

M. His's paper, however, and the remarkable structure of *Epistephium* of M. Kunth, have together given rise to a third hypothesis, whose author, M. Achille Richard,[2] considers an Orchideous flower as generally deprived of the outer series of the perianthium, which is present only in Epistephium. He consequently regards the existing inner series of perianthium, or that to which the labellum belongs, as formed of metamorphosed stamina.

This hypothesis, although apparently sanctioned by the structure of Scitamineæ, I consider untenable; the external additional part in Epistephium, which I have examined, appearing to me rather analogous to the calyculus in some Santalaceæ, in a few Proteaceæ, and perhaps to that of Loranthaceæ.

With reference to the support the hypothesis may [699
derive from the monstrosity described by M. His, I may add that I have met with more than one case of similar conversion into stamina of the inner series of the perianthium, or at least of its two lateral divisions, with a manifest tendency to the same change in the labellum: and in one of these cases, namely *Neottia picta*, in addition to the conversion of the two lateral divisions of the perianthium, the lateral processes of the column were also completely developed.

The next point examined was the composition of the Stigma with the relation of its lobes or divisions to the other parts of the flower, and especially to the supposed compo-

[1] *Journal de Physique*, lxv. (1807), p. 241.
[2] *Mém. de la Soc. d'Hist. Nat. de Paris*, iv, p. 16.

nent parts of the ovarium. On this subject very little information is to be obtained from the writings of botanists, most of whom have contented themselves with describing the stigma as a disk, a fovea glutinosa, a secreting surface, or viscid space in front of the column. The late celebrated Richard, however, who adverts to the occasional existence of two lateral processes of his gynizus, may be supposed to have had more correct notions of its composition: and it may also be observed, that in Curtis's plate of *Ophrys apifera* already referred to, and still more distinctly in Mr. Bauer's figure of *Orchis mascula*, the two lateral lobes are represented as distinct, corresponding very exactly with Haller's description, in 1742, of the stigma in this genus.

The result of my examination of this point satisfied me that Orchideæ have in reality three stigmata, generally more or less confluent, but in some cases manifestly distinct, and two of which are in several instances even furnished with styles of considerable length.

These stigmata are placed opposite to the three outer divisions of the perianthium, and consequently terminate the axes of the supposed component parts of the ovarium, always regarded by me as made up of three simple ovaria
700] united by their ovuliferous margins ; a structure in which the ordinary relation of stigmata to placentæ is that here found.

In Mr. Bauer's 'Illustrations' already referred to, a very different account is given of the composition of the ovarium, which is there said to be formed of six pieces.

This view of its composition seems to be founded on the existence of six vascular cords, on the apparent interruptions in the cellular tissue, and on the singular dehiscence of the capsule. But the mere number of vascular cords, which, being destined to supply all parts of the flower, may be said rather to indicate the divisions of the perianthium than those of the ovarium, cannot be considered as affording an argument of much importance, and, if it were, would equally apply to many other families having trilocular ovaria, as Irideæ ; while the interruptions or inequalities of cellular tissue may be viewed as only the preparation for that dehiscence which, though very remarkable in this order, is in a great degree analogous to that taking place in most Cruciferæ, in several Leguminosæ, and in other families of plants. It may also be objected to Mr. Bauer's view of the composition of ovarium, that the arrangement of the parietal placentæ, which on this hypothesis would occupy the axes of the three alternate component parts, is contrary to every analogy; while the position of the stigmata, if my account should prove to be correct, affords evidence nearly conclusive of the ovarium being formed of only three parts.

In those genera of Orchideæ in which the lateral stamina are perfect, and the middle stamen without anthera, namely, Cypripedium and Apostasia, all these lobes or divisions of stigma are equally developed, are of nearly similar form and texture, and, as I have proved by direct experiment in Cypripedium, are all equally capable of performing the proper function of the organ.

In most other cases the anterior lobe, or that placed [70]
opposite to the perfect stamen, and deriving its vessels from the same cord, manifestly differs both in form and texture from the other two. To this anterior, or upper lobe, as it generally becomes in the expanded flower, the glands always belong to which the pollen masses become attached, but from which they are in all cases originally distinct, as may be proved even in Ophrydeæ.

According to my view, therefore, of the mode of impregnation, its office is essentially different from that of the two lateral lobes or stigmata, which in various degrees of development are always present, and in all cases, when the ovarium is perfect, are capable of performing their proper function.

The greatest development of these lateral stigmata takes place in the tribe of Satyrinæ or Ophrydeæ, as in many species of *Habenaria*, those especially which are found near or within the tropics; and still more remarkably in *Bonatea speciosa*, a plant hardly indeed distinguishable from the same extensive genus.

It would seem that in *Bonatea* the extraordinary development and complete separation of these lateral stigmata,

have effectually concealed their true nature; and accordingly they have uniformly been considered as forming parts or appendages of the labellum, with which indeed their bases cohere. That they are really stigmata, however, I have proved by a careful examination of the tissue of their secreting surface, by the action of the pollen artificially applied to this tissue, by the descent of its tubes, hereafter to be described, along the upper surface of the styles which is destitute of epidermis, and by the consequent enlargement of the ovarium. *Diplomeris* of Mr. Don,[1] which may also be regarded as a species of Habenaria, is another example of nearly the same kind; and the
702] description of stigma which, in 1813, I introduced into the character of *Satyrium*,[2] implies an analogous development in that genus.

On the relative position of stamina and stigmata in the column of an Orchideous plant, it may be remarked that there is hardly an instance of a perfectly developed stamen and stigma placed opposite to each other, and consequently deriving their vessels from the same cord.

For, in the ordinary structure of the family in which only one perfect stamen is produced, the corresponding stigma loses entirely or in great part its proper function, which it recovers, so to speak, in those cases where this stamen becomes imperfect, or is destitute of an anthera: and hence, perhaps, it may be said that to obtain in any case the complete devolopment of the lateral stamina, and, what is of greater importance, to ensure in all cases the perfection of the lateral stigmata, these organs are never placed opposite, but uniformly alternate with each other.

The general conformation of the ovarium, with regard to the number and relative position of the parietal placentæ, and the arrangement of their numerous ovula, has long been well understood. But the early structure and evolution of the unimpregnated ovulum have not yet, as far as I know, been in any degree attended to.

In its gradual development, the ovulum exhibits a series

1 *Prodr. Flor. Nepal.* p. 26.
2 *Ait. Hort. Kew.* ed. 2, vol. v, p. 196.

of changes nearly agreeing with those which M. Mirbel[1] has described and illustrated as taking place in other families.

In the earliest state in which I have examined the ovulum in Orchideæ, it consists merely of a minute papilla projecting from the pulpy surface of the placenta. In the [703 next stage the annular rudiment of the future testa is visible at the base of the papilliform nucleus. The subsequent changes, namely, the enlargement of the testa, the production of a funiculus, which is never vascular, and the curvature or inversion of the whole ovulum, so as to approximate the apex of its nucleus to the surface of the placenta, take place in different genera at different periods with relation to the development of the other parts of the flower. In general when the flower expands, the ovulum will be found in a state and direction proper for receiving the male influence. But in several cases, as in Cypripedium and Epipactis, genera which in many other respects are nearly allied, the ovulum has not completed its inversion, nor is the nucleus entirely covered by its testa until long after expansion, and even after the pollen has been acted on by the stigma, and its tubes have penetrated into the cavity of the ovarium.

The tissue of the perfect stigmata in Orchideæ does not materially differ from that of many other families. In the early state the utriculi composing it are densely approximated, having no fluid interposed. In the more advanced but unimpregnated state, these utriculi enlarge, and are separated from each other by a copious and generally viscid secretion. The channel of the style, or stigma, whose parietes are similarly composed, undergoes the same changes. Both these states are represented in one of Mr. Bauer's plates, who however considers the more advanced stage as subsequent to impregnation.

In the advanced but still unimpregnated state of the ovarium, the upper portions, which are in continuation with the axes of the three placentæ, but do not produce

[1] *Annal. des Sc. Nat.* xvii, p. 302;—and in *Mém. de l'Acad. des Sc. de l'Instit.* ix, p. 212.

ovula, are of a texture somewhat different from that of the greater part of the cavity, but still more obviously different from that of the cavity of the style, being neither apparently
704] secreting nor consisting of similar utriculi. A narrow line of like surfaee is found extending on each side of every placenta nearly as far as it is ovuliferous. The three lines occupying the upper part of the axes, and the six lines marginal to the three placentæ, may, for a reason which will hereafter appear, be called the conducting surfaces of the ovarium.

The female organ, as now described, is in a proper state to be acted upon by the pollen applied to the stigma, and for the transmission of the fecundating matter into the cavity of the ovarium, in a manner and form which I shall presently attempt to explain.

In reflecting on the whole evidence existing in favour of the direct application of the pollen mass to the stigma, and especially on the recent experiments of Professor Treviranus,[1] I could no longer doubt that in this manner impregnation was actually effected in Orchideæ; and the sole difficulty in my mind to its being the only way arose from adverting to a circumstance that must have been remarked by every one who has particularly attended to this family, either in Europe or in tropical regions; namely, that all the capsules of a dense spike are not unfrequently ripened: a fact which at first seems hardly reconcilable with this mode of fecundation, at least on the supposition that the pollen mass is applied to the stigma by insects.

Without going fully into the question at present, I shall here only remark, that in several such cases I have satisfied myself, by actual examination of the stigmata belonging to capsules taken at many different heights in the spike, that pollen, by whatever means, had actually been applied to them.[2]

[1] *Zeitschrift f. Physiol.* ii, p. 225.

[2] It may also be observed, that the same difficulty applies to many other cases of dense inflorescence, as to the female spikes or strobili of Coniferæ, Zamia, and Zea; in all of which the symmetry of the ripe fruit is generally perfect, although partial failures of impregnation might be at least equally expected.

Believing, therefore, this is to be the only mode in [705
which impregnation is effected, I proceeded to examine the immediate changes produced by the application of the pollen masses to the stigma.

From numerous observations and experiments made with this view, chiefly in Satyrinæ or Ophrydeæ, and Arethuseæ, not however confined to these tribes, it was ascertained that the grains of pollen, soon after being applied to the stigma, either in the entire mass or separately, produce tubes or *boyaux* analogous to those first observed in one case by Professor Amici,[1] and afterwards in numerous others, and in many families, by M. Adolphe Brongniart.[2]

In Orchideæ one tube only is emitted from the absolutely simple grain, while the number of tubes generally corresponds with that of the divisions or cells of the compound grain. These tubes are of extreme tenuity, their diameter being generally less than 1-2000th of an inch, and they acquire a great length, even while adhering to the grains producing them. From these, however, they separate generally while still involved in the secretion and mixed with the utriculi of the stigma; and I have never observed an instance of a tube with its grain attached to it lower than the tissue of the stigma. In form they are perfectly cylindrical, or of equal diameter, neither dilated at the apex nor sensibly contracted in any part of their course. I have never found them either branched or jointed; but have frequently observed apparent interruptions in the tube, probably caused by partial coagulations of the contained fluid. Even in their earliest stage, while in length hardly equal to the diameter of the grain, I have not been able to observe them to contain distinct granules in employing a magnifying power of 150. With a [706
power of 300 or 400 indeed, extremely minute and very transparent granular matter may be detected; but such granules are very different from those which have been supposed to belong to the grains of pollen.

As an entire pollen mass is usually applied to the surface

Atti della Soc. Ital. xix, par. 2, p. 254. *Annal. des Sc. Nat.* ii, p. 66.
Annal. des Sc. Nat. xii, p. 34.

of the stigma, and as a great proportion of the mass so applied is acted upon by the fluid in which it is immersed, the tubes produced are generally very numerous, and together form a cord which passes through the channel of the stigma or style.

On reaching the cavity of the ovarium this cord regularly divides into three parts, the divisions being closely applied to those short upper portions of the axes of the valves which are not placentiferous; and at the point where the placenta commences each cord again divides into two branches. These six cords descend along the conducting surfaces already described when speaking of the unimpregnated ovarium, and generally extend as far as the placentæ themselves, with which they are thus placed nearly but perhaps not absolutely in contact.

The cords now described, both general and partial, seem to me to be entirely composed of pollen tubes, certainly without any mixture of the utriculi of the stigma, or, as far as I can ascertain, of the tissue of the conducting surfaces.

In two cases, namely *Ophrys apifera* and *Cypripedium spectabile*, I at one time believed I had seen tubes going off laterally from the partial cords towards the placentæ and mixing with the ovula; but I am not at present entirely satisfied with the exactness of these observations, and I have never been able to detect similar ramifications in any other case.[1]

That the existence of these tubes in the cavity of the ovarium is essential to fecundation in Orchideæ, can hardly be questioned. But the manner in which they operate on,
707] or whether they come actually in contact with, the ovula, are points which still remain undetermined.

I am aware that Professor Amici,[2] who discovered in several plants the remarkable fact of the penetration of the pollen tubes into the cavity of the ovarium, and who regards this economy as being very general, likewise believes that in all cases a pollen tube comes in contact with an

[1] See Additional Observations.
[2] *Annal. des Sc. Nat.* xxi, p. 329.

ovulum. M. Du Petit Thouars also, in his account already quoted of these cords, supposed by him to belong to the stigma of Orchideæ, describes their ultimate ramifications as mixing with the ovula.

I do not however consider myself so far advanced as these observers in this very important point;[1] and what I shall have to adduce on the subject of Asclepiadeæ, makes me hesitate still more to adopt their statements.

I may also remark that in Orchideæ the six cords are to be met with even in the ripe capsule, in which, allowance being made for the effect of pressure, they are not materially reduced in size; and the statement by M. Du Petit Thouars, of the lateral branches separating the ovula into irregular groups, is certainly not altogether correct; these groups being equally distinct before the existence of the cords.

With regard to the question of the origin of the pollen tubes, several arguments might be adduced in favour of M. Brongniart's opinion; which is, that they belong to the inner membrane of the grain, the intimate cohesion of the two membranes being assumed in most cases, and the no less intimate union of the constituent parts of compound grains in some others. That an inner membrane does occasionally exist is manifest in the pollen of several Coniferæ, in which the outer coat regularly bursts and is deciduous; and it will hereafter appear, that the structure in Asclepiadeæ confirms the correctness of this view.

But whatever opinion may be entertained as to the [708
origin of the tube, it can hardly be questioned that its production or growth is a vital action excited in the grain by the application of an external stimulus. The appropriate and most powerful stimulus to this action is no doubt contact, at the proper period, with the secretion or surface of the stigma of the same species. Many facts, however, and among others the existence of hybrid plants, prove that this is not the only stimulus capable of producing the effect; and in Orchideæ I have found that the action in

[1] See Additional Observations.

the pollen of one species may be excited by the stigma of another belonging to a very different tribe.

The elongation of the tubes, so remarkable in this family, and their separation from the grain long before their growth is completed, render it probable that they derive nourishment either from the particles contained in the grain, or from the conducting surfaces with which they are in contact.

The first visible effect of the action of the pollen on the stigma is the enlargement of the ovarium, which, in cases where it was reversed by torsion in the flowering state, generally untwists and resumes its original position.

Of the changes produced in the ovulum consequent to impregnation, the first consists in its enlargement merely; and in the few cases where the nucleus is at this period still partially exposed, it becomes completely covered by the testa, the original apex, but now the lower extremity of which continues open. The next change consists in the disappearance of the nucleus, probably from its acquiring greater transparency, and becoming confluent with the substance of the testa. Soon after, or perhaps simultaneously with, the disappearance of the original nucleus, and while the enlargement of the whole ovulum is gradually proceeding, a minute opaque round speck, generally seated about the middle of the testa, becomes visible. The
709] opaque speck is the commencement of the future embryo. At this period, or until the opaque corpuscle or nucleus has acquired more than half the size it attains in the ripe seed, a thread may be traced from its apex very nearly to the open end of the testa, or as it may be supposed, to the apex of the original nucleus of the unimpregnated ovulum.

This thread consists of a simple series of short cells, in one of which, in a single instance only however, I observed a circulation of very minute granular matter; and in several cases I have been able to distinguish in these cells that granular areola so frequently existing in the cells of Orchideous plants, and to which I shall have occasion hereafter to advert.

The lowermost joint or cell of this thread is probably the original state of what afterwards, from enlargement and deposition of granular matter, becomes the opaque speck or rudiment of the future embryo.

The only appreciable changes taking place in this opaque rudiment of the embryo are its gradual increase in size, and at length its manifest cellular structure.

In the ripe state it forms an ovate or nearly spherical body, consisting, as far as I have been able to ascertain, of a uniform cellular tissue covered by a very thin membrane, the base of which does not exhibit any indication of original attachment at that point; while at the apex the remains of the lower shrivelled joints of the cellular thread are still frequently visible.

This cellular body may be supposed to constitute the Embryo, which would therefore be without albumen, and whose germinating point, judging from analogy, would be its apex, or that extremity where the cellular thread is found; and consequently that corresponding with the apex of the nucleus in the unimpregnated ovulum.

The description here given of the undivided embryo in Orchideous plants as forming the whole body of the nucleus, [710 and consequently being destitute of albumen, agrees with the account first I believe published by M. du Petit Thouars,[1] and very soon after by the late excellent Richard.[2]

The only other remark I have to make on the fructification of this family, is, that the seed itself, as well as its funiculus, is entirely without vessels, and that the funiculus, which in the ripe seed is inserted into the testa close to one side of its open base, can hardly be traced beyond that point.

I shall conclude my observations on Orchideæ with a notice of some points of their general structure, which chiefly relate to the cellular tissue.

In each cell of the epidermis of a great part of this family, especially of those with membranaceous leaves, a single circular areola, generally somewhat more opaque than the membrane of the cell, is observable. This areola, which is more or less distinctly granular, is slightly convex,

[1] *Hist. des Orchid.* p. 19. [2] *Mém. du Mus. d'Hist. Nat.* iv, p. 41.

and although it seems to be on the surface is in reality covered by the outer lamina of the cell. There is no regularity as to its place in the cell; it is not unfrequently however central or nearly so.

As only one areola belongs to each cell, and as in many cases where it exists in the common cells of the epidermis it is also visible in the cutaneous glands or stomata, and in these is always double,—one being on each side of the limb,—it is highly probable that the cutaneous gland is in all cases composed of two cells of peculiar form, the line of union being the longitudinal axis of the disk or pore.

This areola, or nucleus of the cell as perhaps it might be termed, is not confined to the epidermis, being also found not only in the pubescence of the surface, particularly when
711] jointed, as in Cypripedium, but in many cases in the parenchyma or internal cells of the tissue, especially when these are free from the deposition of granular matter.

In the compressed cells of the epidermis the nucleus is in a corresponding degree flattened; but in the internal tissue it is often nearly spherical, more or less firmly adhering to one of the walls, and projecting into the cavity of the cell. In this state it may not unfrequently be found in the substance of the column, and in that of the perianthium.

The nucleus is manifest also in the tissue of the stigma, where, in accordance with the compression of the utriculi, it has an intermediate form, being neither so much flattened as in the epidermis, nor so convex as it is in the internal tissue of the column.

I may here remark, that I am acquainted with one case of apparent exception to the nucleus being solitary in each utriculus or cell, namely in *Bletia Tankervilliæ*.

In the utriculi of the stigma of this plant I have generally, though not always, found a second areola apparently on the surface, and composed of much larger granules than the ordinary nucleus, which is formed of very minute granular matter, and seems to be deep seated.

Mr. Bauer has represented the tissue of the stigma in this species of Bletia, both before and as he believes after impregnation; and in the latter state the utriculi are marked with from one to three areolæ of similar appearance.

The nucleus may even be supposed to exist in the pollen of this family. In the early stages of its formation at least a minute areola is often visible in the simple grain, and in each of the constituent parts or cells of the compound grain. But these areolæ may perhaps rather be considered as merely the points of production of the tubes.

This nucleus of the cell is not confined to Orchideæ, [712
but is equally manifest in many other Monocotyledonous families; and I have even found it, hitherto however in very few cases, in the epidermis of Dicotyledonous plants; though in this primary division it may perhaps be said to exist in the early stages of development of the pollen. Among Monocotyledones the orders in which it is most remarkable are Liliaceæ, Hemerocallideæ, Asphodeleæ, Irideæ, and Commelineæ.

In some plants belonging to this last-mentioned family, especially in *Tradescantia virginica* and several nearly related species, it is uncommonly distinct, not only in the epidermis and in the jointed hairs of the filaments,[1] but in

[1] The jointed hair of the filament in this genus forms one of the most interesting microscopic objects with which I am acquainted, and that in three different ways:

1st. Its surface is marked with extremely fine longitudinal parallel equidistant lines or striæ, whose intervals are equal from about 1-15,000th to 1-20,000th of an inch. It might therefore in some cases be conveniently employed as a micrometer.

2ndly. The nucleus of the joint or cell is very distinct as well as regular in form, and by pressure is easily separated entire from the joint. It then appears to be exactly round, nearly lenticular, and its granular matter is either held together by a coagulated pulp not visibly granular,—or, which may be considered equally probable, by an enveloping membrane. The analogy of this nucleus to that existing in the various stages of development of the cells in which the grains of pollen are formed in the same species, is sufficiently obvious.

3rdly. In the joint when immersed in water, being at the same time freed from air, and consequently made more transparent, a circulation of very minute granular matter is visible to a lens magnifying from 300 to 400 times. This motion of the granular fluid is seldom in one uniform circle, but frequently in several apparently independent threads or currents: and these currents, though often exactly longitudinal and consequently in the direction of the striæ of the membrane, are not unfrequently observed forming various angles with these striæ. The smallest of the threads or streamlets appear to consist of a

713] the tissue of stigma, in the cells of the ovulum even before impregnation, and in all the stages of formation of the grains of pollen, the evolution of which is so remarkable in those species of Tradescantia.[1]

The few indications of the presence of this nucleus, or areola, that I have hitherto met with in the publications of botanists, are chiefly in some figures of epidermis, in the recent works of Meyen and Purkinje, and in one case in M. Adolphe Brongniart's memoir on the structure of leaves. But so little importance seems to be attached to it, that the appearance is not always referred to in the explanations of the figures in which it is represented. Mr. Bauer, however, who has also figured it in the utriculi of the stigma of *Bletia Tankervilliæ,* has more particularly noticed it, and seems to consider it as only visible after impregnation.

714] The second point of structure in Orchideæ to which I shall at present more briefly advert, is the frequent exist-

single series of particles. The course of these currents seems often in some degree affected by the nucleus, towards or from which many of them occasionally tend or appear to proceed. They can hardly, however, be said to be impeded by the nucleus, for they are occasionally observed passing between its surface and that of the cell; a proof that this body does not adhere to both sides of the cavity, and also that the number and various directions of the currents cannot be owing to partial obstructions arising from the unequal compression of the cell.

[1] In the very early stage of the flower-bud of *Tradescantia virginica,* while the antheræ are yet colourless, their loculi are filled with minute lenticular grains, having a transparent flat limb, with a slightly convex and minutely granular semi-opaque disk. This disk is the nucleus of the cell, which probably loses its membrane or limb, and, gradually enlarging, forms in the next stage a grain also lenticular, and which is marked either with only one transparent line dividing it into two equal parts, or with two lines crossing at right angles, and dividing it into four equal parts. In each of the quadrants a small nucleus is visible; and even where one transparent line only is distinguishable, two nuclei may frequently be found in each semicircular division. These nuclei may be readily extracted from the containing grain by pressure, and after separation retain their original form.

In the next stage examined, the greater number of grains consisted of the semicircular divisions already noticed, which had naturally separated, and now contained only one nucleus, which had greatly increased in size.

In the succeeding state the grain apparently consisted of the nucleus of the former stage considerably enlarged, having a regular oval form, a somewhat granular surface, and originally a small nucleus. This oval grain continuing to increase in size, and in the thickness and opacity of its membrane, acquires a pale yellow colour, and is now the perfect grain of pollen.

ence, particularly in the parasitical tribes, of fibrous or spirally striated cells in the parenchyma, especially of the leaves, but also in the white covering of the radical fibres.

In the leaves, they are either short spirally striated cells whose longer diameter is at right angles to the surface, as in *Stelis* and *Pleurothallis*, and whose fibres or striæ are connected by a broader membrane; or, being greatly elongated and running in the direction of the leaf, resemble compound spiral vessels of enormous diameter, aud consisting entirely of the spiral fibres with no visible connecting membrane: the real spiral vessels in the same species being, as they generally are in the family, very slender and simple. In the white covering of the radical fibres the shorter striated cell is met with in many genera, especially I think in Oncidium and Epidendrum, in one species of which they have been remarked and figured by Meyen.[1]

My concluding observation on Orchideæ relates to the very general existence and great abundance, in this family, of Raphides or acicular crystals in almost every part of the cellular tissue.

In each cell where they exist these crystals are arranged in a single fasciculus, which is generally of a square form.

The individual crystals,—which are parallel to each other,—are cylindrical, with no apparent angles, and have short and equally pointed extremities.

The abundance of these fasciculi of crystals in the cellular tissue of the auriculæ of the column or supposed lateral stamina in Orphydeæ, is very remarkable, giving these processes externally a granular appearance, which has been [715
noticed though its cause seems to have been overlooked.

In the recent work of Meyen,[2] also, some examples of these crystals in Orchideæ are given.

[1] *Phytotomie*, tab. 11, f. 1 and 2.

[2] *Phytotomie*.

ASCLEPIADEÆ.

The various statements and conjectures on the structure and functions of the sexual organs in this family were collected, and published in 1811, by the late Baron Jacquin, in a separate volume, entitled, 'Genitalia Asclepiadearum Controversa.'

To this work, up to the period when it appeared, I may refer for a complete history, and to the tenth volume of the Linnean Society's Transactions, along with the first of the Wernerian Natural History Society's Memoirs, published somewhat earlier, for a slight sketch, of the subject.

I shall here therefore only notice such statements as Jacquin has either omitted or imperfectly given, and continue the history to the present time.

In 1763, Adanson correctly describes the stamina in Asclepias as having their filaments united into a tube surrounding the ovaria, their antheræ bilocular and cohering with the base of the stigma, and the pollen of each cell forming a mass composed of confluent grains as in Orchideæ. He is also correct in considering the pentagonal body as the stigma; but he has entirely overlooked its glands and processes, nor does he say anything respecting the manner in which the pollen masses act upon or communicate their fecundating matter to it.

In 1779, Gleichen,[1] although he expressly says that in young flower-buds the pollen masses are distinct from those glands of the pentagonal central body to which they
716] afterwards are attached, yet considers both masses and glands as equally belonging to the anthera, the mass being the receptacle of the pollen. He further states that before the masses unite with the glands they are removed from the cells in which they were lodged, and are found firmly implanted by their sharp edge into the wall of the tube which surrounds the ovaria; that in this state a white

[1] *Microscop. Entd.* p. 73, et seq.

viscid substance hangs to them, which, when highly magnified, appears to consist of very slender tubes containing minute globules; and these tubes with their contents he considers as constituting the early preparation for the formation of pollen. He also asserts that the tops of the styles are not originally connected with the pentagonal body to which the glands belong—the stigma of Adanson, Jacquin, and others; and that therefore the true stigmata are those extremities of the styles on which, he adds, vesicles and threads are observable. And lastly, he supposes that impregnation, which he says is of rare occurrence in this family, does not usually take place until those stigmata have penetrated through the substance of the pentagonal body, and are on a level with its apex; at the same time he is disposed to believe that insects may occasionally assist in this function, by carrying the fecundating matter directly to the stigmata, if I understand him, even before they enter the pentagonal body. His conclusion therefore is, that in Asclepiadeæ impregnation may be effected in two different ways.

This description, in several respects so paradoxical, and of which Jacquin has overlooked some of the most important parts, is too remarkable to be here either omitted or abridged. It is not indeed strictly correct in more than two points, namely, in the pollen masses being originally distinct from the glands, and in the masses, when found implanted in the membrane surrounding the ovarium, having minute tubes filled with granular matter [717
hanging to them. The remaining statements, however, though essentially erroneous, are so far founded in fact, that had Gleichen either opened or rather dilated the opening which must have existed in the pollen mass when these tubes were found hanging to it, and more carefully attended to the state of the other parts of the flower when the mass was seen implanted in the tube, he must necessarily have obtained a correct view of the whole structure, and consequently have greatly advanced—by at least half a century—not only our knowledge of this particular family, but also the general subject of vegetable impregnation.

In 1793, Christian Konrad Sprengel, who adopts the opinion of Jacquin both with respect to the pollen masses and pentagonal stigma, further states, that this stigma has a secreting upper surface or apex, and is formed of two united bodies, each of which conveys to its corresponding ovarium the fecundating matter, consisting of the oily fluid which exudes from the surface of the pollen mass. He also considers insects as here essentially necessary in impregnation, which they effect by extracting, in a manner particularly described, the pollen masses from the cells, and applying them to the apex of the stigma. And lastly, as extraordinary activity of the insect is necessary, or at least advantageous in the performance of this operation, that activity is, according to him, produced by the intoxicating secretion of the nectaria.[1]

In 1809, an essay on Asclepiadeæ was published in the first volume of the Memoirs of the Wernerian Natural History Society, in which one of my principal objects was to establish the opinion, more or less conjectural, of Adanson,
718] Richard, Jussieu, and Schreber, respecting the structure of the stamina and stigma. With this view I appealed to the remarkable fact, that in the early state of the flower-bud the pollen masses are absolutely distinct from the glands and processes of the stigma, to which they in a more advanced stage become attached. This proof of the real origin of parts I then believed to be entirely new. It has, however, been already seen that the fact was noticed by Gleichen, and it will presently appear that it was also well known to another original observer.

In the essay referred to, I had not very minutely examined the texture of the pollen mass, and in true Asclepiadeæ I had failed in ascertaining its real internal structure; not having been then aware of the existence of the included grains of pollen, but believing, until very lately, that the mass in its most advanced state consisted of one

[1] It may here be remarked, that the prevailing form of inflorescence in Asclepiadeæ is well adapted to this economy; for the insect so readily passes from one corolla to another, that it not unfrequently visits every flower of the umbel.

undivided cavity, filled with minute granular matter mixed with an oily fluid; and hence concluded that the fecundating matter was conveyed from the mass through the arm and gland to the stigma.

In the month of April last I saw, for the first time, drawings of several Asclepiadeæ made between 1805 and 1813 by Mr. Bauer, who, aware of the interest I took in this subject, with his accustomed liberality and kindness, offered me any part of them for publication.

Among these drawings, exceeding perhaps in beauty and in the completeness of the details all the other productions with which I am acquainted even of this incomparable artist, an extensive series, exhibiting the gradual development of the parts of the flower in *Asclepias curassavica*, were the most important.

In this series, made in 1805, and commencing when the pollen is just separable in a pulpy mass from its cell, the glands of the undivided stigma being still invisible, the fact of the distinct origins of these parts is very satis- [719
factorily shown, in accordance with my observations in the essay referred to.[1]

But in these drawings Mr. Bauer has gone further than I did, having also represented the internal structure of the pollen mass as cellular; each cell in the flower-bud just before expansion being filled with a grain of pollen, marked with lines indicating its quaternary composition; while in the expanded flower this grain is exhibited as shrivelled, having discharged its contents, which consist of a mixture of an oily fluid and minute granules. From this, the concluding stage of the series, it may be inferred that Mr. Bauer's opinion respecting the mode of impregnation in Asclepiadeæ agrees with that which I had adopted, and

[1] In a flower bud much earlier than the commencoment of Mr. Bauer's series I have found the pistilla to consist merely of two distinct very short semicylindrical bodies, the rudiments no doubt of the future stigma.

In this stage also the antheræ are flat, nearly orbicular or ovate, greenish, rather thick and opaque, but petal-like, with no inequality of surface, or any other appearance of the future cells, which in a somewhat more advanced stage are indicated by two less opaque areolæ, and at the same time the two semicylindrical bodies unite to form the stigma. (Pl. 36, figs. 7—11.)

which, though probably originating with Richard in 1799,[1] and briefly stated by him in 1802,[2] was first distinctly expressed as a conjecture in 1789 by M. de Jussieu.

In 1817, Mr. Stephen Elliott states that he observed, in his *Podostigma*[3]—a genus nearly allied to Asclepias—a fibre or cord extending through the centre of the corpuscular pedicel or attenuated base of the stigma, and communicating from the anthera to the ovarium. He adds, that Dr. Macbride has since seen it in some species of Asclepias.

There can be no doubt that the cord here noticed is of the same nature with that which Gleichen has described in a different state, and of which I shall presently have occasion to speak.

720] In 1824, Professor Link,[4] while he admits the distinct origins of the pollen masses and glands or corpuscula seated on the angles of the stigma, yet considers both these parts as equally belonging to the anthera. In this respect his opinion is identical with that of Gleichen. The pollen mass, he adds, is composed either of a cellular tissue, or manifestly of grains of pollen: the former part of the description being no doubt meant to apply to true Asclepiadeæ, the latter to Periploceæ.

Professor L. C. Treviranus, in 1827,[5] published some observations on this family, in which his account of the structure of the pollen differs in several points from that exhibited in Mr. Bauer's drawings, which he states he had seen three years before this publication.

In *Asclepias curassavica*, the species more particularly examined by Treviranus, he describes the pollen mass as filled with compressed, nearly round but obtusely angular, colourless, simple grains, containing minute granules; the pressure of the external grains, or those in contact with the general covering, giving it the appearance of being cellular.

In speaking of the mode of impregnation, he says, that the pollen mass, at the time when its connection is esta-

[1] *Encycl. Botan.* i, p. 212.
[2] *Bulliard, Dict. de Bot.* ed. 2, p. 56.
[3] *Bot. of Carol. and Georg.* i, p. 327.
[4] *Phil. Bot.* p. 300.
[5] *Zeitsch. f. Physiol.* ii, p. 230.

blished with the process or arm of the gland, which is then very viscid, undergoes manifest changes, from being ventricose and opaque becoming flat, hard, and transparent. These changes he thinks are probably owing to the extraction of its fecundating matter by the process through which it passes to the glands, and by them to the angles of the stigma, whence it may be easily communicated to the styles and ovaria. His opinion, therefore, in every respect agrees with that which originated with Richard and Jussieu, and which I had adopted.

The celebrated traveller and naturalist, Dr. Ehrenberg,
in 1829,[1] has given a very interesting account of the [72]
structure of the pollen masses in Asclepiadeæ, from observations commenced in 1825, and others made in 1828.

In this account he describes the pollen mass as consisting of a proper membrane bursting in a regular manner, the cavity being not cellular but undivided and filled with grains of pollen, each grain having a cauda or cylindrical tube often of great length, and all these tubes being directed towards the point or line of dehiscence. This appendage or cauda he considers analogous to the *boyau* of Amici and Brongniart differing however in its forming an essential part of the grain in Asclepiadeæ; whereas in other families the application of an external stimulus is necessary for its production.

He is entirely silent as to the manner in which these caudate grains communicate with or act upon the stigma; and does not in any case remark,—what must, I think, have been the fact, at least in several of the plants in which this structure was observed, and especially in those with pendulous pollen,—that the mass examined was no longer in the cell of the anthera, but had been removed and probably applied to some part of the stigma.

In the month of July last I examined several species of Asclepias, with reference to Mr. Bauer's drawings and Dr. Ehrenberg's account of the pollen;—the first object, there-

[1] *Linnæa* iv, p. 94.

therefore, was to ascertain the structure of the pollen mass.

[1] Although on this subject my earliest observations essentially agreed with Mr. Bauer's figures of the mass, which represent it as having a subdivided cavity with a grain of pollen in each cell; yet a further examination had led me to adopt the opinion of Treviranus and Ehrenberg, who describe its cavity as being undivided and filled with distinct grains.

722] I was confirmed in this opinion on considering the state of the mass after the production of the pollen tubes; for it appeared very improbable that the cells, unless they were of extreme tenuity, could be either suddenly removed or sufficiently ruptured to admit of the passage of the tubes from its more distant parts to the point or line of dehiscence.

The appearance, however, occasionally met with, of lacerated membranes proceeding, as it seemed, from the

[1] [In the original impression, printed for distribution in October, 1831, the passage from this point down to the paragraph on p. 524 commencing "On the 16th of July," stood as follows. This was replaced in the 'Linnean Transactions' by that which is given in the text.—EDIT.]

"My earliest observations on this subject, made on several species of *Asclepias*, seemed to prove that the mass is cellular, nearly as Mr. Bauer has represented it. But on a further examination I was convinced that it can be termed cellular only in the early stages, in consequence of the state of the grains of pollen which then certainly cohere; while in the more advanced, and especially in the mature state, it is no longer really cellular, the grains being now distinct from each other; sections of the mass, however, whether transverse or longitudinal, still exhibit a cellular appearance.

"These grains, when in this their perfectly developed state, are colourless, nearly round, but slightly and obtusely angular, probably from mutual pressure, much compressed, with an undivided cavity, and no indication of their being composed of four or any other number of united cells. Their membrane is transparent, and has no appearance of being made up of two united coats, and the cavity is filled and rendered opaque by spherical granules of nearly uniform size, with occasionally a few oily particles. In this state no appearance or indication of the tubes or appendages described by Dr. Ehrenberg was found.

"The general covering of the mass, which is of a deep yellow colour and very distinctly areolated, the meshes being angular, and in size as well as in form nearly corresponding with the included grains, may perhaps be considered as the outermost series of cells, whose laminæ are closely applied to each other, as in the epidermis, and their cavity consequently obliterated. They thus form a coat of considerable thickness, necessary for the protection of the grains of pollen, in a mass which is destined to be removed from its original place by an insect, and applied by this agent to a distant part of the same or of a different flower."

margins of the areolæ of the inner surface of the mass, added to the facts which had originally led me to adopt Mr. Bauer's view, determined me to re-examine the subject.

The result of this examination, made on specimens of *Asclepias phytolaccoides* and *purpurascens*, but especially the former, proved that the mass in these species is really cellular in all stages, as Mr. Bauer has represented it in *A. curassavica*, and that in the advanced flower-bud, as in the expanded flower, the cells may be seen, though not without difficulty, after their grains are removed.

The pollen mass in several species of Asclepias, particularly in *Asclepias phytolaccoides*[1] (and in *A. curassavica*, as figured by Mr. Bauer), consists of cells disposed in three series parallel to its sides, the middle series being often more or less interrupted.

The cells of the outer layer of each side have their opposite walls very unequal both in colour and thickness. The outer wall of each of these cells, which is formed by one of the areolæ of the surface, is of a deep yellow colour, nearly opaque, and of such thickness as to prevent external bursting; the inner is of a paler yellow, semi-transparent, and so much thinner as to determine internal rupture, which in these cells, after the production of the tubes, seems to take place without regularity, and to such an extent, that after the removal of the grain the remains of the inner wall are [723
not very readily distinguishable.

Sections of the mass, indeed, both transverse and longitudinal, exhibit an appearance of cellularity; but there is here a source of fallacy, unless the contained grains are also visible in the section; and the best proof of its being cellular is derived from the state of the central or middle series after the bursting of the mass.

The cells of this central layer are of equal thickness throughout, and on the production of the tubes burst in a definite manner towards the convex edge of the mass, and at the same time generally separate from each other. They continue, however, to inclose the grain, or, as it may be

[1] Tab. 35, fig. 8.

considered, the inner membrane of the grain of pollen, whose outer membrane is formed by the cell itself; and the tenacity of this outer membrane is such that it may easily be removed from the inner without further apparent rupture.

These central grains, thus covered by their respective cells, may readily be distinguished, by their pale yellow colour and a certain degree of opacity, from the naked grains or inner membranes, which, like their tubes, are entirely colourless, and transparent.[1]

In Asclepiadeæ, therefore, it may be said that the greatest development of the pollen grain exists; namely a grain having an undivided cavity, whose membranes are entirely distinct, and the pollen tubes of which seem to possess the highest degree of vitality yet met with.

In the perfectly developed state of the pollen mass, the grain, considered as distinct from its outer membrane or containing cell, is nearly round, but slightly and obtusely angular, much compressed, with an undivided cavity, and exhibiting no indication of its being composed of four or
724] any other number of united cells. Its membrane is transparent and colourless, made up of two united coats, and the cavity is filled with spherical granules of nearly uniform size, among which a few oily particles are occasionally observable.[2] In this state no appearance or indication of the tubes or appendages described by Dr. Ehrenberg is found.

On the 16th of July, in repeating my examination of *Asclepias purpurascens*,[3] I observed in several flowers one or more pollen masses removed from their usual place, namely the cell of the anthera, and no longer fixed by the descending arm to the gland of the stigma, but immersed in one of the fissures formed by the projecting alæ of the antheræ, and in most cases separated from the gland, a small portion of the arm or process, generally that only below its flexure, remaining attached to the mass.[4]

[1] Tab. 35, fig. 9.
[2] Tab. 34, fig. 6; and tab. 56, figs. 3 and 13.
[3] Tab. 34.
[4] Tab. 35, figs. 2, 3, 4, and 7.

In the cases now described, the mass, which in general is entirely concealed by the alæ, was so placed in the fissure, that its inner or more convex edge was in contact with the outer wall of the tube formed by the united filaments, and the gibbous part of the edge closely pressed to that point where this tube is joined to the base of the corresponding angle of the stigma.[1]

These masses, at the point of contact, in most cases adhered firmly to the tube or base of the stigma, and on being separated, a white cord or fasciculus of extremely slender threads or tubes, issuing from the gibbous part of the edge, which had then regularly burst, came into view

On laying open the pollen mass,—which in this state was easily done, by first dilating the aperture that gave issue to the cord,—each of the tubes composing it was found to proceed from a grain of pollen. These grains retained nearly their original form, but were become more transparent, and had generally lost a great portion of their granules; and these granules were not often to be found even in the tube, especially after it had acquired considerable [725
length.[2]

Almost every grain in the mass had produced its tube, and the tubes were directed from all parts of it towards the point of dehiscence. In this state the mass had become more convex from the increased bulk of its contents.

The tube so produced from each grain of pollen cannot be said to be emitted from it, but is manifestly an elongation of its membrane. These tubes are transparent, cylindrical, about 1-2000th of an inch in diameter, neither branched nor jointed, with no apparent interruption in their cavity, and when of great length, which they often attain, are frequently without granular matter.

I next proceeded to examine the course of the cord, which in most cases,—and indeed in all where the mass had remained a sufficient length of time in the fissure,—had opened a passage for itself through the membrane, or rather had separated the upper edge of this membrane from

[1] Tab. 34, fig 7.
[2] Tab. 35, figs. 7 and 10; and tab. 34, fig. 12.

the base of the stigma, to which it was before united. Having effected this separation, it was found to proceed along the surface of the base of the stigma in a line exactly opposite to the glands seated on the apex of the same bevelled angle. The cord having passed along the surface of the attenuated base of the stigma until it arrives at its articulation with the two styles, then inclines towards the inner side of the apex of the style nearest to it, and actually introduces itself, wholly or in part, into the hollow of the apex, which in this stage is in some degree exposed.[1] But as the partial separation of the styles from the stigma, then taking place, is not always sufficient for the free admission of the whole cord, a few of the tubes not unfrequently become bent, in some cases even zigzag, doubtless
726] in consequence of the obstacles opposed to them; and such tubes very seldom enter the style, but along with others hang down externally below the joint. This introduction of part of the tubes into the apex of the style is soon followed by a manifest enlargement of the ovarium, and of the style itself, which, in *Asclepias purpurascens*, then exhibits a discoloured blackish line, visible even on the surface of its inner side. On opening the cavity or body of the style in this stage, a fasciculus of tubes was constantly seen passing down the centre, which was originally pulpy, and the walls of the cavity formed by the passage of these tubes were always found indurated and blackened, having every appearance of being absolutely killed.

I have never been able hitherto to follow these tubes further than the commencement of the placenta, where they really appear to terminate.[2] I have not at least yet succeeded in tracing any of them either on the surface or in the substance of the placenta, though with this object I have examined it not only in its first degree of enlargement, but also in some of its more advanced stages.

The same series of appearances, with very slight modifications only, were observed in all the species of Asclepias (not indeed more than seven in number) which I had

[1] Tab. 34, figs. 7—9; and tab. 35, figs. 4 and 10.
[2] Tab. 34, figs. 10 and 11; and tab. 35, figs. 5 and 6.

opportunities of examining during the summer. For in those species in which the pollen mass was not found transferred from its original position to the fissure, and in contact with the base of the style, no doubt by means of insects, it was not difficult to place it there; and in doing so I never failed to obtain the same results.

I now turned my attention to the base of the stigma, expecting to find there such a modification of surface as might serve to account for the rupture and production of the tubes in the mass brought in contact with it. I have, however, in no case been able to observe the slightest [727
appearance of secretion, or any difference whatever in texture, between that part and the general surface of the stigma.

The bursting of the mass in Asclepias is uniformly on the more rounded edge; and this, it may be observed, is the inner edge or margin of the mass, with reference to the cell of the anthera in which it is formed; and I may further remark, that in the only case in which I have hitherto observed dehiscence in an erect pollen mass, namely, in *Hoya carnosa*, it also takes place along the inner margin.

In Asclepias the bursting always commences at the most prominent point of the convex edge, and to this part it is generally confined: it is sometimes, however, found extending through the greater part of its length.

On carefully examining the convex edge, and more particularly its most prominent portion, I have not been able to observe in it any change or peculiarity of texture, or even any obvious difference in the form of the meshes of the reticulated surface. Notwithstanding this apparent want of secretion in the base of the stigma, and of difference of texture in the covering of the mass of pollen at the point where it comes in contact with that organ, it must still be supposed that there is some peculiarity both in the surface of the stigma and in the prominent edge of the mass, on which the effects in question depend.

These effects are indeed very remarkable; the stimulus here supposed to be derived from the surface of the stigma,

and applied to the prominent point of the convex edge of the pollen mass, producing its appropriate action not only in those cells or grains of pollen in immediate contact with that point, but generally in every grain in the mass. But as there are no visible conductors of this stimulus within the mass, it must either be supposed to be propagated from one cell to another, or conveyed from the prominent [728
point of the edge to every other part of the surface of the covering itself.

To ascertain whether contact of the convex edge of the pollen mass with this point of the stigma was absolutely necessary for the rupture of the mass and the production of tubes, I in the first place introduced a mass into the fissure, but with its convex edge outwards. In this position no change whatever took place.

I next removed one of the glands of the angles of the stigma, and applied the convex edge of a mass to the surface thus exposed, which even in this stage—to facilitate the removal of the gland by insects—continues to secrete. In this case, dehiscence and protrusion of pollen tubes did follow, more slowly, however, and less completely, than when brought in contact with the non-secreting base.

On applying the pollen mass of one species of Asclepias to the base of the stigma of another, the usual changes generally took place; but still, as it seemed, less perfectly, and only after a longer interval.

Pollen masses of *Asclepias purpurascens* being applied to the stigma of *Epipactis palustris*, and immersed in its viscid secretion, the dehiscence, contrary to expectation, not only took place, but even more speedily than usual, that is within twenty-four hours. Some of the grains were also found discharged from the mass unchanged, while others, both discharged and still inclosed, had begun to produce tubes.

The greater number of these observations were also made with *A. phytolaccoides*, which, on account of the greater size of its flower, I at first preferred. I found, however, with reference to such experiments, an objection to employing this species, arising from the great excitability,

so to speak, of its mass, which in some cases produced its tubes merely on continued immersion in water. I even found that in this species, in the gradual decay of the [729
flower, where the parts remain soft, the rupture and protrusion of tubes took place while the mass was still in its original position, immersed in the cell of its anthera.[1] The tubes produced in this situation often acquire a great length, but coming, immediately on their protrusion from the mass, in contact with the membrane of the anthera, their course is necessarily altered; and in their new direction, which is generally upwards, they not unfrequently arrive at the top of the cell, or even extend beyond it.

In addition to the several species of Asclepias already referred to, *Cynanchum* (*Vincetoxicum*) *nigrum* is the only plant of this family in which I have observed the whole of the appearances; namely, the rupture of the mass, the production and protrusion of the pollen tubes, their union into a cord, with the course and entrance of this cord into the cavity of the style.

The present essay, therefore, as far as regards this family, might with greater propriety have been entitled, "On the mode of impregnation in the genus Asclepias." It seems, however, allowable to conclude, that in all the genera having pendulous pollen masses, the same economy, slightly modified perhaps in some cases, is likely to be found. But among those with erect pollen masses, there are several in which more considerable differences may be expected. Of this section of the family I have hitherto had the opportunity of submitting only one plant to careful examination, namely, *Hoya carnosa;* and even here my observations are incomplete.

In *Hoya carnosa* I have never found the pollen tubes produced, or masses ruptured, while remaining in their original position; but I have succeeded in producing these effects by bringing them in contact with certain parts of the corona.

The rupture and protrusion of pollen tubes, then, take

[1] Tab. 35, fig. 11.

730] place through the whole length of the inner edge of the mass, which, as in all the genuine species of Hoya, is truncated and pellucid.[1] But I have not yet been able so to place the mass as to produce a cord of tubes communicating with the stigma, nor can I at present conjecture how this is to be effected.

I shall conclude with some observations equally relating to both the families that have been treated of.

It is in the first place deserving of remark, that while Asclepiadeæ and Orchideæ so widely differ in almost every other respect, there should yet be an obvious analogy between them in those points in which they are distinguished from all other Phænogamous plants.

It is unnecessary here to state the numerous and important differences existing between these two families: but it may be of some interest to make a few remarks on their points of agreement or analogy.

These are chiefly two: The first being the presence of
731] an apparently additional part, not met with in other families; the second, the cohesion of the grains of pollen, and their application in masses to the female organ.

With regard to the first peculiarity it may be observed, that there is no real addition made to the number of organs in either family, and that in both families the apparent

[1] In the tubes of *Hoya carnosa* I have been able to confirm Professor Amici's observation with respect to circulation taking place in the *boyaux* of the grains of pollen. In this case the membrane being very transparent, and the granules, before the tube has acquired any considerable length, not being so numerous as to obscure the view of the opposite currents, they were very distinctly seen.

I have also observed circulation in the pollen tubes in a few other cases; especially in *Tradescantia virginica*, in which, while the tube was still very short, the circle partly existing in the tube was completed in the body of the grain. The circular current in grains of pollen before the production of the tube may likewise, in some cases, but not very readily, be distinguished, as in *Lolium perenne*.

It might perhaps be supposed that the molecular motion, which in a former essay I stated I had seen within the body of the grain of pollen, might have been merely an imperfect view of the circulation of granules, and such I am inclined to think it really was in *Lolium perenne*.

I have, however, also very distinctly seen within the membrane of the grain of pollen in some species of Asclepias, vivid oscillatory motion of granules without any appearance of circulation.

addition consists in a modification or production of the stigma; the modified part of which loses the proper function of that organ.

This production of the stigma,—which is generally present, and wanting only in certain Orchideæ, where its place is sometimes supplied by an analogous modification of the male organ,—though differing very remarkably in appearance in the two families, agrees in being originally distinct from the pollen masses, and in the advanced stage becoming firmly attached to them; in adhering but slightly to the point of its formation after the attachment to the pollen takes place; and in being so constructed as to be readily removed by insects from its original position along with the pollen masses.

As to the second point of agreement; namely, the cohesion of the grains of pollen into masses of considerable size, and the application of these masses to the stigma,—it is obviously connected with that which might perhaps be termed a third peculiarity; the apparent necessity for an unusual number of pollen tubes which are to act in concert; in the one family to penetrate to and regularly arrange themselves in the cavity of the ovarium;[1] in the other to open a communication with the stigma, and then to pass along a non-secreting surface, until they arrive at a distant point, where they are to be introduced into the cavity or body of the style.

With respect to the agency of Insects in fecundation in those two orders, there can be no doubt that it is very frequently employed in Orchideæ; at the same time there are evidently cases in that family in which, from the relative [732
position of the organs, the interposition of these agents is not always required. But in those Asclepiadeæ at least that have been fully examined, the absolute necessity for their assistance is manifest.

Two questions still remain.

The first regards the proof of the actual penetration of the pollen tubes into the cavity of the ovarium in both families.

[1] See Additional Observations.

In Asclepiadeæ I shall only observe, that I consider the evidence complete; but in Orchideæ it may be admitted that it is not altogether so satisfactory. Of the descent of pollen tubes through the cavity of the stigma in Orchideæ, the evidence appears to me unquestionable. With respect, however, to the origin of the cords formed of similar tubes, so numerous and so regularly arranged in the cavity of the ovarium, and which are in contact with surfaces not altogether incapable of secretion, it might perhaps be alleged, either that they wholly originate from the supposed conducting surfaces, or that they consist of a mixture derived from both sources.

That mucous threads, or capillary tubes, in most respects similar to pollen tubes, and certainly altogether belonging to the style, exist in some plants, there is no doubt; and such I have observed in Didymocarpus, Ipomopsis, and in Allamanda, before the application of the pollen to the stigma. I am still, however, of opinion, that those found in the cavity of the ovarium in Orchideæ are really derived from the pollen;[1] an opinion which receives some confirmation from the manifest descent of the pollen tubes in the style in many other families, as in several Scrophularinæ, Cistineæ, Viola, and Tradescantia.

The second question is, Whether the granules originally filling the grain of pollen, and which may often be found in the tubes, especially in their nascent state, both in these and in many other families, are the essential agents in the
733] process of fecundation; the tubes being merely the channels conveying them to the organ or surface on which they are destined to act.

The arguments which might be adduced in favour of this, the generally received opinion, would probably be the variety in the form and size of the granules in different plants, with their great uniformity in these respects in the same species, added to the difficulty of conceiving in what manner the tubes themselves can operate. On the other hand, their great diminution in number, or even total disappearance, in Asclepiadeæ and Orchideæ, long before the

[1] See Additional Observations.

tubes have finished their growth, would afford an argument of some weight at least against their essential importance in any case; and it may be added, that in Asclepiadeæ there appears to be no other source of nourishment for the tube until it has penetrated into the style, than these granules. Nor is it necessary to suppose that the tubes themselves act directly, it being even probable that they also contain a fluid or granular matter much more minute than that originally filling the cavity of the grain.[1]

Our knowledge indeed appears to me not yet sufficient to warrant even conjectures as to the form of the immediate agent derived from the male organ, or the manner of its application to the ovulum in the production of that series of changes constituting fecundation. I may, however, be allowed to observe, that at present, with respect to this function, we are at least as far advanced in these two families, hitherto considered so obscure, as we are in any other tribe of Phænogamous plants: and I even venture to add, that in investigating the obscure subject of generation, additional light is perhaps more likely to be derived from a further minute and patient examination of the structure and action of the sexual organs in Asclepiadeæ and Orchideæ, than from that of any other department either of the vegetable or animal kingdom.

[1] See Additional Observations.

734] EXPLANATION OF THE PLATES.

TAB. 30 (34). ASCLEPIAS PURPURASCENS.

Fig. 1. A branch in flower :—natural size.

Fig. 2. An expanded flower, of which two of the foliola coronæ and one of the antheræ are removed :—moderately magnified.

Fig. 3. A front or inner view of an anthera, to show the extent of bursting, particularly with relation to the pollen mass, of which the greater part is included in the non-dehiscent portion :—magnified as fig. 2.

Fig. 4. A pollen mass, more highly magnified, separated from its gland and arm, and divided transversely, to show its cellular structure (first discovered in *Asclepias curassavica* in 1805 by Mr. Bauer), with grains of pollen, their granules, and some drops of an oily fluid.

Fig. 5. A pollen mass entire, with a small portion of the arm adhering to its apex :—magnified as fig. 4.

Fig. 6. A transverse section of a pollen mass, still more highly magnified, in one of the cells of which is seen the single grain (or inner membrane), also separately exhibited to show that it is simple and slightly angular.

Fig. 7. The pistillum with pollen masses, that have burst and protruded their tubes, applied to the base of the stigma, the glands and their arms being removed. The cords formed by the pollen tubes have passed along the corresponding sides of the conical base of the stigma, and have reached the tops of the styles.

Fig. 8. A longitudinal section (more highly magnified) of the conical base of the Stigma with the two styles, to show more distinctly the course of the pollen tubes.

735] Fig. 9. A pollen mass after bursting, with its cord formed of the pollen tubes, entering the apex of the style, which is there lacerated.

Fig. 10. The two Ovaria with their styles, one being somewhat enlarged in consequence of impregnation, and opened longitudinally; exhibiting pollen tubes extending from the apex of the style to the commencement of the placenta.

Fig. 11. The same two ovaria and styles, both opened, to show that in one (the left), which is somewhat smaller, no pollen tubes are contained; the other (the right), which is impregnated, shows the tubes reaching the ovula, but not extending further.

Fig. 12. Two grains of pollen (or rather grains deprived of their outer membranes), with portions of their tubes and contained spheroidal granules; proving that the tubes are extensions of this (the inner) membrane;—very highly magnified.

TAB. 31 (35). ASCLEPIAS PHYTOLACCOIDES.

Fig. 1. An expanded flower (magnified), from which two of the foliola coronæ and one anthera have been removed.

Fig. 2. The complete Pistillum, and on one side two of the antheræ, the membrane formed by the united filaments being cut off a little below the stigma; on the other side, a naked pollen mass applied to the stigma, with its gland and arm adhering.

Fig. 3. A longitudinal section of fig. 2, to show on the left side a pollen mass, with a small portion only of the arm adhering, applied to the base of the stigma, and which, having burst, shows the protrusion of the cord formed by the pollen tubes.

Fig. 4. A longitudinal section of one half of the Stigma and the corres- [736
ponding style transversely cut near the base, showing more distinctly the position of the pollen mass with the protrusion and course of the tubes.

Fig. 5. The Style of fig. 4, laid open lengthways, exhibiting within its cavity and beyond it the pollen tubes reaching the apex of the placenta, a reflected portion of which, with three of its ovula, is also shown.

Fig. 6. An impregnated Pistillum, of which the style is laid open longitudinally, and the placenta, thickly covered with ovula, exposed, to show the descent and course of the pollen tubes.

Fig. 7. A Pollen mass, to the apex of which the base of the arm adheres, with pollen tubes protruding from the point of dehiscence:—more highly magnified.

Fig. 8. A transverse section of a Pollen mass, showing an arrangement of the cells somewhat different from that of *A. purpurascens*, there being here a middle irregular series, the cells of which in some cases appear to separate and cover the grains after the production of the tubes.

Fig. 9. Two grains of pollen with portions of their tubes, very highly magnified, the grain to the left having its outer covering or membrane, which is removed from the grain to the right, and shown separately further to the left.

Fig. 10. A Pollen mass which has burst and protruded its tubes, exhibited as entering the cavity of the style, which is laid open to show the commencement of their descent.

Fig. 11. Two Pollen masses (with their arms and gland), which have burst and protruded their tubes while still inclosed in the cells of the antheræ; [737
this happening in *A. phytolaccoides* in that particular kind of decay mentioned in p. 529 of the text.

TAB. 32 (36).

Fig. 1. Two Pollen masses of *Asclepias purpurascens* with protruded tubes; the only instance met with in which both cords are introduced into the same style.

Fig. 2. A grain of pollen, of the same species, with a portion of its tube; the unusual form probably caused by the pressure of other grains and their tubes.

Fig. 3. A grain of pollen of *Asclepias purpurascens* containing numerous minute granules and two larger drops or globules of an oily fluid.

Figs. 4, 5, & 6. Various combinations of pollen masses of *Asclepias purpurascens*. In these it is supposed that the insect having removed and applied to the stigma some of the masses, has extracted, by means of the arms still adhering to it, other masses with their glands and arms.

A combination of the same kind, different from and more remarkable than any of these, but perhaps not very accurately represented, is given, in his *Microscop. Entdeck.*, tab. 36, fig. 8, by Gleichen, who appears (op. cit. p. 81) to have also met with other combinations, without suspecting in any case the real cause of such apparently anomalous structures.

Fig. 7. A flower bud of *Asclepias curassavica* in the earliest stage in which I was able to distinguish its parts; the unopened corolla in its place with one of the sepala, the other four being exhibited separately:—highly magnified.

738] Fig. 8. The Corolla of fig. 7, opened and in part removed, to show the state of the contained organs; the figure exhibiting two petals hardly cohering at base; within these, two distinct petal-like bodies, alternating with them, and which are the antheræ; and two other smaller bodies, which are the pistilla as yet unconnected.

Fig. 9. An Anthera taken from fig. 8, and more highly magnified, to show that in this early stage it is entirely petal-like, there being no indication of the two cells, of which the first appearance in a somewhat more advanced stage is given at Fig. 10.

Fig. 11. A Petal of fig. 8, more highly magnified.

Fig. 12. The Pistilla of fig. 8, as yet distinct, scarcely at all angular, and with no manifest cavities; so that these two bodies may be regarded as chiefly or entirely the component parts of the stigma.

Fig. 13. Two Grains of pollen taken from the pollen mass of the expanded flower of *Asclepias curassavica*.

ADDITIONAL OBSERVATIONS ON THE MODE OF FECUNDATION IN ORCHIDEÆ. [739

Read June 5th, 1832.

The following additions to the Paper which was communicated to the Society in November last, on the Sexual Organs and Mode of Fecundation in Orchideæ and Asclepiadeæ, relate entirely to the former family.

In the essay itself I had ascertained from the examination of a considerable number of species belonging to different tribes of Orchideæ, that in the expanded flower of this family, however long it had remained in that state, no appearance whatever existed of those tubes which form the mucous cords, either in the tissue of the stigma or in the cavity of the ovarium, anterior to the application of the pollen to the stigma; and that in all cases where pollen had been applied to that organ and enlargement of the ovarium had followed, the mucous cords were to be found.

From these facts I had concluded that the tubes forming the cords were entirely and directly produced from the grains of pollen; and hence I accounted for the cohesion of the pollen into masses, and its frequent application in that state to the stigma.

Some cases, however, in which a few lobules or even grains of pollen only were observed on the stigmata of impregnated flowers, had led me to express myself doubtfully on this point. And since my paper was read, I have had opportunities of making several observations and experiments which prove that the application of a very small portion of a pollen mass to the stigma is sufficient for the production of mucous cords of the ordinary size in the cavity of the ovarium.

My observations on this point and on the gradual pro- [740
duction and descent of these cords have been made chiefly on *Bonatea speciosa*, perhaps the most favourable subject for such experiments in the whole family.

My first observation on Bonatea related to the probability of a single insect impregnating several or even many flowers with one and the same mass of pollen.

To effect this, it is only necessary that the viscidity of the retinaculum or gland with which the pollen mass becomes inseparably connected, and by means of which the mass is removed from its cell and adheres to the insect, should exceed that of the surface of the stigma, and that the viscidity of the stigma should be sufficient to overcome the mutual cohesion of the lobules composing the mass.

These different degrees of viscidity are very manifest in *Bonatea speciosa*, in which, imitating the supposed action of the insect, I have succeeded in impregnating most of the flowers of the spike with a single pollen mass. I believe they exist also in the greater number of Ophrydeæ, as well as in many Neotteæ and Arethuseæ.

But even in Ophrydeæ they are not universally met with, a very remarkable exception existing, I believe, in the whole genus Ophrys, in which the resemblance of the flower to an insect is so striking, and in which also the retinacula, whose viscidity hardly equals that of the stigma, are included and protected by concave processes of the upper lip of that organ.

It may also be remarked, that in the genus Ophrys impregnation is frequently accomplished without the aid of insects, and in general the whole pollen mass is found adhering to the impregnated stigma. Hence it may be conjectured, that the remarkable forms of the flowers in this genus are intended to deter not to attract insects,
741] whose assistance seems to be unnecessary, and the action of which, from the diminished vicidity of the retinaculum, might be injurious. On this subject I will also hazard another remark, that the insect forms in Orchideous flowers, resemble those of the insects belonging to the native country of the plants.

The next object I had in view was to determine the first appearance and progress of the mucous tubes.

My observations on the *origin* of these tubes are not altogether satisfactory.

It appeared, however, in Bonatea, which was also the plant most particularly examined, that they first become visible soon, but not immediately, after the production of the pollen tubes from the lobules or grains of the mass applied to the stigma; and that their earliest appearance is in the tissue of the stigma, in the immediate vicinity of the pollen tubes, from which they are with difficulty distinguishable, and only by their being less manifestly or not at all granular in their surface or contents, and in general having those interruptions in their cavity, which I have termed coagula, and which I have never yet met with in tubes actually adhering to the grain of pollen.

But even these characters, in themselves so minute, might be supposed to depend on a difference in the state of the contents of the pollen tube, after it has quitted the grain producing it. It is possible therefore that the mucous cords may be entirely derived from the pollen, not however by mere elongation of the original pollen tubes, but by an increase in their number, in a manner which I do not attempt to explain.

The only other mode in which these tubes are likely to be generated, is by the action of the pollen tubes on the coagulable fluid, so copiously produced in the stigma at the only period when impregnation is possible.

The obscurity respecting the origin of these mucous
tubes does not, however, extend to their gradual in- [742
crease and progress, both of which may be absolutely
ascertained.

In Bonatea they are, in the first stage of their production, confined to the stigma, with the proper tissue of which they are more or less mixed. Soon after they may be found on the anterior protected surface of the style, at first in small numbers; but gradually increasing, they form a mucous cord of considerable size, in which very few or none of the utriculi of the stigma are observable. This cord, which is originally limited to the style, begins, though sometimes not until several days have elapsed, to appear in the cavity of the ovarium, where it divides and subdivides in the manner I have described in my paper, its descent being gradual until the cords nearly equal the length of the placenta, to which they are parallel and approximated.

That these cords are not in any degree derived from those portions of the walls of the cavity of the ovarium, to which they are closely applied, and which I have termed the conducting surfaces, is manifest from the identity in state of those surfaces before and after the production of the cords.

In Bonatea the first evidence of the action of the pollen consists in the withering of the stigma; a similar decay of the greater part of the style soon follows, and the enlargement of the ovarium generally begins before the withering of the style is completed. When the enlargement of the ovarium is considerable, and the mucous cords are fully[1] formed in its cavity, a corresponding enlargement of the ovula takes place, and the nucleus becomes first visible.

I have no satisfactory observations in Bonatea respecting any tubes going off from these cords and mixing with the ovula; but in *Orchis Morio* I have repeatedly and very clearly observed them scattered in every part of the surface of the placenta, and in not a few cases have been able to
743] trace them into the aperture of the ovulum, to which they adhere with considerable firmness.[2]

At what period they reach the foramen of the testa, whether before or immediately after the first faint appearance of the nucleus, I have not yet been able to determine. That the tubes thus traced to the foramen of the ovulum are of the same nature as those which I have called mucous tubes, and not those directly produced by the pollen, is proved by their exact agreement with the former in every respect, except in their being remarkably and irregularly flexuose, apparently from the numerous obstacles they have to overcome after leaving the cords and beginning to mix with the ovula; for in the cords themselves, where the

[1] ["Carefully" in the original—an obvious error of the press.—Edit.]

[2] Since these additional observations were read, I have found in several other Orchideæ, especially *Habenaria viridis* and *Ophrys apifera*, tubes scattered over the surface of the placenta, and not unfrequently inserted, in like manner, into the apertures of ovula.

course of the tubes is not at all impeded, they are very nearly or altogether straight.

The two most important facts stated in the present communication are; *first*, the production of tubes not directly emitted from the grains of pollen, but apparently generated by them; and, *secondly*, the introduction of one or sometimes more than one of those tubes into the foramen of the ovulum, the point corresponding with the radicle of the future embryo.

The principal points remaining to be examined, and which we may hope, by careful investigation, to ascertain, are the precise state of the ovulum at the moment of its contact with the tube, and the immediate changes consequent to that contact.

Supplementary Note. [744

SINCE the paper on Fecundation in Orchideæ and Asclepiadeæ was read before the Society, and a Pamphlet containing all its more important statements was distributed in the beginning of November, 1831,[1] two essays have appeared on the same subject. The first on both families by M. Adolphe Brongniart, in the numbers of the *Annales des Sciences Naturelles* for October and November, 1831, but which were not published until January and February, 1832; the second, by Dr. Ehrenberg, on Asclepiadeæ alone, in the Transactions of the Royal Academy of Sciences of Berlin, before which it was read in November, 1831.

M. Brongniart's statements respecting ORCHIDEÆ to a great extent agree with those of my essay. They differ, however, in the following important points:

1st, He does not seem to be aware of the operation of insects in the fecundation of this family.

2ndly, He considers the mucous cords in the cavity of the ovarium (first seen by M. du Petit Thouars, with whose observations he seems to be entirely unacquainted),

[1] I may also refer to an excellent abstract of the Paper which appeared on the 1st of December, 1831, in the Philos. Mag. and Annals of Philosophy.

as a continuation of the tissue of the stigma and style, and as existing before the application of the pollen to the female organ.

And 3rdly, He supposes that the male influence reaches the ovula in Orchideæ before the inversion of the nucleus; an opinion founded, as it seems, on his observations on Epipactis, in which, as well as in some other genera of the order, this is the state of the ovulum in the expanded flower.

In ASCLEPIADEÆ M. Brongniart's observations, made chiefly in *Asclepias amœna* and *Gomphocarpus fruticosa*, accord with my statements as far as relates to the applica-
745] tion of the more convex edge of the pollen mass to the base of the stigma, its consequent dehiscence, the protrusion of the pollen tubes, and their penetration into the cavity of the style.

The chief differences are,

1st, His not even suspecting the agency of insects in the fecundation of this family, and particularly in the plants examined by him, in which I have regarded their assistance as absolutely necessary.

2ndly, In his assuming that the pollen mass in these two genera of Asclepiadeæ is ruptured, and comes in contact with the base of the stigma without leaving the cell of the anthera.

3rdly, His conjecturing that the secretion visible in the expanded flower on the angles of the stigma after removing the glands, is absorbed by the glands and conveyed through their arms or processes to the pollen mass, which it excites to the production of pollen tubes.

Dr. Ehrenberg, on the subject of Asclepiadeæ, repeats, with some slight modifications, his former statements quoted in my paper, and illustrates them by figures. In addition, he suspects that the pollen masses (which with Professor Link he regards as the true anthera, and the cells in which they are lodged as processes of the perigonium), are not originally distinct from the glands of the stigma, regarded by him as the filaments of his supposed anthera.

The central pentangular body he considers as the stigma, but he has no observations on the mode in which the pollen is applied to it.

And lastly, His original statement respecting the grains of pollen is so far modified, that he now believes them to be in the early stages without tubes or *boyaux*, which, according to him, make their appearance at the period of impregnation.

SUPPLEMENTARY OBSERVATIONS

ON THE

FECUNDATION

OF

ORCHIDEÆ AND ASCLEPIADEÆ.

BY

ROBERT BROWN.

[*Reprinted from a separate publication for distribution.*]

LONDON.

1833.

SUPPLEMENTARY OBSERVATIONS, &c.

ORCHIDEÆ.

In the observations appended to my Paper on these two Natural Families, printed in the 16th Volume of the 'Linnean Society's Transactions,' and which relate entirely to Orchideæ, it is stated, that in several species of Ophrydeæ the Tubes, produced either directly from the grains of Pollen, or in consequence of their application to the Stigma, were found spread over the surface of the Placentæ, and not unfrequently inserted into the aperture of the Ovula. The correctness of this statement I have confirmed, during the present season, by numerous observations, not only on the same, but also on several other species. Another remarkable appearance observed in some of these species, especially in *Orchis ustulata*, *fusca*, *Morio*, and in *Ophrys apifera*, and which indeed I had before met with, but neglected to mention in my Paper, consists in the elongation and protrusion of the jointed or cellular filament connecting the upper extremity of the Embryo with that of the original nucleus (the Tercine of M. Mirbel).

The Filament so protruded often equals the whole Ovulum in length, and its elongation seems to depend not only on the enlargement of each of the cells or joints, of which the included thread consists, but also on the production of additional joints.

As, however, the Pollen tube is found applied to the aperture of the Ovulum uniformly before either the Embryo or its thread is distinguishable, and as I have never observed the protruded thread of the Ovulum until after the secondary nucleus or Embryo, of which it is a continuation, becomes visible, I consider it as a production subsequent to impregnation.

It is possible, therefore, that the nearly similar tubes which have been observed terminating, as it is supposed, the nucleus of the unimpregnated Ovulum in a few other Families, may in some of these cases be of like origin.

To the observations formerly made on the general structure of Orchideæ, I have here to add,—

1st, That the cells of the testa of the ripe seed are frequently spirally striated, though these cells in the Ovulum before and even for some time after impregnation are absolutely without striæ.

2nd, The Fibrillæ constituting the pubescence frequently produced, and in some cases entirely covering the surface of the aërial roots, as they have been called, of the parasitical portion of the Order, are very remarkable.

These Fibrillæ, which I have examined both in dried and recent specimens of several species, but more particularly in the living state in *Renanthera coccinea*, are simple tubular hairs without joints, and whose apices, by which they adhere when attached to other bodies, are either of the same diameter, or somewhat dilated; and then, as in Renanthera, often more or less lobed.

In their natural state they exhibit, in most cases, hardly any indication of spiral structure; but the membrane, of which they entirely consist, is sufficiently elastic to admit of being extended, and at the same time unrolled, to about twice the length of the Tube. They then form a broad ribbon of equal width throughout, and spirally twisted from right to left,—a direction opposite to that which generally obtains in spiral vessels. It is possible that this may not be the direction of the spire in all cases; it is manifest, however, very generally, if not universally, in Renanthera.

The existence of spiral tubes produced on the surface is probably of very rare occurrence; and among Phænogamous plants I have hitherto met with it only in the hairs of the inner surface of the Corolla of some species of Ceropegia, in the wool enveloping the spines in several species of Mammillaria and Melocactus, and in the Coma of the seed of an Apocyneous plant from Brazil: for the spiral vessels in the seeds of Collomiæ, first observed by Mr. Lindley, and described by him as external, are seated between the two membranes of the testa, as I have long since described those of Casuarina. They differ, however, in direction; being in Collomia transverse or perpendicular, while in Casuarina they are longitudinal, or parallel to the membranes.

ASCLEPIADEÆ.

With regard to this Family, it was remarked, both in the Pamphlet which was distributed in the beginning of November, 1831, and in my Paper in the 'Transactions of the Linnean Society,' published in 1833, that I had [3
never been able to find the Pollen tubes descending lower than the commencement of the ovuliferous portion of the Placenta. But as this was far from satisfactory, especially after the further course of the analogous Tubes in Orchideæ had been ascertained, I determined to re-examine the subject.

For this purpose *Asclepias phytolaccoides* was selected; and on the 12th of the present month I succeeded in tracing the Pollen tubes in that species, not only over the whole ovuliferous surface of the Placenta, but also going off to the Ovula, to a definite point of each of which a single Tube was found in many cases attached.

These observations I have now so frequently repeated, and always with results so exactly similar, that I have great confidence in the correctness of the following statement:

In the newly expanded flower, the Ovulum in *Asclepias phytolaccoides* is nearly obovate, and is compressed in the same direction as the ripe seed, but in a much less degree: its umbilical cord is inserted on the axis of the inner or ventral side, about one fourth from the apex, and a process proceeding from it is continued, though not very distinctly, to the opposite or lower extremity. On the upper and broader end of the Ovulum a deep groove is observable, commencing at its inner margin, which is nearly in contact with the Placenta, and extending through its whole breadth, and somewhat obliquely downwards, so as to terminate at the same height on the outer side of the Ovulum with the upper edge of the ventral umbilical cord. This groove, or that point of it to which the Pollen tube is attached, occupies the place of the Foramen so generally found in the unimpregnated ovulum of Phænogamous plants. In *Asclepias phytolaccoides*, however, and I believe the observation may be extended to every species of the genus, there is certainly no perforation, nor at this period are the coats and nucleus of the Ovulum separable or even distinguishable; and the same apparent simplicity of structure is found even in its earlier stages.

Soon after the Pollen tubes enter the cavity of the Ovarium, even before the Corolla falls off, they may be found spread over the whole ovuliferous surface of the Placenta, which then often becomes of a light brown colour, but never dark brown or black, like the upper non-ovuliferous portion. From the surface of the Placenta the Tubes go off, one to each Ovulum, along the depressed apex of which the Tube passes till it reaches the outer extremity of the groove, where it is invariably inserted. To this point the Tube adheres so firmly, that I am inclined to think it actually penetrates, to some depth at
4] least into the substance of the Ovulum; a fact, however, which I have not yet absolutely ascertained.

Soon after the insertion of the Pollen tube, a change takes place in the appearance of the Ovulum, an internal body or nucleus becoming visible, with the upper attenuated extremity of which the point of insertion accurately corresponds.

The Pollen tube, when thus inserted into the Ovulum, is not always absolutely destitute of granules; in some cases containing a few, which in size and form seem to be identical with those that completely fill it in its nascent state.

But as such granules, at the period of insertion, are either very few in number, or apparently altogether wanting, I am still inclined to consider them rather as furnishing the nourishment of the Tube than as being the essential agents in fecundation; the really active particles in this function being probably much more minute.

These supplementary observations may be concluded with the remark, that although the descent of Tubes derived from the Pollen into the cavity of the Ovarium, and their insertion into that point of the Ovulum where the Radicle of the future Embryo is seated, has been absolutely ascertained in several species of Orchideæ and in one of Asclepiadeæ, and probably will be found in the whole of these two extensive families, yet it does not follow that this descent and insertion of Tubes should be expected to extend to all Phænogamous plants; for among these some structures of the female organ exist, which hardly admit of this œconomy.

LONDON; *July* 31*st*, 1833.

ON THE RELATIVE POSITION OF THE DIVISIONS OF STIGMA AND PARIETAL PLACENTÆ IN THE COMPOUND OVARIUM OF PLANTS.

BY

ROBERT BROWN, F.R. & L.S.

[*Reprinted from 'Plantæ Javanicæ Rariores.' Part II, pp.* 107—112.]

LONDON.

1840.

ON THE DIVISIONS OF STIGMA,[1] &c. [107

To estimate correctly the importance of the relation between the divisions of the Stigma and the parietal placentæ of the compound ovarium, namely, whether when agreeing [108
in number they are placed opposite to or alternate with each other, it is necessary to take into consideration the theoretical view which appears the most probable of the origin or formation of a simple ovarium, and that of the stigma belonging to it, as well as the various kinds and degrees of confluence by which the real nature of both organs, but especially the latter, is so often obscured.

It is at present, I believe, universally agreed to consider a polyspermous legumen as that state of the simple ovarium, which best exemplifies the hypothetical view of the formation of this organ generally adopted; namely, that it consists of the modification of a leaf folded inwards and united by its margins, which in most cases are the only parts of the organ producing ovula; or, at least, where this power of production is not absolutely confined to the margins, it generally commences with or includes them.

The exceptions to the structure as here stated are of two kinds :—

First. Where the whole internal surface of the carpel is equally ovuliferous, which is the case in a few families of very small extent, as *Butomeæ*, *Nymphæaceæ*, and *Lardizabaleæ*.

[1] [Extracted from Mr. Brown's account of *Cyrtandreæ*, given in the second part of Dr. Horsfield's 'Plantæ Javanicæ Rariores,' published in 1840. Separate copies of this note were distributed in December, 1839.—*Edit. Ann. Nat. Hist.*]

Secondly. Where the production of ovula is limited to the external angle of the cell or axis of the leaf supposed to form the carpel.

A case of this kind is found in a portion of one of those families in which the whole surface is generally ovuliferous, namely, in *Hydropeltideæ*, which I have always regarded as merely a section of *Nymphæaceæ*;[1] and from the nature of these differences in placentation, which are more apparent than real, an argument might even be adduced in favour of that opinion.

A placenta apparently limited to the outer angle of the cell also occurs in the greater number of species of *Mesembryanthemum.* As this structure, however, is certainly not without exception in that very natural genus, several species, among which are *Mesembryanthemum crystallinum, cordifolium, papulosum* and *nodiflorum*, having the placenta confined to the internal angle of the cell or margins of the carpel; and as in some of those species in which the outer angle is placentiferous, the production of ovula is not confined to it, but extends to the lower half of the inner angle;—this apparent deviation from ordinary structure may perhaps be explained by assuming cohesion of the inflected portion of the carpel with the wall of the cell;—an hypothesis, in some degree supported by the fact, that in several species the termination of the assumed inflected portion is free and not ovuliferous.

But whatever opinion may be adopted as to the relation of this seemingly anomalous to the ordinary structure, it cannot, as M. Fenzl proposes,[2] be employed as the essential character of a distinct natural family limited to the Linnæan genus *Mesembryanthemum.*

The placenta then of a simple ovarium in its usual state, according to this view, is necessarily double; though by the complete suppression of ovula in one of its two component parts, and their diminished production in the other, the ovarium is not unfrequently reduced to a single ovulum. That such is the origin of the single ovulum is at least

[1] Gen. Rem. in Flinders's Voy. vol. ii, Append. p. 598. (*Anté p.* 74.)
[2] Annal. des Wien. Mus. vol, i, p. 349.

manifest in a monstrosity of *Tropæolum majus*, in which the stamina are converted into pistilla; but the complete action being impeded by the presence of the regular trilocular pistillum, and the two marginal cords of each open ovarium remaining distinct, the origin of the ovulum from one only of these cords is satisfactorily shown.

An ovarium with two or a greater number of cells, whose placentæ project into the cavities more or less from their inner angles, is an organ, the composition of which is sufficiently obvious.

But a compound ovarium may be differently constructed; and, first, instead of each simple organ forming a complete cell by the union of its own margins or adjoining portions of its surface, the corresponding margins or adjoining portions of surface of the proximate component parts may unite together so as to form a parietal placenta, often apparently simple, but in reality double in all cases. This view of the composition of a unilocular ovarium having two or more parietal placentæ is also very generally received. But exceptions, supposed to prevail in whole families, in which the disk and not the margins are placentiferous, have lately been assumed by Professor Lindley, *Orchideæ* and *Orobancheæ* being the examples of this structure to which he more particularly refers.

The accurate determination of this question appears to me of great importance to the theoretical botanist, but the subject will be most advantageously discussed after treating of the origin and modifications of stigmata.

An ovarium less manifestly compound is that in which the centre of the cavity is occupied by a placenta entirely unconnected with its sides; the supposed inflected portions of each component organ, according to the view here adopted, being removed, or reabsorbed so completely in a very [109
early stage of its development as to leave no trace of their existence either on the walls of the cavity or on the surface of the central placenta, which may either be polyspermous, or produce only a smaller and definite number of ovula having a relation to its supposed component parts, or, lastly, in some cases be reduced to a single ovulum.

These are the principal modifications of the compound ovarium when forming a simple series; but it is necessary to observe that both surfaces of the inflected and included portions of the carpels are not unfrequently equally productive of ovula, a structure which is manifest in many *Cyrtandraceæ*, especially *Cyrtandra*, although in several other genera of the same family the production is confined to the inner or upper surface of the margin. In other cases the polyspermous ovuliferous portion or placenta is connected with the inner angle of the cell by a single point only, which may proceed either from the apex or base of the cavity. This modification of structure, though in some families hardly of generic importance, seems to me to assist in explaining the apparently anomalous structures of *Hydnora*, *Rafflesia*, and *Brugmansia*.

On the subject of the origin and type of Stigma, my first observation is, that the style where present can only be regarded as a mere attenuation, in many cases very gradual, of the whole body of the ovarium. Hence the idea naturally suggests itself, that the inner margins of the carpel, which in the lower part are generally ovuliferous, in the upper part perform the different, though in some degree analogous, function of stigma. As the function, however, of this organ implies its being external, and as in different families, genera, and even species, it has to adapt itself to
110] various arrangements of parts destined to act upon it, corresponding modifications of form and position become necessary; hence it is frequently confined to the apex, and very often, especially in the compound ovarium with united styles, appears to be absolutely terminal.

In such cases, as it must always include and be closely approximated to the vascular cord of the axis, it has by some botanists been considered as actually derived from it, which it is, however, only in the same manner as the marginal placentæ are derived from the axis of the carpel. But according to the notion now advanced, each simple pistillum or carpel has necessarily two stigmata, which are to be regarded, not as terminal, but lateral.

That the stigma is always lateral may be inferred from its

being obviously so in many cases; and in one genus at least, *Tasmannia*, it extends nearly the whole length of the ovarium, so as to be commensurate with and placed exactly opposite to the internal polyspermous placenta.

That the stigma is always double appears probable from those cases in which it is either completely developed, as in the greater part of *Gramineæ* where the ovarium is simple; in the compound ovarium in *Urena;* and from those in which the development, though less complete, is still sufficiently obvious, as in many *Euphorbiaceæ* and in several *Irideæ*. This degree of development, however, is comparatively rare, confluence between the two stigmata of each carpel being the more usual structure; and in the compound pistillum a greater degree of confluence often takes place in the stigmata than in the placentæ;—a fact, which in all such cases is obviously connected with adaptation of surface to the more complete performance of function.

Another difference frequently occurs between the mode of confluence of placentæ and stigmata, namely, that in the compound but unilocular ovarium, while the placentæ of the adjoining carpels are united, the stigmata of each carpel are generally confluent. But this rule admits of exceptions, as in *Parnassia*, in many *Cruciferæ*, and in *Papaveraceæ*; in all these cases the stigmata as well as placentæ of the adjoining carpels are confluent, a structure satisfactorily proved in *Cruciferæ* by several cases of monstrosity, in which the stamina are transformed into pistilla; and in *Papaveraceæ* by a series of modifications of structure as well as by a like transformation of stamina.

A similar confluence of stigmata in the compound multilocular pericarpium is of much rarer occurrence; it is found, however, in the majority of *Irideæ*, in which the three stigmata alternate with the cells, and consequently with the placentæ of the trilocular ovarium. That this is the correct view of the composition of the stigmata in *Irideæ* is at least probable from their occasional deep division, and more particularly still from the bifid petal-like styles or stigmata which are opposite to the cells of the ovarium in other genera of the same family, as in *Iris* and *Moræa*. In both these arrangements the adaptation to the performance of function is equally manifest.

If the correctness of these observations be admitted, it follows that characters dependent on the various modifications of stigmata are of less value, both in a systematic point of view as determining the limits of families, and theoretically in ascertaining the true composition of organs, than those derived from the analogous differences in the ovaria or placentæ.

In those cases in which the nature of the composition of the ovarium is doubtful, it may, in the first place, be remarked, that wherever in the compound unilocular pistillum the placentæ are double or two-lobed, it is more probable that such placentæ are derived from two adjoining carpels, and are consequently marginal or submarginal, than that they occupy the disc of one and the same carpel; this being entirely the appearance in many cases where the marginal origin of placentæ is admitted; while in the greater part of those in which the disc is known to be ovuliferous, the ovula are never collected in two distinct masses, being generally scattered equally over the surface.

But the double placentæ are manifest in *Orchideæ*, the principal family in which Mr. Lindley considers the ovula as occupying the disc and not the margins. In this family also the alternation of stigmata with placentæ is that relation which is most usual in compound unilocular ovaria, where the apparent number of stigmata and placentæ is equal; and that in *Orchideæ* each apparent stigma is formed by the confluence of the two stigmata of one and the same carpel, is proved by tracing to their origins their vascular cords, which are found to coalesce with those of the three outer foliola of the perianthium.

This view of the composition of the ovarium in *Orchideæ* is confirmed by finding that it agrees with the ordinary arrangement in monocotyledonous plants; namely, the opposition of the double parietal placentæ to the three inner divisions of perianthium[1], while in *Apostasia* the three placentæ of the trilocular ovarium are opposite to the three outer

[1] Denham, Trav. in Afr. Append. p. 243. (*Anté, p.* 300.)

divisions; and it is further strengthened on considering what takes place in *Scitamineæ*, where the same agreement is found both in the placentæ of the trilocular ovarium, which in this family is the ordinary structure, and in the unilocular, which is the exception.

I am aware that the agreement of *Orchideæ* with the usual relation of parts in Monocotyledones is not admitted by M. Achille Richard, nor by Mr. Lindley, who has adopted his hypothesis respecting the structure of the flower in this family. According to M. Richard, the outer series of perianthium is generally wanting, being found only in one genus, *Epistephium*: the three outer divisions actually existing in the whole order, according to this view, become petals, and the three inner divisions sterile petaloid stamina.

I have some years ago[1] stated several objections to this hypothesis; at present I shall advert to one of those only, considering it as conclusive; namely, the position of the two lateral stamina, which are generally rudimentary, but in some cases perfectly developed, in this family. In several species of *Cypripedium*, which is one of these cases of perfect development, I had then ascertained, by means of numerous transverse sections made at various heights in the column and at its base, that their vascular cords united with those of the two lateral inner divisions of the flower, while that of the third, generally the only perfect stamen, is manifestly opposite to the anterior division of the outer series. The position of stamina, therefore, so far from being regular, as the hypothesis in question considers it, is absolutely without example, two of the inner series being opposite to two of the supposed outer series of stamina.

A very different view respecting the formation of the ovarium in *Orchideæ* is that first advanced by Mr. Bauer and adopted by Mr. Lindley, namely, that it consists of six carpels, of which three, placed opposite to the outer series of perianthium or sepals, are sterile; the remaining three, opposite to the inner series, or petals, being fertile, and bearing their placentæ on their axes or disks.

The chief argument in support of this view is no doubt

[1] Linn. Soc. Trans. vol. xvi, p. 698. (*Antè*, *p*. 501.)

derived from the very remarkable dehiscence of the capsule into six valves. But I have elsewhere pointed out cases where an analogous dehiscence occurs, in which, however, a similar composition has never been supposed to exist: and if the presence of six vascular cords in sections of the ovarium be likewise adduced in favour of the opinion, I may add that I have in the same place remarked that these vascular bundles belong not to the ovarium only, but also to the perianthium and stamina, and are equally observable in other families with adherent ovarium, as *Irideæ*, in which a similar composition has never been inferred.

With regard to the second family, in which Mr. Lindley believes the disk of the carpel to be ovuliferous, namely, *Orobancheæ*, I find no other argument advanced in support of this view than that derived from the bursting of the capsule into two lateral valves; but an opinion founded on dehiscence only may be said to be a mere begging of the question; division through the axis of carpels, especially in the families related to *Orobancheæ*, being nearly as common as separation of their margins. In this family also, as in *Orchideæ*, the placentæ are double, an argument in favour of their submarginal origin: and although, whether the carpels be regarded as lateral, or anterior and posterior, the placentæ are not strictly marginal, yet there are other families where a similar position of placentæ is found, but in which the structure assumed in this hypothesis has never been suspected. As to the supposed affinity of *Orobancheæ* with *Gentianeæ*, which might be adduced in support of this view, as far as it is founded on the assumed agreement of the two orders in the lateral position of their carpels, the argument, even if correct, would hardly be conclusive; for in *Gentianeæ* there is at least one genus having quadrifid and another with quinquefid flowers, in which the carpels are
112] not lateral, but anterior and posterior, as I believe them to be in *Orobancheæ;* nor has it ever been supposed that in *Gentianeæ* the disk or axis is ovuliferous.

In the account now given of the modifications of ovarium and stigma, I have, in conformity with the ordinary language of botanists, employed the term *confluence*, by which, however, is not to be understood the union or cohesion of parts originally distinct, for in the great majority of cases the separation or complete development of these parts from the original cellular and pulpy state has never taken place. But with this explanation the word may still be retained, unless connate should be considered less exceptionable.

I have also assumed that ovula belong to the transformed leaf or carpel, and are not derived from processes of the axis united with it, as several eminent botanists have lately supposed. That the placentæ and ovula really belong to the carpel alone is at least manifest in all cases where stamina are changed into pistilla. To such monstrosities I have long since referred in my earliest observations on the type of the female organ in phænogamous plants,[1] and since more particularly in my paper on *Rafflesia*:[2] the most remarkable instances alluded to in illustration of this point being *Sempervivum tectorum*, *Salix oleifolia*, and *Cochlearia armoracia*, in all of which every gradation between the perfect state of the anthera and its transformation into a complete pistillum is occasionally found.

[1] In Linn. Soc. Trans., vol. xii, p. 89.
[2] Ibid. vol. xiii, p. 212, note. (*Antè*, *p*. 379.)

ON THE PLURALITY AND DEVELOPMENT OF THE EMBRYOS IN THE SEEDS OF CONIFERÆ.

BY

ROBERT BROWN, ESQ., F.R.S., F.L.S.,

AND FOREIGN MEMBER OF THE ACADEMY OF SCIENCES IN THE INSTITUTE OF FRANCE.

[*Reprinted from the 'Annals and Magazine of Natural History,' for May, 1844. Vol. XIII, pp.* 368—374.]

ON THE PLURALITY, &c.[1] [368

THE following short paper on a subject which I intend to treat at greater length, contains a few facts of sufficient interest perhaps to admit of its being received as a communication to the present meeting.

In my observations on the structure of the female flower

[1] Read before the British Association at Edinburgh in August 1834, and published in the Annales des Sciences Naturelles for October 1843. The following abstract was given in the "Report of the Fourth Meeting of the British Association," 1835, pp. 596-7:—"The earliest observations of the author on this subject were made in the summer of 1826, soon after the publication of his remarks on the female flower of *Cycadeæ* and *Coniferæ*. He then found that in several *Coniferæ*, namely, *Pinus Strobus*, *Abies excelsa*, and the common Larch, the plurality of embryos in the impregnated ovulum was equally constant, and their arrangement in the albumen as regular as in *Cycadeæ;* and similar observations made during the present summer on several other species, especially *Pinus sylvestris* and *P. Pinaster*, render it highly probable that the same structure exists in the whole family. The first change which takes place in the impregnated ovulum of the *Coniferæ* examined, is the production or separation of a solid body within the original nucleus. In this inner body, or albumen, several subcylindrical corpuscula, of a somewhat different colour and consistence from the mass of the albumen, seated near its apex and arranged in a circular series, soon become visible. In each of these corpuscula, which are from three to six in number, a single thread or funiculus, consisting of several, generally of four, elongated cells or vessels, with or without transverse septa, originates. The funiculi are not unfrequently ramified, each branch or division terminating in a minute rudiment of an embryo. But as the lateral branches of the funiculi usually consist of a single elongated cell or vessel, while the principal or terminal branch is generally formed of more than one, embryos in *Coniferæ* may originate either in one or in several cells, even in the same funiculus. A similar ramification in the funiculi of the *Cycas circinalis* has been observed by the author. Instances of the occasional introduction of more than one embryo in the seeds of the several plants belonging to other families have long been known, but their constant plurality and regular arrangement have hitherto only been observed in *Cycadeæ* and *Coniferæ*."

in *Cycadeæ* and *Coniferæ*, published in 1826,[1] I endeavoured to prove that in these two families of plants the ovulum was in no stage inclosed in an ovarium, but was exposed directly to the action of the pollen.

In support of this opinion, which has since been generally, though I believe not universally adopted, the exact resemblance between the organ until then termed ovarium in these two families, and the ovulum in other phænogamous plants, was particularly insisted on; and I at the same time referred, though with less confidence, to their agreement in the more important changes consequent to fecundation.

I noticed also the singular fact of the constant plurality of embryos in the impregnated ovula of *Cycadeæ*, and the not unfrequent occurrence of a similar structure in *Coniferæ*. In continuing this investigation, in the course of the same summer in which the essay referred to appeared, it seemed probable, from the examination of several species of the Linnæan genus *Pinus*, namely, *Pinus Abies, Strobus* and *Larix*, that the plurality and regular arrangement of embryos were as constant in *Coniferæ* as in *Cycadeæ;* for in all the species of *Pinus* here referred to, the preparation for the production of several embryos was equally manifest, and the points or areolæ of production were in like manner disposed in a single circular series at the upper extremity of the amnios.

From these observations, which I have since confirmed in the same and also in other species of *Pinus*, an additional and important point of resemblance is established between
369] *Cycadeæ* and *Coniferæ;* and it is worthy of remark, that while the female organ in these two families exists in a simpler form than in other phænogamous plants, the normal state of the impregnated ovulum is much more complex, and might even be considered as compound, or made up of the essential parts of several confluent ovula.

On considering the well-known œconomy of several *Coniferæ*, and especially of the genus *Pinus*, as at present limited, namely, in their requiring (at least) two seasons to ripen their cones, it occurred to me that these plants, from

[1] In the Appendix to Capt. King's Voyage. [*Antè, p.* 453.]

the extreme slowness in the process of maturation, conjoined with the considerable size of their seeds, and also from the striking peculiarity already noticed, were probably the best adapted for an investigation into the origin and successive changes of the vegetable embryo.

With this view chiefly I commenced in the present summer (1834) a series of observations, intending to follow them up from the period when the enlargement of the impregnated cone begins to take place, to its complete maturity at the end of the second or beginning of the third year.

Pinus sylvestris was selected for this purpose, corresponding observations being also made on other species, particularly *Pinaster* and *Strobus;* and although the investigation is necessarily incomplete, the facts already ascertained appear to me of sufficient importance to be submitted to physiological botanists.

In an essay on the organs and mode of fecundation in *Orchideæ* and *Asclepiadeæ*, published in 1831, I have given some account of the earliest changes observable in the impregnated ovulum of the former family; and in noticing the jointed thread, or single series of cells by which the embryo is suspended, I remarked that the terminating cell or joint of this thread is probably the original state of what afterwards, from enlargement, subdivision of its cavity, and deposition of granular matter in its cells, becomes the more manifest rudiment of the future embryo.

I had not indeed actually seen this joint in its supposed earliest state; the following observations on *Pinus*, however, will perhaps be considered as giving additional probability to the conjecture.

But before entering on my account of the origin and development of the embryo in *Pinus*, I shall state briefly the still earlier changes consequent to impregnation that take place in this genus; not only with the view of rendering the account of the embryo itself more readily intelligible, but also in confirmation of the opinion formerly advanced on the nature of the female organ in *Coniferæ* and *Cycadeæ*.

The first and most evident change observable is the production or separation of a distinct body within the nucleus of the ovulum, which, before impregnation, is a solid uniform substance.

370] In this stage the upper extremity of the included body, or amnios, is slightly concave, and has a more or less rough or unequal surface; the inequality being in consequence of the laceration of the cellular tissue, by which it was in its early stage attached to the apex of the original nucleus, or rather to a short cylindrical process arising from it and corresponding in size and form with this concave upper extremity, from which it separates when the amnios has attained its full size.

On this concave upper extremity of the amnios a few minute points of a deeper colour, and disposed in a single circular series, are sometimes observable; in general, however, they are hardly to be distinguished.

Below the concave apex the amnios itself is slightly transparent for about one fourth of its length, the remaining portion being entirely opake.

On dividing the whole longitudinally it is found to consist of a pulpy cellular substance, in which no definite cavity is originally observable; the upper transparent portion is, however, of a looser texture, and on the included embryos becoming manifest, a cavity irregular both in figure and extent is formed in its centre.

But before the embryos themselves or their funiculi become manifest, the areolæ, or portions of the substance destined for their production, are visible.

These areolæ, as I observed them in the common larch in May, 1827, are from three to five in number, of nearly cylindrical form, arranged in a circular or elliptical series, and are seated near the apex, with which they probably communicate by the similarly arranged points of its surface already noticed.

In the amnios of *Pinus sylvestris*, as observed in June and July last, the corresponding parts were found considerably more advanced. In the specimens then examined, the remains of the embryoniferous areolæ, from four to six in number, were still visible, but consisting of conical membranes of a brown colour, presenting their acute apices towards the surface, and at the base seeming to pass gradually into the lighter-coloured pulpy substance of which the mass of the amnios consists.

Corresponding and nearly approximated to each of these conical membranes, a filament, generally of great length, and either entirely simple or giving off a few lateral branches, was found. This filament or funiculus consisted generally of four series of elongated transparent cells or vessels, usually adhering together with firmness, but in some cases readily separable without laceration; and in one of the species examined, *Pinus Pinaster*, the transverse septa of the funiculus were either very obscure or altogether wanting.

The upper extremity of each funiculus was in all cases [371
manifestly thickened and of a depressed spheroidal form; and in each of the four cells or vessels of which it consisted exhibited a small opake areola analogous to the nucleus of the cell, so frequently observable in the tissue of Monocotyledonous plants, and which also exists, though less commonly, in Dicotyledones.

A lacerated and extremely transparent membrane was generally found surrounding and adhering to the thickened origin or head of the funiculus.

In the earliest state examined of *Pinus Pinaster*, the funiculus was found equally transparent through its whole length, and having no appearance of subdivision or any other indication of embryo at its lower extremity. In a somewhat more advanced state of the same plant, as well as in the two other species observed, namely, *Pinus sylvestris* and *Strobus*, the lower extremity of the funiculus was subdivided into short cells, sometimes disposed in a double series, but more frequently with less regularity and in greater numbers, the lowest being in all cases the most minute and also the most opake, from the deposition of granular matter, which is nearly or entirely wanting in the upper part of the cord. This opake granular extremity of the funiculus is evidently the rudiment of an embryo. When the funiculus ramifies, each branch is generally terminated by a similar rudiment, and these lateral embryoniferous branches not unfrequently consist of a single vessel or cell, while the embryo of the trunk or principal branch is as generally derived from more than one.

That each of those opake bodies terminating the trunk and branches of the funiculi are really rudimentary embryos, is proved by tracing them from their absolutely simple state to that in which the divisions of the lower extremity become visible; and those again into the perfect cotyledons.

The results of this investigation in its present incomplete state are, 1st, that the plurality of rudimentary embryos in *Pinus* (and probably in other *Coniferæ*) is not only constant, but much greater than could well have been imagined independent of actual observation ; each impregnated ovulum not only containing several distinct funiculi, but each funiculus being capable of producing several embryos. In the ripe seed, however, it is a rare occurrence to find more than one of these embryos perfected.

2ndly. That an embryo in *Coniferæ* may originate in one or in more than one cell or vessel even in the same cord ; and it also appears that the lower extremity of the funiculus, the seat of the future embryo, is originally in no respect different from the rest of its substance.

The greater part of the appearances now described are represented in the accompanying Plate.

372] *April* 20, 1844.

Postscript.—It is necessary to notice the recent publication of a very important memoir by MM. de Mirbel and Spach on the development of the embryo in *Coniferæ*.[1]

These excellent observers confirm the principal statements of the preceding essay, with the brief abstract of which only they were acquainted.

They have also extended the investigation to *Thuja* and *Taxus*, two genera which I had not examined, and in which, especially in the latter, the structure appears to be re-

[1] Annales des Sc. Nat. 2 série, November 1843.

markably modified; and they have ascertained some points in *Pinus* itself that I had overlooked.

In this memoir M. de Mirbel refers to his early observations on the structure of the seeds of *Cycas* which occur in an essay read before the Academy of Sciences in October 1810, and soon after published in the 'Annales du Muséum.'[1]

These observations and the figures illustrating them clearly prove M. de Mirbel's knowledge of the plurality of embryos in *Cycas* at that period. And in his recent memoir on *Coniferæ* he regards them as giving the earliest notice of that remarkable structure; stating also that my first publication on the same subject was in 1835.

But as the 'Prodromus Floræ Novæ Hollandiæ' was published before M. de Mirbel's essay in the 'Annales du Muséum,' which appears from his references to that work in the essay in question, he must have overlooked the following passages:—

"In Cycadi angulata puncta areæ depressæ apicis seminis totidem canalibus brevibus respondent gelatina homogenea primum repletis et membrana propria instructis, unico quantum observavimus embryonifero, quo augente reliqui mox obliterati sunt."—*Prodr.* p. 347.

"Structura huic omnino similis hactenus absque exemplo nec ulla analoga (nempe embryones plures in distinctis cavitatibus ejusdem albuminis) nisi in Cycadi et nonnunquam in Visco cognita sit."—*Prodr.* p. 307.

I may add, that this structure of *Cycas* was ascertained in living plants on the east and north coasts of New Holland in 1802 and 1803.

The earliest observer of the principal fact, however, was probably the late Aubert du Petit Thouars, who in a dissertation on the structure and affinities of *Cycas* published in 1804,[2] distinctly notices the points on the surface and the corresponding corpuscula within the apex of the albumen, into which corpuscula he hazards the conjecture that the grains [373
of pollen enter and become the future embryos. This, in

[1] Annales du Muséum d'Hist. Nat. tom. xvi, p. 252, tab. 20.
[2] Histoire des Végétaux des Iles d'Afrique, p. 9, tab. 2, *n*.

regard to *Cycas*, might be considered the revival of the general hypothesis advanced by Morland in 1703,[1] and some years afterwards adopted, but without acknowledgment, by C. J. Geoffroy,[2] and which seems to have entirely originated in the discovery by Grew of the existence of a foramen opposite to the radicle of the embryo in the ripe seeds of some Leguminous plants.[3]

But as M. du Petit Thouars had evidently no intention of extending his hypothesis beyond *Cycas* and probably *Zamia*, it can hardly be said to anticipate the general and ingeniously supported theory of Dr. Schleiden, respecting which physiological botanists are at present almost equally divided. On this theory it is not my intention at present to express an opinion; nor did the question of the mode of action of the pollen form any part of my object in the preceding essay. I shall only here remark, that according to the latest statements of Dr. Schleiden with which I am acquainted,[4] although he admits that his investigation is not in all points complete, he seems to have no doubt that his theory of the origin of the vegetable embryo in the pollen tube is applicable to *Coniferæ*. He has in the first place ascertained the existence of my areolæ or corpuscula, which he denominates large cells in the embryo-sac or albumen, in all the European genera of *Coniferæ*;[5] and in *Abies excelsa*, *Taxus baccata*, and *Juniperus Sabina*, he states that he has succeeded in preparing free the whole pollen tubes from the nucleary papillæ to the bottom of the corpuscula. But as (if my observations are correct, and they seem to be confirmed by those of M. de Mirbel) the corpuscula are not developed in *Pinus*, as the genus is at present limited, until the spring or even beginning of summer of the year after flowering, and if Dr. Schleiden's statement be also correct, the pollen must remain inactive for at least twelve months.

The quiescent state of pollen for so long a time is indeed

[1] Philosophical Transactions, vol. xxiii, part 2, n. 287, p. 1474.
[2] Mém. de l'Acad. des Sc. de Paris, 1711, p. 210.
[3] Anat. of Plants, p. 2. [4] Schleiden, Grund. der Bot. 2 Theil, p. 374.
[5] *Op. cit.* pp. 354 et 357.

not altogether improbable on considering the analogous œconomy in several tribes of insects, in some of which the male fluid remains inactive in the female for a still longer period;[1] and in plants, though for a much shorter period, I may refer to *Goodenoviæ*, in which the pollen is applied to the stigma a considerable time before that organ is sufficiently developed to act upon or transmit its influence.[2] But the supposed protracted state of inactivity in the pollen of *Pinus* does not necessarily lead to the adoption of Dr. [374
Schleiden's theory. With respect to *Cycadeæ*, whatever opinion may be adopted as to the precise mode of action of the pollen in that family, it is certain that the mere enlargement of the fruit, the consolidation of albumen, and the complete formation of the corpuscula in its apex are wholly independent of male influence, as I have proved in cases where pollen could not have been applied, namely, in plants both of *Cycas* and *Zamia* (*Encephalartos*) producing female flowers in England at a time when male flowers were not known to exist in the country.

EXPLANATION OF PLATE 33 (VII).

Fig. 1. A scale of the cone of *Pinus sylvestris*, with its winged seeds, one of which is abortive: natural size.

N.B. The remaining figures are more or less magnified.

Fig. 2. An unripe seed, of which the testa, in this state cartilaginous, is cut open, partly removed and thrown back to show the included body, which is the half-ripe original nucleus with its sphacelated apex and the free portion of the inner coat, extending from the apex to about one third of the length of the nucleus, below which it is intimately connected with and inseparable from the outer coat.

Fig. 3. The amnios or albumen, with the coats opened and laid back.

a. The body of the albumen, with its slightly concave upper extremity: in this stage separated from *b*, the apex, which is conical above, below cylindrical, and which was suspended from the top of the original nucleus.

---

[1] Herold. Entwickel. der Schmetterl. &c. 1815, et Siebold in Müller's Archiv, 1837, p. 392.

[2] Append. to Flinders's Austral. p. 561. [*Antè*, *p*. 33.]

Fig. 4. A plan rather than actual representation of a longitudinal section of any one seed examined, but the parts accurately copied from the calyptræform membranes, the funiculi or suspensors, and the nascent embryos of seeds of *Pinus sylvestris*. In this stage the funiculi are distinct from the calyptræform membranes within which they originated.

Fig. 5 is also a plan of the slightly concave apex of the amnios or albumen, with its semitransparent points or pores circularly arranged; in this species (*Pinus sylvestris*) seldom exceeding five, and not unfrequently being only four or even three.

Fig. 6. One of the funiculi or suspensors, with its dilated upper extremity, to which the lacerated remains of a thin transparent membrane adhere: the funiculus itself ramified, each of the two lateral branches consisting of a single elongated tube or cell terminating in a rudimentary embryo: the trunk of the funiculus composed of several (apparently four) tubes or cells terminated by a single embryo, which is already slightly divided, the divisions being the commencement of its cotyledons.

Figs. 7 & 8. Two other funiculi belonging to the same seed less advanced, but both ramified.

Fig. 9. A funiculus of *Pinus Pinaster* with its thickened head, in which the nuclei of its component elongated cells or tubes, and its adhering lacerated membrane are visible. The figure is given particularly to show that in this (the only one observed) there is no opake granular portion of the compound funiculus; in other words, no indication of a nascent embryo.

Fig. 10. A funiculus of *Pinus Abies*, Linn., with its rudimentary embryo and thickened head, still partly inclosed in the calyptræform membrane.

ON THE ORIGIN

AND

MODE OF PROPAGATION

OF THE

GULF-WEED.

BY

ROBERT BROWN, ESQ.,

PRESIDENT OF THE LINNEAN SOCIETY.

[*Reprinted from the 'Proceedings of the Linnean Society.' Vol. II, pp.* 77—80.]

ON THE

ORIGIN AND MODE OF PROPAGATION OF THE GULF-WEED.

Read before the Linnean Society, May 7, 1850.

Read a letter, dated May 19th, 1845, addressed by the President to Admiral Sir Francis Beaufort, for communication to Baron Alexander von Humboldt, "On the Origin and Mode of Propagation of the Gulf-weed." The letter is as follows:—

"My dear Captain Beaufort,—I am vexed to have kept Baron Humboldt's letter so long, and now in returning it, that it should be accompanied by so little satisfactory information on the only one of its queries with which I could have been supposed to deal, namely, that which relates to the origin and mode of propagation of the Gulf-weed.

"On this subject it appears that M. de Humboldt (in his Personal Narrative) first supported the more ancient notion, that the plant, originally fixed, was brought with the stream from the Gulf of Florida, and deposited in what Major Rennell calls the recipient of that stream. More recently, however, Baron Humboldt has adopted the opinion,[1] also held by several travellers, that the Gulf-weed originates and propagates itself where it is now found. To the adoption of this view it appears that he has been led chiefly by the

[1] Histoire de la Géographie du Nouveau Continent, vol. iii, p. 73, and Meyen, Reise, vol. i, p. 36-9.

observations of the late Dr. Meyen, who in the year 1830 passed through a considerable portion of the great band of
78] Gulf-weed, and who ascertained, as he states, from the examination of several thousand specimens, that it was uniformly destitute both of root and fructification; he concludes, therefore, that the plant propagates itself solely by lateral branches; he at the same time denies that it is brought from the Gulf of Florida, as, according to his own observation, it hardly exists in that part of the stream near the great band, though found in extensive masses to the westward. I have here to remark that, as far as relates to the absence of root and fructification, Meyen has only confirmed by actual observation what had been previously stated by several authors, particularly by Mr. Turner (in his 'Historia Fucorum,' vol. i, p. 103, published in 1808), and Agardh (in his 'Species Algarum,' p. 6, published in 1820). But Meyen materially weakens his own argument in stating that he considers the Gulf-weed (*Sargassum bacciferum* of Turner and Agardh), and the *Sargassum natans*, or *vulgare*, specifically distinguished from it by these authors, as one and the same species; adding, that he has observed among the Gulf-weed all the varieties of *Sargassum vulgare* described by Agardh; and finally, that on the coast of Brazil he has found what he regards as the Gulf-weed in fructification. Now, as *Sargassum natans* has been found fixed by a discoid base or root, in the same manner as the other species of the genus, and as according to Meyen the Gulf-weed has been found in fructification, the legitimate conclusion from his statements seems to be, that this plant is merely modified by the peculiar circumstances in which it has so long been placed. I am not, however, disposed to adopt Dr. Meyen's statement that he actually found the true *Sargassum natans*, much less all its supposed varieties, mixed with the Gulf-weed, having reason to believe that at the period of his voyage his practical knowledge of marine submersed Algæ was not sufficient to enable him accurately to distinguish species in that tribe. It is not yet known what other species of *Sargassum* are mixed with the Gulf-weed, what proportion they form of the great band, nor in what state, with respect

to root or fructification, they are found; though, in reference to the questions under discussion, accurate information on these points would be of considerable importance.

"That some mixture of other species probably exists may be inferred even from Dr. Meyen's statement, and indirectly from that of Lieut. Evans, who, in his communication published in Major Rennell's invaluable work on the Currents of the Atlantic, asserts that he found the Gulf-weed in fructification, which he compares with that of Ferns, a statement which would seem to prove merely that he had found along with the Gulf-weed a species of *Sargassum* with dotted leaves, the real fructification of the genus bearing [79
no resemblance to that of Ferns, though to persons slightly acquainted with the subject the arranged dots on the leaves might readily suggest the comparison.

"With regard to the non-existence of roots in the Gulf-weed as a proof of specific distinction, it is to be observed that the genus *Sargassum*, now consisting of about sixty species, is one of the most natural and most readily distinguished of the family *Fucaceæ*, and that there is no reason to believe that any other species of the genus, even those most nearly related to, and some of which have been confounded with it, are originally destitute of roots; though some of them are not unfrequently found both in the fixed and in considerable masses in the floating state, retaining vitality and probably propagating themselves in the same manner (see Forskål, Fl. Ægypt.-Arab., p. 192, n. 52). It is true, indeed, that a *Sargassum*, in every other respect resembling Gulf-weed, has, I believe, not yet been found furnished either with roots or fructification, neither Sloane's nor Browne's evidence on this subject being satisfactory.[1] But the shores of the Gulf of Florida have yet not been sufficiently examined to enable us absolutely to decide that that is not the original source of the plant; and the differences

[1] See Sloane's Jam. i, p. 59. I have examined Sloane's specimens in his Herbarium; they belong to Gulf-weed in its ordinary form, and are alike destitute of root and fructification; hence they are probably those gathered by him in the Atlantic, and not those which he says grew on the rocks on the shores of Jamaica. Browne's assertion to the same effect is probably merely adopted from Sloane.

between the Gulf-weed and some other *Sargassa*, especially *S. natans*, are not such as to prove these two species to be permanently distinct. The most remarkable of these differences consists in the leaves of the Gulf-weed being uniformly destitute of those dots or areolæ so common in the genus *Sargassum*, and which are constantly present in *S. natans*. These dots, in their greatest degree of development, bear a striking resemblance to the perforations or apertures of the imbedded fructification in the genus. But as the receptacles of the fructification, as well as the vesicles, are manifestly metamorphosed leaves; and as the production of fructification is not adapted to the circumstances in which the Gulf-weed is placed, it is not wholly improbable, though this must be regarded as mere hypothesis, that the propagation by lateral branches, continued for ages, may be attended with the entire suppression of these dots.

"That the Gulf-weed of the great band is propagated
80] solely by lateral or axillary ramification, and that in this way it may have extended over the immense space it now occupies, is highly probable, and perhaps may be affirmed absolutely without involving the question of origin, which I consider as still doubtful.

"My conclusion, therefore, is somewhat different from that of Baron Humboldt, to whom I would beg of you to forward these observations, which will prove that I have not been inattentive to his wishes and to your own, though they will at the same time prove that I have had very little original information to communicate."

SOME ACCOUNT

OF

TRIPLOSPORITE,

AN

UNDESCRIBED FOSSIL FRUIT.

BY

ROBERT BROWN, ESQ., D.C.L., F.R.S.,

HONORARY MEMBER OF THE ROYAL SOCIETY OF EDINBURGH AND ROYAL IRISH ACADEMY; PRESIDENT OF THE LINNEAN SOCIETY.

[*Reprinted from the 'Transactions of the Linnean Society.' Vol. XX, pp.* 469—475.]

LONDON.

1851.

SOME ACCOUNT OF AN UNDESCRIBED FOSSIL FRUIT. [469

Read June 15th, 1847.

The following imperfect account of a singularly beautiful and instructive silicified Fossil has been hastily drawn up, to supply in some measure the possible want of any other memoir for the present Meeting.

The remarks which I am enabled to make, from detached memoranda, on so short a notice, will principally serve to explain the accompanying drawings, which I have carefully superintended, and which exhibit a very satisfactory microscopic analysis of its structure, and do great credit to the artistical talent of Mr. George Sowerby, jun.

The only specimen of this fossil known to exist was brought to London in 1843 by M. Roussell, an intelligent dealer in objects of natural history. His account of it was, that it had been in the possession of Baron Roget, an amateur collector in Paris, for about thirty years; that after his death it was brought to public sale with the rest of his collection, but no offer being made nearly equal to the sum he paid for it, which was 600 francs, it was bought in. It was purchased here from M. Roussell jointly by the British Museum, the Marquis of Northampton, and myself, for nearly £30. It seems to have entirely escaped the notice of the naturalists of Paris. Nothing else is known of its history, but from its obvious analogy in structure and in its mineral condition with *Lepidostrobus*, it may be conjectured to belong to the same geological formation.

The specimen is evidently the upper half of a Strobilus

very gradually tapering towards the top. As brought to England it was not quite two inches in length; but a transverse slice, probably of no great thickness, had been removed from it in Paris: the transverse diameter of the lower slices somewhat exceeded the length of the specimen;
470] its surface, which was evidently waterworn, is marked with closely-approximated hexagonal areæ, of which the four lateral sides are nearly twice the length of the upper and lower: these hexagons, which are the waterworn terminations of the bracteæ of the Strobilus, becoming gradually smaller and less distinct towards the top.

A transverse section of the Strobilus exhibits a central axis, from which radii directly proceed, constantly thirteen in number, resembling, when perfect, the spokes of a wheel, but several of them being always more or less incomplete. These radii alternate with an equal number of oblong bodies, also radiating, of a lighter colour, and which are not directly connected with the axis: beyond these twenty-six radiating bodies a double series of somewhat rhomboidal areolæ exist. These appearances not readily indicating the actual structure in the transverse, are satisfactorily explained by the vertical section.

From the vertical section it appears that the Strobilus is formed of a central axis of small diameter compared with the parts proceeding from it, which consist,—

1. Of bracteæ densely approximated and much imbricated; the lower half of each of these stands at right angles to the axis, while the imbricating portion, of about equal length with the lower, and forming an obtuse angle with it, is gradually thickened upwards: these form the spokes and external rhomboidal areæ of the transverse section.

2. Of an equal number of oblong bodies of a lighter colour and more transparent, each of which is adnate and connected by cellular tissue with the upper surface of the supporting bractea. These bodies are sections of *Sporangia* filled with innumerable microscopic *sporules*, originally connected in threes (very rarely in fours), but ultimately separating, as shown in TAB. 35 (XXIV), fig. G.

From this triple composition or union of sporules, which differs from the constant quadruple union in tribes of existing plants, namely, *Ophioglosseæ* and *Lycopodiaceæ*, which, from other points of structure, may be supposed most nearly related to the fossil, I have called it *Triplosporite*, a name which expresses its fossil state, the class or primary division to which it belongs, and its supposed peculiarity of structure.

The structure of the *axis*, which is well preserved in the specimen, distinctly shows, in the arrangement of its [471
vascular bundles, a preparation for the supply of an equal number of bracteæ. These vascular fasciculi are nearly equidistant in a tissue of moderately elongated cells.

The vessels are exclusively scalariform, very closely resembling those of the recent Ferns and *Lycopodiaceæ* ; and among fossils, those of *Psarolites*, *Lepidodendron*, and its supposed fruit, *Lepidostrobus*, as well as several other fossil genera; namely, *Sigillaria*, *Stigmaria*, *Ulodendron*, *Halonia ?* and *Diploxylon*.

The coat of the sporangium appears to be double; the outer layer being densely cellular and opake, the inner less dense, of a lighter colour, and formed of cells but slightly elongated.

On the lower or adnate side of the sporangium this inner layer seems to be continued, in some cases at least, in irregular processes to a considerable depth. I cannot, however, find that the sporules are actually formed in this tissue, but in another of somewhat different appearance and form, of which I have been only able to see the torn remains.

The minute granular bodies which accompany the sporules in the drawing TAB. 35 (XXIV), fig. G, are probably particles of the mother cells, and are neither uniform in size nor outline.

The whole specimen has suffered considerable decay or loss of substance, which is most obvious in the sporangia from their greater transparency, but equally exists in the opake bracteæ, in which radiating crystallization occupies the space of the removed cellular substance.

I cannot at present enter fully into the question of the affinities of *Tripolosporite*. I may remark, however, that in its scalariform vessels it agrees with all the fossil genera supposed to be Acotyledonous. In the structure of its sporangia and sporules it approaches most nearly, among recent tribes, to *Lycopodiaceæ*, and *Ophioglosseæ*; and among fossils, no doubt, to *Lepidostrobus*, and consequently to *Lepidodendron*.

The stem structure of *Lepidodendron*, known to me only in one species, *Lepidodendron Harcourtii*, offers no objection to this view, the vascular arrangement of the axis of its stem bearing a considerable resemblance to that of *Triplosporite*. To the argument derived from an agreement in structure between axis of stem and of strobilus I attach considerable importance, an equal agreement existing both in recent and fossil *Coniferæ*.

472] In conclusion I have to state, that very recently (since the drawings were completed, and as well as the specimens seen by such of my friends as were interested in fossil botany) Dr. Joseph Hooker has detected in the sporangia of a species referred to *Lepidostrobus* sporules, and those also united in threes. There are still, however, characters which appear to me sufficient to distinguish that genus from the fossil here described.

To the brief account here given of *Triplosporite* it is necessary to add a few remarks on some nearly-related fossils, chiefly *Lepidostrobi*, whose structure is now more completely known than it was when that account was submitted to the Society.

On the affinities of *Lepidostrobus* to existing structures, respecting which various opinions have been held, it is unnecessary here to advert to any other than that of M. Brongniart, which is now very generally adopted, namely, that *Lepidostrobus* is the fructification of *Lepidodendron*, and that the existing family most nearly related to *Lepidodendron* is *Lycopodiaceæ*. The same view is in great part adopted in my paper. But I hesitated in absolutely referring *Triplosporite* to *Lepidostrobus*, from the very imperfect knowledge then possessed of the structure of that genus. The specimens of *Lepidostrobus* examined by M. Brongniart were so incomplete, that they suggested to him an erroneous view of the relation of the supposed sporangium to its supporting bractea, and of the contents of the sporangium itself they afforded him no information whatever.

In concluding my account of *Triplosporite,* I noticed the then very recent discovery of spores in an admitted species of *Lepidostrobus* by Dr. Joseph Hooker, who, aware of the interest I took in everything relating to *Triplosporite*, the sections and drawings of which he had seen, communicated to me a section of the specimen in which spores had been observed, but which in other respects was so much altered by decomposition, that it afforded no satisfactory evidence of the mutual relation of the parts of the strobilus. The appearances, however, were such, that I hazarded the opinion of its being generically different from *Triplosporite,* an opinion strengthened by M. Brongniart's account of the origin of the sporangium.

Since the abstract of my paper was printed in the Pro- [473
ceedings of the Society, the second volume of the Memoirs of the Geological Survey of Great Britain has appeared, which contains an article entitled "Remarks on the Structure and Affinities of some *Lepidostrobi*." The principal object of Dr. Hooker, the author of this valuable essay, is from a careful examination of a number of specimens, all more or less incomplete, or in various degrees of decomposition and consequent displacement or absolute abstraction of parts, to ascertain the complete structure or common type of the genus *Lepidostrobus ;* but the type so deduced is in every essential point manifestly exhibited, and in a much more satisfactory manner, by the single specimen of *Triplosporite.* This does not lessen the value of Dr. Hooker's discovery and investigation, but it gives rise to the question whether *Triplosporite,* which he has not at all referred to, and therefore probably considered as not belong-

ing to *Lepidostrobus*, be really distinct from that genus; and although there are still several points of difference remaining, namely, the form of the strobilus in *Triplosporite*, confirmed by a second specimen presently to be noticed, and in *Lepidostrobus* the more limited insertion of sporangium, and the very remarkable difference in the form of the unripe spores, hardly reconcilable with a similar origin to that described in *Triplosporite*, I am upon the whole inclined to reduce my fossil to *Lepidostrobus* until we are, from still more complete specimens of that genus, better able to judge of the value of these differences. The name *Triplosporites*, however, is already adopted, and a correct generic character given, in the second edition of Professor Unger's 'Genera et Species Plantarum Fossilium,' p. 270, published in 1850, who at the date of his preface in 1849 was not aware of Dr. Hooker's essay on *Lepidostrobus*, the character of which he has adopted entirely from M. Brongniart's account.

In October 1849 M. Brongniart showed me a fossil so closely resembling the *Triplosporite*, both in form and size, that at first sight I concluded it was the lower half of the same strobilus. On examination, however, it proved to be of somewhat greater diameter. It was nearly in the same mineral state, except that the crystallizations consequent on loss of substance were rather less numerous; it differed also in the central part of the axis being still more complete; in the bracteæ being more distant and of a slightly different
474] form: but the spores in composition, form, and apparently in size were identical. This specimen had then very recently been received from the Strasburg Museum, but nothing was known of its origin or history.

May 5, 1851.

EXPLANATION OF THE PLATES OF TRIPLOSPORITE.

Tab. 34 (XXIII).

The figures A, B, C, and D are of the natural size.

Fig. A. A portion of the surface of the Strobilus, showing the hexagonal areolæ.

Figs. B & C. Transverse sections, exhibiting different appearances of the bracteæ and sporangia.

Fig. D. A vertical section of fig. A.

The remaining figures, E, F, G and H, are all more or less magnified.

Fig. E. A transverse section of the axis.

Fig. F. A more highly magnified drawing of a portion of fig. E, to show the arrangement and proportion of the vascular and cellular tissues.

Fig. G. A horizontal section of a sporangium, made probably near its origin.

Fig. H. A portion of the outer wall of a sporangium or bractea.

Tab. 35 (XXIV).

All the figures magnified.

Fig. A. A vertical section of the axis, near, but not exactly in the centre, showing the ramifications of the central cord of the axis going to the circumference of the axis, and connected or supported by a loose cellular tissue at *a a*.

Fig. B. A small portion of the axis, from which proceeds a bractea cut vertically through its centre, showing its vascular cord, and bearing on its lower and horizontal half a vertical section of an adnate sporangium, of which the base is cellular, rising irregularly and without spores,—probably a rare occurrence.

Fig. C. A small portion of the axis, to show the scalariform vessels with [475
the slightly elongated surrounding cells.

Fig. D. A similar portion, from the central axis of the bractea of fig. B.

Fig. E. A similar portion, from the line of union between the bractea and sporangium of fig. B.

Fig. F. A small portion of a sporangium, sufficiently magnified to show the arrangement and composition of sporules.

Fig. G. Several sporules, both in their compound and simple state, still more highly magnified, with the minute granular matter which usually accompanies them.

INDEX.

END OF VOL. I.

PRINTED BY J. E. ADLARD, BARTHOLOMEW CLOSE.

Printed in the United States
By Bookmasters